Silicon, Germanium, and Their Alloys

GROWTH, DEFECTS, IMPURITIES, AND NANOCRYSTALS

Silicon, Germanium, and Their Alloys

GROWTH, DEFECTS, IMPURITIES, AND NANOCRYSTALS

Edited by
Gudrun Kissinger
Sergio Pizzini

CRC Press is an imprint of the
Taylor & Francis Group, an **informa** business

CRC Press
Taylor & Francis Group
6000 Broken Sound Parkway NW, Suite 300
Boca Raton, FL 33487-2742

First issued in paperback 2021

ISBN-13: 978-0-367-78365-5 (pbk)
ISBN-13: 978-1-4665-8664-2 (hbk)

Library of Congress Cataloging-in-Publication Data

Silicon, germanium, and their alloys : growth, defects, impurities, and nanocrystals / edited by Gudrun Kissinger and Sergio Pizzini.
pages cm
Includes bibliographical references and index.
ISBN 978-1-4665-8664-2 (hardcover : alk. paper) 1. Germanium alloys. 2. Silicon alloys. I. Kissinger, Gudrun. II. Pizzini, Sergio.

TA480.G4S545 2015
660'.2977--dc23 2014036669

Visit the Taylor & Francis Web site at
http://www.taylorandfrancis.com

and the CRC Press Web site at
http://www.crcpress.com

To Anna
—S. P.

To Wolfgang, Dietmar, and Thomas
—G. K.

Contents

Preface ix
Editors xi
Contributors xiii

Chapter 1 Modern Aspects of Czochralski and Multicrystalline Silicon Crystal Growth 1

Koichi Kakimoto and Bing Gao

Chapter 2 Growth and Characterization of Silicon–Germanium Alloys 23

Ichiro Yonenaga

Chapter 3 Germanium on Silicon: Epitaxy and Applications 61

Daniel Chrastina

Chapter 4 Self-Interstitials in Silicon and Germanium 87

Alexandra Carvalho and Robert Jones

Chapter 5 Vacancies in Si and Ge 119

Eiji Kamiyama, Koji Sueoka, and Jan Vanhellemont

Chapter 6 Self- and Dopant Diffusion in Silicon, Germanium, and Their Alloys 159

Hartmut Bracht

Chapter 7 Hydrogen in Si and Ge 217

Stefan K. Estreicher, Michael Stavola, and Jörg Weber

Chapter 8 Point Defect Complexes in Silicon 255

Edouard V. Monakhov and Bengt G. Svensson

Chapter 9 Defect Delineation in Silicon Materials by Chemical Etching Techniques 289

Bernd O. Kolbesen

Chapter 10 Investigation of Defects and Impurities in Silicon by Infrared and Photoluminescence Spectroscopies 323

Simona Binetti and Adele Sassella

Chapter 11 Device Operation as Crystal Quality Probe 353

Erich Kasper and Wogong Zhang

Chapter 12 Silicon and Germanium Nanocrystals 377

Corrado Spinella and Salvo Mirabella

Index 407

Preface

Silicon, germanium, and silicon–germanium alloys cover the entire microelectronic applications with still undiscovered potentialities.

Silicon is a mature material, but it is still unique for its crucial role in the microelectronic and photovoltaic sectors. Thousands of articles and hundreds of books have been published since the beginning of the microelectronic era, discussing all the issues concerning the preparation and the growth technologies of semiconductor silicon, the role of impurities and of point and extended defects, and the techniques to minimize their effect on silicon as a material and as a substrate for microelectronics and photovoltaic applications.

Albeit the first transistor was made of germanium, germanium lost its importance soon and silicon became the basic material for integrated circuits. During the last decade, germanium experienced, however, a renaissance as a material capable of being integrated in silicon and compound semiconductor technologies. The integration of germanium into silicon integrated circuit (IC) technology is crucial for the on-chip light emission and for optoelectronic systems, and it can act as a bridge to GaAs integration into silicon devices.

Eventually, silicon–germanium alloys cover a technology sector where silicon is in defect, as it enables faster, more efficient devices to be manufactured using smaller, less noisy circuits than conventional silicon permits through greater integration of components onto the chips. The key advantage of silicon–germanium alloys over their rival technologies is their compatibility with mainstream complementary metal-oxide-semiconductor (CMOS) processing. In addition, silicon–germanium alloys provide ultra-high-frequency capability (well over 100 GHz) on the identical silicon platform where baseband, memory, and digital signal processing functions can also be integrated. This is the rationale and the reason why silicon–germanium alloys prominently appear on technology roadmaps around the world and why the alloys attract increasing attention.

Despite the vast knowledge accumulated on silicon, germanium, and their alloys in more than half a century, these materials still demand research, eminently in view of the improvement of knowledge on silicon–germanium alloys and the potentialities of silicon as a substrate for high-efficiency solar cells and for compound semiconductors and the ongoing development of nanodevices based on nanowires and nanodots.

These are the basic reasons for editing this book on these materials, at least partially based on recent achievements in the field of crystal growth, point defects, extended defects, and impurities of silicon and germanium nanocrystals discussed in a series of symposia held under the sponsorship of the European EMRS in the last four years, which was chaired by the editors of this book.

The editors' choice in the selection of chapter titles and authors was facilitated by the eminent scientific personalities of most of the participants and by the topics discussed by them, which covered the entire spectrum of R&D activities in the field of silicon, germanium, and their alloys, and by a common consensus on the issues, which should be brought together as the binding factor of the book.

It was possible, therefore, to focus attention on bulk, thin film, and nanostructured materials growth and characterization problems in the first three chapters, and in Chapter 12, where the importance of theoretical modeling is also strongly emphasized as a key for success. The role of defects and impurities is deeply accounted for in five chapters authored by world-renowned experts, each of them reporting a concerned view of the most recent experimental and theoretical advancements in the field. On the editors' preference, the three remaining chapters deal with issues of key applicative value: chemical etching as a defect identification technique, the spectroscopic analysis of impurities, and the use of devices as tools for the measurement of materials' quality.

The editors are indebted to their colleagues and friends who dedicated consistent time to work on and write the book. The success they wish for the book is entirely their due.

Editors

Gudrun Kissinger earned a PhD in crystallography at the University of Leipzig and worked as a research scientist and project leader in Innovations for High Performance Microelectronics (IHP) in Frankfurt (Oder), Germany, from 1984 to 2001. She developed a technology for gettering heavy metal impurities in defect-engineered and dielectrically insulated high-voltage substrates. She also worked in the development of silicon wafer bonding for silicon on insulator (SOI) substrate technologies. Dr. Gudrun was a guest scientist at the Institute for Microelectronics in Stuttgart in the field of wafer bonding in 1990. After the unification of Germany, the scope of the IHP changed to SiGe technologies for high-frequency applications. Her main activities were first with the development of a graded SiGe buffer with extremely low density of threading dislocations, with failure analysis and gettering of heavy metals in high-frequency SiGe-HBT devices. In the mid-1990s, she started working in the field of as-grown defects in silicon and their evolution during device processing. From 2001 to 2004, she cofounded and joined the start-up company Communicant Semiconductor Technologies AG as executive assistant to the CEO and Vice President of Strategic Planning. After the winding down of the project in 2004, she returned to the IHP, where she is now working as a project leader in industrial cooperation with Siltronic AG.

Her scientific expertise spans from defect generation during silicon crystal growth and device processing to failure analysis and yield in device manufacturing, gettering of metallic impurities in silicon, SOI substrate fabrication, crystal defects in SiGe layer structures and HBT devices, and SiGe graded buffer layers for strained silicon substrates.

Dr. Gudrun has published more than 100 technical papers in international journals and conference proceedings. She has cochaired two SPIE symposia on "In-Line Characterization, Yield, Reliability and Failure Analysis in Wafer Manufacturing" and the European Materials Research Society (EMRS) symposia on "Advanced Silicon for the 21st Century," "Advanced Silicon Materials Research for Electronic and Photovoltaic Applications I–III," and "Materials Research for Group IV Semiconductors: Growth, Characterization, and Technological Developments." To date, she has mainly worked on direct research contracts and projects with semiconductor device manufacturers and with the wafer manufacturing industry. Dr. Gudrun is a member of the Electrochemical Society.

Sergio Pizzini started his scientific career at the Joint Research Centre of the European Commission in Ispra, Italy, and later in Petten, the Netherlands, where he conducted thermodynamic and electrochemical studies on molten fluorides and ionic oxides in cooperation with the Oak Ridge nuclear center in the United States.

After leaving the Commission at the beginning of 1970, he joined the University of Milano as an associate professor, where he started basic studies on solid electrolytes, which resulted as an example in the realization of a prototype of a solid electrolyte sensor for the determination of the stoichiometry of nuclear oxide fuels and of the concentration of oxygen in inert gases. Still maintaining his position at the university, from 1975 to 1979, he worked as a director of the Materials Department of the Corporate Research Centre of Montedison in Novara, where he launched several new R&D activities on advanced materials for electronics, including InP and silicon. In the next few years, as the CEO of a research company, he studied a process for the production of solar-grade silicon and patented a new furnace and a process for the directional solidification of silicon in multicrystalline ingots. In 1982, he left external assignments for the University of Milano and, later, the University of Milano-Bicocca as a full-time professor of physical chemistry.

During this last period of activity, in addition to his teaching and university management duties, he carried out systematic studies on semiconductor silicon, mostly addressed at the understanding of the electronic and optical properties of point and extended defects of Czochralski, multicrystalline, and nanocrystalline silicon in the frame of national and European projects, as local or the European coordinator of joint R&D activities with major Italian companies operating in the sector.

He was chairman or co-chairman of a number of international symposia in the materials science field in Europe, USA, China, and India. Recently, he cochaired the European Materials Research Society (EMRS) 2008, 2010, 2012, and 2014 symposia on Advanced Silicon and Silicon–Germanium Materials.

He is the author of more than 250 technical articles published in peer-reviewed international journals. He has authored or coauthored four books, including *Advanced Silicon Materials for Photovoltaic Applications*. He served as the guest editor for the *Proceedings of the 2007 Gadest Conference on Gettering and Defect Engineering in Semiconductor Technology* and all the *EMRS Symposia on Advanced Silicon Materials*.

Contributors

Simona Binetti
Department of Materials Science
University of Milano-Bicocca
Milano, Italy

Hartmut Bracht
Institute of Materials Physics
University of Münster
Münster, Germany

Alexandra Carvalho
Graphene Research Center
National University of Singapore
Singapore

Daniel Chrastina
L-NESS Politecnico di Milano
Como, Italy

Stefan K. Estreicher
Physics Department
Texas Tech University
Lubbock, Texas

Bing Gao
Research Institute for Applied Mechanics
Kyushu University
Kasuga, Japan

Robert Jones
School of Physics
University of Exeter
Exeter, England, United Kingdom

Koichi Kakimoto
Research Institute for Applied Mechanics
Kyushu University
Kasuga, Japan

Eiji Kamiyama
Global Wafers Japan Co., Ltd.
Kanagawa, Japan

Erich Kasper
University of Stuttgart
IHT
Stuttgart, Germany

Bernd O. Kolbesen
Institute of Inorganic and Analytical Chemistry
Goethe-University of Frankfurt/Main
Frankfurt am Main, Germany

Salvo Mirabella
Institute of Microelectronics and Microsystems – CNR
Catania, Italy

Edouard V. Monakhov
Department of Physics/Center for Materials Science and Nanotechnology
University of Oslo
Oslo, Norway

Adele Sassella
Department of Materials Science
University of Milano-Bicocca
Milano, Italy

Corrado Spinella
Institute of Microelectronics and Microsystems—CNR
Catania, Italy

Michael Stavola
Physics Department
Lehigh University
Bethlehem, Pennsylvania

Koji Sueoka
Department of Communication Engineering
Okayama Prefectural University
Okayama, Japan

Bengt G. Svensson
Department of Physics/Center for Materials Science and Nanotechnology
University of Oslo
Oslo, Norway

Jan Vanhellemont
Department of Solid State Sciences
Ghent University
Ghent, Belgium

Jörg Weber
Institute for Applied Physics
Technical University Dresden
Dresden, Germany

Ichiro Yonenaga
Institute for Materials Research
Tohoku University
Sendai, Japan

Wogong Zhang
University of Stuttgart
IHT
Stuttgart, Germany

1 Modern Aspects of Czochralski and Multicrystalline Silicon Crystal Growth

Koichi Kakimoto and Bing Gao

CONTENTS

1.1 Introduction ... 1
1.2 Light Element Transfer in a CZ Furnace ... 3
1.3 Gas Flow Fields and Temperature Fields in a CZ Furnace ... 5
1.4 Distributions of SiO(*g*) and CO(*g*) in the Gases in a CZ Furnace ... 8
1.5 Distributions of C(*m*) and O(*m*) Atoms in the Melt in a CZ Furnace ... 10
1.6 Impurity Transfer in a DS Furnace ... 11
1.7 Controlling the Heat and Mass Transfer with Magnetic Fields in a CZ Furnace ... 11
1.8 Summary ... 18
Acknowledgment ... 18
References ... 19

1.1 INTRODUCTION

Advanced developments in microelectronic and optoelectronic devices, photovoltaic (PV) cells, and power devices for electric vehicles are required to improve the quality of our life in the information, energy production, and energy-saving fields. Single-crystal silicon, having widely been used in applications such as computers, solar cells, and inverters in power devices, is one of the key materials to realize these goals due to its low-cost and high quality. To realize these improvements, there is the need to grow silicon crystals with an even lower concentration of impurities (including oxygen and carbon), dislocations and point defects in the material grown today.

It is well-known that the Czochralski (CZ)[1] process is a well-established technique for large-scale production of silicon-single crystals. It is used not only for microelectronic applications but also for the manufacturing of PV cells. PV cells based on CZ silicon possess high conversion efficiency and the production costs are relatively low. Another crystal growth process, the unidirectional solidification process (DS),[2] is almost exclusively used for the growth of silicon ingots for PV applications. The production cost of such silicon material is lower than for CZ silicon but the efficiencies of the PV cells usually remain lower than for PV cells on CZ silicon.[3] In both crystal growth methods, the main source of impurities is the quartz crucible in which the silicon charge is melted. The growth process itself is responsible for the introduction of impurities and defects into the silicon ingot.[4]

The mechanism of carbon and oxygen incorporation during CZ and DS growth is similar. However, due to the specific configurations of the DS growth furnaces, where the quartz crucible is fully embedded in a graphite felt insulation, the concentration of oxygen in a crystal grown by the unidirectional solidification method is one order of magnitude smaller than that in a CZ crystal, while the concentration of carbon is two or three orders of magnitude larger than that in a CZ crystal.

While CZ crystals are almost dislocation free, thanks to the use of the necking process,[5] the control of intrinsic point defects (vacancies and interstitials) in the crystals during the growth process still remains a challenge. The accomplishment of this issue is a crucial technological requirement, because vacancies can form voids and self-interstitials can aggregate in dislocation loops during the cooling process after the solidification of the crystals. The voids are known to degrade the gate oxide integrity when employed in VLSI applications and both defects reduce the lifetime of the minority carriers when used in power devices.[6,7] Therefore, the distribution and concentration of point defects need to be carefully controlled during the crystal cooling process by an effective design of the furnace hot zone and active management of the heater power control.

A further challenge concerns the presence of oxygen and carbon, which are the main contaminants of the CZ crystals. The effective control of their concentration is crucial for the production of high-quality crystals. The experimental approach to the solution of this problem is limited by the time, cost and complexity of the analysis, especially when carried out on industrial scale furnaces. Developments in computer technologies allow, instead, the simulation of the global environments of crystal growth and to find solutions capable of improving the purity of the crystals.

The main concern in the simulation of the impurity contamination processes during crystal growth is the transport of impurities from the impurity sources to the crystal. Among the many simulations of impurity transport which have been performed,[8–19] most were local simulations[8–14] that neglected the transport of the impurities in the gas phase. A few of these studies used global simulations,[15–18] but in one of these the presence of both oxygen and carbon in the melt was neglected,[15] while in the others the presence of carbon was neglected for both the gas and the silicon melt.[16–18] There have been no simulations that take both the oxygen and carbon impurities into account in cooling gas and silicon melt.

Similar to the CZ method, contamination with light elements in the unidirectional solidification is governed by the transport of impurities.[19] Carbon is one of the major impurities in multicrystalline silicon ingots and the carbon concentration can affect the density and electrical activity of dislocations.[20] When the concentration of carbon exceeds its solubility limit, it will precipitate to form silicon carbide (SiC) particles, which can cause severe Ohmic shunts in PV cells and result in the nucleation of new grains in silicon ingots.[21] Carbon, oxygen, and SiC particles in a solidified silicon ingot cause significant deterioration of the conversion efficiency in PV cells. If the carbon concentration exceeds 4×10^{16} atoms/cm^3, it will markedly influence the precipitation of oxygen during the thermal annealing of crystals and during device processing of the wafers cut from these crystals.[22–27] Oxygen precipitates are known to act as intrinsic gettering sites for impurities and affect also the mechanical strength of the wafers.[28,29] Therefore, effective control of the carbon concentration in a crystal is required for the production of high-quality crystals.

Experiments[30–33] and numerical simulations [10,17,34–37] have been carried out to find technical solutions capable of improving the purity of crystals. The C and O concentrations in the multicrystalline silicon crystals can be controlled by a specially designed gas flow system.[33] A tungsten crucible cover in a DS furnace can reduce the O concentration to a reasonable value and markedly reduces the C concentration. However, the silicon vapor is very reactive toward tungsten.[29] Therefore, covers made of SiC-coated carbon are preferred.

1.2 LIGHT ELEMENT TRANSFER IN A CZ FURNACE

It is well recognized that the contamination of silicon ingots with oxygen impurities originates from the dissolution of the silica crucible.[10] The dissolved oxygen atoms combine with the silicon atoms in the melt to form gas-phase SiO at the gas–melt interface. The SiO is then transferred by the argon gas flow to all of the graphite components in the furnace, reacting with them to form CO. The formed CO is then transported back to the melt surface by diffusion and/or convection. Finally, the CO is dissolved into the melt and C and O are segregated into the crystal.

The global process of C and O transfer is described by the following set of differential equations:

$$\frac{\partial c_{\mathrm{SiO}}}{\partial t} + \nabla \cdot \left(c_{\mathrm{SiO}} \vec{u}_{\mathrm{Ar}}\right) = \nabla \cdot \left[c_{\mathrm{Ar}} D_{\mathrm{SiO}} \nabla \left(\frac{c_{\mathrm{SiO}}}{c_{\mathrm{Ar}}}\right)\right] \tag{1.1}$$

$$\frac{\partial c_{\mathrm{CO}}}{\partial t} + \nabla \cdot \left(c_{\mathrm{CO}} \vec{u}_{\mathrm{Ar}}\right) = \nabla \cdot \left[c_{\mathrm{Ar}} D_{\mathrm{CO}} \nabla \left(\frac{c_{\mathrm{CO}}}{c_{\mathrm{Ar}}}\right)\right] \tag{1.2}$$

where c_{SiO} is the molar concentration of the SiO vapors, c_{CO} is the molar concentration of the CO gas and c_{Ar} is the molar concentration of the argon gas. The vector $\vec{u}_{\mathrm{Ar}}$ is the flow vector of the argon gas. The diffusivities D_{SiO} and D_{CO} in Equations 1.1 and 1.2 depend on the temperature T (K) and the pressure p (dyn/cm^2)[14,37,38] according to the following equations:

$$D_{\mathrm{SiO}} = a_{\mathrm{SiO}} \frac{T^{1.75}}{p} \tag{1.3}$$

$$D_{\mathrm{CO}} = a_{\mathrm{CO}} \frac{T^{1.75}}{p} \tag{1.4}$$

where $a_{\mathrm{SiO}} = 0.862611$ and $a_{\mathrm{CO}} = 1.79548$ if D, T, and p are measured in cm²/s, K and dyn/cm², respectively. The concentrations of the oxygen and carbon atoms in the melt were modeled using the following equations:

$$\frac{\partial c_{\mathrm{C}}}{\partial t} + \nabla \cdot \left(c_{\mathrm{C}} \vec{u}_{\mathrm{Si}}\right) = \nabla \cdot \left[c_{\mathrm{Si}} D_{\mathrm{C}} \nabla \left(\frac{c_{\mathrm{C}}}{c_{\mathrm{Si}}} \right) \right] \tag{1.5}$$

$$\frac{\partial c_{\mathrm{O}}}{\partial t} + \nabla \cdot \left(c_{\mathrm{O}} \vec{u}_{\mathrm{Si}}\right) = \nabla \cdot \left[c_{\mathrm{Si}} D_{\mathrm{O}} \nabla \left(\frac{c_{\mathrm{O}}}{c_{\mathrm{Si}}} \right) \right] \tag{1.6}$$

where c_{C} is the molar concentration of carbon atoms in the melt, c_{O} is the molar concentration of oxygen atoms in the melt, and c_{Si} is the molar concentration of silicon atoms. $\vec{u}_{\mathrm{Si}}$ is the flow vector of silicon melt. D_{O} and D_{C} are the diffusivities of oxygen and carbon in the melt, respectively, assuming to amount 5.0×10^{-8} m²/s.[38]

The chemistry of the process is the last point to be examined to complete the model of the process starting from the dissolution of the quartz crucible in the silicon melt, which can be described by the following reaction:

$$\mathrm{SiO_2}(s) \leftrightarrow \mathrm{Si}(m) + 2\mathrm{O}(m) \tag{1.7}$$

The index symbol (*s*) denotes the solid, (*m*) denotes the melt, (*g*) denotes the gas, (*l*) denotes the liquid, and (*c*) denotes the crystal.

For a CZ furnace (it also holds for DS), the equilibrium concentration of oxygen atoms c_{O} on the walls of the quartz crucible is expressed as[19]

$$c_{\mathrm{O}} = \frac{b}{1-b} \times 0.5 \times 10^{23} [\mathrm{cm}^{-3}] \tag{1.8}$$

where the parameter $b = 1.32 \times \exp((-7150/T) - 6.99)$, which depends on the temperature of the crucible wall.

The oxygen atoms in the melt are transported to the melt–gas interface by convection or diffusion, where they react with silicon atoms to form SiO vapors.

$$\mathrm{O}(m) + \mathrm{Si}(l) \leftrightarrow \mathrm{SiO}(g) \tag{1.9}$$

The equilibrium relationship between the concentrations c_{SiO} of SiO in the gas phase and that of oxygen in the melt phase $c_{\mathrm{O}(m)}$ is given by the following expression:[15,38]

$$c_{\mathrm{SiO}} = (101{,}325/RT) \frac{c_{\mathrm{O}(m)}}{c_{\mathrm{Si}}} e^{-21{,}000/T + 17.8} \tag{1.10}$$

where R is the universal gas constant (8.314 $J \cdot K^{-1} \cdot mol^{-1}$). When SiO($g$) comes into contact with the hot carbon surfaces embedding the crucible during the gas phase transport, the following reaction occurs:

$$\mathrm{SiO}(g) + 2\mathrm{C}(s) \leftrightarrow \mathrm{CO}(g) + \mathrm{SiC}(s) \tag{1.11}$$

with a free energy of reaction ΔG given by the following expressions:[15,38]

$$\begin{aligned} \Delta G &= -81{,}300 + 3.02T(\mathrm{J/mol}) \quad \text{for } T < 1640\,\mathrm{K} \\ \Delta G &= -22{,}100 - 33.1T(\mathrm{J/mol}) \quad \text{for } 1640\,\mathrm{K} < T < 1687\,\mathrm{K} \\ \Delta G &= -72{,}100 - 3.44T(\mathrm{J/mol}) \quad \text{for } T > 1687\,\mathrm{K} \end{aligned} \tag{1.12}$$

Eventually, one has to account for the transport of CO from the gas phase back to the gas–melt interface by the argon flow. At the gas–melt interface, the CO gas dissolves into the melt according to the reaction

$$\mathrm{CO}(g) \leftrightarrow \mathrm{C}(m) + \mathrm{O}(m) \tag{1.13}$$

and the equilibrium condition is given by[15,38]

$$c_{\mathrm{CO}} = (101{,}325/RT)\frac{c_{\mathrm{O}}}{c_{\mathrm{Si}}}\frac{c_{\mathrm{C}}}{c_{\mathrm{Si}}}e^{-5210/T+14.5} \tag{1.14}$$

We used C(c) = $k_{\mathrm{C_{seg}}}$C(m) and O(c) = $k_{\mathrm{O_{seg}}}$O(m) with $k_{\mathrm{C_{seg}}} = 0.07$[13] and $k_{\mathrm{O_{seg}}} = 0.85$.[39]

1.3 GAS FLOW FIELDS AND TEMPERATURE FIELDS IN A CZ FURNACE

The flow of argon gas through the furnace is considered to be a compressible and axisymmetric flow. The compressible flow solver can accurately simulate the buoyancy-driven flow due to the large density variation in the furnace. Although the flow velocity in this furnace is low, yet the density variation is significant for this buoyancy-driven flow, which is similar as combustion problem. Fully coupled compressible multispecies flow for N species written in general curvilinear coordinates is treated as follows:[40]

$$\frac{\partial \overline{Q}}{\partial t} + \frac{\partial \overline{E}}{\partial \xi} + \frac{\partial \overline{F}}{\partial \eta} = \frac{\partial \overline{E}_v}{\partial \xi} + \frac{\partial \overline{F}_v}{\partial \eta} + \overline{H} + \overline{G} \tag{1.15}$$

where the vectors $\overline{Q}$, $\overline{E}$, $\overline{F}$, $\overline{E}_v$, $\overline{F}_v$, $\overline{H}$, and $\overline{G}$ are defined as

$$\overline{Q} = \frac{y^{\delta}}{J}Q, \overline{E} = \frac{y^{\delta}}{J}(\xi_x E + \xi_y F), \overline{F} = \frac{y^{\delta}}{J}(\eta_x E + \eta_y F), \overline{E}_v = \frac{y^{\delta}}{J}(\xi_x E_v + \xi_y F_v)$$

$$\overline{F}_v = \frac{y^{\delta}}{J}(\eta_x E_v + \eta_y F_v), \overline{H} = \frac{1}{J}H, \overline{G} = \frac{y^{\delta}}{J}G$$

In the above expressions, t, ξ, and η are the time and generalized spatial coordinates, x and y are the axial and radial coordinates, and J is the transformation Jacobian. The symbol δ is an index for two types of governing equations: with $\delta = 0$ the system is two-dimensional and with $\delta = 1$ the system is axisymmetric. The vector H includes axisymmetric terms and rate of change of species due to chemical reaction.[40] The other vectors Q, E, F, E_v, F_v, and G are

$$Q = \left[\rho, \rho u, \rho v, \rho e_t, \rho C_1, \ldots, \rho C_{N-1}\right]^T$$

$$E = \left[\rho u, \rho u^2 + p, \rho uv, (\rho e_t + p)u, \rho u C_1, \ldots, \rho u C_{N-1}\right]^T$$

$$F = \left[\rho v, \rho uv, \rho v^2 + p, (\rho e_t + p)v, \rho v C_1, \ldots, \rho v C_{N-1}\right]^T$$

$$E_v = \left[0, \tau_{xx}, \tau_{xy}, u\tau_{xx} + v\tau_{xy} + q_{xe}, q_{x1}, \ldots, q_{xN-1}\right]^T$$

$$F_v = \left[0, \tau_{xy}, \tau_{yy}, u\tau_{xy} + v\tau_{yy} + q_{ye}, q_{y1}, \ldots, q_{yN-1}\right]^T$$

$$G = \left[0, \rho g_x, \rho g_y, \rho g_x u, \rho g_y v, \ldots, 0\right]^T$$

where ρ, p, u, and v represent the density, pressure, and Cartesian velocity components, e_t is the total internal energy, C_i is the mass fraction of species i, τ_{xx} and τ_{yy} are normal stresses and τ_{xy} is shear stress, g_x and g_y are Cartesian gravity components, and q_{xi} and q_{yi} are the diffusion fluxes of the species. The energy flux $\vec{q}_e$ includes the thermal flux due to the temperature gradient and the thermal flux due to the concentration diffusion, that is,

$$\vec{q}_e = k\nabla T + \rho \sum_{i=1}^{N} \left(h_i D_i \nabla C_i\right) \tag{1.16}$$

where h_i is the thermodynamic enthalpy of species, i, k is the thermal conductivity of mixtures calculated using Wilke's mixing rule,[41] and D_i is the effective binary diffusivity of species i in the gas mixture defined as[40]

$$D_i = (1 - X_i) / \sum_{j \neq i}^{N} \left(X_j / \overline{D}_{ij}\right) \tag{1.17}$$

X_i is the molar fraction of species i, and $\overline{D}_{ij}$ is the binary mass diffusivity between species i and j obtained from the Chapman–Enskog theory.[41]

A second-order total variation diminishing (TVD) scheme is used for convection terms and Davis–Yee symmetric TVD is used for the flux limiter vectors.[42]

An entropy correction term is also introduced to satisfy the entropy condition. A fourth-order Runge–Kutta scheme is used for time marching and a central difference scheme is used for viscous terms. Local time step[42] and low Mach number acceleration technique[43,44] are adopted for efficiency. The domain-decomposition method is used to solve the whole flow field.

The temperatures at all solid surfaces are set to the values computed from the global heat transfer. The velocities at all solid surfaces are set to zero. For the conditions at inlet and outlet, different methods can be used according to our practical tests. For axisymmetric and subsonic inflow, three characteristic lines enter into the domain and, therefore, three analytical boundary conditions are specified. The remaining one is evaluated on the basis of information from the interior points near the inlet. For subsonic outflow, one characteristic line enters into the computation domain from the outside and, thus, one analytical boundary condition is given and others are evaluated from the interior points near the outlet. For the first simulation, inlet flow rate Q is set to 0.5 l/min, inlet static temperature T is set to 350 K, and inlet tangential velocity u is set to 0. The outlet static pressure p is set to 10 Torr. The inner and outer radii of inlets were 9.5 and 15.9 mm, respectively.

The gas flow field inside a typical configuration of a CZ furnace is shown in Figure 1.1 and the temperature field inside the global furnace is shown in Figure 1.2.[38] In the furnace, several re-circulations were formed. There was a small re-circulation above

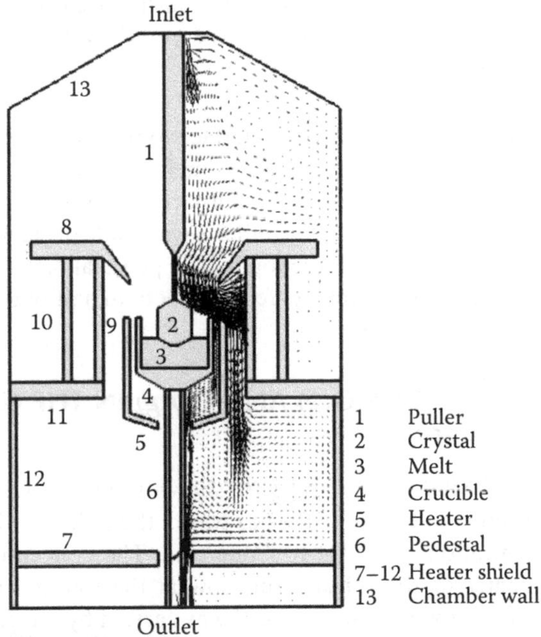

FIGURE 1.1 Configuration of a furnace and the flow rates in the furnace. (Reprinted from *J. Cryst. Growth*, 312, B. Gao, K. Kakimoto. Global simulation of coupled carbon and oxygen transport in a Czochralski furnace for silicon crystal growth, 2972, Copyright 2010, with permission from Elsevier.)

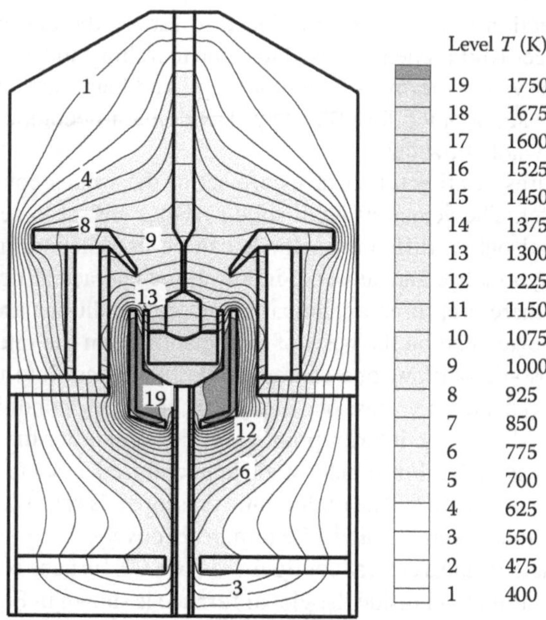

FIGURE 1.2 Temperature contours inside the CZ furnace (spacing 75 K). (Reprinted from *J. Cryst. Growth*, 312, B. Gao, K. Kakimoto. Global simulation of coupled carbon and oxygen transport in a Czochralski furnace for silicon crystal growth, 2972, Copyright 2010, with permission from Elsevier.)

the melt, bringing SiO from the gas–melt interface to the top of the crucible, before taking CO from the top of a crucible to the gas–melt interface.

The concentrations of SiO and CO in the furnace were sensitive to the materials of furnace components. In Figure 1.1, the puller 1 was made of steel and the crucible 4 was made of quartz. The heater 6, the top half of pedestal 5, and the heat shields 7–12 were all made of carbon or graphite felt. The bottom half of the pedestal 5 and the chamber wall 13 were made of steel.

1.4 DISTRIBUTIONS OF SiO(*g*) AND CO(*g*) IN THE GASES IN A CZ FURNACE

The concentrations of SiO(*g*) and CO(*g*) are shown in Figure 1.3a and b, respectively. The concentration of SiO(*g*) in the gas was in the order of 10^{-9} mol/cm^3 with a maximum concentration of ~7.7×10^{-9} mol/cm^3. The concentration of SiO(*g*) was high in the center of the furnace and low at the top and bottom because SiO(*g*) evaporated from the melt surface and was carried by the gas from the center of the furnace to the bottom of the furnace. During transportation in a furnace, the SiO(*g*) reacted with the hot carbon walls and was gradually reduced. The maximum CO concentration shown in Figure 1.3b was approximately 6.0×10^{-10} mol/cm^3, lying in the corners of the furnace, formed by the heat shields labeled 10 and 12 shown in Figure 1.1. This value was one order of magnitude lower

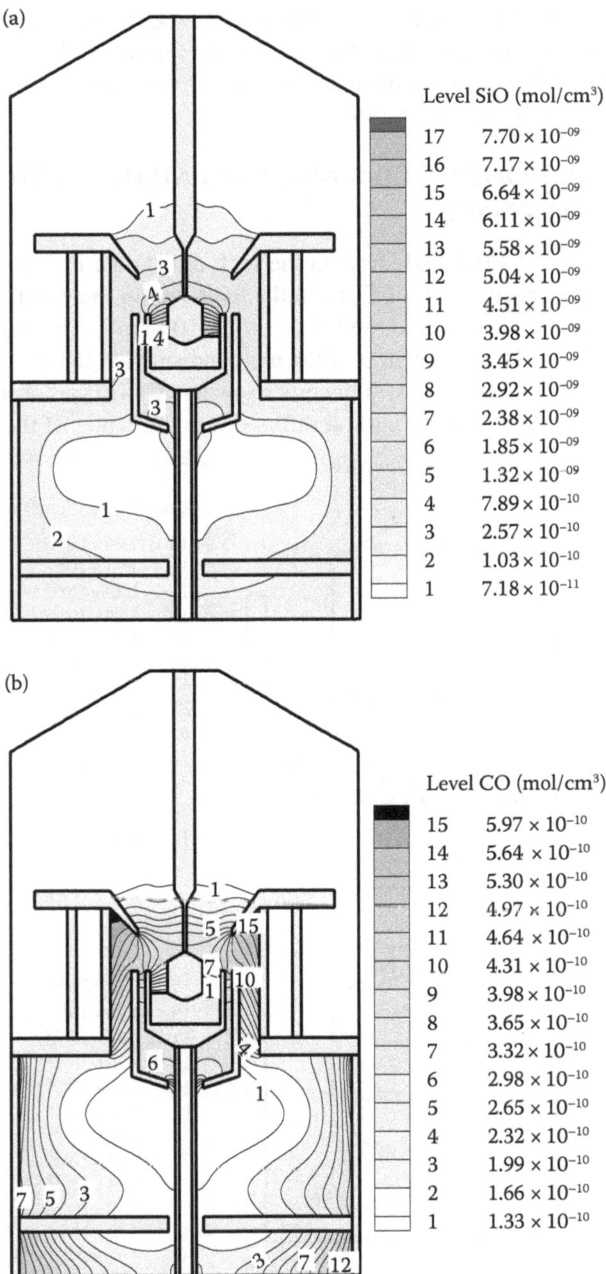

FIGURE 1.3 SiO (a) and CO (b) concentrations in argon gas in the furnace. (Reprinted from *J. Cryst. Growth*, 312, B. Gao, K. Kakimoto. Global simulation of coupled carbon and oxygen transport in a Czochralski furnace for silicon crystal growth, 2972, Copyright 2010, with permission from Elsevier.)

than the maximum SiO concentration. The minimum CO concentration was about 1.3×10^{-10} mol/cm^3 on the melt surface and at the top of the furnace. Similar to the distribution of SiO(*g*), the concentration of CO was large in the center of the furnace, except for inside the crucible.

1.5 DISTRIBUTIONS OF C(*m*) AND O(*m*) ATOMS IN THE MELT IN A CZ FURNACE

The concentrations of C(*m*) and O(*m*) in the melt are shown in Figure 1.4a and b, respectively. The concentration of C(*m*) in the melt was in the order of ~10^{18} cm^{-3}. The average concentration in the melt was ~1.4×10^{18} cm^{-3}. The concentration of C(*m*) in the melt was large at the top of the melt and small at the center of the melt. This distribution was consistent with the process where CO(*g*) was absorbed and dissolved into the melt and then the carbon diffused into the center of the melt.

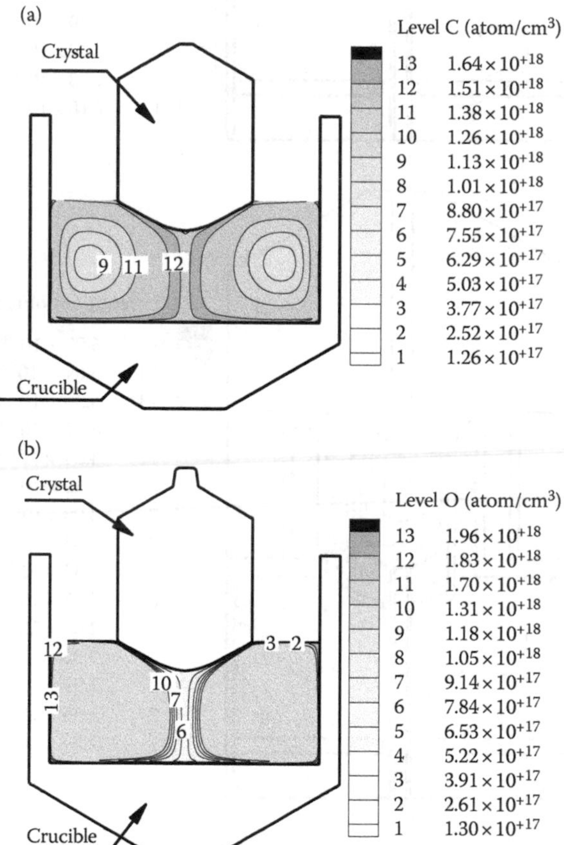

FIGURE 1.4 Carbon (a) and oxygen (b) concentrations in the silicon melt. (Reprinted from *J. Cryst. Growth*, 312, B. Gao, K. Kakimoto. Global simulation of coupled carbon and oxygen transport in a Czochralski furnace for silicon crystal growth, 2972, Copyright 2010, with permission from Elsevier.)

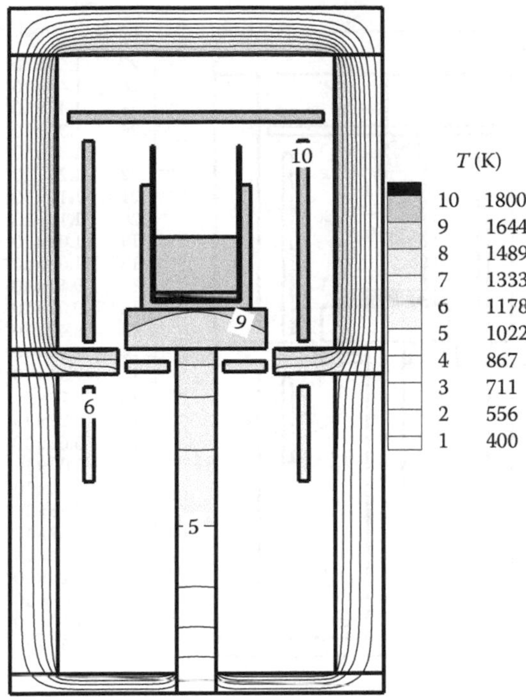

FIGURE 1.5 Temperature contours in the whole furnace (spacing of 54 K).

1.6 IMPURITY TRANSFER IN A DS FURNACE

This process is basically the same as that of the CZ system. Figure 1.5 shows the typical temperature distribution in a unidirectional DS furnace. At high temperatures, the silica crucible (SiO_2) was partially dissolved into the melt, the dissolved oxygen atoms were then transported to the gas–melt interface and evaporated as a SiO vapor. Subsequently, the SiO was carried by the argon gas flow to all of the graphite components, reacting with them to produce gas phase CO. The resultant CO was then transported back to the surface of the melt by either diffusion or convection and then dissolved into the melt. Finally, the C and O atoms were segregated into the crystal. The concentrations of CO and SiO in the furnace are shown in Figure 1.6a and b, respectively. The concentrations of C and O in the melt are shown in Figure 1.7a and b, respectively. Their distributions were similar to those in the CZ furnace.

1.7 CONTROLLING THE HEAT AND MASS TRANSFER WITH MAGNETIC FIELDS IN A CZ FURNACE

The diameters of the crystals should be increased to reduce production costs of integrated circuits. To obtain crystals with large diameters, a crucible with a large diameter is required, but the flow in the melt in large diameter crucibles becomes

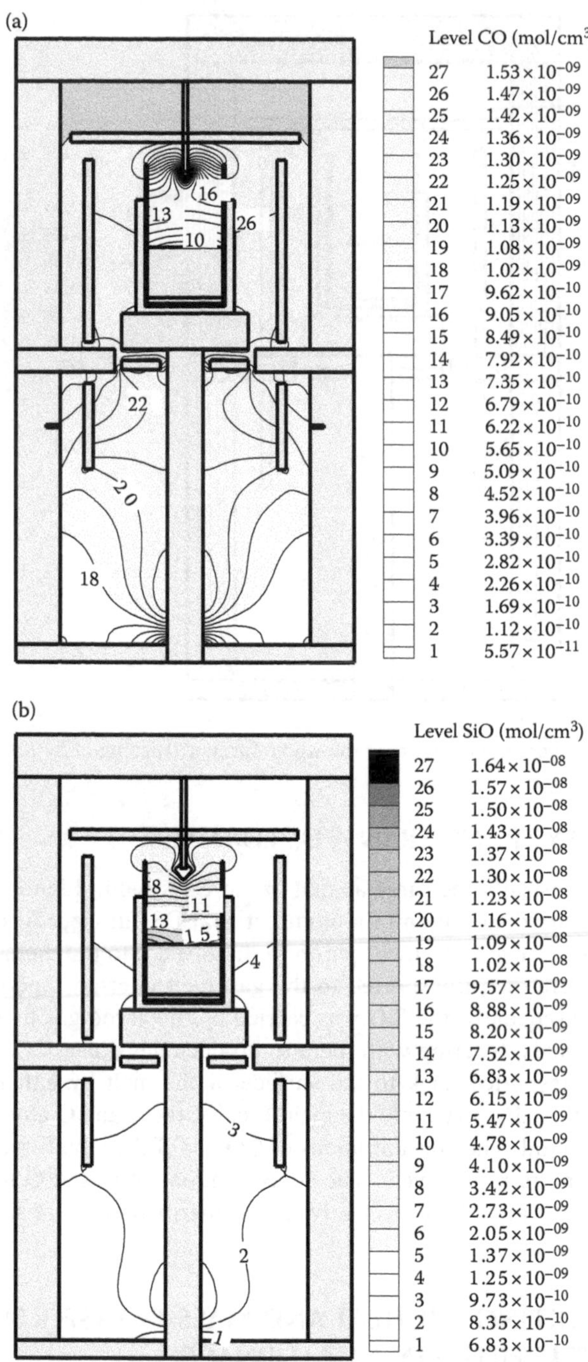

FIGURE 1.6 CO (a) and SiO (b) concentration distributions in the furnace. (Reproduced with permission from B. Gao, S. Nakano, K. Kakimoto, *J. Electrochem. Soc.*, 157, H153. Copyright 2010, The Electrochemical Society.)

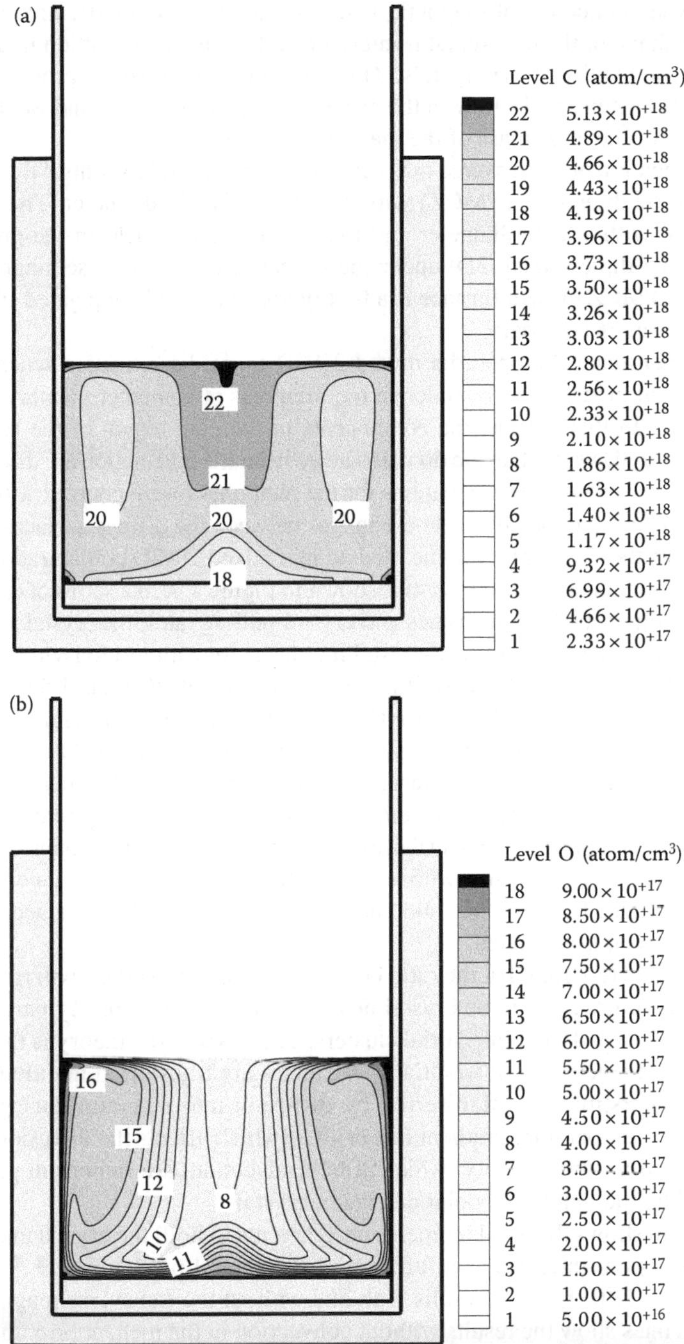

FIGURE 1.7 Carbon (a) and oxygen (b) concentrations in the melt. (Reproduced with permission from B. Gao, S. Nakano, K. Kakimoto, *J. Electrochem. Soc.*, 157, H153. Copyright 2010, The Electrochemical Society.)

unstable. The application of magnetic fields is known to be an effective method to control the shape of the melt–crystal interface and the melt convection in a crucible, improving the quality of the crystals. This method is effective for growing crystals with large diameters, as the flow in the crucible becomes unstable and weakly turbulent because of the large mass of the melt in the crucible.

The application of a transverse magnetic field can be used to control the melt flow in the CZ growth process (TMCZ) with crucibles of large diameter. The melt flow in a crucible with a large diameter and the global thermal field in the growth furnace are three-dimensional (3D) under the influence of a transverse magnetic field. Because the TMCZ growth furnace is a highly nonlinear and conjugated system, 3D global modeling is necessary.

Liu and Kakimoto[45] proposed a mixed 2D/3D space discretization scheme to perform 3D global modeling with moderate requirements of computer memory and computation time. In this scheme, the components in the core region of the furnace are discretized in a 3D way, while the domains away from the core region are discretized in a 2D way. The Navier–Stokes equations for the melt phase were coupled with conductive heat transfers and radiative heat exchanges between the diffuse surfaces, and they are solved together by a finite volume method in a mixed 2D/3D configuration. A local view of the computational grid system is shown in Figure 1.8. In a series of calculations with various magnetic field intensities and crystal-pulling rates, the crystal was rotated at 2 rpm and the shapes of the melt–crystal interfaces were almost axisymmetric.[45]

Figure 1.9 shows the influence of the crystal-pulling rate (0.3 and 1.5 mm/min) on the shape of the melt–crystal interface (Figure 1.9a) and the temperature distribution (Figure 1.9b) in the crystal and the melt.[45] Figure 1.9c shows the shapes of the calculated interfaces with the two different crystal-pulling rates. The magnitude of the magnetic field was set to 0.2 T for both cases. When a low pulling rate was applied to the crystal, the temperature distribution in the melt was less homogeneous and the interface was more convex than with a high pulling rate. This phenomenon is related to the heat balance between the solid and the liquid through the interface where the solidification heat is generated.

Voronkov[46,47] reported that the ratio between the local crystal growth rate (V_g) and the temperature gradient in the crystal near the interface (G) are key parameters in the formation of voids and interstitial clusters. The basis of the theory is the reaction between the vacancies and interstitials. Vacancies are transferred mainly by advection, while interstitials are transferred by diffusion into a crystal during cooling. Therefore, the temperature gradient in a crystal (which affects the diffusion process) and the crystal growth velocity (which affects advection) are important parameters for controlling the number of point defects in crystals.

Figure 1.10 shows the axial temperature gradients in both the crystal and the melt at the melt–crystal interface as a function of the crystal-pulling rate.[48–50] The solid lines in Figure 1.10 show the results with and without the transverse magnetic field, and dashed lines show the results without convection in the melt, approximately corresponding to the case for an infinite magnetic field. The arrows in Figure 1.10 show the contributions from convection in the melt. The values of the axial temperature gradients in the melt and crystal were obtained by averaging over the central area of the interface. These values were not identical, even when the crystal-pulling rate

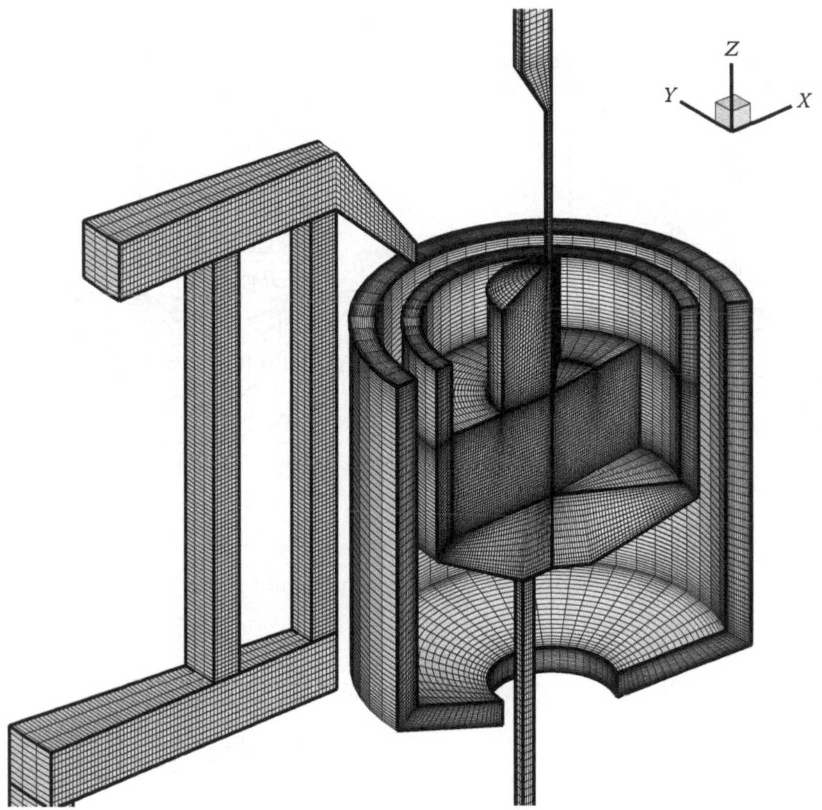

FIGURE 1.8 A local view of the grid calculated at the center domains inside a discrete system of a typical CZ growth furnace. (After *Int. J. Heat Mass Transfer* 48, L. J. Liu, K. Kakimoto. 2005. Partly three-dimensional global modeling of a silicon Czochralski furnace. I. Principles, formulation and implementation of the model, 4481, Copyright 2005 with permission from Elsevier.)

was zero because of the difference in the thermal conductivity between the crystal and the melt. These results showed that the axial temperature gradients in the melt and the crystal near the interface increased with an increasing magnetic field. The temperature gradient near the interface of the crystal increased, while that in the melt decreased with an increasing crystal-pulling rate. Meanwhile, this difference was reduced when the magnetic field intensity had a finite value (including zero) and when there was no melt convection. This difference occurred because of the melt convection, indicating that the contribution of the melt flow was reduced with an increase in the crystal-pulling rate.

When a magnetic field with a large intensity was applied, natural convection in the melt was suppressed, resulting in an inhomogeneous temperature distribution in the melt. Therefore, the temperature gradient in the melt increased with an increasing magnetic field. This was because there was a heat balance between the liquid and the solid at the interface. The temperature gradient in the crystal near the interface also increased. However, the melt convection remained, even when a magnetic field

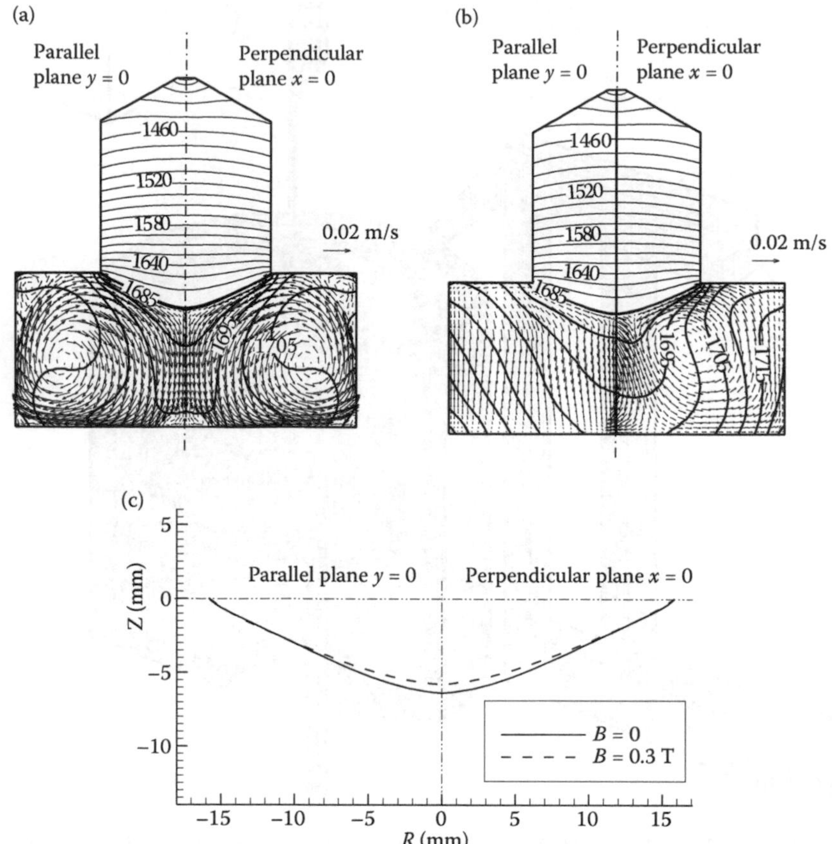

FIGURE 1.9 The melt flow, thermal field, and melt–crystal interface profiles in symmetric planes for $x = 0$ (right half of the plane) and $y = 0$ (left half of the plane) for different crystal-pulling rates of (a) 0.3 and (b) 1.5 mm/min. The isotherms were plotted every 15 K in the crystal and 5 K in the melt. (c) Comparisons between the shapes of the melt–crystal interfaces with different crystal-pulling rates. (After *Cryst. Res. Technol.* 40, L. J. Liu, K. Kakimoto, 3D global analysis CZ-Si growth in transverse magnetic field with rotating crucible and crystal, 347, Copyright 2005, with permission from Elsevier.)

with a relatively large intensity was applied to the melt. As a result, when a relatively large magnetic field of 0.3 T was applied to the system, the temperature gradients near the interface in both the crystal and the melt were very different from those without melt convection, as shown in Figure 1.10.

A larger crystal growth rate always resulted in a lower heater power. The temperature on the side wall of the melt decreased because of the lower heater power. The temperature difference was reduced, becoming more homogeneous in the melt. This led to weaker melt convection caused by a decrease in the thermal buoyant force that was induced by the temperature gradient in the melt. Therefore, increasing the crystal-pulling rate caused both the axial temperature gradient in the melt near the interface and the contribution from melt convection to decrease. However, the heat

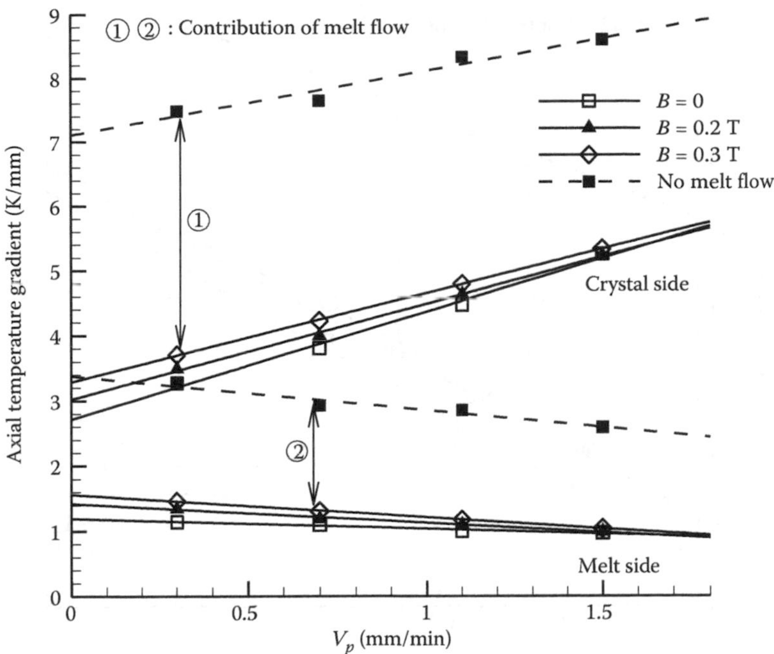

FIGURE 1.10 Axial temperature gradients in a crystal (upper part) and a melt (lower part) near the interface as a function of the crystal-pulling rate for different applied magnetic fields. (After *Cryst. Res. Technol.* 40, L. J. Liu, K. Kakimoto, 3D global analysis CZ-Si growth in transverse magnetic field with rotating crucible and crystal, 347, Copyright 2005, with permission from Elsevier.)

released during solidification at the interface was proportional to the crystal growth rate and was transported away from the interface through the crystal. Thus, a larger axial temperature gradient field was generated in the crystal near the interface when the crystal-pulling rate was increased.

Figure 1.11 shows the interface deflection toward the melt as a function of the ratio V_p/G between the crystal-pulling rate (V_p) and the temperature gradient in the crystal near the interface (G).[47] The values of interface deflection and V_p/G were obtained by averaging the results from the center of the interface. The arrows in Figure 1.11 show the contributions of convection in the melt. The interface moved upward to the crystal with an increase in either the magnetic field intensity or V_p/G. This tendency is consistent with that of the axial temperature gradient in the crystal near the interface, shown in Figure 1.11, because the shape of the interface is determined mainly by the temperature distribution in the crystal close to the interface and the melt convection in the crucible. When a large magnetic field was applied to the system or a faster pulling rate was applied to the crystal, the melt convection was suppressed and the axial temperature gradient in the crystal near the interface increased. The increase of the axial temperature gradient near the interface, which originates from the decrease of the melt convection, causes the upward movement of the melt–crystal interface toward the crystal.

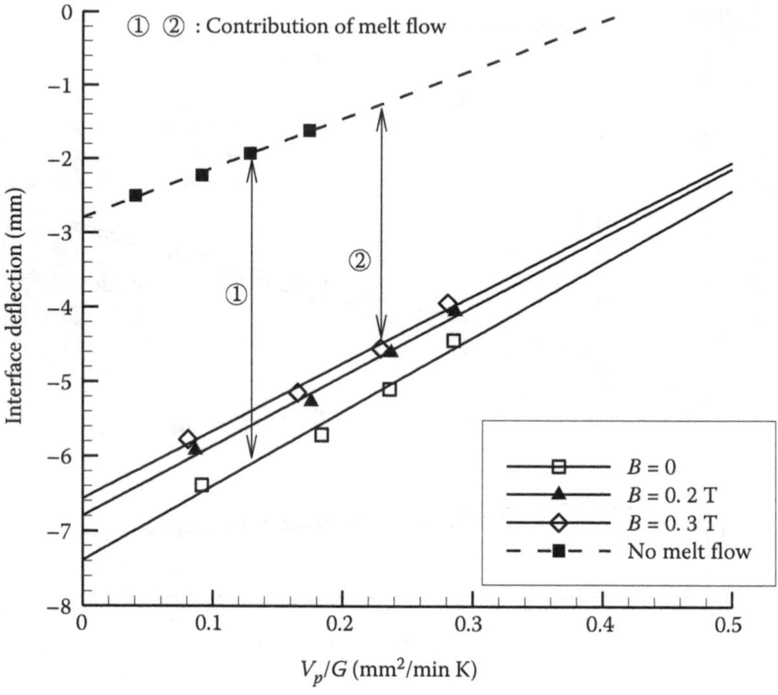

FIGURE 1.11 The interface deflection as a function of the crystal-pulling rate for different applied magnetic fields. (After *Cryst. Res. Technol.* 40, L. J. Liu, K. Kakimoto, 3D global analysis CZ-Si growth in transverse magnetic field with rotating crucible and crystal, 347, Copyright 2005, with permission from Elsevier.)

1.8 SUMMARY

With the help of computer simulation of the global environments of crystal growth techniques, technical solutions can be found that improve the purity of the crystals. The behavior of the oxygen and carbon impurities that are the main light element impurities in silicon crystals can be well described by simulation of the heat and mass transfer in a CZ furnace. In the same way, the models can be applied to the unidirectional solidification process.

Transverse magnetic fields used in a large-scale CZ silicon furnace allow the melt flow to be controlled. In this way, important parameters affecting the formation of intrinsic point defects that means vacancies and interstitials can be modified. A 3D calculation enables predictions of the tendency to form vacancy-rich or interstitial-rich crystals by estimating the ratio between the growth rate and the temperature gradient in the crystals.

ACKNOWLEDGMENT

This work was partly supported by a Grant-in-Aid for Scientific Research (B) 24360012 from the Japanese Ministry of Education, Culture, Sports, Science and Technology.

REFERENCES

1. J. Czochralski. 1917. Ein neues Verfahren zur Messung des Kristallisationsgeschwindigkeit der Metalle. *Z. Physik. Chem.* 92: 219.
2. L. Raabe, O. Patzold, I. Kupka, J. Ehrig, S. Wurzner, M. Stelter. 2011. The effect of graphite components and crucible coating on the behaviour of carbon and oxygen in multicrystalline silicon. *J. Cryst. Growth* 318: 234.
3. W. Koch, A. L. Endrös, D. Franke, C. Häßler, J. P. Kalejs, H. J. Möller. 2003. Bulk crystal growth and wafering for PV. In *Handbook of Photovoltaic Science and Engineering*, eds. A. Luque, S. Hegedus, John Wiley & Sons, Ltd, Europe.
4. G. K. Fraundorf, L. Shive. 1990. Transmission electron microscope study of quartz crucibles used in growth of Czochralski silicon. *J. Cryst. Growth* 102: 157.
5. G. K. Teal, J. B. Little. 1950. Growth of germanium single crystals. *Phys. Rev.* 78: 647.
6. T. G. Digges, R. H. Hopkins, R. G. Seidensticker. 1975. The basis of automatic diameter control utilizing "bright ring" meniscus reflections. *J. Cryst. Growth* 29: 326.
7. M. Itsumi, H. Akiya, T. Ueki, M. Tomita, M. Yamawaki. 1995. The composition of octahedron structures that act as an origin of defects in thermal SiO_2 on Czochralski silicon. *J. Appl. Phys.* 78: 5984.
8. N. Kobayashi. 1991. Oxygen transport under an axial magnetic field in Czochralski silicon growth. *J. Cryst. Growth* 108: 240.
9. H. Hirata, K. Hoshikawa. 1992. Three-dimensional numerical analyses of the effects of a cusp magnetic field on the flows, oxygen transport and heat transfer in a Czochralski silicon melt. *J. Cryst. Growth* 125: 181.
10. T. A. Kinney, R. A. Brown. 1993. Application of turbulence modeling to the integrated hydrodynamic thermal-capillary model of Czochralski crystal growth of silicon. *J. Cryst. Growth* 132: 551.
11. K. Kakimoto, K. W. Yi, M. Eguchi. 1996. Oxygen transfer during single silicon crystal growth in Czochralski system with vertical magnetic fields. *J. Cryst. Growth* 163: 238.
12. K. W. Yi, K. Kakimoto, M. Eguchi, H. Noguchi. 1996. Oxygen transport mechanism in Si melt during single crystal growth in the Czochralski system. *J. Cryst. Growth* 165: 358.
13. M. Watanabe, K. W. Yi, T. Hibiya, K. Kakimoto. 1999. Direct observation and numerical simulation of molten silicon flow during crystal growth under magnetic fields by x-ray radiography and large-scale computation. *Progr. Cryst. Growth Charact. Mater.* 38: 215.
14. C. Reimann, T. Jung, J. Friedrich, G. Müller. 2008. The importance of convective heat and mass transfer for controlling material properties in ingot casting of multi-crystalline-silicon for photovoltaic application. In *Proceedings of the 33rd IEEE Photovoltaic Specialists Conference*, p. 223, The IEEE Xplore Digital Library, San Diego, USA.
15. D. E. Bornside, R. A. Brown, T. Fujiwara, H. Fujiwara, T. Kubo. 1995. The effects of gas-phase convection on carbon contamination of Czochralski-grown silicon. *J. Electrochem. Soc.* 142: 2790.
16. Y. R. Li, M. W. Li, N. Imaishi, Y. Akiyama, T. Tsukada. 2004. Oxygen-transport phenomena in a small silicon Czochralski furnace. *J. Cryst. Growth* 267: 466.
17. A. D. Smirnov, V. V. Kalaev. 2008. Development of oxygen transport model in Czochralski growth of silicon crystals. *J. Cryst. Growth* 310: 2970.
18. A. D. Smirnov, V. V. Kalaev. 2009. Analysis of impurity transport and deposition processes on the furnace elements during Cz silicon growth. *J. Cryst. Growth* 311: 829.
19. H. Matsuo, R. B. Ganesh, S. Nakano, L. J. Liu, Y. Kangawa, K. Arafune, Y. Ohshita, M. Yamaguchi, K. Kakimoto. 2008. Analysis of oxygen incorporation in unidirectionally solidified multicrystalline silicon for solar cells. *J. Cryst. Growth* 310: 2204.
20. S. Pizzini, A. Sandrinelli, M. Beghi, D. Narducci, F. Allegretti, S. Torchio, G. Fabbri, G. P. Ottaviani, F. Demartin, A. Fusi. 1988. Influence of extended defects and native

impurities on the electrical properties of directionally solidified polycrystalline silicon. *J. Electrochem. Soc.* 135: 155.
21. J. Bauer, O. Breitenstein, J. P. Rakotoniaina. 2006. Precipitates and inclusions in block-cast silicon—isolation and electrical characterization. In *Proceedings of the 21st European Photovoltaic Solar Energy Conference*, eds. J. Poortmans, H. Ossenbrink, E. Dunlop, P. Helm, pp. 1115–1118, WIP, Munich, Germany.
22. T. Fukuda, M. Koizuka, A. Ohsawa. 1994. A Czochralski silicon growth technique which reduces carbon impurity down to the order of 10^{14} percubic centimeter. *J. Electrochem. Soc.* 141: 2216.
23. Q. Sun, K. H. Yao, J. Lagowski, H. C. Gatos. 1990. Effect of carbon on oxygen precipitation in silicon. *J. Appl. Phys.* 67: 4313.
24. M. Ogino. 1982. Suppression effect upon oxygen precipitation in silicon by carbon for a two step thermal anneal. *Appl. Phys. Lett.* 41: 847.
25. B. O. Kolbesen. 1983. Aggregation phenomena of point defects in silicon. In *The Electrochem. Soc. Proc. Ser.*, eds. E. Sirtl, J. Goorissen, vol. 83–4, pp.155–175, Pennington, NJ: Electrochemical Society.
26. U. Goesele. 1986. Oxygen, carbon, hydrogen and nitrogen in crystalline silicon. In *Mater. Res. Soc. Symp. Proc.*, eds. J. C. Mikkelsen, Jr. S. P. Peaton, J. W. Corbett, S. J. Pennycook, MRS, Pittsburgh, PA, p. 419.
27. F. Shimura. 1989. *Semiconductor Silicon Crystal Technology*. Academic Press, New York, p.148.
28. T. Abe. 1985. Silicon materials. In *VLSI Electronics Microstructure Science Series*, eds. N. G. Einspruch, H. Huff, Academic Press, Inc., New York, vol. 12, p.3.
29. U. Goesele, T. Y. Tan. 1982. Oxygen diffusion and thermal donor formation in silicon. *Appl. Phys. A* 28: 79.
30. N. Machida, Y. Suzuki, K. Abe, N. Ono, M. Kida, Y. Shimizu. 1998. The effects of argon gas flow rate and furnace pressure on oxygen concentration in Czochralski-grown silicon crystals. *J. Cryst. Growth* 186: 362.
31. N. Machida, K. Hoshikawa, Y. Shimizu. 2000. The effects of argon gas flow rate and furnace pressure on oxygen concentration in Czochralski silicon single crystals grown in a transverse magnetic field. *J. Cryst. Growth* 210: 532.
32. T. Inada, T. Fujii, T. Kikuta, T. Fukuda. 1987. Growth of semi-insulating GaAs crystals with low carbon concentration using pyrolytic boron nitride coated graphite. *Appl. Phys. Lett.* 50: 143.
33. C. Reimann, J. Friedrich, M. Dietrich. 2009. Device and method for preparing crystalline bodies by directional solidification. Patent WO2009100694A1, Europe.
34. B. Gao, S. Nakano, K. Kakimoto. 2011. Influence of reaction between silica crucible and graphite susceptor on impurities of multicrystalline silicon in a unidirectional solidification furnace. *J. Cryst. Growth* 314: 239.
35. H. Hirata, K. Hoshikawa. 1992. Three-dimensional numerical analyses of the effects of a cusp magnetic field on the flows, oxygen transport and heat transfer in a Czochralski silicon melt. *J. Cryst. Growth* 125: 181.
36. B. Gao, S. Nakano, K. Kakimoto. 2010. Global simulation of coupled carbon and oxygen transport in a unidirectional solidification furnace for solar cells. *J. Electrochem. Soc.* 157: H153.
37. B. Gao, X. J. Chen, S. Nakano, K. Kakimoto. 2010. Crystal growth of high-purity multicrystalline silicon using a unidirectional solidification furnace for solar cells. *J. Cryst. Growth* 312: 1572.
38. B. Gao, K. Kakimoto. 2010. Global simulation of coupled carbon and oxygen transport in a Czochralski furnace for silicon crystal growth. *J. Cryst. Growth* 312: 2972.
39. K. Hoshikawa, X. Huang. 2000. Oxygen transportation during Czochralski silicon crystal growth. *Mater. Sci. Eng. B* 72: 73.

40. J. S. Shun, K. H. Chen, Y. Choi. 1993. A coupled implicit method for chemical non-equilibrium flows at all speeds. *J. Comput. Phys.* 106: 306.
41. R. C. Reid, J. M. Prausnitz, T. K. Sherwood. 1987. *The Properties of Gases and Liquids*, 3rd ed., McGraw-Hill, New York.
42. K. A. Hoffmann, S. T. Chiang. 2000. *Comput. Fluid Dyn.* II: 69.
43. I. Mary, P. Sagaut, M. Deville. An algorithm for low Mach number unsteady flows. 2000. *Comput. Fluids* 29: 119.
44. P. Jenny, B. Müller. 1999. Convergence acceleration for computing steady-state compressible flow at low Mach numbers. *Comput. Fluids* 28: 951.
45. L. J. Liu, K. Kakimoto. 2005. Partly three-dimensional global modeling of a silicon Czochralski furnace. I. Principles, formulation and implementation of the model. *Int. J. Heat Mass Transfer* 48: 4481.
46. V. V. Voronkov. 1982. The mechanism of swirl defects formation in silicon. *J. Cryst. Growth* 59: 625.
47. V. V. Voronkov, R. Falster. 2002. Intrinsic point defects and impurities in silicon crystal growth. *J. Electrochem. Soc.* 149: 167.
48. L. J. Liu, K. Kakimoto. 2005. Partly three-dimensional global modeling of a silicon Czochralski furnace.II. Model application: Analysis of a silicon Czochralski furnace in a transverse magnetic field. *Int. J. Heat Mass Transfer* 48: 4492.
49. L. J. Liu, S. Nakano, K. Kakimoto. 2005. An analysis of temperature distribution near the melt-crystal interface in silicon Czochralski growth with a transverse magnetic field. *J. Cryst. Growth* 282: 49.
50. L. J. Liu, K. Kakimoto. 2005. 3D global analysis CZ-Si growth in transverse magnetic field with rotating crucible and crystal. *Cryst. Res. Technol.* 40: 347.

2 Growth and Characterization of Silicon–Germanium Alloys

Ichiro Yonenaga

CONTENTS

2.1 Introduction 23
2.2 Crystal Growth 25
  2.2.1 Crystallinity of Grown Crystals 26
  2.2.2 Grown-in Defects 27
2.3 Key Phenomena in Crystal Growth of SiGe Alloys 28
  2.3.1 Heteroseeding 28
  2.3.2 Critical Growth Velocity 30
  2.3.3 Spatial Variation of the Composition of SiGe Alloys 30
  2.3.4 Distribution Coefficient of Impurities 33
2.4 Fundamental Properties 36
  2.4.1 Local Atomic Structure 36
  2.4.2 Carrier Transport 38
  2.4.3 Thermal Conductivity 41
  2.4.4 Seebeck Coefficient 43
2.5 Impurities and Defects 44
  2.5.1 Oxygen 44
  2.5.2 Hydrogen 46
  2.5.3 Dislocations 48
  2.5.4 Mechanical Strength 49
2.6 Applications 52
2.7 Summary 53
Acknowledgments 54
References 55

2.1 INTRODUCTION

Silicon–germanium (Si_xGe_{1-x} or germanium–silicon $Ge_{1-x}Si_x$) alloys, where x indicates the mole fraction of silicon, form a continuous series of solid solutions crystallizing with the diamond structure. The lattice parameters of the two elements

(0.564 nm of pure Ge and 0.534 nm of pure Si) differ by 4% and the lattice parameters of the alloys closely follow Vegard's law in the entire range of compositions, while the bandgap changes from 0.72 eV (Ge) to 1.2 eV (Si). SiGe alloys are important for both microelectronic and optoelectronic devices and for various functional materials in view of the potential for bandgap and lattice parameter engineering they offer. In fact, alloying Si and Ge leads to various unique effects on fundamental properties, absent in the component materials Si and Ge.

Applications of SiGe alloys can be divided into two categories,[1] depending on the morphology of the material. Thin films grown on crystalline substrates by various epitaxial techniques are suitable for applications in high-performance electronic and optoelectronic devices as heterostructure bipolar transistors (HBTs), modulation-doped field-effect transistors (MO DFETs), buried channel metal-oxide semiconductor (CMOS) transistors, optical devices working at 1.3–1.55 μm wavelengths, avalanche multiplication photo diodes (APDs), and quantum wells (QWs) and strained layer superlattices. Films of SiGe alloys can also be used as a buffer layer between a GaAs layer and Si substrate as a means of soft lattice matching.

Bulk crystals are used in thermoelectric power converters installed in deep-space spacecrafts such as Voyager, Galileo, and Cassini-Hyugens. Other possible applications include solar cells, neutron and x-ray monochromators, lattice-matched substrates for the heteroepitaxial growth of various compounds, including for SiGe epitaxial growth instead of Si, photodetectors for 1–1.5 μm, high-speed temperature sensors, γ-ray detectors, Hall effect transducers, and so on. Moreover, the local variation of the bandgap and lattice parameter, dependent on the local alloy composition, leads to the potential application of SiGe as a functionally graded material.

For such functional applications as well as for further technological applications, it is essential to gain a deep insight into the native and intrinsic physical properties of bulk crystals and thin films. For example, the growth of thin films on Si substrates is inevitably associated with the presence of misfit dislocations. Such dislocations affect the electrical and optical properties of the alloy and limit its technological applications, while strain at the film/substrate interface hinders its application as electronic and optoelectronic devices.

The literature has reported a large number of results concerning the growth of single crystals of SiGe alloys using various growth techniques.[2–24] However, the growth of bulk single crystals of SiGe alloys is difficult.

Figure 2.1 shows the phase diagram of the SiGe system.[25] The solidus and liquidus curves are widely separated. The equilibrium distribution coefficient, $k = x_s/x_l$ (x_s and x_l being the solidus and liquidus concentration of silicon), is as large as 6 for Ge-rich solutions and about 2.2 at a temperature intermediate between the melting points of the elements. This large gap between the liquidus and the solidus and the difference in the physical properties of the constituent elements, such as density, lattice parameters, and melting temperatures, make the crystal growth of SiGe alloys very difficult.

The present author has attempted the Czochralski (CZ) growth of SiGe bulk crystals of large size in the whole composition range $0 < x < 1$ and succeeded in growing full single crystals of SiGe alloys of large size within the composition

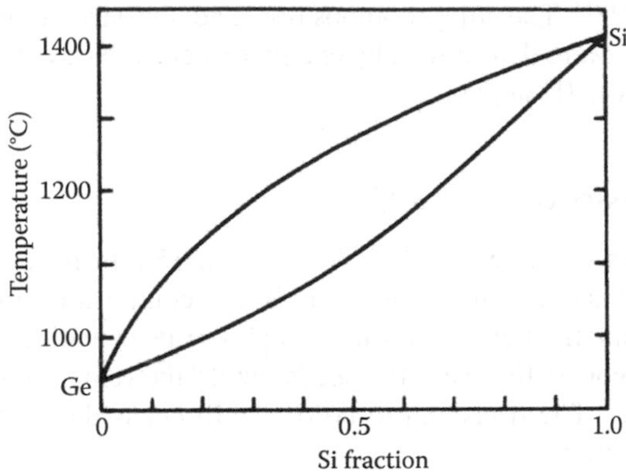

FIGURE 2.1 Phase diagram of the $Ge_{1-x}Si_x$ system. (Reprinted from R. W. Olesinski and J. C. Abbaschian. 1984. The Ge-Si (germanium–silicon) system. *Bulletin of Alloy Phase Diagrams* 5: 180. Copyright 1984 by ASM Publication. With permission.)

ranges of $0 < x < 0.15$ and $0.73 < x < 15$.[9–11,13,19] Heavily impurity-doped SiGe crystals have also been grown for application in thermoelectric devices.[15,16] It should be noted, however, that success in growing alloy materials requires an extensive knowledge of the physical properties that are involved in the process and the start-up of additional experimental and theoretical research activities concerning still unexplored aspects.

This chapter reviews the most relevant aspects of CZ growth of bulk crystals of SiGe alloys with a discussion of their intriguing solidification problems, and reports some typical physical properties of the alloys, such as their atomistic bonding structure of imperfect Pauling type and their mechanical, electrical, optical, and spectroscopical properties. Some specific properties, such as electron and hole mobilities, thermal conductivity, oxygen local vibration, hydrogen-like muonium, and so forth, which concern the overall behavior of germanium or silicon, are also emphasized.

2.2 CRYSTAL GROWTH

Bulk crystals of Si_xGe_{1-x} alloys in the whole composition range were grown by the CZ technique using a graphite resistance heater. Mixtures of Si and Ge in the desired proportions were charged into a crucible with a diameter of 50 mm and a depth of 50 mm within a graphite susceptor. Seeds prepared from a Si or Ge crystal oriented parallel to [111] or [001] and $4.5 \times 4.5 \times 45$ mm^3 in size were used to start the growth of the crystals. The growth was carried out in a flowing gas atmosphere of high-purity Ar with a flux of 150 L/h under a pressure of 1 atm with a low pulling rate ranging from 0.5 to 8 mm/h. The rotational rate of the crystal was 7.5 rpm, whereas that of the crucible was 2 rpm in the opposite direction. Also, impurity-doped crystals were grown from a melt to which certain kinds of impurities were added. Detailed procedures and conditions of SiGe crystal growth have been described elsewhere.[5,9–11,13,15,16,19] The alloy composition and the structural homogeneity of grown SiGe boules were determined by energy dispersive x-ray (EDX) spectroscopy using a JEOL JXA-8600 apparatus.

2.2.1 Crystallinity of Grown Crystals

Full single crystals of large size (i.e., larger than 15 mm in diameter and longer than 30 mm) were successfully grown within the composition ranges $0 < x < 0.15$ and $0.73 < x < 1$. Small single crystalline samples of intermediate composition were obtained in the ingot section near the seeds, while the remaining part of the ingot was polycrystalline. The transition to polycrystallinity is due to the occurrence of constitutional supercooling.[10,19,23]

Figure 2.2 shows several undoped SiGe alloy crystals grown in these studies. The starting composition of the crystals, that is, that for a solidification fraction $g \approx 0$, was $x_0 \approx 0.06$, 0.15, 0.50, 0.73, and 0.85, in the order from the top to the bottom in the figure.[9,13,19,22] Their size was 15–25 mm in diameter and more than 40 mm in length. Four {111} facets on the body outside of the boules confirm the growth of a full single crystalline material except the boule at the center of the figure. The central boule in Figure 2.2c was grown from a melt with an initial Si content $x_{m0} = 15.4$ at% at a very low pulling rate of 0.5 mm/h. The grown crystal was 13 mm in length and had a maximum diameter of 10 mm. The composition of the crystal from the top to the tail was determined to be almost constant as $Si_{0.50}Ge_{0.50}$ by EDX. Apparently,

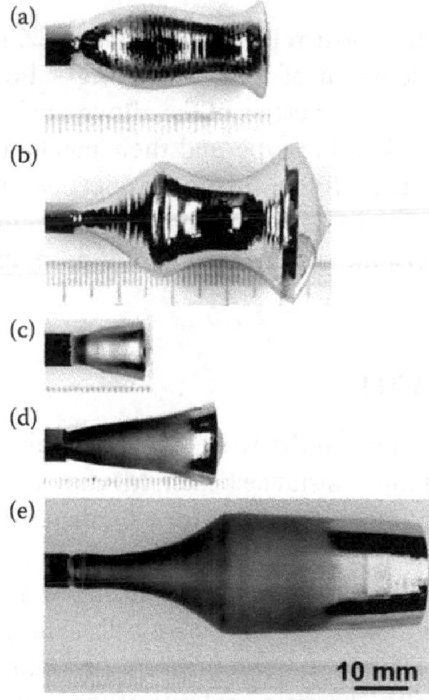

FIGURE 2.2 Si_xGe_{1-x} alloys grown by the Czochralski technique with a Si seed crystal with the [001] orientation. The compositions at the growth starting positions are $x_0 \approx 0.06$, 0.15, 0.50, 0.73, and 0.85 from the top (a) to the bottom (e), respectively.[9,13,19,22]

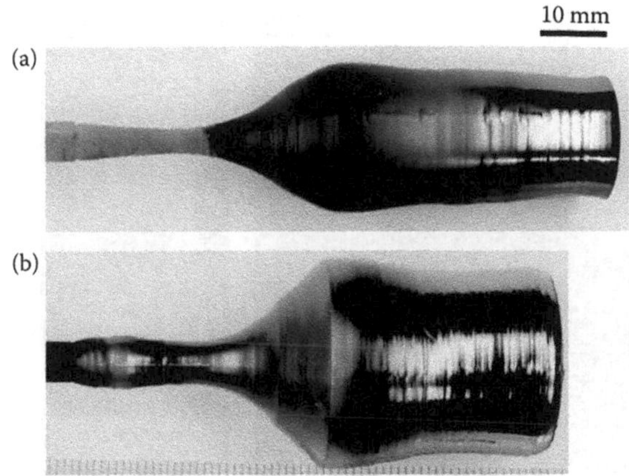

FIGURE 2.3 Boules of single crystal of (a) B- and (b) P-doped $Si_{0.93}Ge_{0.07}$ alloys.

the crystal has four {111} facets in the top part close to the Si seed and a few twins in the middle-tail part.

Similarly, CZ growth of *n*- and *p*-type impurity-doped SiGe alloys was attempted. Single crystals of Si-rich Si_xGe_{1-x} alloys ($0.8 < x < 1$) heavily doped with electrically active impurities (boron, gallium, phosphorus, and so forth) at a concentration up to 10^{20} cm^{-3} were successfully grown.[15,16] Figures 2.3a and 2.3b show B- and P-doped single crystals, respectively, of $Si_{0.93}Ge_{0.07}$ alloys. The carrier concentration of the boules was 5×10^{20} and 6×10^{19} cm^{-3}, respectively, measured by Hall effect measurements.

2.2.2 Grown-in Defects

Figure 2.4a shows an x-ray topographic image of a $Si_{0.73}Ge_{0.27}$ crystal.[19] In the image taken perpendicular to the growth axis, fine concentric circles, the so-called striations, are apparent. Such striations are commonly observed in highly impurity-doped crystals and are due to compositional variation of the order of 10^{-4}–10^{-3} in atomic fraction, which can be sensitively detected by x-ray topography. Figure 2.4b shows an x-ray topographic image of a longitudinal cross section of a Si seed/$Si_{0.98}Ge_{0.02}$ crystal interface.[19] Striations develop homogeneously, and reveal a convex shape of the melt–crystal interface from the seed–crystal interface to the tail of the boule.

Many dislocations are generated at the seed–crystal interface of SiGe alloys. As seen in Figure 2.4b, there are many straight misfit dislocations with the Burgers vector parallel to the seed–crystal interface, running parallel to the growth direction from the seed–crystal interface. Such dislocations are present when the lattice mismatch $f\,(=\Delta a/a = (a_{SiGe} - a_{Si})/a_{Si})$, where a_{SiGe} and a_{Si} are lattice parameters of the SiGe crystal and Si seed, respectively, is higher than 10^{-5}.[5,10,19] The critical mismatch is around 8×10^{-4} for $Si_{0.98}Ge_{0.02}$ alloys.

Besides misfit dislocations, many curved dislocations, inclined against the interface, are generated from the crystal periphery due to the thermal stress during the seeding stage. The density of grown-in dislocations in the undoped and highly impurity-doped alloys was in the range 10^3–10^5 cm^{-2}, strongly dependent on alloy

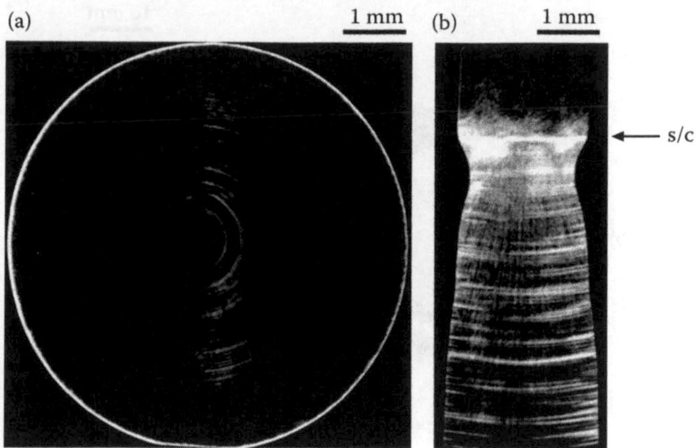

FIGURE 2.4 (a) X-ray topographic image of a $Si_{0.73}Ge_{0.27}$ crystal. The plate is perpendicular to the growth axis. (b) X-ray topographic image of a $Si_{0.98}Ge_{0.02}$. The plate is parallel to the growth axis. s/c shows the seed–crystal interface. (Reprinted from *Journal of Crystal Growth* 275, I. Yonenaga, Growth and fundamental properties of SiGe bulk crystals, 91, Copyright 2005, with permission from Elsevier BV.)

composition. The generation process is controlled by the magnitude of the misfit strain between the alloy and seed, the temperature, the temperature gradient, and the mobility of dislocations.[10,19]

Twins are often observed in Si-rich and intermediate Si content SiGe alloys, even under conditions of a pulling rate lower than the critical growth velocity (see Section 2.3.2). The stacking fault energy of SiGe alloy determined in transmission electron microscopic studies is 55–61 mJ/m^2, intermediate between those of Si and Ge.[26] Therefore, it is assumed that twin generation in SiGe alloys should not be ascribed to the stacking fault energy but rather related to high strain/stress originating from the axial and radial composition or from temperature gradients due to their spatial variations.

2.3 KEY PHENOMENA IN CRYSTAL GROWTH OF SiGe ALLOYS

2.3.1 Heteroseeding

In general, growth of bulk crystals of SiGe alloys with an intended orientation inevitably requires the use of a Si or Ge single crystal as seed material, that is, a heteroseeding process. Accordingly, alloy composition changes abruptly across the interface, along a distance less than 10 μm in the growth direction, as seen in Figure 2.5.

Interestingly, a single crystal of Si_xGe_{1-x} can be grown using a Si seed even from an initially pure Ge melt, though the crystal becomes a dilute SiGe alloy due to Si segregating from the melted Si seed. This implies that a Si seed could be used as a seed material for the growth of SiGe alloys in the whole composition range $0 < x < 1$.[27] However, the high density of misfit dislocations due to the difference of

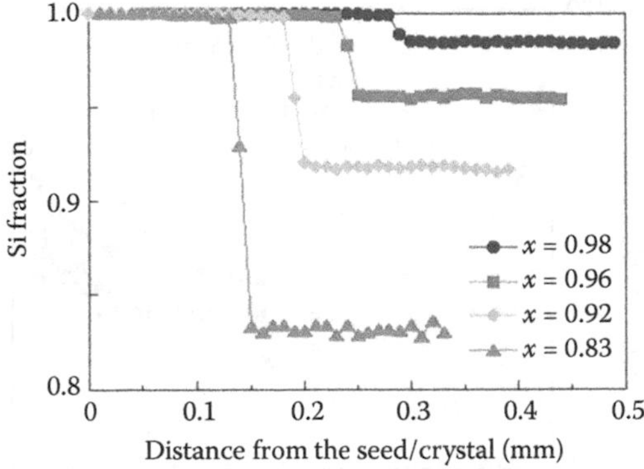

FIGURE 2.5 Abrupt variation of Si fraction at the seed–crystal interface.

the lattice parameter between a Si seed and the crystal is responsible of the difficulties encountered in the crystal growth of SiGe alloys, especially in the intermediate composition range.

Dash-necking with a 3–4 mm neck is generally recognized as a standard method to eliminate dislocations generated during the seeding stage of crystal growth. Figure 2.6a shows a boule of $Si_{0.95}Ge_{0.05}$ grown with adopting Dash-necking, though the elimination of misfit dislocations running straight parallel to the growth direction is not yet complete.[19] Then, a boule of $Si_{0.95}Ge_{0.05}$ in Figure 2.6b was grown with a SiGe seed of the same composition obtained in a preceding growth batch and Dash-necking.[19] In this case, the density of grown-in dislocations was extremely low, as seen in Figure 2.6c.

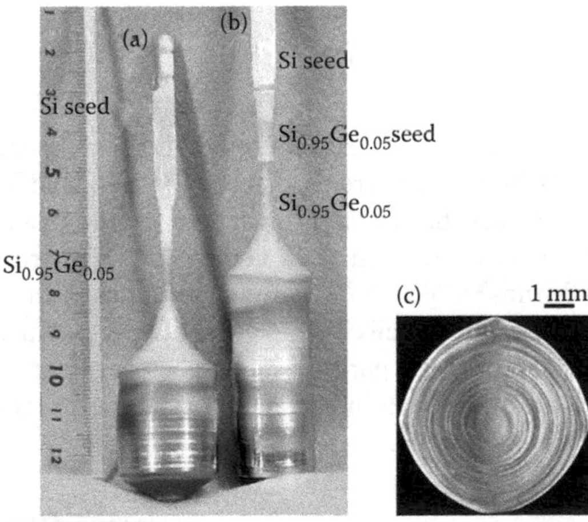

FIGURE 2.6 SiGe boules (a) grown with Dash-necking and (b) grown with a SiGe seed with the same composition. (c) X-ray topographic image of the crystal (b). The plate is perpendicular to the growth axis. (Reprinted from *Journal of Crystal Growth* 275, I. Yonenaga, Growth and fundamental properties of SiGe bulk crystals, 91, Copyright 2005, with permission from Elsevier BV.)

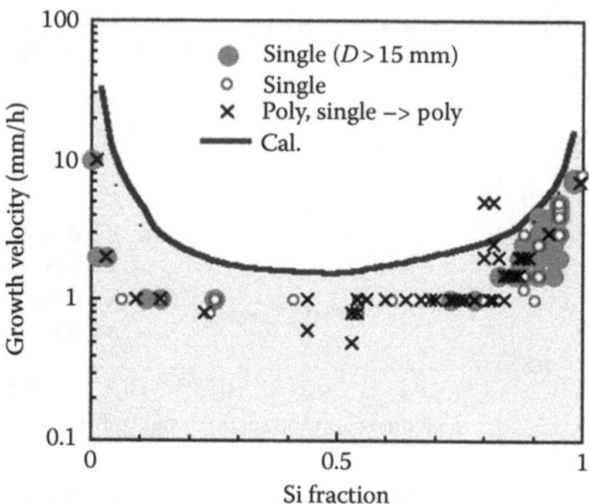

FIGURE 2.7 Growth velocity of SiGe crystals as a function of Si fraction. Solid line shows the estimated critical growth rates for a temperature gradient of 40 K/cm. Solid and open circles illustrate the successful growth of large size (larger than 15 mm in diameter) and small single crystals, respectively. Crosses show occurrence of polycrystallization from the initiation of growth and during growth. (Reprinted from *Journal of Crystal Growth* 275, I. Yonenaga, Growth and fundamental properties of SiGe bulk crystals, 91, Copyright 2005, with permission from Elsevier BV.)

2.3.2 Critical Growth Velocity

The critical growth velocity v_c for the start-up of polycrystalline growth caused by constitutional supercooling can be estimated as a function of the composition x by the following equation:

$$v_c = D\nabla\theta k/\nabla T_l x(k-1), \tag{2.1}$$

where k is the segregation coefficient, D the diffusion coefficient [$D = (30-24x) \times 10^{-5}$ cm^2/s], $\nabla\theta$ the temperature gradient of the melt, and ∇T_l the slope of the liquidus.[28] Figure 2.7 shows the critical growth velocity estimated using a temperature gradient of 40 K/cm as a function of the composition. Experimental results from our previous growth runs[5,10,11,13,15,19,23,27,29] are superimposed in the figure. Indeed, the estimated critical growth velocity is somewhat higher than that arising from the experimental results. For practical purposes, it is necessary to adopt very low pulling rates for full single crystal growth in consideration of the phenomena that will be described in the following section.

2.3.3 Spatial Variation of the Composition of SiGe Alloys

For any SiGe alloy, the Si content gradually decreases along the pulling direction, as can be seen in the phase diagram of Figure 2.1 because of the strong segregation mechanism. In such a situation, Si atoms in the melt are preferentially transferred to the crystal during the growth.[10,19,30] Such compositional variation of a grown SiGe crystal along the pulling direction can be described satisfactorily by the following equation:

$$x(g) = x_{m0}k(1 - g)^{k-1}, \tag{2.2}$$

where $x(g)$ is the Si content in a SiGe ingot grown at a solidification fraction g, and x_{m0} is the initial Si content in the melt, according to the Pfann relation, valid for complete melt mixing conditions.[31] It should be noted that, in SiGe growth, the gravity effect caused by a large difference in the densities of Ge and Si should be additionally taken into account.[10,19]

Figure 2.8a shows a SiGe boule, whose starting composition x_0 was 0.89.[23] The grown crystal has a 15 mm cone-shaped shoulder, a body 33 mm in length, and a maximum diameter of 25 mm. Externally, the boule is single crystalline along the [001] direction, since four crystal lines, relating to {111} facets, appeared clearly on the side surface up to the tail. Figure 2.8b shows the variation of Si fraction x in the grown crystal along the growth axis, quantitatively analyzed by means of EDX spectroscopy. The Si fraction changes abruptly across the seed–crystal interface and decreases gradually from $x = 0.89$ to 0.82 up to the tail along the growth axis. No transient change of the composition was observed at the seed–crystal interface. The variation of the Si atomic fraction in the grown crystal in Figure 2.8b is well fitted by Equation 2.2. The distribution coefficient of Ge in the SiGe alloy was determined to be 0.51.[30]

Figure 2.9a shows an x-ray topographic image of a longitudinal cross section of the crystal.[23] There is a high density of grown-in dislocations (10^3–10^5 cm^{-2}) at the

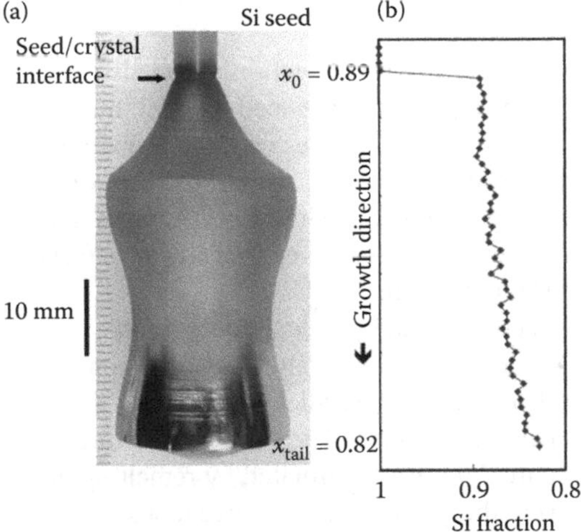

FIGURE 2.8 (a) A grown $Si_{0.89}Ge_{0.11}$ crystal. (b) Variation of Si fraction x in the grown crystal along the growth axis analyzed by means of EDX spectroscopy. (Reprinted from *Journal of Crystal Growth* 312, I. Yonenaga et al. Cellular structures in Czochralski-grown SiGe bulk crystal, 1065, Copyright 2010, with permission from Elsevier BV.)

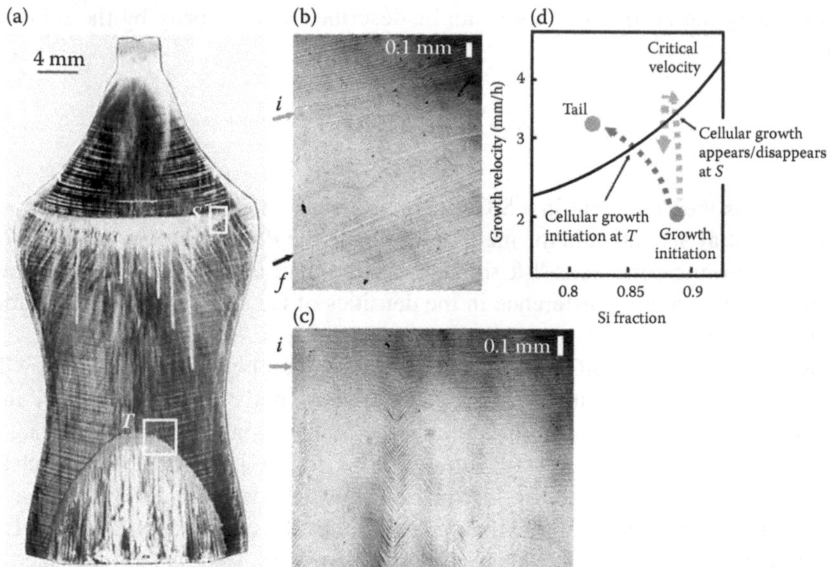

FIGURE 2.9 (a) X-ray topographic image of the grown SiGe. Image of the plate parallel to the growth axis. (b) and (c) show striation patterns observed in the regions ***S*** and ***T***, respectively, in (a). Arrows labeled by *i* and *f* mean initiation and termination of cellular growth, respectively. (d) Schematic view of the occurrence of the cellular structure based on the relation between critical growth velocity for the occurrence of constitutional supercooling and alloy composition. (Reprinted from *Journal of Crystal Growth* 312, I. Yonenaga et al. Cellular structures in Czochralski-grown SiGe bulk crystal, 1065, Copyright 2010, with permission from Elsevier BV.)

seed–crystal interface and in the whole section. It can also be seen that growth striations homogeneously developed from the seed–crystal interface to the tail except in the regions starting at ***S*** and ***T***. In these regions, the growth interface is strongly disturbed and also many vertical white lines are present. Above the end of the disturbed region ***S***, some defects still remain as white linear images, and the disturbed region ***T*** expands in size with an increasing density of linear defects and finally occupies almost the entire crystal at the tail.

These features well correspond to optical micrographic observations of regions ***S*** and ***T*** in Figures 2.9b and 2.9c, respectively.[23] In correspondence with these regions, the regular striations become wavy parallel to the traces of (111) planes, showing the initiation of a typical cellular growth. The cellular structure in the crystal periphery of the region ***S*** in Figure 2.9b was terminated by remelting due to some increase of melt temperature during the growth, and homogeneous growth restarted with regular striations. However, in the region starting at position ***T***, the cellular structure continued toward the crystal tail.

The observed occurrence and development of cellular structures starting at positions ***S*** and ***T*** are related to constitutional supercooling. We can speculate that the cellular structure appears and disappears in the following ways, as shown in

Figure 2.9d. First, the cellular structure generation at ***S*** may be attributed to a variation in the local growth velocity at a certain stage. In the growth stage, with expansion of the shoulder part of the crystal, the controlling temperature was intentionally reduced, which led to enhancement of the effective growth velocity. As a result, the effective growth velocity exceeded the critical growth velocity at that time. The effective growth velocity was then reduced by heat-up of the melt/crystal. Thus, the furnace temperature should be strictly controlled to avoid remarkable expansion of a crystal diameter in the growth stage of the shoulder part formation of the crystal.

However, the development of the cellular structure at ***T*** can be understood as follows. The Si fraction of the growing alloy decreases gradually and the effective pulling rate simultaneously increases with the progress of growth, as shown in Figure 2.9d. Thus, the Si fraction and the local growth velocity at the growth interface at ***T*** may cross the critical relation of the growth velocity versus the composition of Figure 2.7, resulting in the initiation of cellular growth. That is fatal to further satisfactory growth if the effective pulling rate is not reduced or kept constant during the growth.

2.3.4 Distribution Coefficient of Impurities

The knowledge of the distribution/segregation of electrically active impurities (dopants) during crystal growth is important in terms of design of semiconductor devices. Thus, the distribution coefficients of various dopants were determined from the variation of their concentrations along the direction of growth of high-quality single crystals of SiGe alloys, mainly of Si-rich alloys.

Figure 2.10 shows the variation of carrier concentrations plotted against the solidification ratio g of heavily impurity-doped Si_xGe_{1-x} alloys ($0.89 < x < 0.96$).[30] For any doped SiGe crystals, the carrier concentration gradually increased along the pulling direction. Depending on the dopant species, the carrier concentration varied from 3×10^{16} to 1.1×10^{20} cm^{-3}. Under the assumption that almost all the dopants are ionized as suppliers of carriers (electrons or holes) in SiGe alloys, the concentration of a dopant and its variation in the alloys were determined. B and P were easily incorporated into SiGe alloys at relatively high concentrations, similar to those incorporated into pure Si crystals. The concentration of In was the lowest among the dopant species investigated in the present study. In the figure, for purposes of comparison, the variation of the concentration of interstitially dissolved oxygen (O) atoms in the undoped SiGe alloys is superimposed. Here, it should be noted that the variations of Si fraction and dopant concentration in any grown Si_xGe_{1-x} alloy crystals against the solidification ratio g do not show a diffusion-controlled profile. Thus, it could be understood that the condition of complete mixing of dopant impurities in the melt was well realized.

The compositional variation of impurities can be analyzed using Equation 2.2[31] to obtain the effective distribution coefficient, k_{eff}, of various dopants for undoped and doped SiGe alloys without any effects of the difference of the initial concentration of melts. In the analysis, the evaluated values are effective distribution coefficients, which might be close to the equilibrium coefficients under conditions of a low growth velocity of 2–5 mm/h.

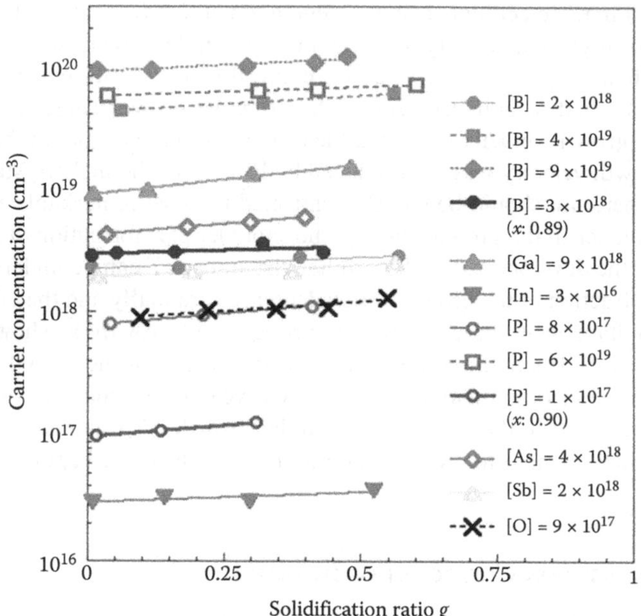

FIGURE 2.10 Concentration of various dopants plotted against the solidification ratio g in SiGe crystals grown from melts with an initial content of $x_{m0} = 0.90$ and 0.79. The concentration variation of O impurity along the direction of growth in undoped SiGe is also superimposed. Numerals show the dopant/impurity concentration per cm^{-3} at the position of growth initiation.[30]

The evaluated distribution coefficients of various dopants, including Ge and O, are shown in Table 2.1, in comparison with those in pure Si and Ge reported in the literature.[32] The covalent radius of the dopants atoms is also shown for purposes of comparison.[33]

Figure 2.11 shows the evaluated effective distribution coefficients of various dopants plotted against the Si fraction together with data on Ge-rich SiGe reported by other groups.[12,34,35] In the figure, Ga, In, and Sb, with a larger atomic size than that of a Si atom, show a large distribution coefficient in Si_xGe_{1-x} alloys ($0.89 < x < 0.96$) in comparison with those in Ge or Si. However, the distribution coefficients of B, P, and As, all with an atomic size smaller than or similar to Si, in SiGe at the concentration $\approx 10^{18}$ cm^{-3} are shown to present values comparable to those in pure Si. Remarkably, the distribution coefficient of B in SiGe decreases with increasing concentration, similar to the results in Si reported by Taishi et al.[36] while the distribution coefficient of P in SiGe increases with increasing concentration. Anyway, it is difficult to deduce a linear interpolation of the distribution coefficients of the dopants between the values reported in Si and Ge.

The observed variation of the dopant segregation in SiGe alloys can be attributed to elastic interaction originating in the difference of the covalent radius, as shown in Table 2.1. Additionally, the lattice of SiGe alloys expands with increasing Ge

TABLE 2.1
Segregation Coefficient of Impurities in Si_xGe_{1-x} ($0.89 < x < 0.96$) in Comparison with Those in Pure Si and Ge[32]

			in SiGe	in SiGe	in SiGe (Present Work)		Covalent
Impurity	in Ge	in Si	(Dold et al.[12,35])	(Kamornik et al.[34])	$x = 0.93–0.96$	$x = 0.89–0.90$	Radius (nm)[33]
B	15	0.8			0.71 [2×10^{18}] 0.58 [4×10^{19}] 0.61 [9×10^{19}]	0.80 [3×10^{18}]	0.088 (−25%)
Ga	0.087	0.008	0.075 {0.035}	0.3 {0.15} (3×10^{19})	0.18 [9×10^{18}]		0.126 (+8%)
In	0.01	0.0004	0.00086 {0.11} 0.0005 {0.06}		0.72 [3×10^{16}]		0.144 (+23%)
P	0.08	0.35	0.095 {0.05}		0.31 [8×10^{17}] 0.76 [6×10^{19}]	0.30 [1×10^{17}]	0.110 (−6%)
As	0.02	0.3	0.05 {0.045}		0.33 [4×10^{18}]		0.118 (+1%)
Sb	0.003	0.023	0.008 {0.13} 0.0045 {0.045}		0.69 [2×10^{18}]		0.136 (+16%)
O		1.4			0.54 [9×10^{17}]		
Ge		0.33			0.51 (undoped)		0.122 (+4%)
Si							0.117

Note: Those in Ge-rich SiGe reported by Kamornik et al.[34] and Dold et al.[12,35] are also shown. Numerals in { } brackets show the Si fraction in Ge-rich SiGe and those in [] show the dopant concentration per cm^{-3} at the position of growth initiation in Si-rich SiGe investigated.

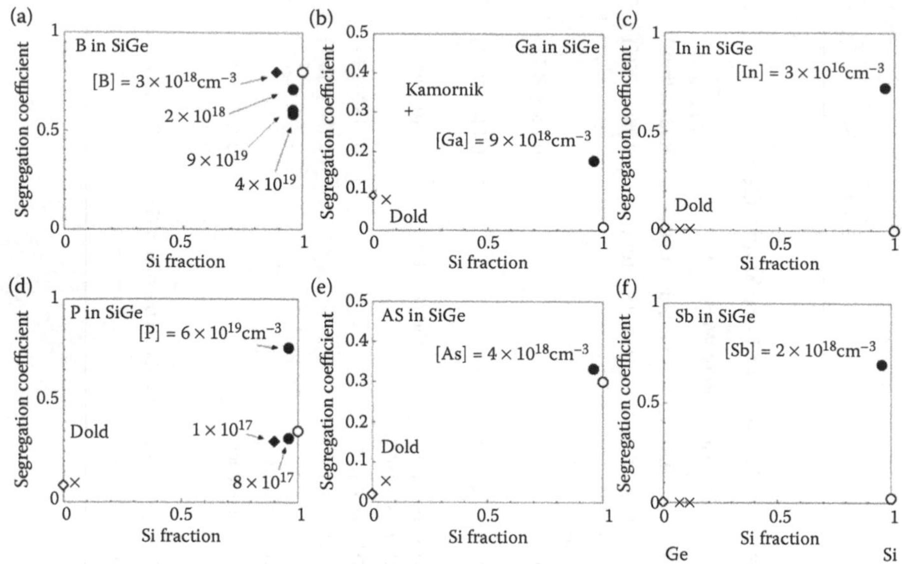

FIGURE 2.11 Evaluated effective segregation coefficients of (a) B, (b) Ga, (c) In, (d) P, (e) As, and (f) Sb together with those in Ge-rich SiGe reported by Kamornik et al.[34] and Dold et al.[12,35]. Values of the segregation coefficients of dopants in Ge and Si as reported in the literature[32] are also superimposed. Numerals show the dopant concentration per cm^{-3} at the position of growth initiation.[30]

content since the covalent radius of Ge ($r_{Ge} = 0.122$ nm) is larger than that of Si ($r_{Si} = 0.117$ nm). Incorporation of Ge into Si (SiGe) favors the stability of Ga–Si, In–Si, or Sb–Si pairs in the crystal, due to a lower strain energy onset associated with their formation, in comparison with that present in pure Si. Similarly, P, As, and especially B may reduce the total energy through a mutual strain compensation by being neighbors to Ge atoms. In the process, the segregation of P can increase with increasing concentration, but an increase of the concentration of B with a misfit strain much larger than P may lead to an increase in total strain energy and result in a suppression of B segregation into the SiGe.

2.4 FUNDAMENTAL PROPERTIES

Success in the growth of high-quality bulk single crystals of SiGe alloys opened insights into basic properties of alloy semiconductors. In the following section, some typical properties are shown.

2.4.1 Local Atomic Structure

The composition dependence of the structure of atomic bonds in SiGe alloys is quite interesting, as it does not follow the Vegard's model,[37] different from the observation that the macroscopical lattice parameters follow the Vegard's law from pure Ge to pure Si. The local atomic structure in bulk Si_xGe_{1-x} alloys in the whole composition range $0 < x < 1$ was investigated by x-ray absorption fine-structure (XAFS) spectroscopy at 11 K and room temperature (RT) and *ab initio* calculation.[38,39]

Figure 2.12 shows the coordination numbers of the nearest-neighbors Si and Ge (first shell) around Si and Ge atoms derived from the XAFS data as a function of alloy composition.* The XAFS results are in good agreement across the whole composition range with a random site occupancy of Si and Ge atoms, shown by the dashed lines. The sum of *N*-Ge(Ge) + *N*-Si(Ge) and also of *N*-Ge(Si) + *N*-Si(Si) is always four in the whole composition range. Diffraction studies of SiGe thin films grown by various epitaxial methods have indicated the existence of ordering[40,41] and the tendency of clustering toward Ge−Ge dimers within Si-rich SiGe thin films. The ordering parameter $[0.25N_{Ge} - (1{-}x)]/x$ estimated from the Ge coordination number N_{Ge} around the Ge atoms in bulk Si_xGe_{1-x} ($0.44 < x < 0.82$) is 0.06–0.11, far from 1, which means that there is no preferential ordering of the Ge−Ge dimers across the whole composition range in the bulk SiGe alloys. These results show a random substitution site occupancy of Si and Ge atoms in the whole composition of SiGe binary alloys with a diamond cubic structure.

The variation of the bond length of nearest-neighbor atoms Ge−Ge, Ge−Si, and Si−Si in SiGe alloys against the alloy composition† is shown in Figure 2.13.[38,39] The uncertainty of the derived bond lengths due to the experimental technique and analysis is less than ±0.0015 nm. Even so, it seems that the Ge−Ge, Ge−Si (Si−Ge),

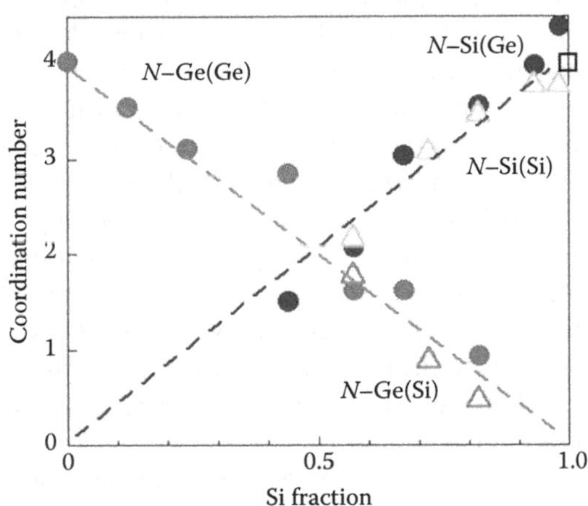

FIGURE 2.12 Ge and Si coordination numbers around Ge and Si atoms derived from XAFS data as a function of Si fraction. "*N*-Si(Ge)" means the number of Si atoms around a centered Ge. The symbol □ is for pure Si crystal. The dashed lines show the coordination numbers predicted from a random mixture of Si and Ge atoms.[39]

* *N*-Ge(Ge) and *N*-Si(Ge) are the coordination numbers of Ge and Si around Ge atoms, respectively, while *N*-Ge(Si) and *N*-Si(Si) are those around Si atoms.

† Here, the Ge−Ge bond length (Ge → Ge) and Ge−Si bond length (Ge → Si) mean that around Ge atoms, whereas Si−Ge bond length (Si → Ge) and Si−Si bond length (Si → Si) mean that around Si.

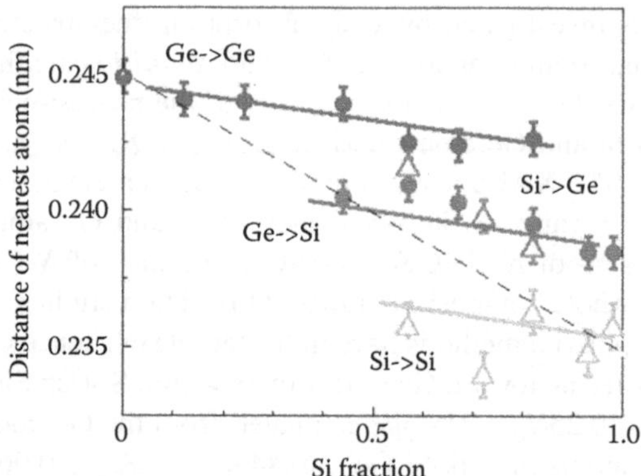

FIGURE 2.13 Ge−Ge bond length (Ge-> Ge), Ge−Si bond length (Ge-> Si), Si−Ge bond length (Si-> Ge), and Si−Si bond length (Si-> Si) in SiGe as a function of Si fraction.[39] The solid lines are the predicted dependence of the bond lengths from *ab initio* electronic structure calculation.[42] The thin dashed line is Vegard's law.

and Si−Si bond lengths in the SiGe alloys are distinctly different and parallel with each other and vary in a linear fashion as a function of the atomic fraction of Si. The Ge−Si (Si−Ge) bond length is 0.240–0.241 nm, corresponding to the sum of the Si and Ge covalent radii. The measured bond length versus composition relations agree rather well with those theoretically estimated from *ab initio* electronic structure calculations[42] that are shown by the solid lines in Figure 2.13. Although the lattice constants follow a linear behavior from Ge to Si according to Vegard's model,[43] the composition dependence of the atomic bond lengths is closer to that of Pauling's model[44] rather than that of an average virtual crystal approximation.

According to the model by Cai and Thorpe,[45] from the slope of the bond length dependencies on the composition, the topological rigidity parameter (implying how a central atom bonded to four neighbor atoms pushes the nearest-neighbor atoms away) is estimated to be 0.60 in agreement with values of 0.63–0.70 reported experimentally[46–48] and found theoretically[49,50] for the Ge−Ge bonds in SiGe. Thus, it can be concluded that both bond lengths and bond angles change with the composition, as schematically shown in Figure 2.14.

Results of an x-ray fluorescence holography study, which includes an analysis of the second neighbor atoms, suggest that such local strain accommodation of SiGe solid solutions may decrease in the range of the second neighbor atoms.[51] It should be noted that SiGe is a suitable model of a disordered material, where there is no restriction for the atomic positions in the sublattices, differing from the ternary or quaternary semiconductors as GaInAs,[52] ZnCdTe,[53] GaInPAs,[54] and so on.

2.4.2 Carrier Transport

In general, the electrical conductivity decreases from pure Ge to Si with increasing Si content in undoped SiGe alloys. The decrease is due to the decrease of intrinsic carrier concentration determined by the increase of the bandgap energy.

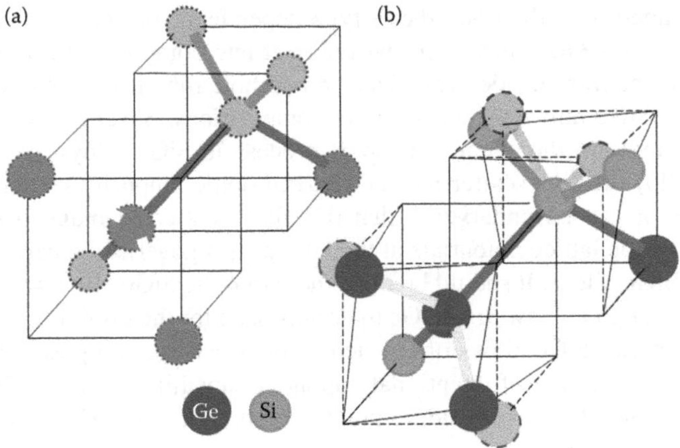

FIGURE 2.14 Schematic of the local atomistic structure of SiGe. (a) The diamond structure and (b) SiGe alloy.

Figure 2.15 shows the dependence of the electron and hole mobilities (μ_e and μ_h, respectively), determined by Hall effect measurements using the Van de Pauw method at RT, on the composition of nominally undoped SiGe alloys (with a carrier concentration higher than 4×10^{15} cm^{-3}) at RT.[22] As seen in the figure, the hole mobility in single crystals is higher than that in polycrystals (at least around 20%) indicating a strong influence of grain boundaries on carrier transport[16,55–57] as previously deduced theoretically.[58] Grain boundaries in SiGe have an electrically distorted region ≈30 μm thick with an energy barrier of ≈0.4 eV.[55] The hole mobility in single crystals of SiGe alloys is much lower than that in Si and Ge and shows a

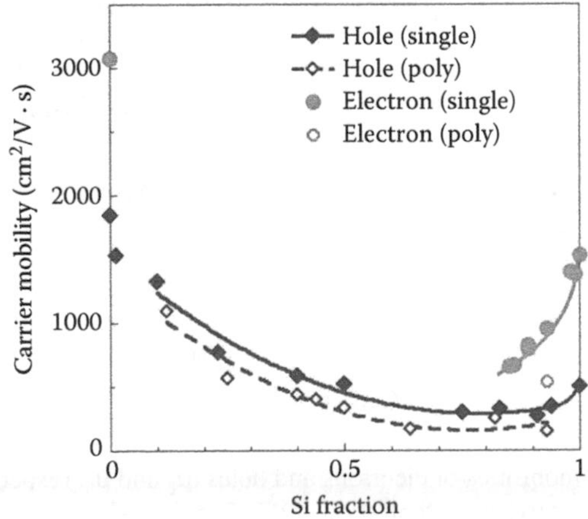

FIGURE 2.15 Compositional dependences of electron and hole Hall mobilities (μ_e and μ_h, respectively) in single crystal and polycrystalline SiGe alloys at RT.[22]

typical U-shaped, so-called Nordheim-type dependence on the Si (or Ge) content with a minimum ≈340 cm²/V · s in the intermediate composition around 0.6–0.8, somewhat on the Si-rich side. The electron and hole mobilities are comparable to each other in SiGe alloys of intermediate compositions. Several scattering mechanisms contribute to the carrier transport process in SiGe alloys: carrier-phonon scattering, alloy disorder scattering, and charged (ionic) impurity scattering, among others. Thus, it can be understood that the alloy scattering induced by disorder fluctuations of the lattice potentials at the lattice sites governs the carrier transport in undoped SiGe alloys. It should be noted here that the SiGe alloy was electrically *p*-type even though the raw Si and Ge materials used for the growth were originally *n*-type. In general, SiGe alloys of Ge-rich or intermediate composition tend to be *p*-type except in the case of intentional doping. Meanwhile, Si-rich SiGe alloys are *n*-type with some effects of residual point defects induced during crystal growth and subsequent cooling.

Figure 2.16 shows the relationship between electron and hole Hall mobilities (μ_e and μ_h, respectively) and carrier concentration at RT in SiGe crystals ($0.93 < x < 0.96$) doped with various impurities (P, As, Sb, B, In).[59] The same figure also displays with a line the relationships between μ_e (P and As dopants) and μ_h (B dopant) and carrier concentration in Si reported in the literature.[60] As shown in the former paragraph, the values of μ_e and μ_h in undoped or lightly impurity-doped SiGe at a carrier concentration less than 10^{17} cm^{-3} are lower than those in Si. With an increase of carrier concentration, the values of μ_e and μ_h decrease due to charged (ionic) impurity scattering, similar to that in Si. The decrease in μ_e with increasing carrier concentration is much larger than the decrease in μ_h. The values of μ_e and μ_h in heavily impurity-doped SiGe at carrier concentrations higher than 10^{18} cm^{-3} are comparable with

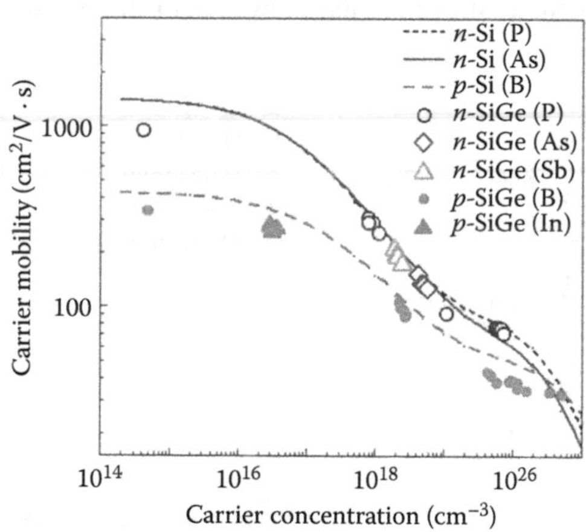

FIGURE 2.16 Hall mobilities of electrons and holes (μ_e and μ_h, respectively) in SiGe crystals plotted against the carrier concentration at RT. The lines show μ_e and μ_h in Si as reported by Masetti et al.[60] (Reprinted with permission from I. Yonenaga. 2006. *Japanese Journal of Applied Physics Part 1* 45: 2678. Copyright 2006 by The Japan Society of Applied Physics.)

those in Si, except that the values of μ_h are somewhat lower than those in Si at carrier concentrations of 10^{19}–10^{20} cm^{-3}. No systematic effect of different dopant species on the relations between carrier mobility and concentration can be detected explicitly. In addition, the effect of alloy scattering becomes unclear in heavily impurity-doped SiGe due to the predominant contribution of charged impurity scattering. This is supported by the fact that both the Hall mobilities of electrons and holes in the SiGe with carrier concentrations around 10^{19}–10^{20} cm^{-3} show a temperature dependence as T^{-n}, with $n \approx 1$, different from $n \approx 3$ observed in undoped SiGe, up to elevated temperatures.[61] Based on these results, the so-called Irvin's curves, for practical use, could be applied to SiGe alloys within the composition range $0.93 < x < 1$.[59]

2.4.3 Thermal Conductivity

Thermal conductivity of SiGe alloys shows a typical U-shaped dependence on the alloy composition, with a minimum in the intermediate composition at around 0.5–0.7.[62,63] Doping with electrically active impurities leads to a slight decrease in thermal conductivity. However, by heavy doping, the thermal conductivity becomes higher than that in undoped materials, due to the electronic contribution to thermal conductivity.[16,57]

Figure 2.17 shows the dependence of thermal conductivity at RT and 600°C on the composition of undoped SiGe alloys.[22] The thermal conductivity of undoped SiGe is around 6 and 3.5 W/m · K at RT and 600°C, respectively. The specific heat increases monotonically from Ge to Si. Thermal conductivity of undoped SiGe alloys decreases weakly with the increase of temperature. In the whole temperature range investigated, the thermal conductivity of the alloys is much lower than that of Si and Ge and decreases more weakly than in the case of Si and Ge.

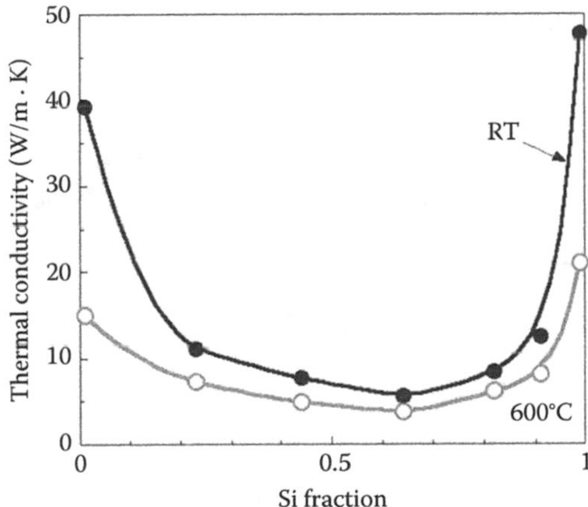

FIGURE 2.17 Compositional dependences of thermal conductivity of undoped SiGe at RT and 600°C. (Reprinted with permission from I. Yonenaga. 2008. *Materia* 47: 3 (in Japanese). Copyright 2008 by The Japan Institute of Metals and Materials.)

The temperature dependence can be described by a T^m law, where $m \approx -0.40$.[16] n-type doping of SiGe alloys results in a slight decrease of thermal conductivity. However, excess doping, for example, for a concentration of P higher than 5×10^{19} cm^{-3}, leads to a slight increase, probably due to the electronic contribution to thermal conductivity. Thus, the fact that thermal conductivity depends strongly on the alloy composition can be explained by phonon scattering effects originating in the lattice distortion at point defects in an alloy semiconductor.[64,65]

Figure 2.18 shows a comparison of the thermal conductivity of $Si_{0.4}Ge_{0.6}$ and of Ge crystals at low temperatures.[66] In analogy to the results at elevated temperatures, the mixing of Si in Ge induces a strong reduction of the thermal conductivity, which is much more remarkable than that of the isotopic mixture in natural Ge from ^{70}Ge. This feature is quite interesting from the viewpoint of the phonon scattering of typically disordered materials, that is, glasses.[66,67] SiGe alloys crystallize in the diamond structure and show a dynamic disordered character, the origin of which is that Si and Ge are randomly substituted in the alloys, as shown in Section 2.4.1, and that their masses are very different ($m_{Ge}/m_{Si} \approx 2.6$).[68] The ordered structure gives sharp Bragg reflections around which acoustic vibration branches and plane-wave phonons are then good approximate eigenmodes, whereas the plane-wave acoustic phonons of short wavelength are scattered in the alloys, owing to the random substitution. That is, a considerable vibrational disorder may appear at high momentum exchange and phonon frequency. This is an interesting model for glasses, with phonons that are accessible to experimental investigation, contrary to real glasses in which the possibilities of direct observation of the high-frequency transverse acoustic waves are extremely limited. Currently, SiGe alloys are being studied by high-resolution neutron spectroscopy to elucidate the fundamental mechanism of high-momentum

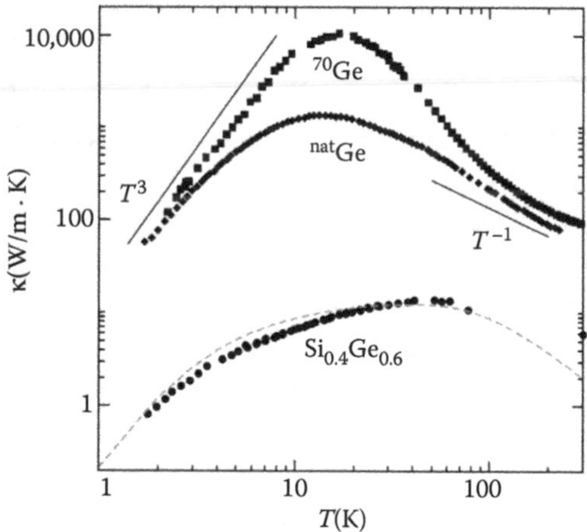

FIGURE 2.18 Thermal conductivity of the sample $Si_{0.4}Ge_{0.6}$ compared to Ge. The line is calculated according to Callaway's model (*Physical Review* **113** (1959) 1046). (Reprinted from *Physica B* 350, A. Béraud et al. Disorder-induced broadening of transverse acoustic phonons in Si_xGe_{1-x} mixed crystals, 254, Copyright 2004, with permission from Elsevier BV.)

exchange acoustic phonons of disordered matter, related to the low-temperature thermal properties.

2.4.4 Seebeck Coefficient

SiGe alloys are well known as materials for thermoelectric power generators at elevated temperatures, 800–1200 K. In fact, SiGe thermoelectric devices prepared from polycrystals have been successfully used as power converters for decay heat of $^{238}PuO_2$ radioisotopes in deep-space spacecraft probes (Voyager, Galileo, etc.).

In undoped SiGe alloys, the magnitude of the Seebeck coefficient is very high at RT and decreases rapidly with increasing temperature due to thermal excitation of carriers across the energy gap. At elevated temperatures (600°C), the Seebeck coefficient of undoped Si_xGe_{1-x} alloys shows a maximum at around $x \approx 0.8$, which is, in fact, the mean composition of the SiGe polycrystalline alloys used for thermoelectric applications.[69] The Seebeck coefficient is positive and negative in *p*-type and *n*-type SiGe alloys, respectively, and is almost constant, with an absolute value of 300–400 μV/K, in the temperature range from RT to 600°C in SiGe alloys suitably doped with impurities at a concentration higher than 10^{18} cm^{-3}.

Figure 2.19 shows the absolute value of the Seebeck coefficient of impurity-doped SiGe alloys at 600°C plotted against the electrical conductivity.[57] The Seebeck coefficient of the SiGe alloys decreases with an increase of doping and then of the electrical conductivity in the range 3×10^3–10^5 (ohm · m)$^{-1}$, which indicates superposition of the charged impurity scattering into the phonon scattering of the alloys, as theoretically supposed.[70]

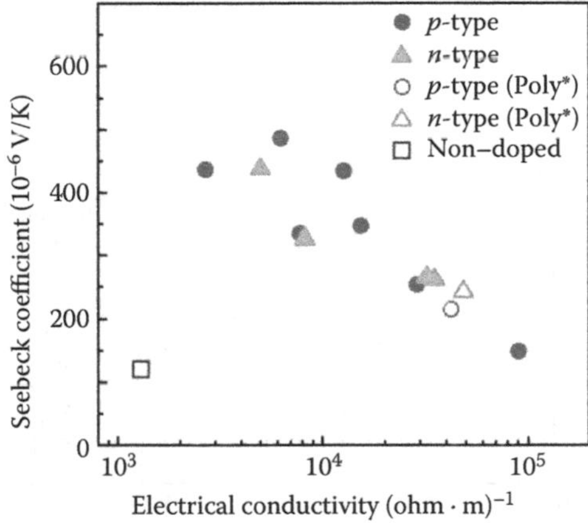

FIGURE 2.19 Relation between the absolute Seebeck coefficient of SiGe alloys versus the electrical conductivity at 600°C. The data on polycrystals $Si_{0.7}Ge_{0.3}$ by Dismukes et al.[63] are superimposed. (Reprinted with permission from I. Yonenaga, T. Akashi, and T. Goto. 2002. *Materials Research Society Symposium Proceedings* 691: G7.4.1. Copyright 2002 by Materials Research Society.)

2.5 IMPURITIES AND DEFECTS

2.5.1 Oxygen

Interstitially dissolved oxygen impurities in CZ-grown SiGe alloys was investigated by infrared (IR) absorption spectroscopy.[71]

A large amount of oxygen atoms, up to around 10^{18} cm^{-3}, are incorporated into the grown SiGe alloys, especially in Si-rich SiGe alloys, due to the chemical reaction between a Si melt and quartz crucible,[11] similar to that occurring in the growth of conventional Si crystals. It is well known that an interstitially dissolved oxygen atom (O_i) occupies a bond-centered position in Si and behaves as a Si−O_i−Si quasi-molecule, leading to local vibration at 1106 cm^{-1} at RT due to the ν_3 antisymmetric stretching mode.[72,73] Similarly, in Ge, oxygen occupies a bond-centered position site, resulting in local vibration of Ge−O_i−Ge at 855 cm^{-1},[72,74,75] but normally the concentration of O_i is more than two orders of magnitude lower in Ge than in Si.[71]

Figure 2.20 shows the infrared absorption spectra of the grown Si_xGe_{1-x} alloys of various Si content x, obtained at RT.[71] In the whole composition range $0 < x < 1$, a remarkable and broad absorption peak was observed at the position of 1106 cm^{-1} which originates from the antisymmetric stretching vibration of interstitially dissolved oxygen atoms, that is, of the Si−O_i−Si quasi-molecule. An absorption coefficient $\alpha = 3$–4 cm^{-1} was measured in Si-rich SiGe alloys, showing that a large number of interstitial oxygen atoms are dissolved in the alloy. The peak intensity decreases with a decrease in the Si content x and then the peak almost disappears in the alloys with x lower than 0.23. In addition, with decreasing x, the peak position shifts to the low wave number side. No peak at 855 cm^{-1}, originating by Ge−O_i−Ge vibration, was detected, even in Ge-rich SiGe alloys. Moreover, there is no peak in the wave number range 855–1106 cm^{-1}. Also, at 5 K, no other peaks are observed in the above-mentioned wave number range, although the main peak originating from

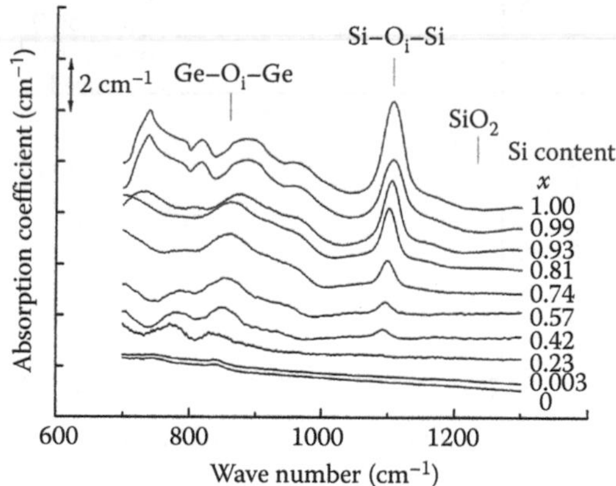

FIGURE 2.20 Infrared absorption spectra of the as-grown Si_xGe_{1-x} alloys of various Si content x, measured at RT. (Reprinted from *Physica B* 308–310, I. Yonenaga, M. Nonaka, and N. Fukata. Interstitial oxygen in GeSi alloys, 539, Copyright 2001, with permission from Elsevier BV.)

$Si-O_i-Si$ vibration locates to the high-frequency side. Thus, it seems that IR emissions originating from $Si-O_i-Ge$ and $Ge-O_i-Ge$ quasi-molecules do not occur in SiGe alloys. In addition, no absorption at 1225 cm^{-1} is observed, showing the absence of SiO_2 precipitates.

The total number of oxygen atoms in the SiGe alloys, evaluated from the peak intensity of the IR spectrum, corresponds well to the total oxygen concentration determined by secondary ion mass spectroscopy (SIMS). It should be noticed that the oxygen concentration shows an x^2 dependence against the Si content x in the alloy.[71] The results show that the oxygen concentration depends strongly on the ratio of Si−Si bonds in the SiGe alloy with respect to the composition if the Si and Ge atoms occupy the lattice sites randomly. In fact, as described in Section 2.4.1, the XAFS investigations show the random substitutional site occupancy of Si and Ge atoms and no preferential ordering of the Ge−Ge dimers across the complete composition range.[38] Thus, it is known that an oxygen atom preferentially occupies the bond-centered position of Si−Si bonds, forming a $Si-O_i-Si$ quasi-molecule. The above-mentioned relationship between oxygen solubility and Si content in the alloy probably depends on the fact that oxygen atoms have a stronger interaction with Si atoms than with Ge ones. Indeed, this is supported by the larger absolute value of the standard Gibbs free energy of formation of SiO_2 than that of GeO_2.[76,77]

A shift of the absorption peak around 1106 cm^{-1}, as seen in Figure 2.20, is clearly observed when plotted against the Si content x in Figure 2.21. The amount of peak shift does not change linearly from Si to Ge, but increases drastically from Si and Si-rich SiGe ($x \sim 1$) to ~0.5 and then changes gradually to Ge-rich SiGe ($x \sim 0$). The deviation from the linear change becomes a maximum at $x \sim 0.5$. For the sake of comparison, previous results on Si-rich SiGe alloys at 5 K or RT reported by Yamada-Kaneta et al.,[78] Wauters and Clauws,[79] and Humlícek et al.[80] are superimposed in the figure. The feature can be understood as follows:

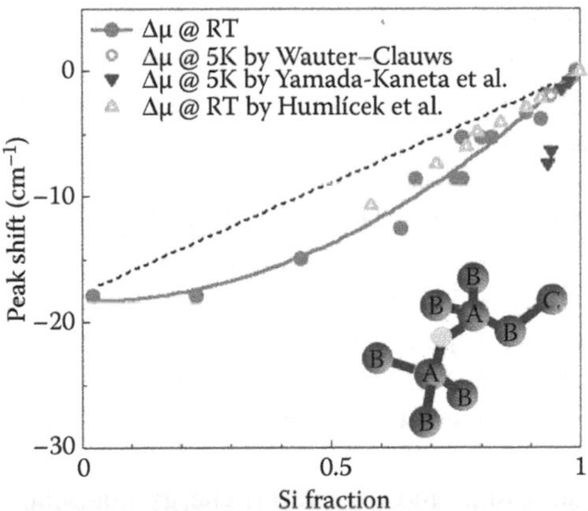

FIGURE 2.21 Shift of the 1106 cm^{-1} peak as a function of Si content x together with those reported by Yamada-Kaneta et al.,[78] Wauters and Clauws,[79] and Humlícek et al.[80] The inset shows atomic configuration around $Si-O_i-Si$ bond.

First, the absolute increase of oxygen atoms in the bond-centered position of a Si-Si bond with the decrease of the Si content in the alloy results in a linear expansion of the bond length, resulting in the linear lowering of the vibration peak of a $Si-O_i-Si$ quasi-molecule as theoretically derived by Coutinho et al.[81] Second, the $Si-O_i-Si$ quasi-molecule has 6 nearest neighbors and 12 next-nearest neighbors, as shown in the inset of Figure 2.21. If the heavier Ge atom substitutes for the Si atom on a neighbor or next-nearest neighbor, the $Si-O_i-Si$ quasi-molecule may be perturbed in the vibration frequency resulting in a downward shift. The probability of Ge substitution for Si or Si substitution for Ge in Si- or Ge-rich SiGe alloys, respectively, depends on the Si content x according to the binomial distribution. Thus, the vibration frequency is perturbed in its maximum at $x \sim 0.5$.

2.5.2 Hydrogen

Hydrogen defect states, lattice sites, and behavior were investigated for SiGe alloys in the full composition range from Si to Ge according to observations of its muonium analogue.[82–85] Hydrogen is a common impurity, affecting the electronic and structural properties of many semiconductors. It can bind to defects or other impurities, eliminating their electrical activity, known as passivation. Thus, efforts are being expended to clarify and predict its behavior. However, direct study of isolated hydrogen in semiconductors may be very difficult or impossible for its high reactivity and mobility. The light hydrogen-like atom muonium, composed by a positive muon and an electron, is electronically almost identical to hydrogen and can provide information on its electronic structure and states.[86]

It is known that, in Si and Ge, a muonium forms two states at low temperatures: an immobile bond-centered (BC) species Mu_{BC} and a highly mobile tetrahedral (T)-site center Mu_T. As seen in Figure 2.22, these two types of muonium show quite different features depending on the alloy composition.[83,84] The hyperfine parameter of Mu_{BC} shows a linear variation with the composition over the full alloy range. As described in Section 2.4.1, there are random site occupancies of Si and Ge atoms. Also, the Si−Si, Si−Ge, and Ge−Ge bond lengths are different and vary linearly with alloy composition. Therefore, implanted muons adopting an immobile, BC position experience a random selection of bonding environments, which shows an overall linear variation with alloy composition. However, the hyperfine parameter of Mu_T shows a very nonlinear behavior with most changes from hyperfine values near that of Si occurring above 80% Ge content. The minimum of conduction band changes from Δ to L at a Ge content of ~85%. The T-site to T-site hopping rate is much faster in Ge than in Si. The expected path between adjacent T-sites is through the center of the puckered six-member ring separating adjacent tetrahedral cages. This opening is physically larger in Ge than in Si since the distance from the center of the ring to the six nearest host atoms is 0.232 and 0.224 nm, respectively. Furthermore, the electronic charge distribution and the overall energy landscape within which a Mu atom moves are considerably flatter for Ge compared with Si.[82,83]

Figure 2.23 shows a band alignment diagram of Mu, and likely H, in SiGe across the full alloy range. The Mu[+/–] level is pinned at 4.45 ± 0.04 eV below vacuum.

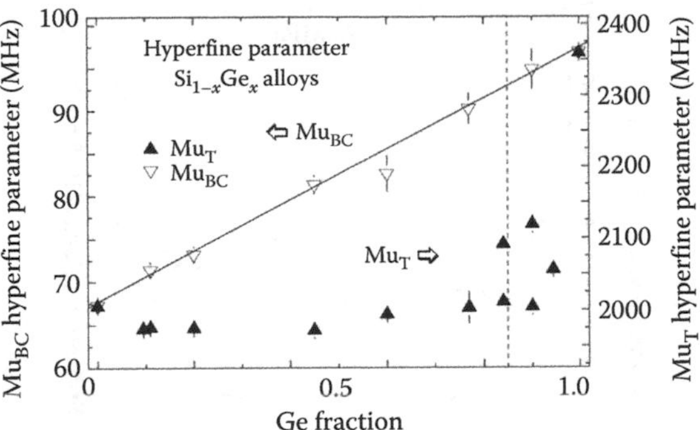

FIGURE 2.22 Variation with alloy composition of the average value of the isotropic component of the Mu_{BC} hyperfine parameter, with straight line fit, together with the Mu_T hyperfine parameter (Mu_T values are for 50 K except $x = 0.45$ and 0.77, which are 70 and 75 K, respectively). (Reprinted from *Physica B* 404, B. R. Carroll et al. Muonium acceptor states in high Ge $Si_{1-x}Ge_x$ alloys, 812, Copyright 2009, with permission from Elsevier BV.)

The Mu[+/–] donor lies deep in the bandgap across the full alloy composition range. The ionization energy of the Mu_{BC}[+/0] acceptor shows a maximum near the Δ to L crossover at $x = 0.85$, whereas the Mu_T[0/–] is deep in pure Si and becomes valence-band resonant for Ge-rich alloys over $x \sim 0.9$.[85] The results are inconsistent with the claims[87] that only H^- is thermodynamically stable for any Fermi energy since H[+/–] is valence-band resonant in Ge. However, the present work is consistent with the deep level transient spectroscopy (DLTS), results[88] for the H^+ level in Ge and the lack of a DLTS acceptor signal.

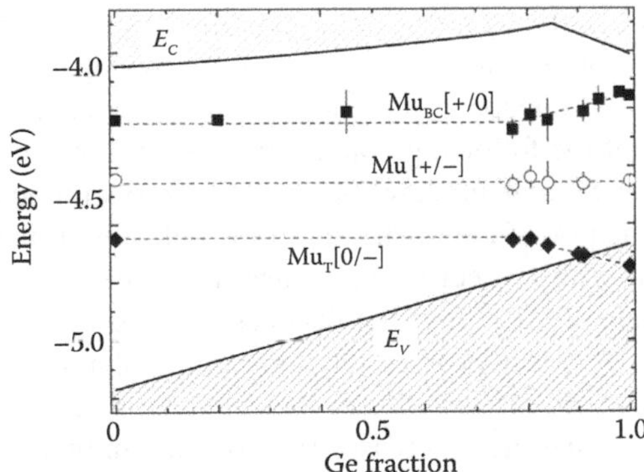

FIGURE 2.23 Band alignment diagram of donor and acceptor levels in $Si_{1-x}Ge_x$. The energy scale is set by accepted electron affinities for Si and Ge with the conduction band varying smoothly between the two. (Reprinted with permission from B. R. Carroll et al. 2010. *Physical Review B* 82: 205205. Copyright 2010 by the American Physical Society.)

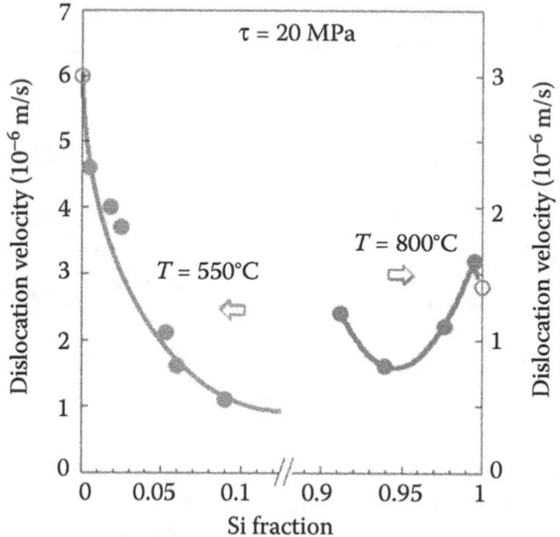

FIGURE 2.24 Velocities of 60° dislocations in the Ge-rich and Si-rich SiGe alloys at 550°C and 800°C, respectively, under a shear stress 20 MPa as dependent on the Si content. (Reprinted with permission from I. Yonenaga. 2013. Dislocation dynamics in SiGe alloys. *Journal of Physics: Conference Series* 471: 012002. Copyright 2013 by IOP Publishing Limited.)

2.5.3 Dislocations

A unique feature of SiGe alloys, that arises from the physics of the alloying process itself, has been observed in their mechanical behavior, determined by measuring the velocities of dislocations in Si_xGe_{1-x} alloys of composition range $0 < x < 0.08$ and $0.922 < x < 1$.

The specimens were stressed at elevated temperature by three-point bending in a vacuum. Displacements of dislocations generated preferentially from a scratch pre-drawn on the surface under stressing were measured by the etch pit technique.[89–91] In SiGe alloys of the composition range $0 < x < 0.08$, the logarithm of the velocity of dislocations is linear with respect to the logarithm of the stress with approximately the same slope except the following feature: the velocity of 60° dislocations in the SiGe alloys of higher Si content shows a break at stress around 8 MPa, depending on the Si content. The plot of the logarithm of the velocity of dislocations in Si-rich SiGe alloys is linear with respect to the logarithm of the stress at temperatures of 750–850°C. The slope of the plot of the dislocation velocity versus the stress in the SiGe alloy with $x = 0.996$ is approximately the same as that in Si. However, in SiGe alloys of $x = 0.922–0.979$, the velocity of dislocations is zero under a stress lower than the threshold one and then increases rapidly with an increase in stress beyond the threshold stress, as reported by Iunin et al.[92] The threshold stress for dislocation generation from a scratch increases with a decrease in the Si content.

Figure 2.24 displays the dependence of the velocity of 60° dislocations in the Ge-rich and Si-rich SiGe alloys at 550°C and 800°C, respectively, under a shear stress of 20 MPa, on the Si content.[91] In the Ge-rich SiGe alloys of composition range $0 < x < 0.08$, the dislocation velocity decreases monotonically with increasing

TABLE 2.2
Magnitudes of v_0, m, and Q for 60° Dislocations in Si_xGe_{1-x} and Pure Ge and Si[91]

Crystal		v_0 (m/s)	m	Q (eV)
Ge		2.9×10^2	1.7	1.62 ± 0.05
Si_xGe_{1-x}	$x = 0.016$	4.6×10^2	1.7	1.68
	$x = 0.047$	2.8×10^2	1.7	1.68
	$x = 0.080$	2.3×10^2	1.6	1.7
	$x = 0.922$	1.2×10^1	3.3	2.2
	$x = 0.946$	9.4×10^1	2.1	2.3
	$x = 0.979$	2.1×10^2	1.9	2.3
	$x = 0.996$	1.4×10^4	1.0	2.4
Si		1.0×10^4	1.0	2.4

Si content, reaching about one-seventh of that in pure Ge at $x = 0.08$. However, in the composition range $0.922 < x < 1$ (Si-rich SiGe), the dislocation velocity first increases, then decreases, and again increases with decreasing Si content in the temperature range 750–850°C. The dislocation velocity in SiGe alloy with the Si content $x = 0.996$ is found to be higher than that in pure Si.

The velocities of dislocations in the SiGe alloys investigated in these studies are expressed in a similar way as those in Ge, Si, and other semiconductors as functions of the stress τ and temperature T by the following empirical equation[93–95]:

$$v = v_0(\tau/\tau_0)^m \exp(-Q/k_B T), \tau_0 = 1\text{MPa}, \quad (2.3)$$

where k_B is the Boltzmann constant. The experimentally determined magnitudes of v_0, m, and Q in SiGe, pure Ge and Si are listed in Table 2.2.[91] Here, it should be noted that dislocations in SiGe alloys show a typical recombination-enhanced dislocation motion under irradiation by an electron beam.[96,97]

2.5.4 Mechanical Strength

The mechanical strength was investigated in Si_xGe_{1-x} alloys for the composition ranges $0 < x < 0.4$ and $0.95 < x < 1$. Rectangular parallelepiped specimens were compressed under a constant strain rate at elevated temperatures.[29,90,91,98] Stress–strain curves of Si-rich SiGe alloys of the composition range $0.95 < x < 1$ are similar to those of pure Si at temperatures in the range of 800–1000°C, characterized by a stress drop followed by an increase in the stress with strain, as commonly seen in semiconductors such as Si, Ge, GaAs, and so forth.[94,95,99] The upper and lower yield stresses and flow stress increase with a decrease in Si content. Similarly, stress–strain curves of Ge-rich SiGe alloys at low temperatures are characterized by a stress drop followed by an increase in the stress with respect to strain, but no stress drop is observed at high temperatures, differing from those of the Si-rich SiGe alloys.

Noticeably, SiGe alloys with $x > 0.10$ exhibit much higher levels of the yield and flow stresses than pure Ge and Si.[29,90,98]

Figure 2.25 shows the temperature dependence of the upper yield stresses of various SiGe alloys, as well as those of Si and Ge for comparison. In a case where there is no stress drop after yielding, the yield stresses are plotted.[29,90,98] The yield stresses in Si and Ge decrease with an increase in the temperature. Such dependence of yield stress τ_y is described as a function of strain rate $\dot{\varepsilon}$ and temperature T by the following empirical equation:

$$\tau_y = A\dot{\varepsilon}^{1/n}\exp(-U/k_B T), \tag{2.4}$$

where A, n, and U are constants.[90,91,98] The Ge-rich SiGe alloys show similar reduction of the yield stresses with increasing temperatures in the low-temperature region. Their dependence on the temperature becomes weak in the high-temperature region and finally becomes nearly constant. The temperature-insensitive range expands toward the low-temperature side with an increase of the Si content x. The yield stress of the alloys is higher at high temperatures with increasing Si content up to 0.4. Typically, the yield stresses of the alloys $x = 0.4$ are temperature-insensitive in the range investigated. The yield stress of the Si-rich SiGe alloy $x = 0.99$ is the same as, or slightly lower than, that of Si and the temperature dependence of the yield stress is similar to that of Si. With a decrease in the Si content to $x = 0.95$, the yield stress increases and the temperature dependence of the yield stress becomes weaker.

Figure 2.26 shows the composition dependence of the yield stress of the SiGe alloys at 800°C and 900°C. The yield stress increases with increasing Si content in the composition range $x = 0$–0.4, and with decreasing Si content from $x = 1$ to $x = 0.95$. Typically, the yield stresses of the Ge-rich SiGe alloys with $x > 0.10$ are

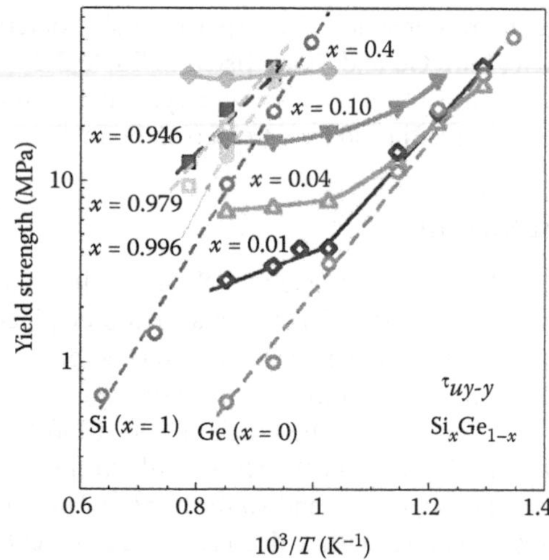

FIGURE 2.25 Yield stresses of the SiGe alloys with various compositions plotted against the reciprocal temperature for deformation under a strain rate of 1.8×10^{-4} s^{-1}.

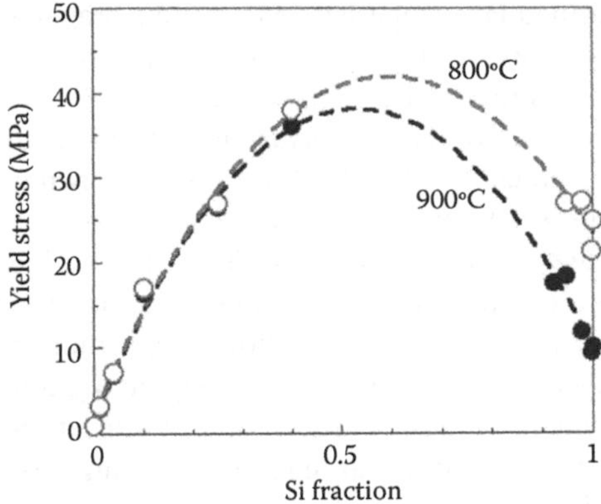

FIGURE 2.26 Composition dependence of the yield stress of the SiGe alloys at 800°C and 900°C. (Reprinted from *Journal of Crystal Growth* 275, I. Yonenaga, Growth and fundamental properties of SiGe bulk crystals, 91, Copyright 2005, with permission from Elsevier BV.)

much higher than that of pure Si. Over the whole composition range of the SiGe alloys, the yield stress shows a maximum at $x \sim 0.5$, dependent on the composition as proportional to $x(1 - x)$.[19,29,98]

SiGe demonstrates the character of an alloy only at high temperatures. The velocity of isolated dislocations in SiGe alloys with compositions close to Si and Ge shows only small differences from that in Si and Ge. Thus, the difference in the dislocation mobility among various compositions of SiGe alloys may not lead to a drastic difference in the mechanical strength of the alloys. Here, it may be noted that the bulk modulus of SiGe alloys was evaluated to increase linearly with alloy composition.[100] Also, dislocations induced by plastic deformation are dissociated into Shockley partial dislocations bounding an intrinsic stacking fault. The intrinsic stacking fault energy decreases from 61 to 55 ± 10 mJ/m^2 with increasing Si content, intermediate between those of Si and Ge.[26]

There are two components of the flow stress of a crystal under deformation. One is the effective stress by which dislocations move at a certain velocity against the intrinsic resistance (Peierls potential) via thermal activation. The other is the athermal or slightly temperature-dependent stress below which dislocations cannot move. If alloying results in a drastic increase in the Peierls potential and reduces the dislocation velocity, the yield stress drop may be found to be more remarkable in SiGe alloys than in Ge. In addition, the strengthening effect caused by alloying should be less remarkable with an increase of temperature. As shown in Figure 2.25, the yield stresses of the alloys, remarkable within the intermediate composition, are temperature-independent at elevated temperatures. Thus, the observed variation in the yield stress against the temperature in the SiGe alloys indicates that the SiGe alloys have an athermal stress that does not exist in other elemental and compound semiconductors. Reasonably, the athermal stress can be related to the alloying effect with a maximum at $x \approx 0.5$.

The origin of athermal stress in alloys may result due to several causes, as discussed for GaAsP and InAsP alloys.[101,102] First, the short-range order of the $L1_1$ (CuPt-type) structure found in strained layer superlattice thin films prepared by molecular beam epitaxy[40,41] can lead to an extra stress of athermal nature since the motion of a dislocation destroys the short-range order along its slip plane.[103] However, there has been no report of detection of an ordered structure in bulk SiGe alloys by transmission electron microscopic studies[104] and by XAFS study, as shown in Section 2.4.1 where the ordering parameter was found to range around 0.06–0.11.[38,39]

Second, a long-range stress field may be developed by local fluctuation of the alloy composition in a crystal. Since the covalent radius of Ge is larger than that of Si by about 4%, local fluctuation of the alloy composition in the crystal, causing the development of Si- or Ge-enriched regions, may induce a long-range stress field that cannot be overcome thermally by dislocations. Dislocations in SiGe alloys may move by a repeat bowing out process around the long-range stress fields.

Third, the dynamic development of a solute atmosphere around a dislocation during deformation at high temperatures leads to additional stress for releasing the dislocation from the solute atmosphere. Indeed, many fine serrations on the stress–strain curve in the deformation under a strain rate as low as 1.8×10^{-5} s^{-1} at 900°C were observed.[98] Such a characteristic is known as the Portevin–LeChatelier phenomenon, interpreted as being a repeated process of locking and releasing of dislocations.[105] In addition, though the width of dissociated dislocations formed by plastic deformation remains constant, photoluminescence (PL) studies have shown a variation of composition around induced dislocations by annealing.[106,107] Although the process of dislocation release from its solute atmosphere is a thermally activated one, the development of a solute atmosphere around the dislocation is enhanced at higher temperature. Thus, the contributions of these effects to the flow stress compensate each other and may give rise to a temperature-insensitive resistance to the dislocation motion, apparently looking like an athermal stress. Either one or both the local fluctuation of alloy composition and/or the dynamic development of a solute atmosphere around the dislocations may suppress the dynamic activity of dislocations, resulting in the strengthening of bulk SiGe alloys at elevated temperatures.

2.6 APPLICATIONS

Strain engineering in group-IV heterostructures plays an important role in achieving further improvements in electronic and optical properties, since high carrier mobilities have been achieved in tensile-strained Si[108] and compressive-strained Ge.[109,110] In general, strains are introduced by epitaxial growth on a SiGe virtual substrate consisting of strain-relaxed SiGe film grown on Si. However, misfit dislocations are inevitably introduced due to the lattice mismatch between SiGe and Si, leading to structural imperfections such as fluctuations in the orientation and strain.[110] Utilization of a bulk crystal of SiGe as a substrate is a potential option for high-quality strain-controlled quantum structures in lattice-matched epitaxy.

A pair of QWs of tensile-strained Si (nominal coverage of 11 monolayers (ML)) and compressively strained Ge (nominal coverage of 3 ML) among 250-nm-thick $Si_{0.915}Ge_{0.085}$ barriers was grown on (100)-oriented Si-rich Si_xGe_{1-x} ($x = 0.915$) wafers prepared from a CZ-grown crystal.[111] In addition to a single pair of QWs, samples with 10 pairs of QWs were also grown. The line width of x-ray rocking curve of the

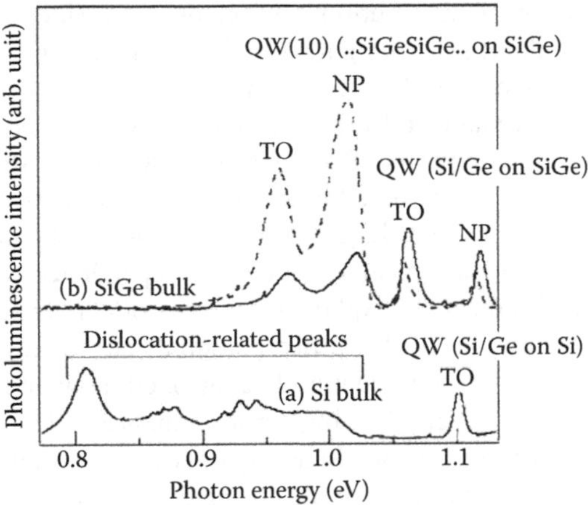

FIGURE 2.27 PL spectra of strained QWs with nominal structures of a pair of strained Si/strained Ge QWs among SiGe barriers grown on (a) Si and (b) SiGe solid lines. The dotted line shows a spectrum from 10 pairs of QWs grown on SiGe at 9 K. (Reprinted with permission from N. Usami et al. 2007. Application of Czochralski-grown SiGe bulk crystal as a substrate for luminescent strained quantum wells. *Applied Physics Letters* 90: 181914. Copyright 2007, American Institute of Physics.)

SiGe (100) substrate was comparable to that of the Si (100) substrate, and no peak splitting was observed in the line scan all over the substrate.

Figure 2.27 shows the PL spectra of the strained QWs with nominal structures of a pair of strained Si/strained Ge QWs between SiGe barriers grown on Si and SiGe.[112] Also, a spectrum of 10 pairs of QWs grown on SiGe was superimposed. Remarkably, the PL spectrum from the sample grown on SiGe (100) exhibits peaks from excitons confined in QWs without any deep levels related to dislocation luminescence in contrast to that grown on Si (100).

2.7 SUMMARY

The basic features of CZ growth of bulk crystals of Si_xGe_{1-x} alloys and their intrinsic properties were reviewed taking into careful account the basic physics of alloy solidification and alloying mechanisms. It is shown that full single crystals of SiGe alloys could be grown in the whole range of compositions, although some difficulties still remain for the growth of large-size SiGe ingots in the intermediate composition range, even with adoption of extremely low pulling velocity.

The experimentally derived criterion relative to the composition dependence of the pulling rate for the growth of single crystal or polycrystals is in good agreement with the classical model of critical growth velocity associated with constitutional supercooling and development of cellular structures. The variation of composition of the grown crystal along the growth direction is explained by solute segregation with the equilibrium coefficient from the melt of complete mixing. Dash-necking with a seed of composition identical to that of the crystal is effective for the growth of dislocation-free crystals.

The large difference of distribution coefficients of various impurities in SiGe alloys with respect to those pertaining to pure Si and Ge may be attributed to elastic interaction due to the size misfit between the relevant atoms.

Considering more fundamental aspects, SiGe alloys may be treated as typical disordered solids with the imperfect Pauling type of structure. Most of the atomistic strain in SiGe alloys is accommodated by changes of both the bond length and bond angle with the alloy composition. Electron and hole mobilities, comparable to each other in SiGe alloys, show a typical $x(1-x)$-type dependence with a minimum of around $x = 0.6$–0.8, which can be explained in terms of carrier scattering produced by the disordered fluctuations of the lattice potentials at the lattice sites. Similarly, thermal conductivity was found to be mainly controlled by phonon scattering due to a distortion of the crystal lattice, showing a minimum at $x = 0.5$–0.7.

Oxygen atoms, incorporated into SiGe crystals as a consequence of the reaction among the Si melt and quartz crucible during the growth process, preferentially occupy a bond-centered position between Si atoms to make a $Si-O_i-Si$ quasi-molecule, leading to a typical optical absorption peak at 1106 cm^{-1}. Muonium, a pseudo-isotope of hydrogen, has a donor level across the alloy system, while the acceptor level is deep in Si and crosses into the valence band. This feature explains the differences of hydrogen behavior between Si and Ge.

Eventually, the mechanical strength of the alloys becomes temperature-insensitive at elevated temperatures and shows a compositional dependence of the $x(1-x)$ type over the whole composition range. Built-in stress fields related to local fluctuation of the alloy composition, together with the dynamic development of a solute atmosphere around dislocations, seem to suppress the activities of dislocations and bring about a hardening of SiGe alloys.

The possibility of growing high-quality bulk crystals of Si_xGe_{1-x} alloys in the full composition range has allowed the start-up of systematic studies on fundamental physics of these kinds of solid solution. Up to now, the effect of various and unique alloying phenomena on the structural and electrical properties and on defects in SiGe has been elucidated in terms of random mixing of the elemental constituents of the solution. This knowledge might be essential for the development of germanium–tin (GeSn)[113] and silicon–tin (SiSn) alloys,[114] exhibiting direct-bandgap features. It is also expected that the results of these studies may lead to further refinement of our knowledge concerning the intrinsic properties of Si and Ge.

ACKNOWLEDGMENTS

The author expresses gratitude to A. Matsui, M. Nonaka, T. Ayuzawa, T. R. Mchedlidze, M. Sakurai, T. Akashi, N. Fukata, T. Goto, D. Shindo, and N. Usami of Tohoku University, P. J. C. King of Rutherford Appleton Laboratory, B. R. Carroll and R. L. Litchti of Texas Tech University, A. Béraud and E. Courtens of Université of Montpellier, M. Werner, M. Bartsch, and U. Messerschmidt of Max-Planck-Institut für Mikrostrukturphysik, E. R. Weber of Fraunhofer Institut Solare Energiesysteme, and J. Wollweber of Leibniz-Institut für Kristallzüchtung for their assistance in conducting the present research. The author is grateful to K. Sumino of Tohoku University and T. Abe of Shin-Etsu Handotai for their encouragement.

REFERENCES

1. I. Yonenaga. 2001. Si_xGe_{1-x} bulk crystals, in *The Encyclopedia of Materials: Science and Technology*, eds. K. H. J. Buschow, R. W. Cahn, M. C. Flemings, B. Ilschner, E. J. Kramer, and S. Mahajan. Amsterdam: Elsevier Science B. V. p. 8647.
2. A. Dahlen, A. Fattah, G. Hanke, and E. Karthaus. 1994. Bridgman and Czochralski growth of Ge-Si alloy crystals. *Crystal Research and Technology* 29: 187.
3. M. Kürten and J. Schilz. 1994. Czochralski growth of Si_xGe_{1-x} single crystals. *Journal of Crystal Growth* 139: 1.
4. J. Schilz and V. N. Romanenko. 1995. Bulk growth of silicon-germanium solid solutions. *Journal of Materials Science: Materials in Electronics* 6: 265.
5. I. Yonenaga, A. Matsui, S. Tozawa, K. Sumino, and T. Fukuda. 1995. Czochralski growth of $Ge_{1-x}Si_x$ alloy crystals. *Journal of Crystal Growth* 154: 275.
6. J. Wollweber, D. Schulz, and W. Schröder. 1996. Extremely reduced dislocation density in Si_xGe_{1-x} single crystals grown by the float zone technique. *Journal of Crystal Growth* 158: 166.
7. N. V. Abrosimov, S. N. Rossolenko, V. Alex, A. Gerhardt, and W. Schröder. 1996. Single crystal growth of $Si_{1-x}Ge_x$ by the Czochralski technique. *Journal of Crystal Growth* 166: 657.
8. K. Kadokura and Y. Takano. 1997. Germanium-silicon single crystal growth using an encapsulant in a silica ampoule. *Journal of Crystal Growth* 171: 56.
9. I. Yonenaga. 1997. Dislocation velocities in GeSi bulk alloys. *Materials Research Society Symposium Proceedings* 442: 337.
10. A. Matsui, I. Yonenaga, and K. Sumino. 1998. Czochralski growth of bulk crystals of $Ge_{1-x}Si_x$ alloys. *Journal of Crystal Growth* 183: 109.
11. I. Yonenaga and M. Nonaka. 1998. Czochralski growth of bulk crystals of $Ge_{1-x}Si_x$ alloys—II: Si-rich alloys. *Journal of Crystal Growth* 191: 393.
12. P. Dold, A. Barz, S. Recha, K. Pressel, M. Franz, and K. W. Benz. 1998. Growth and characterization of $Ge_{1-x}Si_x$ ($x \leq 10$ at%) single crystals. *Journal of Crystal Growth* 192: 125.
13. I. Yonenaga. 1999. Czochralski growth of GeSi bulk alloy crystals. *Journal of Crystal Growth* 198/199: 404.
14. Y. Azuma, N. Usami, T. Ujihara, G. Sazaki, Y. Murakami, S. Miyashita, K. Fujiwara, and K. Nakajima. 2001. Growth of SiGe bulk crystal with uniform composition by directly controlling the growth temperature at the crystal–melt interface using in situ monitoring system. *Journal of Crystal Growth* 224: 204.
15. I. Yonenaga. 2001. Czochralski growth of heavily impurity doped crystals of GeSi alloys. *Journal of Crystal Growth* 226: 47.
16. I. Yonenaga, T. Akashi, and T. Goto. 2001. Thermal and electrical properties of Czochralski grown GeSi single crystals. *Journal of Physics and Chemistry of Solids* 62: 1313.
17. T. A. Campbell, M. Schweizer, P. Dold, A. Cröll, and K. W. Benz. 2001. Float zone growth and characterization of $Ge_{1-x}Si_x$ ($x \leq 10$ at%) single crystals. *Journal of Crystal Growth* 226: 231.
18. M. P. Volz, M. Schweizer, N. Kaiser, S. D. Cobb, L. Vujisic, S. Motakef, and F. R. Szofran. 2002. Bridgman growth of detached GeSi crystals. *Journal of Crystal Growth* 237–239: 1844.
19. I. Yonenaga. 2005. Growth and fundamental properties of SiGe bulk crystals. *Journal of Crystal Growth* 275: 91.
20. M. Yildiz, S. Dost, and B. Lent. 2005. Growth of bulk SiGe single crystals by liquid phase diffusion. *Journal of Crystal Growth* 280: 151.
21. H. Miyata, S. Adachi, Y. Ogata, T. Tsuru, Y. Muramatsu, K. Kinoshita, O. Odawara, and S. Yoda. 2007. Crystallographic investigation of homogeneous SiGe single crystals grown by the traveling liquidus-zone method. *Journal of Crystal Growth* 303: 607.

22. I. Yonenaga. 2008. Growth and fundamental properties of SiGe bulk crystals. *Materia* 47: 3 (in Japanese).
23. I. Yonenaga, T. Taishi, Y. Ohno, and Y. Tokumoto. 2010. Cellular structures in Czochralski-grown SiGe bulk crystal. *Journal of Crystal Growth* 312: 1065.
24. A. Dario, H. O. Sicim, and E. Balikci. 2012. A comparative study on the growth of germanium–silicon single crystals grown by the vertical Bridgman and axial heat processing techniques. *Journal of Crystal Growth* 351: 1.
25. R. W. Olesinski and G. J. Abbaschian. 1984. The Ge-Si (germanium-silicon) system. *Bulletin of Alloy Phase Diagrams* 5: 180.
26. I. Yonenaga, S.-H. Lim, and D. Shindo. 2000. Dislocation dissociation and stacking-fault energies in $Ge_{1-x}Si_x$ alloys. *Philosophical Magazine Letters* 80: 193.
27. I. Yonenaga and Y. Murakami. 1998. Segregation during the seeding process in the Czochralski growth of GeSi alloys. *Journal of Crystal Growth* 191: 399.
28. W. A. Tiller, K. A. Jackson, J. W. Rutter, and B. Chalmers. 1953. The redistribution of solute atoms during the solidification of metals. *Acta Metallurgica* 1: 428.
29. I. Yonenaga. 1999. Growth and mechanical properties of GeSi bulk crystals. *Journal of Materials Science: Materials in Electronics* 10: 329.
30. I. Yonenaga and T. Ayuzawa. 2006. Segregation coefficients of various dopants in Si_xGe_{1-x} ($0.93 < x < 0.96$) single crystals. *Journal of Crystal Growth* 297: 14.
31. W. G. Pfann. 1952. Principles of zone-melting. *Journal of Metals* 4: 747.
32. D. T. J. Hurle. 1994. *Handbook of Crystal Growth.* Vol. 2b. Amsterdam: North-Holland.
33. J. C. Phillips. 1973. *Bonds and Bands in Semiconductors.* New York: Academic Press.
34. E. K. Kamornik, E. D. Nensberg, G. V. Nikitina, A. G. Orlov, and V. N. Romanenko. 1968. Distribution coefficients in certain semiconductor materials. *Inorganic Materials* 4: 667.
35. A. Barz, P. Dold, U. Kerat, S. Recha, K. W. Benz, M. Franz, and K. Pressel. 1998. Germanium-rich SiGe bulk single crystals grown by the vertical Bridgman method and by zone melting. *Journal of Vacuum Science and Technology B* 16: 1627.
36. T. Taishi, X. Huang, M. Kubota, T. Kajigaya, T. Fukami, and K. Hoshikawa. 1999. Heavily boron-doped silicon single crystal growth: Boron segregation. *Japanese Journal of Applied Physics Part 2* 38: L223.
37. J. P. Dismukes, L. Ekstrom, and R. J. Paff. 1964. Lattice parameter and density in germanium-silicon alloys. *The Journal of Physical Chemistry* 68: 3021.
38. I. Yonenaga and M. Sakurai. 2001. Bond lengths in $Ge_{1-x}Si_x$ crystalline alloys grown by the Czochralski method. *Physical Review B* 64: 113206.
39. I. Yonenaga, M. Sakurai, M. H. F. Sluiter, and Y. Kawazoe. 2005. Atomic-scale structure of Si_xGe_{1-x} solid solutions. *Journal of Metastable and Nanocrystalline Materials* 24–25: 523.
40. A. Ourmazd and J. C. Bean. 1985. Observation of order-disorder transitions in strained-semiconductor systems. *Physical Review Letters* 55: 765.
41. J. Z. Tischler, J. D. Budai, D. E. Jesson, G. Eres, P. Zschack, J.-M. Baribeau, and D. C. Houghton. 1995. Ordered structures in Si_xGe_{1-x} alloy thin films. *Physical Review B* 51: 10947.
42. M. H. F. Sluiter and Y. Kawazoe. 2001. Bondlengths and phase stability of silicon-germanium alloys under pressure. *Materials Transactions* 42: 2201.
43. L. Vegard. 1921. Die Konstitution der Mischkristalle und die Raumfüllung der Atome. *Zeitshrift für Physik* 5: 17.
44. L. Pauling. 1967. *The Nature of the Chemical Bond.* Ithaca (NY): Cornell University Press.
45. Y. Cai and M. F. Thorpe. 1992. Length mismatch in random semiconductor alloys. II. Structural characterization of pseudobinaries. *Physical Review B* 46: 15879.
46. D. B. Aldrich, R. J. Nemanich, and D. E. Sayers. 1994. Bond-length relaxation in $Si_{1-x}Ge_x$ alloys. *Physical Review B* 50: 15026.

47. J. C. Woicik, K. E. Miyano, C. A. King, R. W. Johnson, J. G. Pellegrino, T.-L. Lee, and Z. H. Lu. 1998. Phase-correct bond lengths in crystalline Ge_xSi_{1-x} alloys. *Physical Review B* 57: 14592.
48. J. C. Aubry, T. Tyliszczak, A. P. Hitchcock, J.-M. Baribeau, and T. E. Jackman. 1999. First-shell bond lengths in Si_xGe_{1-x} crystalline alloys. *Physical Review B* 59: 12872.
49. N. Mousseau and M. F. Thorpe. 1993. Structural model for crystalline and amorphous Si-Ge alloys. *Physical Review B* 48: 5172.
50. P. Venezuela, G. M. Dalpian, A. J. R. da Silva, and A. Fazzio. 2001. *Ab initio* determination of the atomistic structure of Si_xGe_{1-x} alloy. *Physical Review B* 64: 193202.
51. K. Hayashi, Y. Takahashi, I. Yonenaga, and E. Matsubara. 2004. X-ray fluorescence holography study on $Si_{1-x}Ge_x$ single crystal. *Materials Transactions* 45: 1994.
52. J. C. Mikkelsen, Jr. and J. B. Boyce. 1982. Atomic-scale structure of random solid solution: Extended x-ray-absorption fine-structure study of $Ga_{1-x}In_xAs$. *Physical Review Letters* 49: 1412.
53. A. Balzarotti. 1987. Local bonding and thermodynamic properties of II-VI pseudobinary alloys by EXAFS, in *Ternary and Multinary Compounds*, eds. S. K. Deb and A. Zunger. Pittsburg: Materials Research Society. p. 333.
54. H. Oyanagi, Y. Takeda, T. Matsushita, T. Ishiguro, T. Yao, and A. Sasaki. 1988. Structural studies of (Ga, In)(As, P) alloys and $(In\ As)_m(Ga\ As)_n$ strained-layer superlattices by fluorescence-detected EXAFS. *Superlattices and Microstructures* 4: 413.
55. T. R. Mchedlidze and I. Yonenaga. 1996. Hall effect in anisotropic Si_xGe_{1-x} polycrystals. *Japanese Journal of Applied Physics Part 1* 35: 652.
56. T. R. Mchedlidze and I. Yonenaga. 1997. Hall effect measurements on Si_xGe_{1-x} bulk alloys. *Materials Research Society Symposium Proceedings* 442: 381.
57. I. Yonenaga, T. Akashi, and T. Goto. 2002. Thermal and electrical properties of Czochralski grown GeSi single crystals. *Materials Research Society Symposium Proceedings* 691: G7.4.1.
58. O. Ka. 1994. Electrical transport in polycrystalline semiconductors. *Solid State Phenomena* 37–38: 201.
59. I. Yonenaga. 2006. Carrier mobility and resistivity of n- and p-type Si_xGe_{1-x} $(0.93 < x < 0.96)$ single crystals. *Japanese Journal of Applied Physics Part 1* 45: 2678.
60. G. Masetti, M. Severi, and S. Solmi. 1983. Modeling of carrier mobility against carrier concentration in arsenic-, phosphorus-, and boron-doped silicon. *IEEE Transactions on Electron Devices* 30: 764.
61. I. Yonenaga, W. J. Li, T. Akashi, T. Ayuzawa, and T. Goto. 2005. Temperature dependence of electron and hole mobilities in heavily impurity-doped SiGe single crystals. *Journal of Applied Physics* 98: 063702.
62. B. Abeles. 1963. Lattice thermal conductivity of disordered semiconductor alloys at high temperatures. *Physical Review* 131: 1906.
63. J. P. Dismukes, L. Ekstrom, E. F. Steigmeier, I. Kudman, and D. S. Beers. 1964. Thermal and electrical properties of heavily doped Ge-Si alloys up to 1300°K. *Journal of Applied Physics* 35: 2899.
64. M. S. Abrahams, R. Braunstein, and F. D. Rosi. 1959. Thermal, electrical and optical properties of (In, Ga)As alloys. *Journal of Physics and Chemistry of Solids* 10: 204.
65. E. F. Steigmeier and B. Abeles. 1964. Scattering of phonons by electrons in germanium-silicon alloys. *Physical Review* 136: A1149.
66. A. Béraud, J. Kulda, I. Yonenaga, M. Foret, B. Salce, and E. Courtens. 2004. Disorder-induced broadening of transverse acoustic phonons in Si_xGe_{1-x} mixed crystals. *Physica B* 350: 254.
67. E. Rat, B. Hehlen, J. Kulda, I. Yonenaga, H. Casalta, E. Courtens, M. Foret, and R. Vacher. 2000. Disorder broadening of the acoustic branches in Si_xGe_{1-x} mixed crystals. *Physica B* 276–278: 429.

68. J. Kulda, Y. Ishii, and S. Katano. 1995. Dynamic structure analysis applied to Si and Ge. *Physica B* 213–214: 427.
69. O. Yamashita and N. Sadatomi. 2000. Thermoelectric properties of $Si_{1-x}Ge_x$ ($x \leq 0.10$) with alloy and dopant segregations. *Journal of Applied Physics* 88: 245.
70. G. A. Slack and M. S. Hussain. 1991. The maximum possible conversion efficiency of silicon-germanium thermoelectric generators. *Journal of Applied Physics* 70: 2694.
71. I. Yonenaga, M. Nonaka, and N. Fukata. 2001. Interstitial oxygen in GeSi alloys. *Physica B* 308–310: 539.
72. W. Kaiser, P. H. Keck, and C. F. Lange. 1956. Infrared absorption and oxygen content in silicon and germanium. *Physical Review* 101: 1264.
73. H. J. Hrostowski and R. H. Kaiser. 1957. Infrared absorption of oxygen in silicon. *Physical Review* 107: 966.
74. W. Kaiser and C. D. Thurmond. 1961. Solubility of oxygen in germanium. *Journal of Applied Physics* 32: 115.
75. P. Clauws. 1996. Oxygen related defects in germanium. *Materials Science and Engineering B* 36: 213.
76. T. Taishi, H. Ise, Y. Murao, T. Osawa, M. Suezawa, Y. Tokumoto, Y. Ohno, K. Hoshikawa, and I. Yonenaga. 2010. Czochralski-growth of germanium crystals containing high concentrations of oxygen impurities. *Journal of Crystal Growth* 312: 2783.
77. I. Barin and O. Knacke. 1977. *Thermochemical Properties of Inorganic Substances*. Berlin: Springer.
78. H. Yamada-Kaneta, C. Kaneta, and T. Ogawa. 1993. Infrared absorption by interstitial oxygen in germanium-doped silicon crystals. *Physical Review B* 47: 9338.
79. D. Wauters and P. Clauws. 1997. Ge content dependence of the infrared spectrum of interstitial oxygen in crystalline Si-Ge. *Materials Science Forum* 258–263: 103.
80. J. Humlícek, R. Stoudek, and A. Dubroka. 2006. Infrared vibrations of interstitial oxygen in silicon-rich SiGe alloys. *Physica B* 376–377: 212.
81. J. Coutinho, R. Jones, P. R. Briddon, and S. Öberg. 2000. Oxygen and dioxygen centers in Si and Ge: Density-functional calculations. *Physical Review B* 62: 10824.
82. P. J. C. King, R. L. Lichti, and I. Yonenaga. 2003. Hydrogen behaviour in bulk $Si_{1-x}Ge_x$ alloys as modelled by muonium. *Physica B* 340–342: 835.
83. P. J. C. King, R. L. Lichti, S. P. Cottrell, I. Yonenaga, and B. Hitti. 2005. Characterization of hydrogen-like states in bulk $Si_{1-x}Ge_x$ alloys through muonium observations. *Journal of Physics: Condensed Matter* 17: 4567.
84. B. R. Carroll, R. L. Lichti, P. J. C. King, Y. G. Celebi, I. Yonenaga, and K. H. Chow. 2009. Muonium acceptor states in high Ge $Si_{1-x}Ge_x$ alloys. *Physica B* 404: 812.
85. B. R. Carroll, R. L. Lichti, P. J. C. King, Y. G. Celebi, I. Yonenaga, and K. H. Chow. 2010. Muonium defect levels in Czochralski-grown silicon-germanium alloys. *Physical Review B* 82: 205205.
86. S. F. J. Cox, R. L. Lichti, J. S. Lord, E. A. Davis, R. C. Vilão, J. M. Gil, T. D. Veal, and Y. G. Celebi. 2013. The first 25 years of semiconductor muonics at ISIS, modelling the electrical activity of hydrogen in inorganic semiconductors and high-κ dielectrics. *Physica Scripta* 88: 068503.
87. C. G. Van de Walle and J. Neugebauer. 2003. Universal alignment of hydrogen levels in semiconductors, insulators and solutions. *Nature (London)* 423: 626.
88. L. Dobaczewski, K. Bonde Nielsen, N. Zangenberg, B. Bech Nielsen, A. R. Peaker, and V. P. Markevich. 2004. Donor level of bond-center hydrogen in germanium. *Physical Review B* 69: 245207.
89. I. Yonenaga and K. Sumino. 1996. Dislocation velocity in GeSi alloy. *Applied Physics Letters* 69: 1264.
90. I. Yonenaga. 1999. Dislocation velocities and mechanical strength of bulk GeSi crystals. *Physica Status Solidi* (a) 171: 41.

91. I. Yonenaga. 2013. Dislocation dynamics in SiGe alloys. *Journal of Physics: Conference Series* 471: 012002.
92. Yu. L. Iunin, V. I. Orlov, D. V. Dyachenko-Dekov, N. V. Abrosimov, S. N. Rossolenko, and W. Schröder. 1997. Ge concentration effect on the dislocation mobility in the bulk SiGe alloy single crystals. *Solid State Phenomena* 57–58: 419.
93. K. Sumino. 1994. Mechanical behaviour of semiconductors, in *Handbook on Semiconductors*, Vol. 3, ed. S. Mahajan. Amsterdam: Elsevier Science B. V. p. 73.
94. I. Yonenaga. 1997. Mechanical properties and dislocation dynamics in III-V compounds. *Journal de Physisque III (France)* 7: 1435.
95. I. Yonenaga. 2005. Hardness, yield strength, and dislocation velocity in elemental and compound semiconductors. *Materials Transactions* 46: 1979.
96. K. Maeda and S. Takeuchi. 1996. Enhancement of dislocation mobility in semiconducting crystals by electronic excitation, in *Dislocations in Solids*, Vol. 10, eds. F. R. N. Nabarro and M. S. Duesbery. Amsterdam: Elsevier Science B. V. p. 443.
97. I. Yonenaga, M. Werner, M. Bartsch, U. Messerschmidt, and E. R. Weber. 1999. Recombination-enhanced dislocation motion in SiGe and Ge. *Physica Status Solidi (a)* 171: 35.
98. I. Yonenaga and K. Sumino. 1996. Mechanical strength of GeSi alloy. *Journal of Applied Physics* 80: 3244.
99. I. Yonenaga and K. Sumino. 1992. Mechanical properties and dislocation dynamics of III–V compound semiconductors. *Physica Status Solidi (a)* 131: 663.
100. M. H. F. Sluiter and I. Yonenaga, unpublished work.
101. I. Yonenaga, K. Sumino, G. Izawa, H. Watanabe, and J. Matsui. 1989. Mechanical property and dislocation dynamics of GaAsP alloy semiconductor. *Journal of Materials Research* 4: 361.
102. I. Yonenaga and K. Sumino. 1989. Mechanical property of III-V alloy semiconductors, in *8th Symposium Record on Alloy Semiconductor Physics and Electronics*, ed. A. Sasaki. Kyoto: Organization of Special Project Research on Alloy Semiconductor Physics and Electronics. p. 187.
103. J. C. Fisher. 1954. On the strength of solid solution alloys. *Acta Metallurgica* 2: 9.
104. D. Stenkamp and W. Jäger. 1992. Dislocations and their dissociation in Si_xGe_{1-x} alloys. *Philosophical Magazine A* 65: 1369.
105. I. Yonenaga and K. Sumino. 1992. Impurity effects on the mechanical behavior of GaAs crystals. *Journal of Applied Physics* 71: 4249.
106. K. Tanaka, M. Suezawa, and I. Yonenaga. 1996. Photoluminescence spectra of deformed Si-Ge alloy. *Journal of Applied Physics* 80: 6991.
107. Y. Ohno, Y. Tokumoto, H. Taneichi, I. Yonenaga, K. Togase, and S. R. Nishitani. 2012. Interaction of dopant atoms with stacking faults in silicon. *Physica B* 407: 3006.
108. F. Schäffler. 1997. High-mobility Si and Ge structures. *Semiconductor Science and Technology* 12: 1515.
109. Y. H. Xie, D. Monroe, E. A. Fitzgerald, P. J. Silverman, F. A. Thiel, and G. P. Watson. 1993. Very high mobility two-dimensional hole gas in $Si/Ge_xSi_{1-x}/Ge$ structures grown by molecular beam epitaxy. *Applied Physics Letters* 63: 2263.
110. M. Myronov, T. Irisawa, O. A. Mironov, S. Koh, Y. Shiraki, T. E. Whall, and E. H. C. Parker. 2002. Extremely high room-temperature two-dimensional hole gas mobility in $Ge/Si_{0.33}Ge_{0.67}/Si(001)$ p-type modulation-doped heterostructures. *Applied Physics Letters* 80: 3117.
111. N. Usami, Y. Nose, K. Fujiwara, and K. Nakajima. 2006. Suppression of structural imperfection in strained Si by utilizing SiGe bulk substrate. *Applied Physics Letters* 88: 221912.
112. N. Usami, R. Nihei, I. Yonenaga, Y. Nose, and K. Nakajima. 2007. Application of Czochralski-grown SiGe bulk crystal as a substrate for luminescent strained quantum wells. *Applied Physics Letters* 90: 181914.

113. Y. Murao, T. Taishi, K. Kutsukake, Y. Tokumoto, Y. Ohno, and I. Yonenaga. 2012. Growth of dilute GeSn alloys. *Extended Abstracts of the Seventh International Workshop on Modeling of Crystal Growth (IWMCG-7)*, Taihei, October 28–31, 2012: p. 130.
114. I. Yonenaga, T. Taishi, K. Inoue, R. Gotoh, K. Kutsukake, Y. Tokumoto, and Y. Ohno. 2014. Czochralski growth of heavily tin-doped Si crystals. *Journal of Crystal Growth* 395: 94.

3 Germanium on Silicon
Epitaxy and Applications

Daniel Chrastina

CONTENTS

3.1 Introduction 61
3.2 Historical Results on Germanium as a Semiconductor 62
3.3 Epitaxy of Germanium 62
3.4 Strained Ge QWs 65
3.5 Multiple QWs 67
  3.5.1 Thermoelectric Devices 67
  3.5.2 Optoelectronic Devices 68
3.6 Thick Ge Layers for Light Emission 70
3.7 Ge Deposition on Structured Substrates 70
3.8 Beyond Ge 72
References 74

3.1 INTRODUCTION

Germanium is attracting increasing interest due to the similarity of its band structure to III–V semiconductors, and its compatibility with Si-based microelectronics. High *p*-type electrical mobility in strained Ge channels has been demonstrated, but recent results also demonstrate the possible optoelectronic applications for both thin-strained Ge quantum wells (QWs) and thick bulk-like Ge layers (Chaisakul et al. 2014b; Rouifed et al. 2014; Camacho-Aguilera et al. 2012).

Modulation-doped strained Ge QWs have demonstrated extremely high hole mobilities at low temperature, and at room temperature, the mobility matches or exceeds the mobility of intrinsic bulk Ge (Xie et al. 1993; von Känel et al. 2002; Rössner et al. 2004; Dobbie et al. 2012; Hassan et al. 2014; Mironov et al. 2014).

A stack of several strained Ge QWs can be employed to exploit the quasi-direct bandstructure of Ge, in which, the Γ-point minimum is only 150 meV above the indirect minimum. This leads to efficient absorption in the region of the technologically important wavelength of 1.5 μm, which can be modulated by means of the quantum-confined Stark effect (QCSE) in which a transverse electric field brings confined electron and hole states closer together in energy. Direct-gap photo- and electroluminescence have been observed both at low temperatures and at room temperature (Bonfanti et al. 2008; Gatti et al. 2011; Wu et al. 2012; Cecchi et al. 2014).

Another method of enhancing direct-gap effects is to apply tensile strain to bulk-like Ge layers, which lowers both the direct and indirect minima while reducing the energy gap between them. *n*-Type doping can then be used to saturate the indirect states such that a significant density of electrons is thermally excited to the direct minimum (Carroll et al. 2012; Liu et al. 2012).

Ge layers more than a few micrometers thick cannot be grown directly on Si substrates without catastrophic cracking of the material due to the thermal expansion mismatch between Ge and Si. However, recent advances in the growth of microcrystals on patterned substrates allow extremely thick layers to be realized, for applications such as x-ray detectors (Falub et al. 2012; Kreiliger et al. 2014).

In this chapter, epitaxial growth of Ge and Ge-rich SiGe is discussed with a particular focus on low-energy plasma-enhanced chemical vapor deposition (LEPECVD), as well as conventional chemical vapor deposition (CVD) and molecular-beam epitaxy (MBE) approaches.

3.2 HISTORICAL RESULTS ON GERMANIUM AS A SEMICONDUCTOR

While silicon dominates the microelectronics industry, the first transistor was realized in germanium (Bardeen and Brattain 1948).* Germanium is undergoing a resurgence as a possible high-mobility replacement for mainstream Si metal-oxide-semiconductor field-effect transistor (MOSFET) technology, since of all the semiconductor materials, Ge exhibits the highest *p*-type mobility (Cuttriss 1961; Pillarisetty 2011) and it is also extremely interesting for optoelectronic devices due to its almost GaAs-like bandstructure (Cardona and Pollak 1966; Elder et al. 2011).

In pure Ge, the Γ-point conduction band minimum is only 150 meV above the overall *L*-point minimum at room temperature (Cardona and Pollak 1966; Lautenschlager et al. 1985) (Figure 3.1), leading to a much sharper absorption edge as compared to even Ge-rich SiGe alloys (Braunstein et al. 1958). Doped material was studied optically (Newman 1953a,b; Pankove 1960; Sommers 1961; Spitzer et al. 1961; Haas 1962; Pankove and Aigrain 1962; Fowler et al. 1962; Kline et al. 1968; Benoit à la Guillaume and Cernogora 1969; Stern 1971; Lukeš and Humlíček 1972; Contreras et al. 1983; Viña and Cardona 1986), demonstrating photoluminescence for both *n*-type and *p*-type bulk crystals (Figure 3.2) (van Driel et al. 1976; Nakamura 1977; Rentzsch and Shlimak 1977; Klingenstein and Schweizer 1978; Wagner et al. 1983; Wagner and Viña 1984) with the possibility of lasing already under discussion (van Driel et al. 1976). The direct bandgap energy of Ge at room temperature (800 meV) corresponds to a telecommunications-relevant wavelength of 1.5 μm, increasing to about 900 meV at low temperature (Allen and Cardona 1983).

3.3 EPITAXY OF GERMANIUM

Epitaxial growth of Ge on Si, as part of the general scope of epitaxial SiGe alloy growth processes, opens up new possibilities in terms of physics of heterostructures

* It is mentioned in a footnote that "the effect has been found with both silicon and germanium."

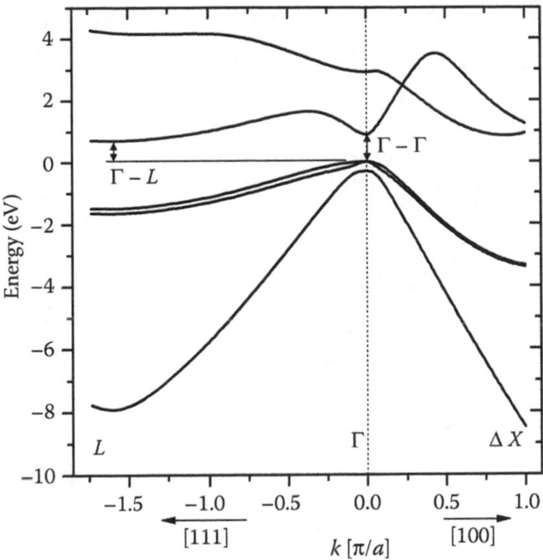

FIGURE 3.1 Energy bands of germanium calculated by the **k** · **p** method in the [111] and [100] directions of **k** space, based on the method of Elder et al. (2011). With thanks to G. Isella.

and also in terms of applications, since the properties of Ge can be accessed while exploiting the massive industrialized resources devoted to the production of cheap high-quality silicon wafers. Thick Ge layers (of the order of 1 μm) can be exploited for optoelectronic applications (Colace et al. 1997, 1999, 2006; Luan et al. 2001; Famà et al. 2002; Isella et al. 2007; Osmond et al. 2009) or as a buffer layer between a silicon wafer and a GaAs-based layer stack. For example, a GaAs-based solar cell structure can be epitaxially grown on the so-called "virtual substrate" (VS) that features only a relatively thin layer of Ge on a Si substrate, rather than using an expensive, fragile, and heavy bulk Ge substrate (Ginige et al. 2006). Similarly, lasers have been fabricated from III–V semiconductor materials epitaxially deposited on a Ge VS (Chriqui et al. 2003; Groenert et al. 2003). These applications exploit the fact that the lattice parameter of Ge is similar to that of GaAs (Dismukes et al. 1964b; Gehrsitz et al. 1999).

In these cases, the VS was realized by growing a thick, graded buffer layer. This takes the form of a layer in which the Ge content is raised from zero (i.e., pure Si) to 100% (i.e., pure Ge) over a thickness of several micrometers, in a process that can obviously be generalized to SiGe alloys of any Ge content (Fitzgerald et al. 1991, 1992, 1999). Thick-graded buffers have been shown to feature low-defect densities (Thomas et al. 2003; Marchionna et al. 2006), but the thickness of the buffer is an impediment to integration and the large quantity of material consumed during growth leads to increased processing time and cost. Therefore, strategies that allow thinner high-quality Ge films to be deposited epitaxially on Si without an intermediate graded layer have been studied.

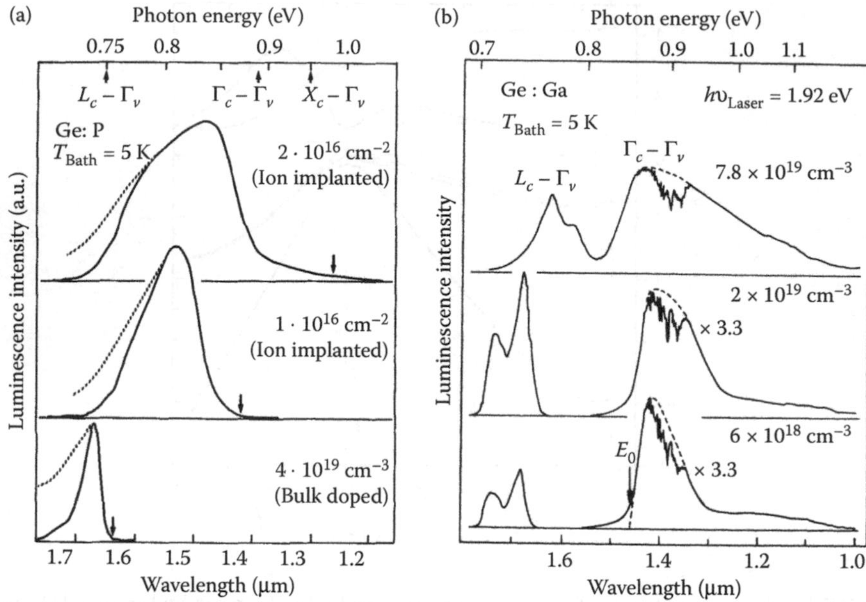

FIGURE 3.2 (a) Photoluminescence of bulk-doped and ion-implanted Ge:P samples. The implantation of 1×10^{16} cm^{-2} corresponds to about 2–3 × 10^{20} cm^{-3} and that of 2×10^{16} cm^{-2} corresponds to about 5×10^{20} cm^{-3} carrier concentration. The arrows indicate the high-energy edge $E_{G,1}$ of the luminescence band. The drawn lines are spectra not corrected for detector response. The dashed lines indicate the luminescence line shape after applying this correction. (b) Photoluminescence spectra of *p*-type Ge for various hole concentrations, showing indirect ($L_C \rightarrow \Gamma_V$) luminescence and direct ($\Gamma_C \rightarrow \Gamma_V$) gap emission. The spectra were recorded using a Ge photodiode. The indirect luminescence spectrum is distorted by the cutoff in detector sensitivity at 1.75 μm. The sharp structure at ≃1.4 μm is caused by atmospheric water absorption. Spectral resolution was 24 Å. ((a) From Wagner, J., A. Compaan, and A. Axmann. *J. Phys. (Paris) Colloq. 44*(C5), 61–64. Copyright 1983, EDP Sciences. Reprinted with permission. (b) Reprinted with permission from Wagner, J. and L. Viña. Radiative recombination in heavily doped *p*-type germanium. *Phys. Rev. B 30*(12), 7030–7036. 1984. Copyright 1984 by the American Physical Society.)

Alternatively, thin Ge layers (of the order of 10 nm) can be exploited as part of a SiGe heterostructure, forming QWs for the confinement of holes and electrons addressed at electronic or optoelectronic applications (Kuo et al. 2005; Tsujino et al. 2006; Bonfanti et al. 2008; Chrastina et al. 2008). In this case, it is necessary to interpose a Ge-rich buffer layer between the active structure and the silicon substrate. The growth of a thin Ge layer directly on Si instead tends to lead to the formation of a layer of SiGe islands (Eaglesham and Cerullo 1990; Rastelli et al. 2001). Such SiGe islands have been extensively studied, and considered for electronic, optoelectronic, and thermoelectric applications (Brehm et al. 2008, 2009; Stoffel et al. 2008, 2009; Bollani et al. 2010a,b, 2012; Pernot et al. 2010; Vanacore et al. 2010a,b, 2014).

Epitaxy of Ge on Si is usually realized by CVD (Koester et al. 2000; Hartmann et al. 2004, 2005, 2008, 2009, 2010, 2012; Liu et al. 2004a; Rouvière et al. 2005; Xu et al. 2013a) or by MBE (Eaglesham and Cerullo 1990, 1991; Xie et al. 1993; Engelhardt et al. 1994; Hackbarth et al. 2000; Madhavi et al. 2001; Irisawa et al. 2002; Liu et al. 2004b). The development of direct Ge/Si VSs brings advantages to both methods. In the case of MBE, the feasibility of the growth of very thick layers is limited by the amount of solid source material within the growth system and the necessity of opening the growth chamber to reload the sources. CVD is based on gas sources and therefore avoids this issue, but depending on the pressure and temperature regime used, growth can be extremely slow and very inefficient, with complications arising from the reactions between Si and Ge during growth of SiGe alloys (Meyerson et al. 1988; Garone et al. 1990; Racanelli and Greve 1990; Jang and Reif 1991; Robbins et al. 1991; Meyerson 2000; Hartmann et al. 2002; Zhang et al. 2002).

Furthermore, the differing thermal expansion coefficients of Si and Ge can lead to wafer bowing and cracking for very thick Ge or GaAs/Ge layers (typically more than 3–5 μm) (Slack and Bartram 1975; Colombo et al. 2006, 2007; Falub et al. 2012) even if the induced tensile strain in the Ge layer can lead to useful optical properties as will be discussed below (Liu et al. 2004a; Süess et al. 2013). The efficiency and growth rate during SiGe epitaxy can be enhanced by means of a low-energy plasma, and the use of low-energy plasma-enhanced CVD (LEPECVD) has, in fact, been shown to produce high-quality material at the research–lab scale (Rosenblad et al. 1998, 2000; Kummer et al. 2002; Isella et al. 2004). LEPECVD was developed to produce high-quality epitaxial SiGe alloys across the whole composition range, with growth rates for pure Si or Ge of around 5 nms^{-1} independent of substrate temperature (Enciso Aguilar et al. 2004; Chrastina et al. 2005a; Marchionna et al. 2006). The advantage of such efficient high-speed deposition is that thick buffer layers (Marchionna et al. 2006) and active layers (Bonfanti et al. 2008, 2009; Cecchi et al. 2013a) can be deposited in reasonable times in a cost-effective manner; high-speed growth at low temperature also extends the parameter space available for the strain relaxation of buffer layers so that thin high-quality buffers can also be developed (Chrastina et al. 2004, 2005b, 2014). The plasma density and gas flows can be reduced to achieve growth rates of around 0.5–1.0 nms^{-1} for the deposition of QW structures (Rössner et al. 2004; Bonfanti et al. 2008; Chrastina et al. 2013). LEPECVD has also been used for the deposition of nanocrystalline Si for photovoltaic applications (Pizzini et al. 2006).

3.4 STRAINED Ge QWs

Ge-rich graded SiGe VSs have been used to realize high-mobility p-type modulation-doped QWs (*p*-MODQWs) that feature strained Ge channels (Xie et al. 1993; von Känel et al. 2002; Chrastina et al. 2003; Rössner et al. 2004), as shown in Figures 3.3 and 3.4. Thinner VSs of various designs give similarly high room-temperature mobilities around 3000 $cm^2\ V^{-1}\ s^{-1}$ (as compared to the bulk Ge hole mobility of 1800 $cm^2\ V^{-1}\ s^{-1}$) but relatively low low-temperature mobility (Irisawa et al. 2002, 2003). However, a strained Ge channel grown on a new VS design that grades

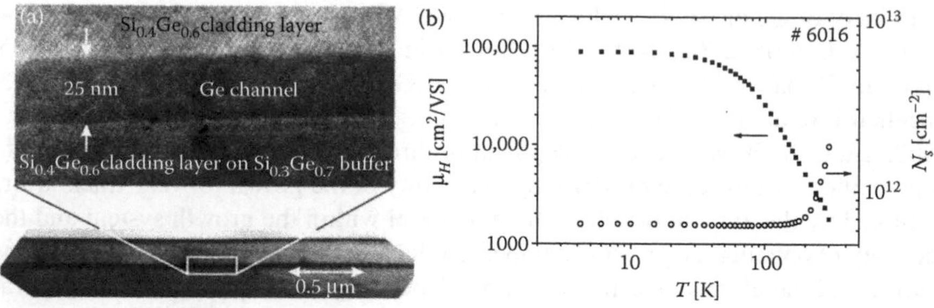

FIGURE 3.3 (a) High-resolution transmission-electron microscopy image of a cross section through a Ge p-MODQW. The cladding layers on both sides of the QW appear lighter than the VS because of the lower Ge content of 60% compared to the 70% of the VS. (b) Temperature dependence of the mobility and the carrier density of a p-MODQW sample grown by LEPECVD. (Reprinted with permission from von Känel, H. et al., Very high hole mobilities in modulation-doped Ge quantum wells grown by low-energy plasma enhanced chemical vapor deposition. *Appl. Phys. Lett. 80*(16), 2922–2924. Copyright 2002, AIP Publishing, LLC.)

backward from a relaxed Ge/Si layer to the desired SiGe concentration (Shah et al. 2010) has recently demonstrated a low-temperature mobility of 1 million $cm^2\ V^{-1}\ s^{-1}$ (Dobbie et al. 2012; Mironov et al. 2014).

In contrast to these electrical results, the structural properties of thin Ge/SiGe layers can also be exploited to create *physical* channels for nanofluidics (Sordan et al. 2009).

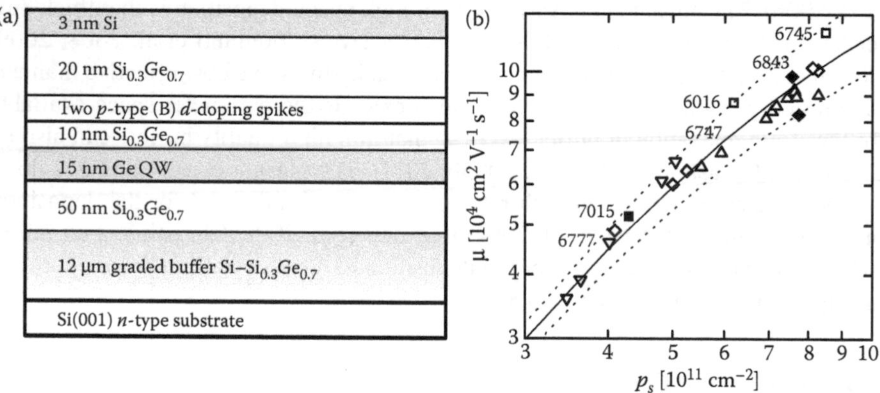

FIGURE 3.4 (a) The structure of a p-MODQW featuring a strained Ge channel. (b) Mobility at 2 K for p-MODQWs grown by LEPECVD. Different sheet densities were induced in some samples by exploiting the effects of annealing on the pinning of surface states. Open symbols are results from square samples, and filled symbols are results from Hall bars. The solid and dotted lines represent calculated mobilities taking into account local and remote impurity scattering and interface roughness. (Reprinted with permission from Rössner, B. et al. Scattering mechanisms in highmobility strained Ge channels. *Appl. Phys. Lett. 84*(16), 3058–3060, 2004. Copyright 2004, AIP Publishing, LLC.)

3.5 MULTIPLE QWs

3.5.1 Thermoelectric Devices

The use of many strained Ge channels in parallel, in the so-called multiple quantum-well (MQW) structure, offers a new route toward thermoelectric materials, based on both lateral (i.e., along the Ge channels) and vertical (i.e., perpendicular to the Ge channels) electrical and thermal transport (Watling and Paul 2011; Paul et al. 2012, 2013; Cecchi et al. 2013a,b; Chrastina et al. 2013; Ferre Llin et al. 2013a,b; Samarelli et al. 2013a,b; Chen et al. 2014). A lateral-transport Hall bar structure is shown in Figure 3.5, with a schematic as an inset. Thermoelectric performance requires a material that is a good electrical conductor while being a poor thermal conductor. Metals are poor choices, since the dense population of mobile charge carriers can also carry thermal energy according to the Wiedemann–Franz law (Wiedemann and Franz 1853). Insulators are also obviously poor choices due to the lack of electrical conductivity, while heat can still be conducted by phonons. Apart from the much lower mobile charge density in a semiconductor with respect to a metal, SiGe also presents a lower phonon-mediated thermal conductivity due to its random alloy nature (Dismukes et al. 1964a; Hackbarth et al. 2003).* Furthermore, structures with reduced dimensionality such as QWs and nanowires also lead to increased Seebeck coefficients due to asymmetry in the density of states across the Fermi level

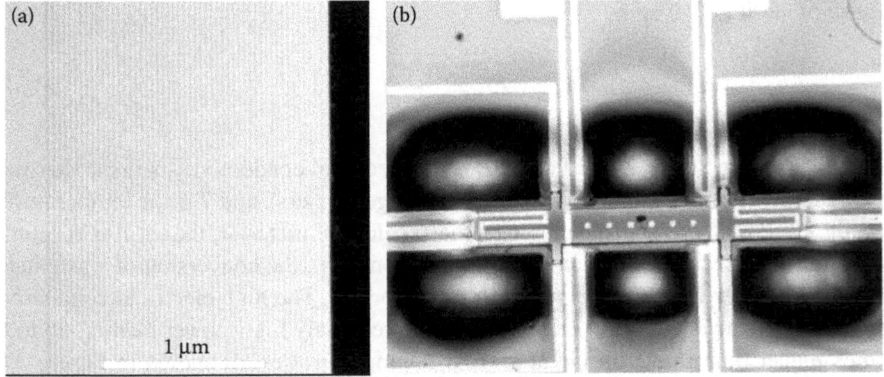

FIGURE 3.5 (a) A transmission-electron microscopy image of the sample structure showing the strain-symmetrized MQW stack grown on top of a 1-µm-thick $Si_{1-y}Ge_y$-relaxed buffer on silicon-on-insulator (SOI) substrate. (b) Optical microscopy picture of a standard device used for the thermoelectric characterization of the MQWs, showing the suspended structure and the metallization required for the measurements. The device features resistive heaters, thermometers, and tilted lateral contacts to connect all the QWs in parallel. The square dots along the Hall bar are used as reference points during thermal atomic-force microscope (AFM) measurements. (Reprinted from *Thin Solid Films* 543, D. Chrastina et al., Ge/SiGe superlattices for nanostructured thermoelectric modules, 153–156, Copyright 2013, with permission from Elsevier.)

* The SiGe material in Dismukes et al. (1964a) is almost certainly polycrystalline, which would be expected to greatly reduce its thermal conductivity with respect to single-crystal epitaxial layers.

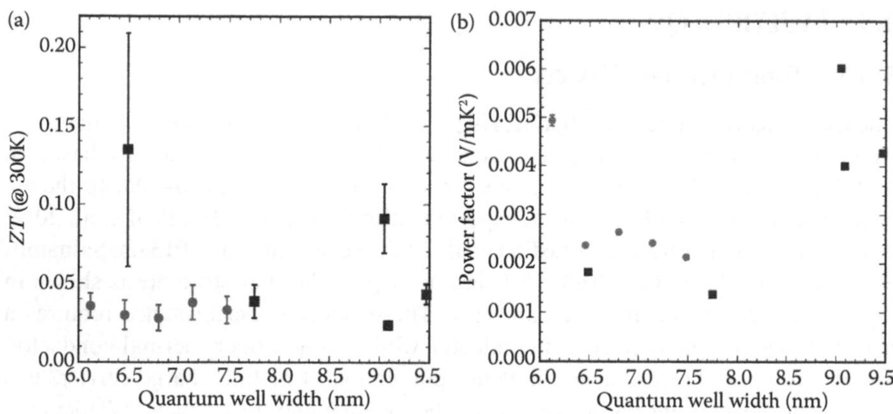

FIGURE 3.6 (a) Thermoelectric figure of merit ZT measured at 300 K on devices as shown in Figure 3.5 on application of heating power via lithographically patterned heating filaments. The different symbols represent two slightly different device designs. (b) Thermoelectric power factors $\alpha^2\sigma$ measured at 300 K. (Reprinted with permission from Samarelli, A. et al., The thermoelectric properties of Ge/SiGe modulation doped superlattices. *J. Appl. Phys. 113*(23), 233704, 2013. Copyright 2013, AIP Publishing, LLC.)

(Dresselhaus et al. 2007). These requirements are summarized in the thermoelectric figure of merit ZT given by

$$ZT = \frac{\alpha^2\sigma}{\kappa}T \tag{3.1}$$

in which α is the Seebeck coefficient, σ is the electrical conductivity, κ is the thermal conductivity (both from phonons and from charge carriers), and T is the temperature. While ZT describes the efficiency of a thermoelectric material, the term $\alpha^2\sigma$ represents the "power factor" that in some applications (i.e., the conversion of waste heat into useful electrical power) is much more important. The high electrical conductivity of MQW stacks based on Ge QWs leads to relatively high-power factors, up to 6 times higher than thin-film Ge at similar doping densities, as shown in Figure 3.6 (Rowe 2006; Samarelli et al. 2013a,b).

3.5.2 Optoelectronic Devices

MQW structures are also under investigation for optoelectronic applications (Kuo et al. 2005; Tsujino et al. 2006; Bonfanti et al. 2008; Lange et al. 2009, 2010; Chaisakul et al. 2010, 2011b,c, 2012, 2014; Gatti et al. 2011; Köster et al. 2011a,b; Giorgioni et al. 2012; Rouifed et al. 2012; Wu et al. 2012; Rouifed et al. 2014). The QCSE was demonstrated in a Ge MQW structure as schematically shown in Figure 3.7 (Kuo et al. 2005). Figure 3.7a is a schematic diagram of a p–i–n diode with an MQW active region, in which light from a monochromator is incident on the top surface (i.e., in a "surface-normal" configuration) and on the open area inside the rectangular frame top electrode. Strong QCSE is seen in Figure 3.7b as a shift of the

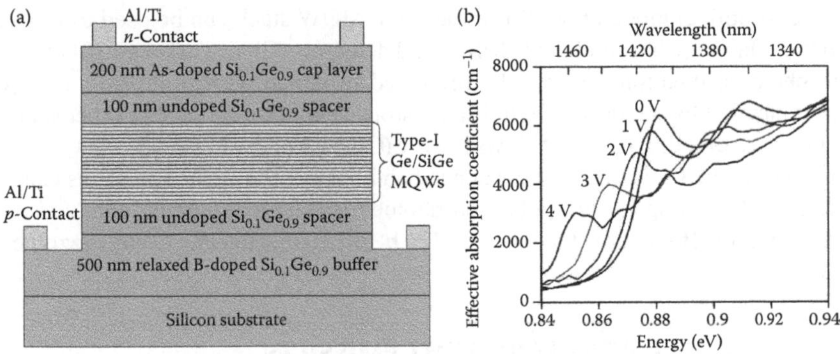

FIGURE 3.7 (a) Schematic diagram of a *p–i–n* diode with an MQW-active region. The cross-sectional view shows the structure of strained Ge/SiGe MQWs grown on silicon on relaxed SiGe direct buffers (not to scale). In the measurements, light from a monochromator is incident on the top surface. (b) Effective absorption coefficient spectra. Strong QCSE is seen as a shift of the absorption peak at room temperature with reverse bias from zero to 4 V. (Reprinted with permission from Macmillan Publishers Ltd. *Nature*, Kuo, Y.-H. et al., 2005. Strong quantum-confined Stark effect in germanium quantum-well structures on silicon. *437*(7063), 1334–1336, copyright 2005.)

absorption peak at room temperature with reverse bias from zero to 4 V. Direct-gap photoluminescence from the Ge QWs was demonstrated in a similar structure at low temperature (Bonfanti et al. 2008) and then at room temperature (Gatti et al. 2011). Ultrafast transient gain was also observed at room temperature (Köster et al. 2011a) (Figure 3.8), which makes MQW structures interesting for light emission as well as

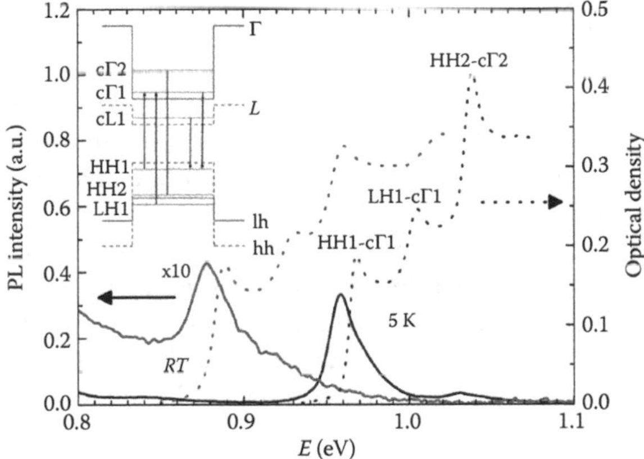

FIGURE 3.8 PL (full lines) and absorption (dashed lines) spectra measured at *RT* and at 5 K. Inset: Conduction and valence band-edge profiles and electron- and hole-confined states (energies not to scale); the transitions visible in the absorption and *PL* spectra are also shown. (Reprinted with permission from Gatti, E. et al., Room temperature photoluminescence of Ge multiple quantum wells with Ge-rich barriers. *Appl. Phys. Lett. 98*(3), 031106, 2011. Copyright 2011, AIP Publishing, LLC.)

for electro-optical modulation. Since the same MQW stack can be used to fabricate both modulators (Chaisakul et al. 2014b) and detectors (Chaisakul et al. 2011a), complete photonic interconnects can be invisaged in which waveguiding in a $Si_{1-x}Ge_x$ layer is enabled by the refractive index constrast (Humlíček et al. 1992) between the $Si_{1-x}Ge_x$ core and the $Si_{1-y}Ge_y$ VS, with $x > y$ (Chaisakul et al. 2014a).

Strain-induced splitting of the valence band heavy- and light-hole states can also be exploited, leading to spin-polarized photoemission on excitation with circularly polarized light (Bottegoni et al. 2011, 2012a,b, 2013; Cecchi et al. 2011; Pezzoli et al. 2013).

3.6 THICK Ge LAYERS FOR LIGHT EMISSION

One of the most active topics in the field of Ge/Si is the possibility of establishing a laser based on the tensile biaxial strain, which reduces the Γ–L offset (Hoshina et al. 2009), and heavy n-type doping to fill L states and promote thermal excitation of electrons to the Γ-point. This possibility was presented theoretically (Liu et al. 2007; Sun et al. 2010), and photo- and electroluminescence were first demonstrated (Hu et al. 2009, 2011; Kurdi et al. 2009; Sun et al. 2009a,b), followed by the report of lasing from an optically pumped Ge waveguide on Si (Liu et al. 2010). This result provoked further interest in heavily doped n-type Ge (Cai et al. 2012), especially in terms of the realization of electrical carrier injection (de Kersauson et al. 2010; Aldaghri et al. 2012; Gallacher et al. 2012a,b). However, such heavy doping should lead to strong free-carrier absorption, which increases the power required to achieve lasing; some calculations call into question whether lasing could ever be achieved at all (Carroll et al. 2012; Dutt et al. 2012; Wang et al. 2013; Geiger et al. 2014). Other workers sought ways in which the tensile strain could be increased beyond that induced by thermal annealing (Capellini et al. 2012, 2013; Ghrib et al. 2012; Jain et al. 2012; Süess et al. 2012; Boucaud et al. 2013; Boztug et al. 2013; Chang and Cheng 2013; Süess et al. 2013; Velha et al. 2013) as well as ways in which the structure could be isolated from defects (Shah et al. 2012, 2014), but nevertheless, an electrically pumped Ge/Si laser was eventually reported (Camacho-Aguilera et al. 2012) as shown in Figures 3.9 and 3.10.

While some microstructuring approaches have given controversial results (e.g., improved emission could be due to increased heating of free-standing structures (Jain et al. 2012, 2013; Boucaud et al. 2013)), the approach shown in Figure 3.11 has demonstrated clear strain-induced enhancement and shift to longer wavelengths of the direct-gap photoluminescence in Ge, as shown in Figure 3.12 (Süess et al. 2013; Etzelstorfer et al. 2014).

3.7 Ge DEPOSITION ON STRUCTURED SUBSTRATES

While this chapter has so far discussed blanket growth over the entire wafers, the waveguides studied for laser emission were in fact grown selectively in oxide windows (Liu et al. 2010; Camacho-Aguilera et al. 2012). Similar approaches can be used to minimize the defectivity in epitaxial layers, by either epitaxial necking

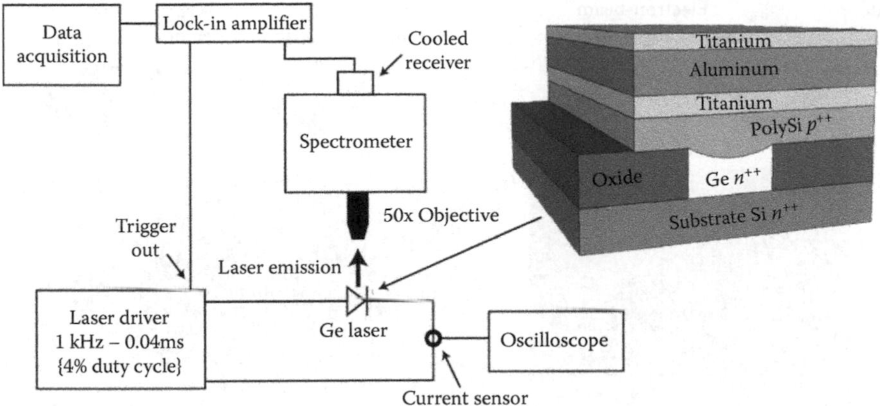

FIGURE 3.9 Schematic of the measurement setup and device structure for the first demonstration of an electrically pumped Ge laser. (From Camacho-Aguilera, R. E. et al. 2012. An electrically pumped germanium laser. *Opt. Express 20*(10), 11316–11320. Copyright 2012. With permission of Optical Society of America.)

(Langdo et al. 2000) or aspect ratio trapping (Bai et al. 2007; Park et al. 2007). A new method of threading dislocation density reduction has recently been proposed, driven by the need to realize extremely thick Ge/Si layers for x-ray or particle detectors as well as for multijunction solar cells, based on LEPECVD epitaxy of Ge on prepatterned micrometer-scale Si pillars (Falub et al. 2012, 2013, 2014; Isa et al. 2013; Kreiliger et al. 2013, 2014; Marzegalli et al. 2013; Pezzoli et al. 2014; Meduňa

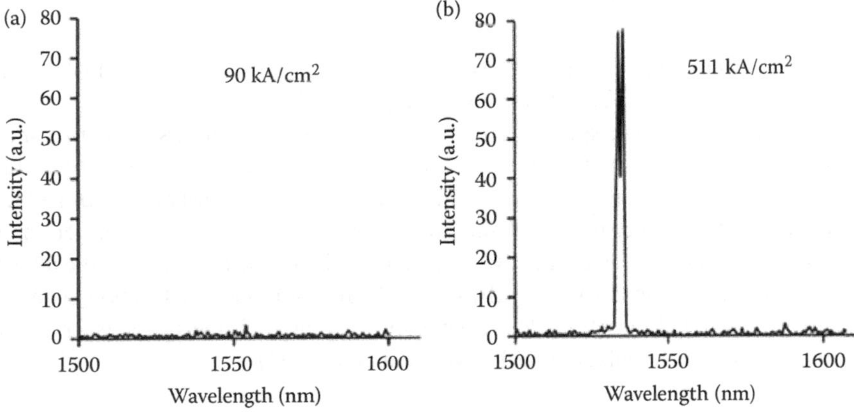

FIGURE 3.10 Ge laser emission spectrum before (a) and after (b) threshold. The cavity length of the waveguide is 333 μm and the waveguide height is about 100 nm. Current injection employed pulse widths of 50 μs at 800 Hz and 15°C. The detector spectral resolution was 1.2 nm. (From Camacho-Aguilera, R. E. et al., 2012. An electrically pumped germanium laser. *Opt. Express 20*(10), 11316–11320. Copyright 2012. With permission of Optical Society of America.)

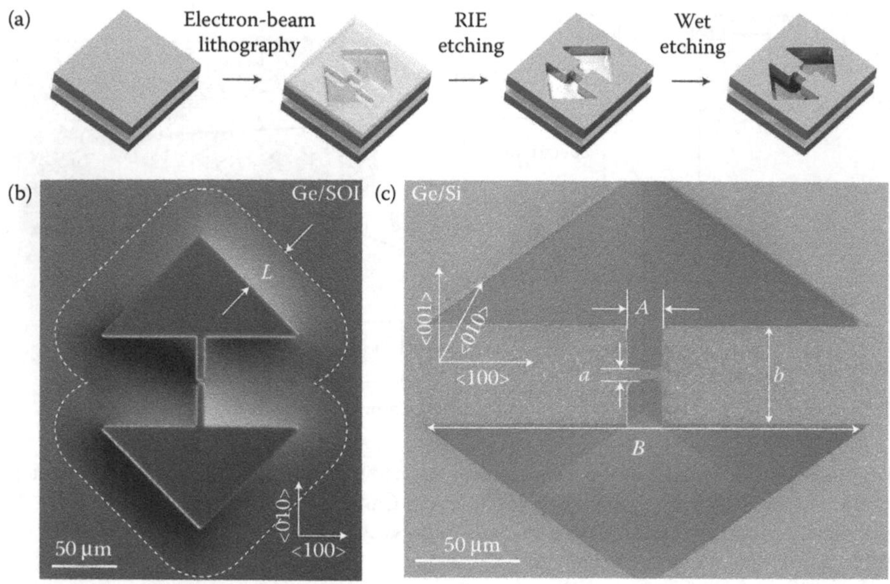

FIGURE 3.11 Suspended microbridges fabricated from epitaxial Ge layers on SOI substrates, in which, the constricted cross-sectional area leads to a multiplication of the tensile strain resulting from postgrowth annealing. (a) Schematics of the process flow used to fabricate suspended and constricted structures. (b) Differential interference contrast light-microscopy image of a Ge/SOI structure. The depth of the etching beneath the frame (*L*) is highlighted by the outer dotted line shown in white. (c) Tilted scanning electron microscopy (SEM) image of a Ge/Si structure showing how a suspended structure of width *b* (length *B*) narrows to width *a* (length *A*). (Reprinted by permission from Macmillan Publishers Ltd. Süess, M. J. et al. 2013. Analysis of enhanced light emission from highly strained germanium micro bridges. *Nat. Photonics* 7(6), 466–472, copyright 2013.)

et al. 2014). An overview of the process (which can also be extended to III–V materials on Si (Taboada et al. 2014)) is shown in Figure 3.13.

In this approach, high-speed vertical growth of Ge on Si pillars leads to a shadowing effect, reducing the lateral growth rate of the Ge towers such that for certain combinations of substrate temperature and growth rate, the towers remain isolated from each other. Further tuning of the growth parameters can lead to variations of the crystal facets at the top of the tower, which facilitates the reduction of the density of threading dislocations (Marzegalli et al. 2013). X-ray nanodiffraction-scanning measurements (Chahine et al. 2014) have shown that each Ge tower is crystallographically perfect (Falub et al. 2013; Meduňa et al. 2014).

3.8 BEYOND Ge

Pure germanium presents the largest lattice parameter available within the SiGe alloy system, and thus, there is no SiGe VS that could simply allow the overgrowth of a tensile Ge layer. However, this possibility is instead provided by the addition of a small amount of tin, and such SiGeSn alloys not only allow the realization of tensile

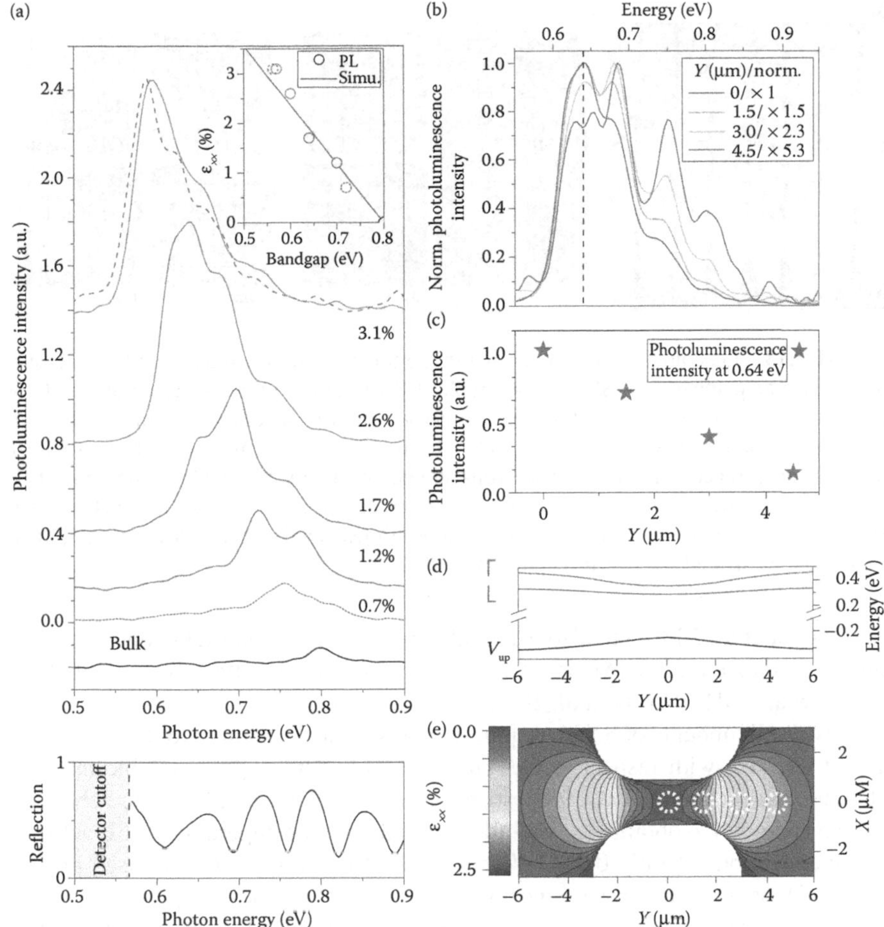

FIGURE 3.12 μPL spectra measured at room temperature for different suspended constricted Ge structures and excitation positions. (a) μPL spectra taken from structures with increasing longitudinal strain up to 3.1% and excitation in the center of the constriction. The μPL spectrum from Ge bulk is plotted for comparison. The curves are offset for clarity. (b) Normalized μPL spectra measured with the excitation laser scanning along the *y*-axis of a Ge constricted structure with 2.6% strain. (c) Although the onset remains constant for different measurement positions *y*, the photoluminescence intensity measured at 0.64 eV decreases at increasing distances from the center of the constriction ($y = 0$). (d) Position-dependent band-edge energies of the structure measured in (b), where Γ, *L*, and V_{up} refer to the conduction band minima at Γ and *L*, and the upper valence band maximum, respectively. (e) Finite-element method (FEM) strain map (ε_{xx}) of the structure shown in (b). (Reprinted by permission from Macmillan Publishers Ltd. Süess, M. J. et al. 2013. Analysis of enhanced light emission from highly strained germanium micro bridges. *Nat. Photonics* 7(6), 466–472, copyright 2013.)

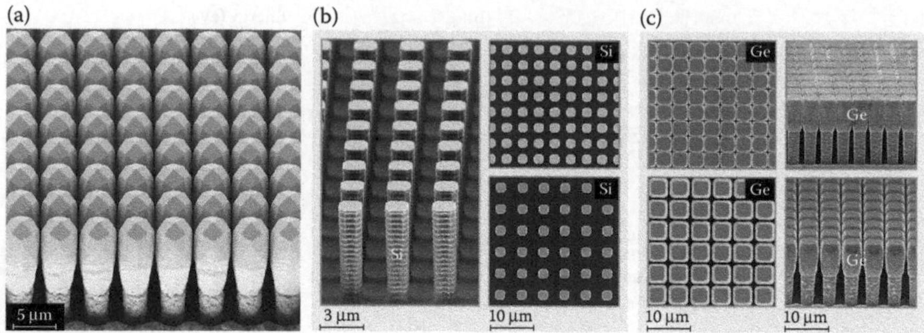

FIGURE 3.13 Ge epitaxial crystals grown in the form of towers on patterned Si(001) substrates. (a) Perspective-view SEM micrograph of 8-μm-tall Ge towers grown at 560°C on patterned Si wafers with 8-μm-tall and 2 μm-wide pillars, spaced by 2 μm trenches. (b) Perspective- and top-view SEM micrographs of 8-μm-tall pillars micromachined on Si(001) wafers by deep reactive ion etching. (c) Top- and perspective-view SEM micrographs of Ge towers grown at 415°C by LEPECVD on the Si pillars depicted in (b). (From Falub, C. V. et al. 2014. Scaling hetero-epitaxy from layers to three-dimensional crystals. *Science 335*(6074), 1330–1334. Reprinted with permission of AAAS.)

Ge (Roucka et al. 2005) but also eventually become direct-gap semiconductors at a suitable Sn concentration (Mäder et al. 1989; Soref and Perry 1991; Soref et al. 2007; Roucka et al. 2011; Werner et al. 2011; Xu et al. 2013a,b).

Only small amounts of Sn are required to make drastic changes to the band structure of $Ge_{1-y}Sn_y$ with respect to Ge, but Sn is only very slightly soluble in Ge or SiGe in general (Mäder et al. 1989); so, even such small concentrations are challenging. An approach to obtain $Ge_{1-y}Sn_y$ alloys based on low-temperature MBE achieves Sn concentrations of only 0.5% (Werner et al. 2011), while a CVD-based process using SnD_4 as a precursor gas (since SnH_4 is unstable) reaches $y \sim 20\%$ (Soref et al. 2007). A quantity of 10% Sn content has been shown to lead to direct-gap emission at around 2.2 μm (Xu et al. 2013a,b).

REFERENCES

Aldaghri, O., Z. Ikonić, and R. W. Kelsall. 2012. Optimum strain configurations for carrier injection in near infrared Ge lasers. *J. Appl. Phys. 111*(5), 053106.

Allen, P. B. and M. Cardona. 1983. Temperature dependence of the direct gap of Si and Ge. *Phys. Rev. B 27*(8), 4760–4769.

Bai, J., J.-S. Park, Z. Cheng, M. Curtin, B. Adekore, M. Carroll, A. Lochtefeld, and M. Dudley. 2007. Study of the defect elimination mechanisms in aspect ratio trapping Ge growth. *Appl. Phys. Lett. 90*(10), 101902.

Bardeen, J. and W. H. Brattain. 1948. The transistor, a semi-conductor triode. *Phys. Rev. 74*(2), 230–231.

Benoit à la Guillaume, C. and J. Cernogora. 1969. Radiative recombination in highly doped germanium. *Phys. Stat. Sol. (b) 35*(2), 599–612.

Bollani, M., E. Bonera, D. Chrastina, A. Fedorov, V. Montuori, A. Picco, A. Tagliaferri, G. Vanacore, and R. Sordan. 2010b. Ordered arrays of SiGe islands from low-energy PECVD. *Nanoscale Res. Lett. 5*(12), 1917–1920.

Bollani, M., D. Chrastina, A. Fedorov, R. Sordan, A. Picco, and E. Bonera. 2010a. Ge-rich islands grown on patterned Si substrates by low-energy plasma-enhanced chemical vapour deposition. *Nanotechnology 21*(47), 475302.

Bollani, M., D. Chrastina, V. Montuori, D. Terziotti, E. Bonera, G. M. Vanacore, A. Tagliaferri, R. Sordan, C. Spinella, and G. Nicotra. 2012. Homogeneity of Ge-rich nanostructures as characterized by chemical etching and transmission electron microscopy. *Nanotechnology 23*(4), 045302.

Bonfanti, M., E. Grilli, M. Guzzi, D. Chrastina, G. Isella, H. von Känel, and H. Sigg. 2009. Direct-gap related optical transitions in Ge/SiGe quantum wells with Ge-rich barriers. *Physica E 41*(6), 972–975.

Bonfanti, M., E. Grilli, M. Guzzi, M. Virgilio, G. Grosso, D. Chrastina, G. Isella, H. von Känel, and A. Neels. 2008. Optical transitions in Ge/SiGe multiple quantum wells with Ge-rich barriers. *Phys. Rev. B 78*(4), 041407(R).

Bottegoni, F., A. Ferrari, S. Cecchi, M. Finazzi, F. Ciccacci, and G. Isella. 2013. Photoinduced inverse spin Hall effect in Pt/Ge(001) at room temperature. *Appl. Phys. Lett. 102*(15), 152411.

Bottegoni, F., A. Ferrari, G. Isella, S. Cecchi, M. Marcon, D. Chrastina, G. Trezzi, and F. Ciccacci. 2012a. Ge/SiGe heterostructures as emitters of polarized electrons. *J. Appl. Phys. 111*(6), 063916.

Bottegoni, F., A. Ferrari, G. Isella, M. Finazzi, and F. Ciccacci. 2012b. Enhanced orbital mixing in the valence band of strained germanium. *Phys. Rev. B 85*(24), 245312.

Bottegoni, F., G. Isella, S. Cecchi, and F. Ciccacci. 2011. Spin polarized photoemission from strained Ge epilayers. *Appl. Phys. Lett. 98*(24), 242107.

Boucaud, P., M. El Kurdi, S. Sauvage, M. de Kersauson, A. Ghrib, and X. Checoury. 2013. Light emission from strained germanium. *Nat. Photonics 7*(3), 162.

Boztug, C., J. R. Sánchez-Pérez, F. F. Sudradjat, R. B. Jacobson, D. M. Paskiewicz, M. G. Lagally, and R. Paiella. 2013. Tensilely strained germanium nanomembranes as infrared optical gain media. *Small 9*(4), 622–630.

Braunstein, R., A. R. Moore, and F. Herman. 1958. Intrinsic optical absorption in germanium–silicon alloys. *Phys. Rev. 109*(3), 695–710.

Brehm, M., M. Grydlik, H. Lichtenberger, T. Fromherz, N. Hrauda, W. Jantsch, F. Schäffler, and G. Bauer. 2008. Quantitative determination of Ge profiles across SiGe wetting layers on Si (001). *Appl. Phys. Lett. 93*(12), 121901.

Brehm, M., T. Suzuki, T. Fromherz, Z. Zhong, N. Hrauda, F. Hackl, J. Stangl, F. Schäffler, and G. Bauer. 2009. Combined structural and photoluminescence study of SiGe islands on Si substrates: Comparison with realistic energy level calculations. *New J. Phys. 11*(6), 063021.

Cai, Y., R. Camacho-Aguilera, J. T. Bessette, L. C. Kimerling, and J. Michel. 2012. High phosphorous doped germanium: Dopant diffusion and modeling. *J. Appl. Phys. 112*(3), 034509.

Camacho-Aguilera, R. E., Y. Cai, N. Patel, J. T. Bessette, M. Romagnoli, L. C. Kimerling, and J. Michel. 2012. An electrically pumped germanium laser. *Opt. Express 20*(10), 11316–11320.

Capellini, G., M. De Seta, P. Zaumseil, G. Kozlowski, and T. Schroeder. 2012. High temperature x-ray diffraction measurements on Ge/Si(001) heterostructures: A study on the residual tensile strain. *J. Appl. Phys. 111*(7), 073518.

Capellini, G., G. Kozlowski, Y. Yamamoto, M. Lisker, C. Wenger, G. Niu, P. Zaumseil, et al. 2013. Strain analysis in SiN/Ge microstructures obtained via Si-complementary metal oxide semiconductor compatible approach. *J. Appl. Phys. 113*(1), 013513.

Cardona, M. and F. H. Pollak. 1966. Energy-band structure of germanium and silicon: The $\mathbf{k} \cdot \mathbf{p}$ method. *Phys. Rev. 142*(2), 530–543.

Carroll, L., P. Friedli, S. Neuenschwander, H. Sigg, S. Cecchi, F. Isa, D. Chrastina, G. Isella, Y. Fedoryshyn, and J. Faist. 2012. Direct-gap gain and optical absorption in germanium correlated to the density of photoexcited carriers, doping, and strain. *Phys. Rev. Lett. 109*(5), 057402.

Cecchi, S., F. Bottegoni, A. Ferrari, D. Chrastina, G. Isella, and F. Ciccacci. 2011. Spin polarized photoemission from strained Ge epilayers grown by low-energy plasma-enhanced CVD (LEPECVD). In *IEEE 8th International Conference Group IV Photonics*, London, UK, pp. 83–85.

Cecchi, S., T. Etzelstorfer, E. Müller, A. Samarelli, L. F. Llin, D. Chrastina, G. Isella, J. Stangl, J. M. R. Weaver, P. Dobson, and D. J. Paul. 2013a. Ge/SiGe superlattices for thermoelectric devices grown by low-energy plasma-enhanced chemical vapor deposition. *J. Electron. Mater. 42*(7), 2030–2045.

Cecchi, S., T. Etzelstorfer, E. Müller, A. Samarelli, L. F. Llin, D. Chrastina, G. Isella, J. Stangl, and D. J. Paul. 2013b. Ge/SiGe superlattices for thermoelectric energy conversion devices. *J. Mater. Sci. 48*(7), 2829–2835.

Cecchi, S., E. Gatti, D. Chrastina, J. Frigerio, E. Müller Gubler, D. J. Paul, M. Guzzi, and G. Isella. 2014. Thin SiGe virtual substrates for Ge heterostructure integration on silicon. *J. Appl. Phys. 115*(9), 093502.

Chahine, G. A., M.-I. Richard, R. A. Homs-Regojo, T. N. Tran-Caliste, D. Carbone, V. L. R. Jaques, R. Grifone, et al. 2014. Imaging of strain and lattice orientation by quick scanning x-ray microscopy combined with three-dimensional reciprocal space mapping. *J. Appl. Cryst. 47*(2), 762–769.

Chaisakul, P., D. Marris-Morini, J. Frigerio, D. Chrastina, M.-S. Rouifed, S. Cecchi, P. Crozat, G. Isella, and L. Vivien. 2014a. Integration of Ge quantum well photonic interconnections on silicon substrates. *Nat. Photonics 8*(6), 482–488.

Chaisakul, P., D. Marris-Morini, G. Isella, D. Chrastina, N. Izard, X. L. Roux, S. Edmond, J.-R. Coudevylle, and L. Vivien. 2011b. Room temperature direct gap electroluminescence from Ge/$Si_{0.15}$ $Ge_{0.85}$ multiple quantum well waveguide. *Appl. Phys. Lett. 99*(14), 141106.

Chaisakul, P., D. Marris-Morini, G. Isella, D. Chrastina, M.-S. Rouifed, X. L. Roux, S. Edmond, E. Cassan, J.-R. Coudevylle, and L. Vivien. 2011a. 10-Gb/s Ge/SiGe multiple quantum-well waveguide photodetector. *IEEE Photonic Tech. L. 23*(20), 1430–1432.

Chaisakul, P., D. Marris-Morini, G. Isella, D. Chrastina, X. L. Roux, S. Edmond, J.-R. Coudevylle, E. Cassan, and L. Vivien. 2011c. Polarization dependence of quantum-confined Stark effect in Ge/SiGe quantum well planar waveguides. *Optics Lett. 36*(10), 1794–1796.

Chaisakul, P., D. Marris-Morini, G. Isella, D. Chrastina, X. L. Roux, E. Gatti, S. Edmond, J. Osmond, E. Cassan, and L. Vivien. 2010. Quantum-confined Stark effect measurements in Ge/SiGe quantum well structures. *Optics Lett. 35*(17), 2913–2915.

Chaisakul, P., D. Marris-Morini, M.-S. Rouifed, J. Frigerio, D. Chrastina, J.-R. Coudevylle, X. Le Roux, S. Edmond, G. Isella, and L. Vivien. 2014b. Recent progress in GeSi electro-absorption modulators. *Sci. Technol. Adv. Mat. 15*(1), 014601.

Chaisakul, P., D. Marris-Morini, M.-S. Rouifed, G. Isella, D. Chrastina, J. Frigerio, X. Le Roux, S. Edmond, J.-R. Coudevylle, and L. Vivien. 2012. 23 GHz Ge/SiGe multiple quantum well electro-absorption modulator. *Opt. Express 20*(3), 3219–3224.

Chang, G.-E. and H. H. Cheng. 2013. Optical gain of germanium infrared lasers on different crystal orientations. *J. Phys. D: Appl. Phys. 46*(6), 065103.

Chen, P., J. J. Zhang, J. P. Feser, F. Pezzoli, O. Moutanabbir, S. Cecchi, G. Isella, T. Gemming, S. Baunack, G. Chen, O. G. Schmidt, and A. Rastelli. 2014. Thermal transport through short-period SiGe nanodot superlattices. *J. Appl. Phys. 115*(4), 044312.

Chrastina, D., S. Cecchi, J. P. Hague, J. Frigerio, A. Samarelli, L. Ferre-Llin, D. J. Paul, E. Müller, T. Etzelstorfer, J. Stangl, and G. Isella. 2013. Ge/SiGe superlattices for nanostructured thermoelectric modules. *Thin Solid Films 543*, 153–156.

Chrastina, D., J. P. Hague, and D. R. Leadley. 2003. Application of Bryan's algorithm to the mobility spectrum analysis of semiconductor devices. *J. Appl. Phys. 94*(10), 6583–6590.

Chrastina, D., G. Isella, M. Bollani, B. Rössner, E. Müller, T. Hackbarth, E. Wintersberger, Z. Zhong, J. Stangl, and H. von Känel. 2005b. Thin relaxed SiGe virtual substrates grown by low-energy plasma-enhanced chemical vapor deposition. *J. Cryst. Growth 281*(2–4), 281–289.

Chrastina, D., G. Isella, B. Rössner, M. Bollani, E. Müller, T. Hackbarth, and H. von Känel. 2004. High quality SiGe electronic material grown by low energy plasma enhanced chemical vapour deposition. *Thin Solid Films 459*(1–2), 37–40.

Chrastina, D., B. Rössner, G. Isella, H. von Känel, J. P. Hague, T. Hackbarth, H.-J. Herzog, K.-H. Hieber, and U. König. 2005a. LEPECVD—A production technique for SiGe MOSFETs and MODFETs. In E. Zschech, C. Whelan, and T. Mikolajick (eds.), *Materials for Information Technology*, London, Springer. pp. 17–29.

Chrastina, D., A. Neels, M. Bonfanti, M. Virgilio, G. Isella, E. Grilli, M. Guzzi, G. Grosso, H. Sigg, and H. von Känel. 2008. Ge/SiGe multiple quantum wells for optical applications. In *IEEE 5th International Conference Group IV Photonics*, Sorrento, Italy, pp. 194–196.

Chriqui, Y., G. Saint-Girons, S. Bouchoule, J.-M. Moisons, G. Isella, H. von Kaenel, and I. Sagnes. 2003. Room temperature laser operation of strained InGaAs/GaAs structure monolithically grown by MOCVD on LE-PECVD Ge/Si virtual substrate. *Electron. Lett. 39*(23), 1658–1659.

Colace, L., M. Balbi, G. Masini, G. Assanto, H.-C. Luan, and L. C. Kimerling. 2006. Ge on Si *p–i–n* photodiodes operating at 10 GBit/s. *Appl. Phys. Lett. 88*(10), 101111.

Colace, L., G. Masini, and G. Assanto. 1999. Ge-on-Si approaches to the detection of near-infrared light. *IEEE J. Quantum Elect. 35*(12), 1843–1852.

Colace, L., G. Masini, F. Galluzzi, G. Assanto, G. Capellini, L. D. Gaspare, and F. Evangelisti. 1997. Ge/Si (001) photodetector for near infrared light. *Solid State Phenom. 54,* 55–58.

Colombo, D., E. Grilli, M. Guzzi, S. Sanguinetti, A. Fedorov, H. von Känel, and G. Isella. 2006. Study of thermal strain relaxation in GaAs grown on Ge/Si substrates. *J. Lumin. 121*(2), 375–378.

Colombo, D., E. Grilli, M. Guzzi, S. Sanguinetti, S. Marchionna, M. Bonfanti, A. Fedorov, H. von Känel, G. Isella, and E. Müller. 2007. Analysis of strain relaxation by microcracks in epitaxial GaAs grown on Ge/Si substrates. *J. Appl. Phys. 101*(10), 103519.

Contreras, G., A. Compaan, J. Wagner, M. Cardona, and A. Axmann. 1983. The $e_1 - e_1 + \delta_1$ transitions in bulk grown and in implanted laser annealed heavily doped germanium: Luminescence. *J. Phys. (Paris) Colloq. 44*(C5), C5-55–C5-59.

Cuttriss, D. B. 1961. Relation between surface concentration and average conductivity in diffused layers in germanium. *Bell System Tech. J. 40*(2), 509–521.

de Kersauson, M., R. Jakomin, M. El Kurdi, G. Beaudoin, N. Zerounian, F. Aniel, S. Sauvage, I. Sagnes, and P. Boucaud. 2010. Direct and indirect band gap room temperature electroluminescence of Ge diodes. *J. Appl. Phys. 108*(2), 023105.

Dismukes, J. P., L. Ekstrom, and R. J. Paff. 1964b. Lattice parameter and density in germanium–silicon alloys. *J. Phys. Chem. 68*(10), 3021–3027.

Dismukes, J. P., L. Ekstrom, E. F. Steigmeier, I. Kudman, and D. S. Beers. 1964a. Thermal and electrical properties of heavily doped Ge–Si alloys up to 1300°K. *J. Appl. Phys. 35*(10), 2899–2907.

Dobbie, A., M. Myronov, R. J. H. Morris, A. H. A. Hassan, M. J. Prest, V. A. Shah, E. H. C. Parker, T. E. Whall, and D. R. Leadley. 2012. Ultra-high hole mobility exceeding one million in a strained germanium quantum well. *Appl. Phys. Lett. 101*(17), 172108.

Dresselhaus, M. S., G. Chen, M. Y. Tang, R. G. Yang, H. Lee, D. Z. Wang, Z. F. Ren, J.-P. Fleurial, and P. Gogna. 2007. New directions for low-dimensional thermoelectric materials. *Adv. Mater. 19*(8), 1043–1053.

Dutt, B. R., D. S. Sukhdeo, D. Nam, B. N. Vulovic, Z. Yuan, and K. C. Saraswat. 2012. Roadmap to an efficient germanium-on-silicon laser: Strain vs. n-type doping. *IEEE Photonics J. 4*(5), 2002–2009.

Eaglesham, D. J. and M. Cerullo. 1990. Dislocation-free Stranski–Krastanow growth of Ge on Si(100). *Phys. Rev. Lett. 64*(16), 1943–1946.

Eaglesham, D. J. and M. Cerullo. 1991. Low-temperature growth of Ge on Si(001). *Appl. Phys. Lett. 58*(20), 2276–2278.

Elder, W. J., R. M. Ward, and J. Zhang. 2011. Double-group formulation of $\mathbf{k} \cdot \mathbf{p}$ theory for cubic crystals. *Phys. Rev. B 83*(16), 165210.

Enciso Aguilar, M., M. Rodriguez, N. Zerounian, F. Aniel, T. Hackbarth, H.-J. Herzog, U. König, et al. 2004. Strained Si HFETs for microwave applications: State-of-the-art and further approaches. *Solid State Electron. 48*(8), 1443–1452.

Engelhardt, C. M., D. Többen, M. Aschauer, F. Schäffler, G. Abstreiter, and E. Gornik. 1994. High mobility 2-D hole gases in strained Ge channels on Si substrates studied by magnetotransport and cyclotron resonance. *Solid State Electron. 37*(4–6), 949–952.

Etzelstorfer, T., M. J. Süess, G. L. Schiefler, V. L. R. Jacques, D. Carbone, D. Chrastina, G. Isella, R. Spolenak, J. Stangl, H. Sigg, and A. Diaz. 2014. Scanning x-ray strain microscopy of inhomogeneously strained Ge micro-bridges. *J. Synchrotron Rad. 21*(1), 111–118.

Falub, C. V., H. von Känel, F. Isa, R. Bergamaschini, A. Marzegalli, D. Chrastina, G. Isella, E. Müller, P. Niedermann, and L. Miglio. 2012. Scaling hetero-epitaxy from layers to three-dimensional crystals. *Science* 3356074, 1330–1334.

Falub, C. V., T. Kreiliger, F. Isa, A. G. Taboada, M. Meduňa, F. Pezzoli, R. Bergamaschini, et al. 2014. 3D heteroepitaxy of mismatched semiconductors on silicon. *Thin Solid Films 557*, 42–49.

Falub, C. V., M. Meduňa, D. Chrastina, F. Isa, A. Marzegalli, T. Kreiliger, A. G. Taboada, G. Isella, L. Miglio, A. Dommann, and H. von Känel. 2013. Perfect crystals grown from imperfect interfaces. *Sci. Rep. 3,* 2276.

Famà, S., L. Colace, G. Masini, G. Assanto, and H.-C. Luan. 2002. High performance germanium-on-silicon detectors for optical communications. *Appl. Phys. Lett. 81*(4), 586–588.

Ferre Llin, L., A. Samarelli, S. Cecchi, T. Etzelstorfer, E. Müller Gubler, D. Chrastina, G. Isella, et al. 2013a. The cross-plane thermoelectric properties of p-Ge/$Si_{0.5}Ge_{0.5}$ superlattices. *Appl. Phys. Lett. 103*(14), 143507.

Ferre Llin, L., A. Samarelli, Y. Zhang, J. M. R. Weaver, P. Dobson, S. Cecchi, D. Chrastina, et al. 2013b. Thermal conductivity measurement methods of for SiGe thermoelectric materials. *J. Electron. Mater. 42*(7), 2376–2380.

Fitzgerald, E. A., A. Y. Kim, M. T. Currie, T. A. Langdo, G. Taraschi, and M. T. Bulsara. 1999. Dislocation dynamics in relaxed graded composition semiconductors. *Mat. Sci. Eng. B 67,* 53.

Fitzgerald, E. A., Y.-H. Xie, M. L. Green, D. Brasen, A. R. Kortan, J. Michel, Y.-J. Mii, and B. E. Weir. 1991. Totally relaxed Ge_xSi_{1-x} layers with low threading dislocation densities grown on Si substrates. *Appl. Phys. Lett. 59*(7), 811–813.

Fitzgerald, E. A., Y.-H. Xie, D. Monroe, P. J. Silverman, J. M. Kuo, A. R. Kortan, F. A. Thiel, and B. E. Weir. 1992. Relaxed Ge_xSi_{1-x} structures for III–V integration with Si and high mobility two-dimensional electron gases in Si. *J. Vac. Sci. Technol. B 10*(4), 1807–1819.

Fowler, A. B., W. E. Howard, and G. E. Brock. 1962. Optical properties of heavily doped compensated germanium. *Phys. Rev. 128*(4), 1664–1667.

Gallacher, K., P. Velha, D. J. Paul, S. Cecchi, J. Frigerio, D. Chrastina, and G. Isella. 2012b. 1.55 μm direct bandgap electroluminescence from strained n-Ge quantum wells grown on Si substrates. *Appl. Phys. Lett. 101*(21), 211101.

Gallacher, K., P. Velha, D. J. Paul, J. Frigerio, D. Chrastina, and G. Isella. 2012a. 1.55 μm electroluminescence from strained n-Ge quantum wells on silicon substrates. In *IEEE 9th International Conference Group IV Photonics*, San Diego, California, pp. 81–83.

Garone, P. M., J. C. Sturm, P. V. Schwartz, S. A. Schwarz, and B. J. Wilkens. 1990. Silicon vapor phase epitaxial growth catalysis by the presence of germane. *Appl. Phys. Lett. 56*(13), 1275–1277.

Gatti, E., E. Grilli, M. Guzzi, D. Chrastina, G. Isella, and H. von Känel. 2011. Room temperature photoluminescence of Ge multiple quantum wells with Ge-rich barriers. *Appl. Phys. Lett.* *98*(3), 031106.

Gehrsitz, S., H. Sigg, N. Herres, K. Bachem, K. Köhler, and F. K. Reinhart. 1999. Compositional dependence of the elastic constants and the lattice parameter of $Al_x Ga_{1-x} As$. *Phys. Rev. B* *60*(16), 11601–11610.

Geiger, R., J. Frigerio, M. J. Süess, D. Chrastina, G. Isella, R. Spolenak, J. Faist, and H. Sigg. 2014. Excess carrier lifetimes in Ge layers on Si. *Appl. Phys. Lett.* *104*(6), 062106.

Ghrib, A., M. de Kersauson, M. El Kurdi, R. Jakomin, G. Beaudoin, S. Sauvage, G. Fishman, G. Ndong, M. Chaigneau, R. Ossikovski, I. Sagnes, and P. Boucaud. 2012. Control of tensile strain in germanium waveguides through silicon nitride layers. *Appl. Phys. Lett.* *100*(20), 201104.

Ginige, R., B. Corbett, M. Modreanu, C. Barrett, J. Hilgarth, G. Isella, D. Chrastina, and H. von Känel. 2006. Characterization of Ge-on-Si virtual substrates and single junction GaAs solar cells. *Semicond. Sci. Technol.* *21*(6), 775–780.

Giorgioni, A., E. Gatti, E. Grilli, A. Chernikov, S. Chatterjee, D. Chrastina, G. Isella, and M. Guzzi. 2012. Photoluminescence decay of direct and indirect transitions in Ge/SiGe multiple quantum wells. *J. Appl. Phys.* *111*(1), 013501.

Groenert, M. E., C. W. Leitz, A. J. Pitera, V. Yang, H. Lee, R. J. Ram, and E. A. Fitzgerald. 2003. Monolithic integration of room-temperature cw GaAs/AlGaAs lasers on Si substrates via relaxed graded GeSi buffer layers. *J. Appl. Phys.* *93*(1), 362–367.

Haas, C. 1962. Infrared absorption in heavily-doped *n*-type germanium. *Phys. Rev.* *125*(6), 1965–1971.

Hackbarth, T., H.-J. Herzog, K.-H. Hieber, U. König, M. Bollani, D. Chrastina, and H. von Känel. 2003. Reduced self-heating in Si/SiGe field-effect transistors on thin virtual substrates prepared by low-energy plasma-enhanced chemical vapor deposition. *Appl. Phys. Lett.* *83*(26), 5464–5466.

Hackbarth, T., H.-J. Herzog, M. Zeuner, G. Höck, E. A. Fitzgerald, M. Bulsara, C. Rosenblad, and H. von Känel. 2000. Alternatives to thick MBE-grown relaxed SiGe buffers. *Thin Solid Films* *369*(1–2), 148–151.

Hartmann, J. M., A. Addadie, J. P. Barnes, J. M. Fédéli, T. Billon, and L. Vivien. 2010. Impact of the H_2 anneal on the structural and optical properties of thin and thick Ge layers on Si; low temperature surface passivation of Ge by Si. *J. Cryst. Growth* *312*(4), 532–541.

Hartmann, J. M., A. Abbadie, N. Cherkashin, H. Grampeix, and L. Clavelier. 2009. Epitaxial growth of Ge thick layers on nominal and 6° off Si(001); Ge surface passivation by Si. *Semicond. Sci. Technol.* *24*(5), 055002.

Hartmann, J. M., A. Abbadie, A. M. Papon, P. Holliger, G. Rolland, T. Billon, J. M. Fédéli, M. Rouvière, L. Vivien, and S. Laval. 2004. Reduced pressure–chemical vapor deposition of Ge thick layers on Si(001) for 1.3–1.55 μm photodetection. *J. Appl. Phys.* *95*(10), 5905–5913.

Hartmann, J. M., J. P. Barnes, M. Veillerot, J. M. Fédéli, Q. Benoit, A. La Guillaume, and V. Calvo. 2012. Structural, electrical and optical properties of *in-situ* phosphorous-doped Ge layers. *J. Cryst. Growth* *347*(1), 37–44.

Hartmann, J. M., J.-F. Damelncourt, Y. Bogumilowicz, P. Holliger, G. Rolland, and T. Billon. 2005. Reduced pressure–chemical vapor deposition of intrinsic and doped Ge layers on Si(001) for microelectronic and optoelectronic purposes. *J. Cryst. Growth* *274*(1–2), 90–99.

Hartmann, J. M., V. Loup, G. Rolland, P. Holliger, F. Laugier, C. Vannuffel, and M. N. Séméria. 2002. SiGe growth kinetics and doping in reduced pressure–chemical vapor deposition. *J. Cryst. Growth* *236*(1–3), 10–20.

Hartmann, J. M., A. M. Papon, V. Destefanis, and T. Billon. 2008. Reduced pressure chemical vapor deposition of Ge thick layers on Si(001), Si(011) and Si(111). *J. Cryst. Growth* *310*(24), 5287–5296.

Hassan, A. H. A., R. J. H. Morris, O. A. Mironov, R. Beanland, D. Walker, S. Huband, A. Dobbie, M. Myronov, and D. R. Leadley. 2014. Anisotropy in the hole mobility measured along the [110] and [$\bar{1}$10] orientations in a strained Ge quantum well. *Appl. Phys. Lett. 104*(13), 132108.

Hoshina, Y., K. Iwasaki, A. Yamada, and M. Konagai. 2009. First-principles analysis of indirect-to-direct band gap transition of Ge under tensile strain. *Jpn. J. Appl. Phys. 48*(4), 04C125.

Hu, M., K. P. Giapis, J. V. Goicochea, X. Zhang, and D. Poulikakos. 2011. Significant reduction of thermal conductivity in Si/Ge core–shell nanowires. *Nano Lett. 11*(2), 618–623.

Hu, W., B. Cheng, C. Xue, H. Xue, S. Su, A. Bai, L. Luo, Y. Yu, and Q. Wang. 2009. Electroluminescence from Ge on Si substrate at room temperature. *Appl. Phys. Lett. 95*(9), 092102.

Humlíček, J., A. Röseler, T. Zettler, M. G. Kekoua, and E. V. Khoutsishvili. 1992. Infrared refractive index of germanium–silicon alloy crystals. *Appl. Optics 31*(1), 90–94.

Irisawa, T., S. Koh, K. Nakagawa, and Y. Shiraki. 2003. Growth of SiGe/Ge/SiGe heterostructures with ultrahigh hole mobility and their device application. *J. Cryst. Growth 251*(1–4), 670–675.

Irisawa, T., S. Tokumitsu, T. Hattori, K. Nakagawa, S. Koh, and Y. Shiraki. 2002. Ultrahigh room-temperature hole hall and effective mobility in $Si_{0.3}Ge_{0.7}$/Ge/$Si_{0.3}Ge_{0.7}$ heterostructures. *Appl. Phys. Lett. 81*(5), 847–849.

Isa, F., A. Marzegalli, A. G. Taboada, C. V. Falub, G. Isella, F. Montalenti, H. von Känel, and L. Miglio. 2013. Onset of vertical threading dislocations in $Si_{1-x}Ge_x$/Si(001) at a critical Ge concentration. *APL Mater. 1*(5), 052109.

Isella, G., D. Chrastina, B. Rössner, T. Hackbarth, H.-J. Herzog, U. König, and H. von Känel. 2004. Low-energy plasma-enhanced chemical vapor deposition for strained Si and Ge heterostructures and devices. *Solid State Electron. 48*(8), 1317–1323.

Isella, G., J. Osmond, M. Kummer, R. Kaufmann, and H. von Känel. 2007. Heterojunction photodiodes fabricated from Ge/Si (100) layers grown by low-energy plasma-enhanced CVD. *Semicond. Sci. Technol. 22*(1), S26–S28.

Jain, J. R., A. Hryciw, T. M. Baer, D. A. B. Miller, M. L. Brongersma, and R. T. Howe. 2012. A micromachining-based technology for enhancing germanium light emission via tensile strain. *Nat. Photonics 6*(6), 398–405.

Jain, J. R., A. Hryciw, T. M. Baer, D. A. B. Miller, M. L. Brongersma, and R. T. Howe. 2013. Light emission from strained germanium. *Nat. Photonics 7*(3), 162–163.

Jang, S.-M. and R. Reif. 1991. Temperature dependence of $Si_{1-x}Ge_x$ epitaxial growth using very low pressure chemical vapor deposition. *Appl. Phys. Lett. 59*(24), 3162–3164.

Kline, J. S., F. H. Pollak, and M. Cardona. 1968. Electroreflectance in Ge–Si alloys. *Helv. Phys. Acta 41*(6–7), 968–977.

Klingenstein, W. and H. Schweizer. 1978. Direct gap recombination in germanium at high excitation level and low temperature. *Solid State Electron. 21*(11–12), 1371–1374.

Koester, S. J., R. Hammond, and J. O. Chu. 2000. Extremely high transconductance Ge/$Si_{0.4}Ge_{0.6}$ p-MODFET's grown by UHV–CVD. *IEEE Electr. Device Lett. 21*(3), 110–112.

Köster, N. S., C. Lange, K. Kolata, S. Chatterjee, M. Schäfer, M. Kira, S. W. Koch, et al. 2011a. Ultrafast transient gain in Ge/SiGe quantum wells. *Phys. Stat. Sol. (c) 8*(4), 1109–1112.

Köster, N. S., K. Kolata, R. Woscholski, C. Lange, G. Isella, D. Chrastina, H. von Känel, and S. Chatterjee. 2011b. Giant dynamical Stark shift in germanium quantum wells. *Appl. Phys. Lett. 98*(16), 161103.

Kreiliger, T., C. V. Falub, F. Isa, G. Isella, D. Chrastina, R. Bergamaschini, A. Marzegalli, et al. 2014. Epitaxial Ge-crystal arrays for x-ray detection. *J. Instrum. 9*(3), C03019.

Kreiliger, T., C. V. Falub, A. G. Taboada, F. Isa, S. Cecchi, R. Kaufmann, P. Niedermann, et al. 2013. Individual heterojunctions of 3D germanium crystals on silicon CMOS for monolithically integrated x-ray detector. *Phys. Stat. Sol. (a) 211*(1), 131–135.

Kummer, M., C. Rosenblad, A. Dommann, T. Hackbarth, G. Höck, M. Zeuner, E. Müller, and H. von Känel. 2002. Low energy plasma enhanced chemical vapor deposition. *Mat. Sci. Eng. B 89*(1–3), 288–295.

Kuo, Y.-H., Y. K. Lee, Y. Ge, S. Ren, J. E. Roth, T. I. Kamins, D. A. B. Miller, and J. S. Harris. 2005. Strong quantum-confined Stark effect in germanium quantum-well structures on silicon. *Nature 437* (7063), 1334–1336.

Kurdi, M. E., T. Kociniewski, T.-P. Ngo, J. Boulmer, D. Débarre, P. Boucaud, J. F. Damlencourt, O. Kermarrec, and D. Bensahel. 2009. Enhanced photoluminescence of heavily *n*-doped germanium. *Appl. Phys. Lett. 94*(19), 191107.

Langdo, T. A., C. W. Leitz, M. T. Currie, E. A. Fitzgerald, A. Lochtefeld, and D. A. Antoniadis. 2000. High quality Ge on Si by epitaxial necking. *Appl. Phys. Lett. 76*(25), 3700–3702.

Lange, C., N. S. Köster, S. Chatterjee, H. Sigg, D. Chrastina, G. Isella, H. von Känel, B. Kunert, and W. Stolz. 2010. Comparison of ultrafast carrier thermalization in Ga_x In_{1-x} As and Ge quantum wells. *Phys. Rev. B 81*(4), 045320.

Lange, C., N. S. Köster, S. Chatterjee, H. Sigg, D. Chrastina, G. Isella, H. von Känel, M. Schäfer, M. Kira, and S. W. Koch. 2009. Ultrafast nonlinear optical response of photoexcited Ge/SiGe quantum wells: Evidence for a femtosecond transient population inversion. *Phys. Rev. B 79*(20), 201306(R).

Lautenschlager, P., P. B. Allen, and M. Cardona. 1985. Temperature dependence of band gaps in Si and Ge. *Phys. Rev. B 31*(4), 2163–2171.

Liu, J., D. D. Cannon, K. Wada, Y. Ishikawa, D. T. Danielson, S. Jongthammanurak, J. Michel, and L. C. Kimerling. 2004a. Deformation potential constants of biaxially tensile stressed Ge epitaxial films on Si(100). *Phys. Rev. B 70*(15), 155309.

Liu, J., H. J. Kim, O. Hul'ko, Y. H. Xie, S. Sahni, P. Bandaru, and E. Yablonovitch. 2004b. Ge films grown on Si substrates by molecular-beam epitaxy below 450°C. *J. Appl. Phys. 96*(1), 916–918.

Liu, J., X. Sun, R. Camacho-Aguilera, L. Kimerling, and J. Michel. 2010. A Ge-on-Si laser operating at room temperature. *Optics Lett. 35*(5), 679–681.

Liu, J., X. Sun, D. Pan, X. Wang, L. C. Kimerling, T. L. Koch, and J. Michel. 2007. Tensile-strained, n-type Ge as a gain medium for monolithic laser integration on Si. *Opt. Expr. 15*(18), 11272–11277.

Liu, Z., W. Hu, C. Li, Y. Li, C. Xue, C. Li, Y. Zho, B. Cheng, and Q. Wang. 2012. Room temperature direct-bandgap electroluminescence from n-type strain-compensated Ge/SiGe multiple quantum wells. *Appl. Phys. Lett. 101*(23), 231108.

Luan, H.-C., K. Wada, L. C. Kimerling, G. Masini, L. Colace, and G. Assanto. 2001. High efficiency photodetectors based on high quality epitaxial germanium grown on silicon substrates. *Opt. Mater. 17*(1–2), 71–73.

Lukeš, F. and J. Humlíček. 1972. Electroreflectance of heavily doped *n*-type and *p*-type germanium near the direct energy gap. *Phys. Rev. B 6*(2), 521–533.

Mäder, K. A., A. Baldereschi, and H. von Känel. 1989. Band structure and instability of Ge_{1-x} Sn_x. *Solid State Commun. 69*(12), 1123–1126.

Madhavi, S., V. Venkataraman, and Y. H. Xie. 2001. High room-temperature hole mobility in $Ge_{0.7}$ $Si_{0.3}$/Ge/$Ge_{0.7}$ $Si_{0.3}$ modulation-doped heterostructures. *J. Appl. Phys. 89*(4), 2497–2499.

Marchionna, S., A. Virtuani, M. Acciarri, G. Isella, and H. von Kaenel. 2006. Defect imaging of SiGe strain relaxed buffers grown by LEPECVD. *Mat. Sci. Semicond. Process. 9*(4–5), 802–805.

Marzegalli, A., F. Isa, H. Groiss, E. Müller, C. V. Falub, A. G. Taboada, P. Niedermann, et al. 2013. Unexpected dominance of vertical dislocations in high-misfit Ge/Si(001) films and their elimination by deep substrate patterning. *Adv. Mater. 25*(32), 4408–4412.

Meduňa, M., C. V. Falub, F. Isa, D. Chrastina, T. Kreiliger, G. Isella, A. G. Taboada, P. Niedermann, and H. von Känel. 2014. X-ray nano-diffraction on epitaxial crystals. *Quantum Matter 3*(4), 290–296.

Meyerson, B. S. 2000. Low-temperature Si and SiGe epitaxy by ultrahigh-vacuum/chemical vapor deposition: Process fundamentals. *IBM J. Res. Develop. 44*(1/2), 132–141.

Meyerson, B. S., K. J. Uram, and F. K. LeGoues. 1988. Cooperative growth phenomena in silicon/germanium low-temperature epitaxy. *Appl. Phys. Lett. 53*(25), 2555–2557.

Mironov, O. A., A. H. A. Hassan, R. J. H. Morris, A. Dobbie, M. Uhlarz, D. Chrastina, J. P. Hague, et al. 2014. Ultra high hole mobilities in a pure strained Ge quantum well. *Thin Solid Films 557,* 329–333.

Nakamura, A. 1977. Luminescence from electron–hole drops in heavily doped n-type germanium. *J. Phys. Soc. Jpn. 43*(2), 529–537.

Newman, R. 1953a. Optical studies of injected carriers. I. Infrared absorption in germanium. *Phys. Rev. 91*(6), 1311–1312.

Newman, R. 1953b. Optical studies of injected carriers. II. Recombination radiation in germanium. *Phys. Rev. 91*(6), 1313–1314.

Osmond, J., G. Isella, D. Chrastina, R. Kaufmann, M. Acciarri, and H. von Känel. 2009. Ultra low dark current Ge/Si(100) photodiodes with low thermal budget. *Appl. Phys. Lett. 94*(20), 201106.

Pankove, J. I. 1960. Optical absorption by degenerate germanium. *Phys. Rev. Lett. 4*(9), 454–455.

Pankove, J. I. and P. Aigrain. 1962. Optical absorption of arsenic-doped degenerate germanium. *Phys. Rev. 126*(3), 956–962.

Park, J.-S., J. Bai, M. Curtin, B. Adekore, M. Carroll, and A. Lochtefeld. 2007. Defect reduction of selective Ge epitaxy in trenches on Si(001) substrates. *Appl. Phys. Lett. 90*(5), 052113.

Paul, D. J., A. Samarelli, L. Ferre-Llin, J. R. Watling, Y. Zhang, J. M. R. Weaver, P. S. Dobson, et al. 2012. Si/SiGe nanoscale engineered thermoelectric materials for energy harvesting. In *Proceedings of the 12th IEEE International Conference on Nanotechnology (IEEE-NANO) 2012,* Birmingham, UK, pp. 1–5.

Paul, D. J., A. Samarelli, L. Ferre Llin, Y. Zhang, J. M. R. Weaver, P. S. Dobson, S. Cecchi, et al. 2013. Si/SiGe thermoelectric generators. *ECS Trans. 50*(9), 959–963.

Pernot, G., M. Stoffel, I. Savic, F. Pezzoli, P. Chen, G. Savelli, A. Jacquot, J. et al. 2010. Precise control of thermal conductivity at the nanoscale through individual phonon-scattering barriers. *Nat. Mater. 9*(6), 491–495.

Pezzoli, F., F. Isa, G. Isella, C. V. Falub, T. Kreiliger, M. Salvalaglio, R. Bergamaschini et al. 2014. Ge crystals on Si show their light. *Phys. Rev. App. 1*(4), 044005.

Pezzoli, F., L. Qing, A. Giorgioni, G. I. E. Grilli, M. Guzzi, and H. Dery. 2013. Spin and energy relaxation in germanium studied by spin-polarized direct-gap photoluminescence. *Phys. Rev. B 88*(4), 045204.

Pillarisetty, R. 2011. Academic and industry research progress in germanium nanodevices. *Nature 479* (7373), 324–328.

Pizzini, S., M. Acciarri, S. Binetti, D. Cavalcoli, A. Cavallini, D. Chrastina, L. Colombo, et al. 2006. Nanocrystalline silicon films as multifunctional material for optoelectronic and photovoltaic applications. *Mat. Sci. Eng. B 134*(2–3), 118–124.

Racanelli, M. and D. W. Greve. 1990. Temperature dependence of growth of $Ge_x Si_{1-x}$ by ultrahigh vacuum chemical vapor deposition. *Appl. Phys. Lett. 56*(25), 2524–2526.

Rastelli, A., M. Kummer, and H. von Känel. 2001. Reversible shape evolution of Ge islands on Si(001). *Phys. Rev. Lett. 87*(25), 256101.

Rentzsch, R. and I. S. Shlimak. 1977. Photoluminescence of heavily doped and compensated germanium. *Phys. Stat. Sol. (a) 43*(1), 231–238.

Robbins, D. J., J. L. Glasper, A. G. Cullis, and W. Y. Leong. 1991. A model for heterogeneous growth of $Si_{1-x} Ge_x$ films from hydrides. *J. Appl. Phys. 69*(6), 3729–3732.

Rosenblad, C., H. R. Deller, M. Döbeli, E. Müller, and H. von Känel. 1998. Low-temperature heteroepitaxy by LEPECVD. *Thin Solid Films 318*(1–2), 11–14.

Rosenblad, C., H. von Känel, M. Kummer, A. Dommann, and E. Müller. 2000. A plasma process for ultrafast deposition of SiGe graded buffer layers. *Appl. Phys. Lett. 76*(4), 427–429.

Rössner, B., D. Chrastina, G. Isella, and H. von Känel. 2004. Scattering mechanisms in high-mobility strained Ge channels. *Appl. Phys. Lett. 84*(16), 3058–3060.

Roucka, R., J. Matthews, R. T. Beeler, J. Tolle, J. Kouvetakis, and J. Menéndez. 2011. Direct gap electroluminescence from Si/$Ge_{1-y}Sn_y$ *p–i–n* heterostructure diodes. *Appl. Phys. Lett. 98*(6), 061109.

Roucka, R., J. Tolle, C. Cook, A. V. G. Chizmeshya, J. Kouvetakis, V. D'Costa, J. Menendez, Z. D. Chen, and S. Zollner. 2005. Versatile buffer layer architectures based on $Ge_{1-x}Sn_x$ alloys. *Appl. Phys. Lett. 86*(19), 191912.

Rouifed, M.-S., P. Chaisakul, D. Marris-Morini, J. Frigerio, G. Isella, D. Chrastina, S. Edmond, X. Le Roux, J.-R. Coudevylle, and L. Vivien. 2012. Quantum-confined Stark effect at 1.3 μm in Ge/$Si_{0.35}Ge_{0.65}$ quantum-well structure. *Optics Lett. 37*(19), 3960–3962.

Rouifed, M.-S., D. Marris-Morini, P. Chaisakul, J. Frigerio, G. Isella, D. Chrastina, S. Edmond, et al. 2014. Advances toward Ge/SiGe quantum-well waveguide modulators at 1.3 μm. *IEEE J. Sel. Top. Quant. 20*(4), 3400207.

Rouvière, M., L. Vivien, X. L. Roux, J. Mangeney, P. Crozat, C. Hoarau, E. Cassan, et al. 2005. Ultrahigh speed germanium-on-silicon-on-insulator photodetectors for 1.31 and 1.55 μm operation. *Appl. Phys. Lett. 87*(23), 231109.

Rowe, D. M. (Ed.). 2006. *Thermoelectrics Handbook: Macro to Nano*. CRC Taylor & Francis, Boca Raton, Florida.

Samarelli, A., L. Ferre Llin, S. Cecchi, J. Frigerio, T. Etzelstorfer, E. Müller, Y. Zhang, et al. 2013a. The thermoelectric properties of Ge/SiGe modulation doped superlattices. *J. Appl. Phys. 113*(23), 233704.

Samarelli, A., L. Ferre Llin, Y. Zhang, J. M. R. Weaver, P. Dobson, S. Cecchi, D. Chrastina, et al. 2013b. Power factor characterization of Ge/SiGe thermoelectric superlattices at 300 K. *J. Electron. Mater. 42*(7), 1449–1453.

Shah, V. A., A. Dobbie, M. Myronov, and D. R. Leadley. 2010. Reverse graded SiGe/Ge/Si buffers for high-composition virtual substrates. *J. Appl. Phys. 107*(6), 064304.

Shah, V. A., M. Myronov, C. Wongwanitwatana, L. Bawden, M. J. Prest, J. S. Richardson-Bullock, S. Rhead, E. H. C. Parker, T. E. Whall, and D. R. Leadley. 2012. Electrical isolation of dislocations in Ge layers on Si(001) substrates through CMOS-compatible suspended structures. *Sci. Technol. Adv. Mat. 13*(5), 055002.

Shah, V. A., S. D. Rhead, J. E. Halpin, O. Trushkevych, E. Chávez-Ángel, A. Shchepetov, V. Kachkanov, et al. 2014. High quality single crystal Ge nano-membranes for opto-electronic integrated circuitry. *J. Appl. Phys. 115*(14), 144307.

Slack, G. A. and S. F. Bartram. 1975. Thermal expansion of some diamond like crystals. *J. Appl. Phys. 46*(1), 89–98.

Sommers, Jr., H. S. 1961. Degenerate germanium. II. Band gap and carrier recombination. *Phys. Rev. 124*(4), 1101–1110.

Sordan, R., A. Miranda, F. Traversi, D. Colombo, D. Chrastina, G. Isella, M. Masserini, L. Miglio, K. Kern, and K. Balasubramanian. 2009. Vertical arrays of nanofluidic channels fabricated without nanolithography. *Lab Chip 9*(11), 1556–1560.

Soref, R., J. Kouvetakis, J. Tolle, J. Menendez, and V. D'Costa. 2007. Advances in SiGeSn technology. *J. Mater. Res. 22*(12), 3281–3291.

Soref, R. A. and C. H. Perry. 1991. Predicted band gap of the new semiconductor SiGeSn. *J. Appl. Phys. 69*(1), 539–541.

Spitzer, W. G., F. A. Trumbore, and R. A. Logan. 1961. Properties of heavily doped *n*-type germanium. *J. Appl. Phys. 32*(10), 1822–1830.

Stern, F. 1971. Optical absorption edge of compensated germanium. *Phys. Rev. B 3*(10), 3559–3560.

Stoffel, M., A. Malachias, T. Merdzhanova, F. Cavallo, G. Isella, D. Chrastina, H. von Känel, A. Rastelli, and O. G. Schmidt. 2008. SiGe wet chemical etchants with high compositional selectivity and low strain sensitivity. *Semicond. Sci. Technol. 23*(8), 085021.

Stoffel, M., A. Malachias, A. Rastelli, T. H. Metzger, and O. G. Schmidt. 2009. Composition and strain in SiGe/Si(001) "nanorings" revealed by combined x-ray and selective wet chemical etching methods. *Appl. Phys. Lett. 94*(25), 253114.

Süess, M. J., L. Carroll, H. Sigg, A. Diaz, D. Chrastina, G. Isella, E. Müller, and R. Spolenak. 2012. Tensile strained Ge quantum wells on Si substrate: Post-growth annealing versus low temperature re-growth. *Mat. Sci. Eng. B 177*(10), 696–699.

Süess, M. J., R. Geiger, R. A. Minamisawa, G. Schiefler, J. Frigerio, D. Chrastina, G. Isella, R. Spolenak, J. Faist, and H. Sigg. 2013. Analysis of enhanced light emission from highly strained germanium micro bridges. *Nat. Photonics 7*(6), 466–472.

Sun, X., J. Liu, L. C. Kimerling, and J. Michel. 2009a. Direct gap photoluminescence of *n*-type tensile-strained Ge-on-Si. *Appl. Phys. Lett. 95*(1), 011911.

Sun, X., J. Liu, L. C. Kimerling, and J. Michel. 2009b. Room-temperature direct bandgap electroluminescence from Ge-on-Si light-emitting diodes. *Optics Lett. 34*(8), 1198–1200.

Sun, X., J. Liu, L. C. Kimerling, and J. Michel. 2010. Toward a germanium laser for integrated silicon photonics. *IEEE J. Sel. Top. Quant. 16*(1), 124–131.

Taboada, A. G., T. Kreiliger, C. V. Falub, F. Isa, M. Salvalaglio, L. Wewior, D. Fustor, et al. 2014. Strain relaxation of GaAs/Ge crystals on patterned Si substrates. *Appl. Phys. Lett. 104*(2), 022112.

Thomas, S. G., S. Bharatan, R. E. Jones, R. Thoma, T. Zirkle, N. V. Edwards, R. Liu, et al. 2003. Structural characterization of thick, high-quality epitaxial Ge on Si substrates grown by low-energy plasma-enhanced chemical vapor deposition. *J. Electron. Mater. 32*(9), 976–980.

Tsujino, S., H. Sigg, G. Mussler, D. Chrastina, and H. von Känel. 2006. Photocurrent and transmission spectroscopy of direct-gap interband transitions in Ge/SiGe quantum wells. *Appl. Phys. Lett. 89*(26), 262119.

van Driel, H. M., A. Elci, J. S. Bessey, and M. O. Scully. 1976. Photoluminescence spectra of germanium at high excitation intensities. *Solid State Commun. 20*(9), 837–840.

Vanacore, G. M., M. Zani, M. Bollani, E. Bonera, G. Nicotra, J. Osmond, G. Capellini, G. Isella, and A. Tagliaferri. 2014. Monitoring the kinetic evolution of self-assembled SiGe islands grown by Ge surface thermal diffusion from a local source. *Nanotechnology 25*(13), 135606.

Vanacore, G. M., M. Zani, M. Bollani, D. Colombo, G. Isella, J. Osmond, R. Sordan, and A. Tagliaferri. 2010b. Size evolution of ordered SiGe islands grown by surface thermal diffusion on pit-patterned Si(100) surface. *Nanoscale Res. Lett. 5*(12), 1921–1925.

Vanacore, G. M., M. Zani, G. Isella, J. Osmond, M. Bollani, and A. Tagliaferri. 2010a. Quantitative investigation of the influence of carbon surfactant on Ge surface diffusion and island nucleation on Si(100). *Phys. Rev. B 82*(12), 125456.

Velha, P., K. F. Gallacher, D. C. Dumas, D. J. Paul, M. Myronov, and D. R. Leadley. 2013. Long wavelength ≥1.9 μm germanium for optoelectronics using process induced strain. *ECS Trans. 50*(9), 779–782.

Viña, L. and M. Cardona. 1986. Optical properties of pure and ultraheavily doped germanium: Theory and experiment. *Phys. Rev. B 34*(4), 2586–2597.

von Känel, H., M. Kummer, G. Isella, E. Müller, and T. Hackbarth. 2002. Very high hole mobilities in modulation-doped Ge quantum wells grown by low-energy plasma enhanced chemical vapor deposition. *Appl. Phys. Lett. 80*(16), 2922–2924.

Wagner, J., A. Compaan, and A. Axmann. 1983. Photoluminescence in heavily doped Si and Ge. *J. Phys. (Paris) Colloq. 44*(C5), 61–64.

Wagner, J. and L. Viña. 1984. Radiative recombination in heavily doped p-type germanium. *Phys. Rev. B 30*(12), 7030–7036.

Wang, X., H. Li, R. Camacho-Aguilera, Y. Cai, L. C. Kimerling, J. Michel, and J. Liu. 2013. Infrared absorption of n-type tensile-strained Ge-on-Si. *Optics Lett. 38*(10), 652–654.

Watling, J. R. and D. J. Paul. 2011. A study of the impact of dislocations on the thermoelectric properties of quantum wells in the Si/SiGe materials system. *J. Appl. Phys.* 110(11), 114508.

Werner, J., M. Oehme, M. Schmid, M. Kaschel, A. Schirmer, E. Kasper, and J. Schulze. 2011. Germanium–tin p–i–n photodetectors integrated on silicon grown by molecular beam epitaxy. *Appl. Phys. Lett.* 98(6), 061108.

Wiedemann, G. and R. Franz. 1853. Über die Wärmeleitungsfähigkeit der Metalle (On the thermal conductivity of metals). *Ann. Phys. 165*(8), 497–531.

Wu, P. H., D. Dumcenco, Y. S. Huang, H. P. Hsu, C. H. Lai, T. Y. Lin, D. Chrastina, G. Isella, E. Gatti, and K. K. Tiong. 2012. Above-room-temperature photoluminescence from a strain-compensated Ge/$Si_{0.15}Ge_{0.85}$ multiple-quantum-well structure. *Appl. Phys. Lett.* 100(14), 141905.

Xie, Y. H., D. Monroe, E. A. Fitzgerald, P. J. Silverman, F. A. Thiel, and G. P. Watson. 1993. Very high mobility two-dimensional hole gas in Si/Ge_xSi_{1-x}/Ge structures grown by molecular beam epitaxy. *Appl. Phys. Lett. 63*(16), 2263–2264.

Xu, C., R. T. Beeler, L. Jiang, G. Grzybowski, A. V. G. Chizmeshya, J. Menéndez, and J. Kouvetakis. 2013a. New strategies for Ge-on-Si materials and devices using non-conventional hydride chemistries: The tetragermane case. *Semicond. Sci. Technol. 28*(10), 105001.

Xu, C., L. Jiang, J. Kouvetakis, and J. Menéndez. 2013b. Optical properties of $Ge_{1-x-y}Si_xSn_y$ alloys with $y > x$: Direct bandgaps beyond 1550 nm. *Appl. Phys. Lett. 103*(7), 072111.

Zhang, J., N. J. Woods, G. Breton, R. W. Price, A. D. Hartell, G. S. Lau, R. Liu, A. T. S. Wee, and E. S. Tok. 2002. Growth mechanisms in thin film epitaxy of Si/SiGe from hydrides. *Mat. Sci. Eng. B 89*(1–3), 399–405.

4 Self-Interstitials in Silicon and Germanium

Alexandra Carvalho and Robert Jones

CONTENTS

4.1 Introduction 88
4.2 Self-Interstitial in Silicon 88
  4.2.1 Defect Configurations and Energetics 88
  4.2.2 Electrical Levels 89
  4.2.3 Thermally Activated Diffusion 91
    4.2.3.1 Excitation-Enhanced Diffusion 93
4.3 Self-Interstitial in Germanium 94
  4.3.1 Defect Configurations and Energetics 94
  4.3.2 Electrical Levels 95
  4.3.3 Thermally Activated Diffusion 96
  4.3.4 Excitation-Enhanced Diffusion 96
4.4 Morphs 96
4.5 Radiation Damage and Frenkel Pairs 97
  4.5.1 *p*-Type Silicon 97
  4.5.2 *n*-Type Silicon 102
  4.5.3 *n*-Type Germanium 102
    4.5.3.1 Deep-Level Transient Spectroscopy 105
  4.5.4 *p*-Type Germanium 105
  4.5.5 High-Purity Ge 108
  4.5.6 From *n*-Type to *p*-Type Ge: Monitoring *I* Trapping with PAC Spectroscopy 109
4.6 Aggregation of Self-Interstitials 110
  4.6.1 Silicon 110
  4.6.2 Di-Interstitial 111
  4.6.3 Tri-Interstitial 111
  4.6.4 Tetra-Interstitial 113
4.7 Germanium 113
  4.7.1 Di-Interstitial 114
  4.7.2 Tri-Interstitial and Larger Defects 114
4.8 Summary 115
References 115

4.1 INTRODUCTION

Self-interstitials and vacancies, the primary intrinsic defects, are invariably present in silicon and germanium crystals. Usually formed thermally, during growth annealing or following oxidation, they can also be introduced in higher concentrations by irradiation with fast electrons, γ-rays, neutrons and heavier particles, deformation, or other damaging treatments. Highly mobile and reactive, the self-interstitial takes part in many defect reactions in a wide temperature range. In silicon, it is a mediator for the diffusion of impurities, and is responsible for the transient enhanced diffusion (TED) of boron during the first step following impurity implantation,[1] as well as for forming compensating defect centers. Thus, knowing the properties of this evasive defect is essential in controlling and understanding the evolution of point and extended defects in electronic and solar cell materials.

Their extraordinary mobility and reactivity, even at low temperatures, make them very difficult to be detected and identified experimentally. Studies of radiation damage in Si and Ge started in the late 1940s,[2] and inspired some of the concepts of defect physics that are still used today. However, it was not until recently that first-principles modeling provided an independent backing to the interpretation of experimental results and a unified view of the problem. In Si, the theoretical studies of the properties of isolated self-interstitials have been innumerous (see, e.g., References 3–16). In parallel, powerful techniques such as electron paramagnetic resonance (EPR) spectroscopy and deep-level transient spectroscopy (DLTS) have contributed to a great degree of understanding of the vacancy, divacancy, vacancy–impurity, and interstitial–impurity complexes,[17–19] and mostly indirectly, of the silicon self-interstitial.

Curiously, in germanium, where the self-interstitial has a comparatively higher formation energy and plays a minor role in dopant diffusion and self-diffusion, it has been detected directly or indirectly by a number of techniques, including perturbed angular $\gamma - \gamma$ correlation (PAC) spectroscopy,[20] DLTS,[21,22] and even electron microscopy imaging.[23] It has also been studied using density functional theory.[24–27] Hence, it is perhaps one of the best characterized defects in semiconductors.

In this chapter, we review the properties of the self-interstitials in Si and Ge, as unveiled by recent theoretical calculations and experiments. We then show how the latest theoretical understanding provides a unifying picture of the phenomena behind the collection of qualitative and quantitative data from decades of low-temperature irradiation experiments.

4.2 SELF-INTERSTITIAL IN SILICON

4.2.1 Defect Configurations and Energetics

A variety of models can be conceived for the self-interstitial. The most important are the split-interstitial (or "dumbbell") configurations and caged configurations. The split-interstitial configurations are obtained by replacing a silicon atom at the lattice site by a pair of interstitial atoms aligned along a particular direction, while caged self-interstitials are placed in the cavities of the silicon lattice, at or close to one of

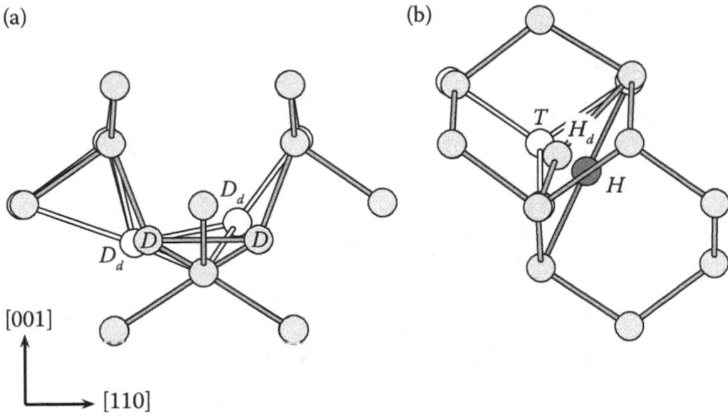

FIGURE 4.1 Charge-dependent structures of the self-interstitial: (a) split-interstitial in a C_{2v} split-interstitial (D) and in a distorted split-interstitial (D_d) configurations, in gray and white, respectively. (b) Caged interstitial at the undistorted hexagonal (H) site, at a ⟨111⟩-displaced hexagonal site (H_d) and at a tetrahedral site, represented in light gray, white, and dark gray, respectively. (Reprinted from *Mater. Sci. Eng. B*, 159–160, Jones, R. et al. 112, Copyright 2009, with permission from Elsevier.)

the high-symmetry sites. Those are the tetrahedral (T) interstitial site, characterized by four nearest neighbors at a distance $\sqrt{3}/4a_0$ from the center and six neighbors at a distance of $a_0/2$; and the puckered hexagonal (H) site, the midpoint between two nearest-neighbor T sites. At the T site, the distance of the interstitial to the nearest lattice atoms is maximized. Other extended self-interstitial configurations are believed to exist at high temperature only (see Section 4.4).

Many of the interesting properties of silicon self-interstitials arise from the fact that they assume different configurations depending on the charge state. In the neutral charge state, the lowest energy structure is a [110]-aligned dumbbell. There is also a metastable configuration where the interstitial is at a caged position between the H and the T sites,[15,25,28] which we label H_d (Figure 4.1), where the subscript "d" stands for distorted.

Since the energy difference between the dumbly and caged interstitials in the neutral charge state is very small, there is some variance in the literature as for which is the lowest in energy (see, e.g., References 4–15, 29). Sometimes differences arise from the fact that the symmetry distortion is not taken into account.

In the positive charge state, it is generally agreed that the [110] dumbbell is not stable. The equilibrium configuration is H_d. The positively charged H_d self-interstitial is 0.7 Å away from the high-symmetry H site, closer to the T site than the neutral self-interstitial, which is displaced by only 0.3–0.5 Å from the H site. Finally, in the double-positive charge state, T is the only stable site found to date (Figure 4.2).

4.2.2 Electrical Levels

The self-interstitials are electrically active, and can exist in three charge states (Figure 4.2). The caged interstitial is a double donor. Table 4.1 shows the ionization

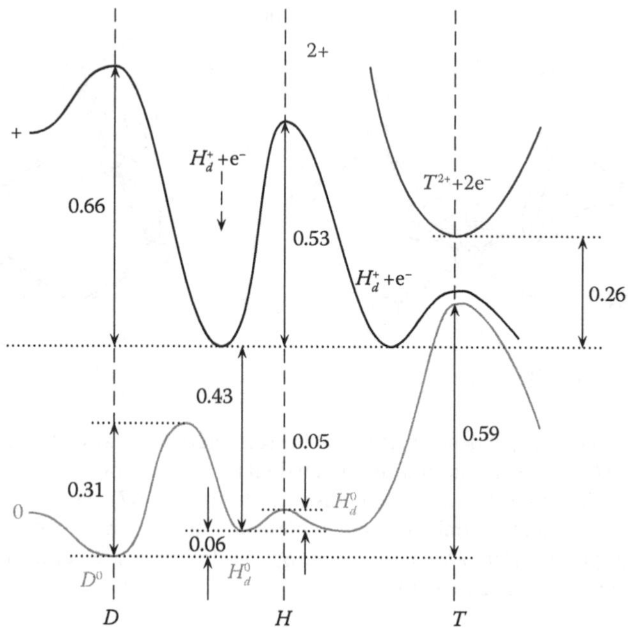

FIGURE 4.2 Proposed configuration-coordinate diagram for the silicon self-interstitial. (Reprinted from *Mater. Sci. Eng. B*, 159–160, Jones, R. et al. Briddon 112, Copyright 2009, with permission from Elsevier.)

levels of the self-interstitial obtained using density functional theory with the local density approximation.[27] The transition between the neutral and the single-positive charge state, in their respective ground state structures, noted (0/+), involves a change between the dumbbell interstitial and the distorted hexagonal interstitial, and it is approximately at midgap. Disregarding excitonic effects, that is equivalent to

TABLE 4.1

Electrical Levels (eV) of the Self-Interstitial Defects in Si and Ge according to Theory (DFT-LDA Calculations in Clusters) and Experiment

			Si		Ge	
Theory	Configuration	Reference	(0/+)	(+/2+)	(0/+)	(+/2+)
	D/D_d	29	$E_c - 0.49$[a]			
	$H>/H_d/T$	27, 29	$E_c - 0.43$	$E_c - 0.26$	$E_c - 0.24$	$E_c - 0.08$
Experiment	E1 (DLTS)	30	$E_c - 0.39$			
	P1 (EPR)	31			$E_c - 0.2$ or	$E_v + 0.2$
	Q2 (PAC)	20			$E_c - 0.04$	
	M2, M3 (DLTS)	21			$E_c - 0.12$	$E_c - 0.05$
	(DLTS)	22			$E_c - 0.24$	$E_c - 0.08$

[a] This transition from the neutral to the positive charge state should be irreversible.

the energy difference between I_D^0 and $I_{H_d}^+$, plus an electron in the conduction band. The (0/+) transition between $I_{H_d}^0$ and $I_{H_d}^+$ is closer to the conduction band, by a small energy difference, which corresponds to the energy difference between $I_{H_d}^0$ and I_D^0. The transition between the single- and the double-positive charged states, (+/2+), involves a relaxation between H_d and T with no associated energy barrier, and it is also close to the conduction band.

There is another level associated with the dumbbell interstitial. The capture of a hole leaves the I_D defect unstable by removing the energy barrier separating $I_{D_d}^+$ and $I_{H_d}^+$; thus, the transformation to be considered is $I_D + h^+ \rightarrow I_{H_d}^+$ with the corresponding donor level at $E_c - 0.49$ eV. Note that the subsequent emission of a hole leaves the defect in a different configuration, since the capture of an electron by $I_{H_d}^+$ leaves the defect as $I_{H_d}^0$. The D structure has a (0/+) level close to E_v, which increases the probability of hole capture.

Experimentally, it has been proposed that the E1 DLTS center with a donor level at $E_c - 0.39$ eV is the I_T self-interstitial.[13,30] This is consistent with the density functional theory calculations giving $E(2\ +/+) = E_c - 0.26$ eV and $E(0/+) = E_c - 0.43$ eV. The properties of this center will be considered in detail in the Section 4.5.1.

4.2.3 Thermally Activated Diffusion

Long-range migration paths can be decomposed into single barrier jumps between the energy minimum points. These elementary steps comprise transformations between different interstitial structures (Figure 4.3a), defect reorientations (Figure 4.3b), and jumps between equivalent configurations (Figure 4.3c and d). As expected, the diffusion paths are very distinct for different charge states[11,32]:

In the neutral charge state, starting on the most stable structure, D, the lowest energy path involves a succession of transformations $D \xrightarrow{\text{step 1}} H_d \xrightarrow{\text{step 2}} H_d' \xrightarrow{\text{step 1}} D$ and its dominant energy barrier, on step 1, is only 0.31 eV high. Another path with the split-interstitial in the $(\bar{1}10)$ plane[10] was found to be higher in energy.[29] The reorientation of H_d passing through the H site (path 2) is very easy requiring only ~0.05 eV of activation energy. However, as the T site is relatively high in energy (0.59 eV), the zigzag migration through the interstitial cages is not likely. Thus, the rate-limiting energy barrier for the lowest-energy path is 0.31 eV (Table 4.2).

In the positive charge state, the migration of I_H^+ requires a succession of jumps through the H sites. Long-range migration has to produce a net displacement to the vicinity of a distinct T cage. This can be accomplished via a succession of the reorientations 2 and 3 shown in Figure 4.3b. The transformation $H_d \xrightarrow{\text{step 3}} H_d'$ is a reorientation inside the interstitial cage, in the neighborhood of the T site, and requires very little energy. The dominant energy for the long-range diffusion is given by step 2, a jump over the 0.53 eV energy barrier centered at the undistorted H site (Table 4.2).

In the double positive charge state, the self-interstitial is less mobile, requiring an activation energy of 1.46 eV to move between T sites (step 5 in Figure 4.3d). The most likely diffusion path is, therefore, a zigzag along the $\langle 111 \rangle$. An alternative that

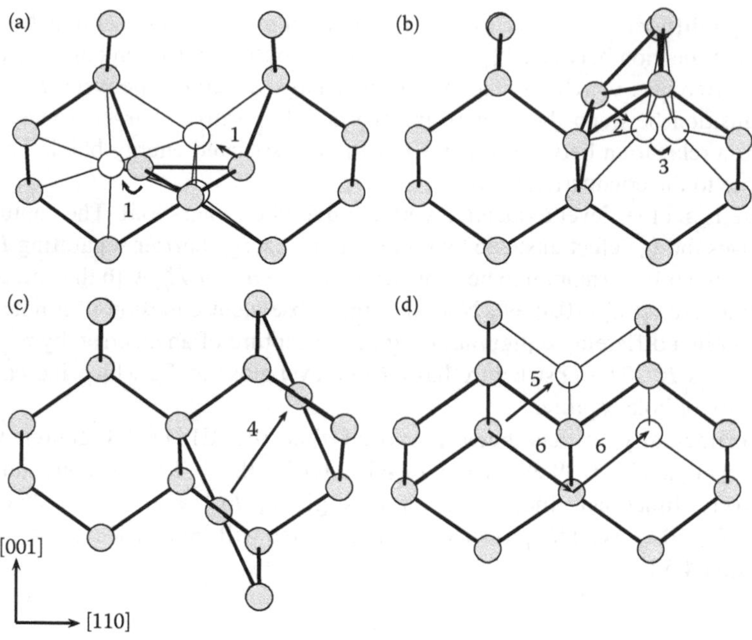

FIGURE 4.3 Structural transformations involved in the migration of Si and Ge self-interstitials, from Reference 32. (a) Reconfiguration of the neutral defect (step 1) from a split-interstitial structure to the nearest H site; (b) reorientation of I_{H_d}, through the undistorted H site (step 2), and reorientation inside the cage (step 3); (c) ⟨111⟩ migration of I_H (step 4) through the T site (Ge only); and (d) migration of I_T^{2+} along the ⟨111⟩ chain (step 5) or by knock-on (step 6). (The source of the material from Carvalho, A. C. Janke et al. 2008, *J. Phys. Condens. Matter* **20**:135220, © Institute of Physics and IOP Publishing Limited 2008, is acknowledged.)

cannot be ruled out is that I^{2+} effectively migrates when it temporarily traps electrons, benefiting from the much lower 0.53 eV barrier in the I^+ state. If the solid is in a state near to thermal equilibrium depending on the lifetime of the $I_{H_d}^+$ charge state, the trapping of an electron from the conduction band by a double positively charged interstitial followed by the fast migration of $I_{H_d}^+$ may become dominant over the pure thermal diffusion of I_T^{2+}. If the motion of I^{2+} is enhanced by charge state change to I^+, one may expect that the relevant diffusion constant will be an average of the diffusion constants of I^{2+} and I^+ weighted by the fraction of time for which each is occupied (assuming that the electronic excitation and the migration jumps are independent events). In a simplistic model, as $\exp(W_{\text{mig}}^{2+}/(kT)) \gg \exp(W_{\text{mig}^+}/(kT))$, the activation energy would be $0.53 + [E^{(2+/+)} - E_F]$. However, it is possible that the activation energies combine in a complex way, or that the passage through the saddle point involves further charge state changes. Unfortunately, such excited state processes are very difficult to model.

The activation energies for each charge state correspond to the rate-limiting energy barrier (Table 4.2).

TABLE 4.2
Lowest Calculated Migration Energies (eV) and Respective Diffusion Paths of the Self-Interstitials in Si and Ge without Charge State Change

Silicon

Charge	Suggested mechanism	W_{mig}
0	$D \xrightarrow{\text{step 1}} H_d \xrightarrow{\text{step 2}} H'_d \xrightarrow{\text{step 1}} D'$	0.31
+	$H_d \xrightarrow{\text{step 2}} H'_d \xrightarrow{\text{step 3}} H''_d$	0.53
2+	$T \xrightarrow{\text{step 5}} T'$	1.46

Germanium

Charge	Suggested mechanism	W_{mig}
0	$D \xrightarrow{\text{step 1}} H \xrightarrow{\text{step 1}} D'$	0.53
+	$H_d \xrightarrow{\text{step 2}} H'_d \xrightarrow{\text{step 3}} H''_d$	0.29
2+	$T \xrightarrow{\text{step 5}} T'$	1.23

Note: When more than one atomic transformation is involved, W_{mig} is taken to be the dominant energy barrier. Primes (′) indicate the same defect in different positions (Figure 4.3).

4.2.3.1 Excitation-Enhanced Diffusion

Under excitation with light or irradiation, long-range migration can take place nearly athermaly through the Bourgoin–Corbett mechanism and/or phonon-aided energy release mechanisms.[33,34]

The Bourgoin–Corbett mechanisms require that the potential energy for the two charge states of the defect is such that the saddle point for one is the minimum for the other and vice versa. The migration is then mediated by the capture and release of charge carriers. The configuration-coordinate diagram of the self-interstitial suggests that this is possible: the (sequential) trapping of two electrons by I_T^{2+} leaves the I_T interstitial unstable, and it tends to relax first to the $I_{H_d}^{+}$ configuration where a barrier of 0.25 eV exists to the I_D site, or is possibly carried to an I_D site. The I_D^0 configuration has a gap donor level at $E_v + 0.49$ eV, and in principle can trap a hole leading to $I_{H_d}^{+}$. Following this, a second hole trapping can switch the defect to a T site that can be different from the starting point.[34]

Phonon-aided energy release diffusion mechanisms take advantage of the energy released on the relaxation following carrier capture. The phonons are either emitted one by one in a single-phonon-assisted transition between electron levels (providing that the defect has a system of Rydberg states with separation close to the phonon energy), or all emitted in one transition between electronic energy levels. In both cases, the phonon energy is equivalent to heat and can enhance diffusion.[33]

4.3 SELF-INTERSTITIAL IN GERMANIUM

4.3.1 Defect Configurations and Energetics

The basic defect configurations of the self-interstitial in Ge are the same as described earlier for Si. However, the relative energies are very different (Figure 4.4).

The main difference is that in the neutral charge state, the caged interstitial is stable at the high-symmetry hexagonal site (H) rather than at a distorted (H_d) configuration; however, it is separated by a nearly vanishing energy barrier (~0.03 eV) from the most stable D configuration.

A second difference is that in the positive charge state, there are two stable configurations, a distorted (D_d) structure with mirror symmetry (Figure 4.1) and a H_d configuration where the self-interstitital is displaced by about 0.8 Å along the ⟨111⟩ from the H site.*

Germanium self-interstitials are among the first defects to be imaged directly using aberration-corrected electron microscopy (Figure 4.5).[23] The self-interstitials

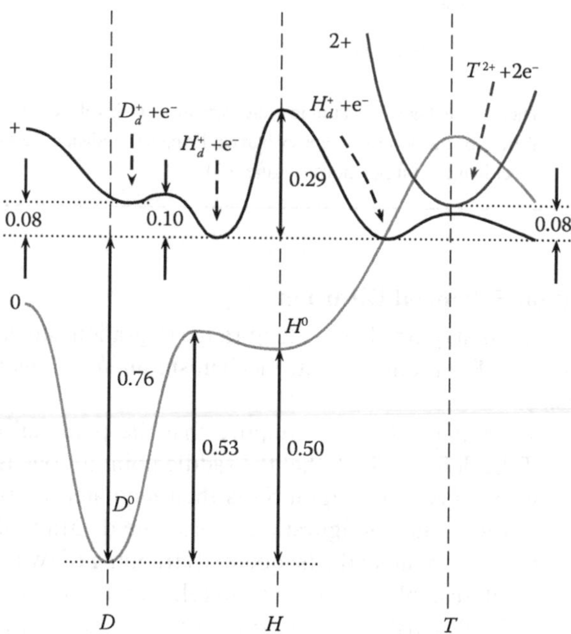

FIGURE 4.4 Proposed configuration-coordinate diagram for the germanium self-interstitial for electrons. (Reprinted with permission from Carvalho, A. et al. 2007. *Phys. Rev. Lett.* 99:175502. Copyright 2008 by the American Physical Society.)

* Spiewak et al. find instead a displaced H_d structure also for the neutral charge state of the defect,[35] as happens in silicon.[28] According to Reference 23, the stabilization of the neutral hexagonal self-interstitial is an effect of the quantum confinement, and is only observed in cluster and slab models. However, Rinke et al. found the H site to be more stable in supercells when using GW,[16] the distortion in LDA supercell calculations may be a spurious effect of the vanishing bandgap in Ge supercells.

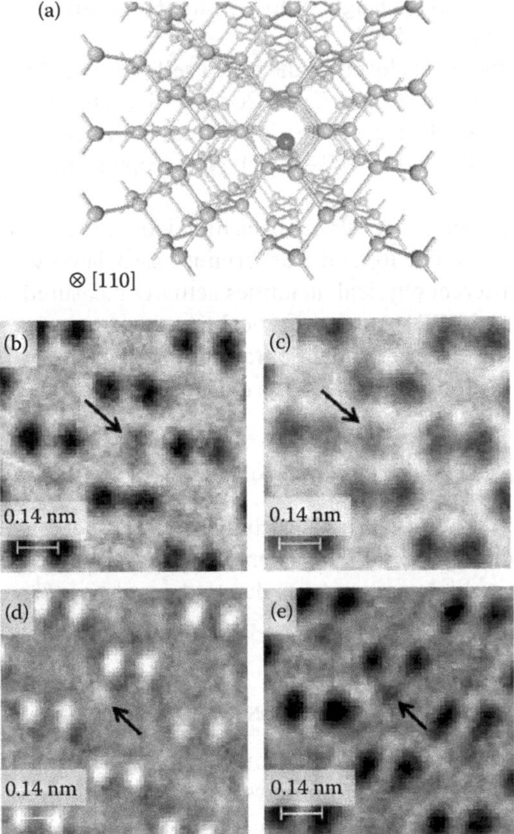

FIGURE 4.5 Self-interstitial in germanium close to a [110] surface, as observed by aberration-corrected transmission-electron microscopy, from Reference 23. (a) Model of a neutral *S* interstitial obtained in a Ge slab; (b, c) Aberration-corrected images of I_T interstitials; (d) I_H or bond-centered interstitial; and (e) self-interstitial at an off-center site. (Reprinted with permission from Alloyeau, D. et al. 2009. *Phys. Rev. B* 80:014114. Copyright 2009 by the American Physical Society.)

observed were predominantly caged self-interstitials in positions close to the *H* and *T* sites, including an additional configuration *S* stabilized by the proximity of the (110) surface, as shown in Figure 4.5a (theory) and Figure 4.5e (experiment).

4.3.2 Electrical Levels

In Ge, as in Si, the self-interstitial can exist in three charge states, 0, +, and +2. The most important difference is that in Ge, there are two distinct defect forms: one is electrically inactive and, therefore, always remains neutral (I_D), and the other, the caged interstitial, is a double donor. The split-interstitial in Ge is stable in both the neutral and positive charge states, but has no (0/+) level in the gap. Recall that in Si the same split-interstitial structure becomes unstable in the positive charge state, relaxing spontaneously to the neighborhood of the H site, and the resulting (0/+) level is deep in the gap, at $E_c - 0.5$ eV.

The position of the (+/2+) levels of the caged self-interstitial is similar in Si and in Ge, at about $E_c - 0.1$ eV in Ge and $E_c - 0.26$ eV in Si (Table 4.1). When reconfiguration between different defect structures is allowed, both defects are negative-U with the donor level lying below the (+/2+) level. No acceptor level was found for any of the defects.

Experimentally, a range of levels has been found for the germanium self-interstitial (Table 4.1). This can be due to (i) the uncertainty associated with the experimental methods, (ii) the different physical quantities actually measured by each experiment (e.g., optically excited EPR measures vertical transition energies, whereas electrical measurements measure thermodynamic transition levels), and (iii) the multiple configurations and transitions available to the defect.

4.3.3 Thermally Activated Diffusion

In the case of Ge, the fastest-diffusing charge state is the single positive, with an activation energy of only 0.3 eV. This is precisely the charge state that is thermodynamically unstable at high temperatures due to the negative-U ordering of the D/H_d (0/+) and H_d/T (+/2+) levels.

4.3.4 Excitation-Enhanced Diffusion

A possible athermal Bourgoin–Corbett migration mechanism for the germanium self-interstitial can be conceived as follows: the equilibrium position for I^{2+} is the T site, but if it traps two electrons, the I_T^0 structure is no longer stable. The double electron capture is then followed by a spontaneous relaxation to one of the two neighboring H sites. Then, a subsequent trapping of two holes would leave the self-interstitial in the I_H^{2+} state, forcing a relaxation to one of the neighboring T sites. This provides a possible process for athermal migration in the presence of excess free carriers, as during irradiation or under other source of excitation,[34] which may be in the origin of the radiation annealing and light-induced annealing at cryogenic temperatures (<5 K) observed in Ge.[36–41]

4.4 MORPHS

There has been some controversy as to whether there is an "extended" form of the self-interstitial, a complex defect similar to an amorphous pocket with an excess atom, also named morph.[36] This idea follows a similar idea that there is an extended form, or morph, of the vacancy.[37] According to Cowern et al.,[36] this model justifies the existence of two diffusion regimes for boron in Si and Ge.

High-temperature defect forms with these characteristics are very difficult to detect by direct methods. Moreover, there are some difficulties in conciliating this model with some of the self-diffusion and impurity diffusion data, which will be further discussed in Chapter 5. In parallel, a recent analysis by Watkins concludes that the low- and high-temperature diffusion processes are identical and there is no large entropy term associated with V migration.[38] In conditions of supersaturation of self-interstitials, Caliste and Pochet[39] suggested, based on first-principles calculations, that for a high concentration of vacancies, the effective diffusion enthalpy of vacancies and divacancies is non-Arrhenius, providing an alternative explanation for nonequilibrium diffusion experiments. Clearly, further studies are necessary for a clearer understanding of the behavior of the self-interstitials at high temperatures.

4.5 RADIATION DAMAGE AND FRENKEL PAIRS

While the conductivity of metals is hardly modified upon irradiation with high-energy particles or γ-rays, the electrical properties of semiconductors may change by orders of magnitude.[19,40,41] These changes are controllable and reproducible, making irradiated material a preferred source of information about intrinsic point defects. Early studies of radiation damage in Si and Ge were a valuable source of information on the properties of Frenkel pairs (i.e., I–V pairs), and a trigger for theoretical studies of defect physics.*

The differences between the effects of near-threshold electron irradiation in silicon and germanium are striking. While the conductivity of n-Ge is drastically changed, that of p-Ge is hardly affected; in silicon, the situation is inverted, and the introduction rates of radiation defects in p-type material are higher by orders of magnitude than in n-type material (Tables 4.3 and 4.4).[17,42–44] This asymmetry can be explained by the different properties of the self-interstitial and vacancy in the two materials. The dumbbell and caged interstitials have different relative stabilities. In parallel, the vacancy takes five different charge states (V^{2+}, V^{+}, V^{0}, V^{-}, and V^{2-}) in Si, but has no donor level in Ge (see Chapter 5).

We now consider in turn the effects of electron or gamma irradiation at low temperatures in Si and Ge, paying attention to some of the defect signatures that have been assigned to isolate self-interstitials.

4.5.1 *p*-TYPE SILICON

In p-Si, $I^{2+} - V^{2+}$ are expected to form during or following low-temperature irradiation. As I^{+} is repelled by V^{2+} due to the long-ranged Coulomb interaction, recombination of I with V is unlikely, since I will be preferentially trapped at acceptors (such as B^{-} or Al^{-}), or possibly at oxygen or carbon. Consequently, electron paramagnetic resonance experiments have shown that electron irradiation of p-type Si results in the formation of boron interstitials even at temperatures as low as 4 K, readily forming complexes with B, Al, C, O, and other impurities[18] (Figure 4.6). The self-interstitial seems to disappear completely during e-irradiation of p-type material at 4.2 K, since the density of interstitial–impurity complexes created roughly equals

* Here, we will concentrate on low-dose regime, where the number N_d of collisions that give rise to atomic displacements per unit volume and unit time is proportional to the flux ϕ of colliding particles. The ratio $\sigma = N_d/(N_{Si}\phi)$, where N_{Si} is the concentration of Si nuclei, is the *cross section* for introduction of defects.

TABLE 4.3
Carrier Removal Rates (r_c), Measured by Electrical Techniques, along with Defect Introduction Rates (r_d), Measured by Other Methods, Following *e*-Irradiation of *n*- and *p*-Type Silicon

	Method	Material	E_{irr} (eV)	T_{irr} (K)	Light/Dark	r_c or r_d (cm^{-1})	Reference
***p*-Si**							
Sivo and Klontz	Electrical measurement	Degenerate, Fz	1	7–90		1.1	43
Sivo and Klontz	Electrical measurement	Degenerate, Cz	1	7–90		0.70–0.74	43
Watkins et al.	DLTS		2.4	4.2		0.1[a]	45
Watkins et al.	EPR		1.5	≤40	Light	0.06[b]	44
Zillgen and Ehrhart	X-ray diffraction	Lightly-doped, Fz and Cz	2.5	4		~1[c]	46
***n*-Si**							
Emtsev et al.	Electrical measurement	Highly doped, Cz	2.5	4.5		0.15	47
Emtsev et al.	Electrical measurement	Highly doped, Cz	2.5	297		4	47
Sivo and Klontz	Electrical measurement	Degenerate, Cz and Fz	1	10–55	Light or dark	0.02	43
Sivo and Klontz	Electrical measurement	Degenerate, Cz	1	90	Light or dark	0.4	43
Sivo and Klontz	Electrical measurement	Degenerate, Fz	1	90	Light or dark	0.9	43
Watkins et al.	EPR	Moderately doped	3.0	4.2	Light	~0.001[d]	45

Note: E_{irr} and T_{irr} represent, respectively, the energy of the electron beam and the temperature of the sample during irradiation.

[a] Introduction rate for *V*, detected by its $E_v + 0.13$ eV DLTS level.

[b] Introduction rate for *V*, observed as V^+.

[c] Introduction rate for FPs.

[d] Introduction rate for *V*, observed as V^-.

TABLE 4.4
Carrier Removal Rates (r_c), Measured by Electrical Techniques, along with Defect Introduction Rates (r_d), Measured by Other Methods, Following e-Irradiation of *n*- and *p*-Type Germanium

	Method	Material	E_{irr} (eV)	T_{irr} (K)	Light/Dark	r_c or r_d (cm^{-1})	Reference
***p*-Ge**							
Flanagan	Electrical measurement	Lightly doped	1.0	10	Dark	3×10^{-4}	48
Flanagan	Electrical measurement	Lightly doped	1.0	10	Long illumination	~0.2–0.6	48
MacKay and Klontz	Electrical measurement	Lightly doped	1.1	4.2		$\sim 10^{-2}$	49
Trueblood	EPR	Undoped	4.5	77	Light	1.6[a]	31
Whitehouse	Electrical measurement	Degenerate	4.5	4.2	Dark	0.4	50
Whitehouse	Electrical measurement	Degenerate	1.1	4.2	Dark	Immeasurable	50
***n*-Ge**							
Brown et al.	Electrical measurement	Lightly doped	1	79		~0.3	51
Calcott and Mackay	Electrical measurement	Lightly doped	4	4.2		~9	49
Calcott and Mackay	Electrical measurement	Lightly doped	1	4.2		0.7–1.1	49
Ehrhart and Zillgen	X-ray diffraction	High-purity	2.5	4		~3[b]	52
Hyatt and Koehler	Electrical measurement	Lightly doped	1.1	5, 10, 15		1.9	53
MacKay and Klontz	Electrical measurement	Lightly-doped	1.1	4.2		2	49
Whitehouse	Electrical measurement	Degenerate	4.5	4.2	Dark	10	50

Note: E_{irr} and T_{irr} represent, respectively, the energy of the electron beam and the temperature of the sample during irradiation.
[a] Introduction rate of P1.
[b] Introduction rate of FPs.

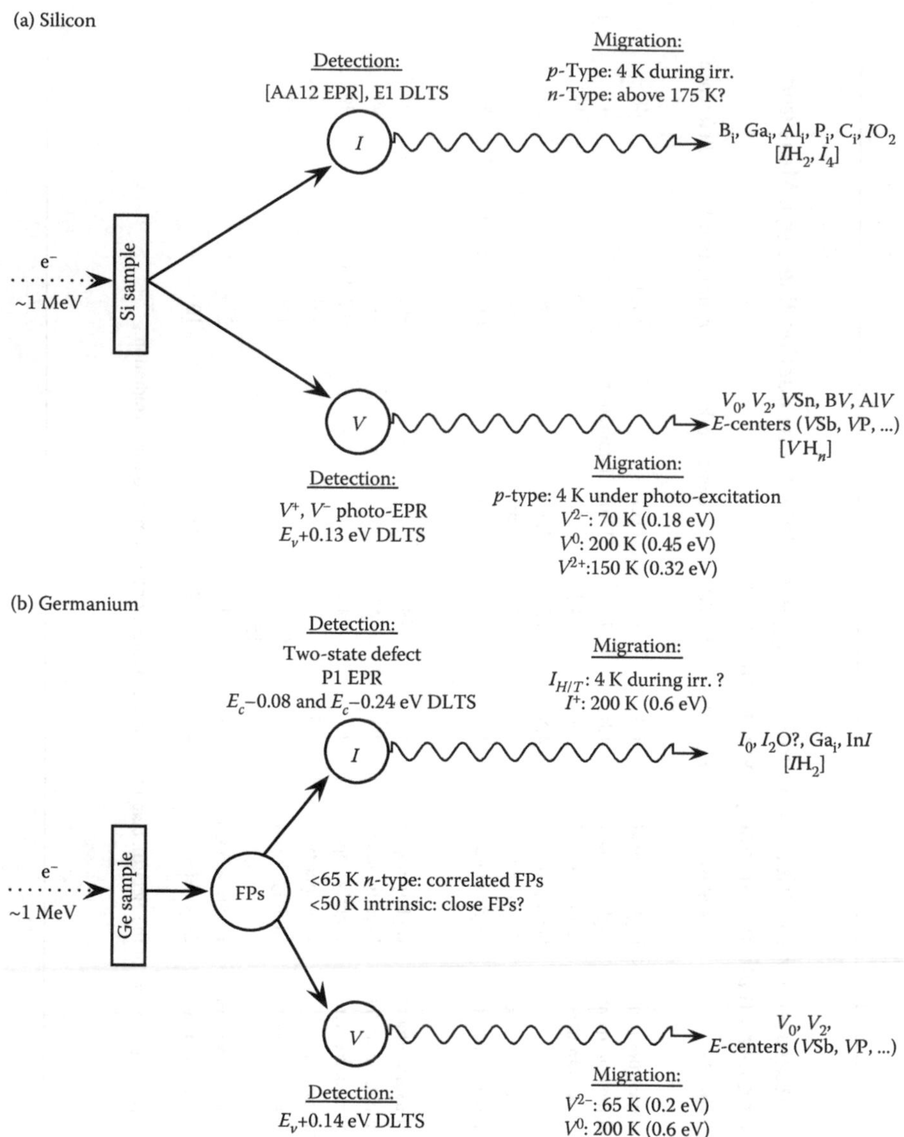

FIGURE 4.6 Summary of the defects introduced by *e*-irradiation in (a) Si and (b) Ge. Some products of proton or ion implantation are given in square brackets. References for defects in Ge are given in the text; for Si, see Reference 18 and references therein. Uncertainties are not shown. The diagram is given as an aid to the reader, but it should be noted that many assignments are tentative, and the list is not exhaustive.

that of vacancies.[18] Everything indicates then that during 1 MeV electron irradiation, I may undertake long-range diffusion via an athermal or nearly athermal excitation-enhanced process.[7,18]

The resulting B_i, Al_i, and C_i have donor levels, compensating the activity of the chemical dopants. As the vacancy is also a double donor, the displacement of a single Si atom results, in average, in the removal of four charge carriers.

Despite its reactivity, in intrinsic or p-type Si irradiated with protons or alpha-particles, there are two defect centers that have been connected to a self-interstitial at a caged interstitial site: the AA12 EPR center and the E1 DLTS center.[13,30] Similar features have not, to our knowledge, been reported in electron or γ-irradiated material. AA12 is an isotropic center showing a resolved hyperfine interaction with a single Si atom,[30] and was correlated with the DLTS E1 center with a level at $E_c - 0.39$ eV. The thermal annealing of both centers was found to take place between 270 and 350 K, simultaneous to the appearance of C_i and Al_i.[54] Alternatively, they can be annealed at 77 K under carrier injection conditions, resulting in the production of an oxygen-related defect in Czochralski samples.[54] A double-positive charge state could have preserved the isolated self-interstitial by reducing its mobility and simultaneously discouraging it to bind to other self-interstitials. However, it is noteworthy that the 30 MeV proton irradiation doses used were very large (up to 1.5×10^{15} cm^{-2}). This would have created defect densities of the orders of magnitude larger than the electron irradiation, which could have converted the initially p-type Si into intrinsic material and provided the conditions for significant defect clustering to occur (Palmer, D. W. 2008, private communication). The question raised then is whether E1 could be a more complex intrinsic defect. Unfortunately, the authors do not report that the introduction rates of E1 and AA12 are the same.

Nevertheless, all we can say is that the known properties of the E1 DLTS center are consistent with the self-interstitial. Since, at the temperature of the measurements (77 K), it is unlikely that the defect would be able to overcome the 0.31 eV barrier from I_D^0 to $I_{H_d}^0$ (Figure 4.2), this level should be compared with the donor levels of the caged interstitial, $E(2+/+) = E_c - 0.26$ eV and $E(0/+) = E_c - 0.43$ eV according to the theoretical calculations,[27] both consistent with the measured value given a 0.2 eV uncertainty.

What is surprising is the correlation between the E1 and the AA12 EPR center, assigned to I_T^+. The AA12 center has an unusual T_d symmetry with hyperfine interaction on one Si atom, compatible with an assignment to I^+ at a T site. This is consistent with our calculations if at the temperature of measurement (77 K) $I_{H_d}^+$ is able to reorientate around the T site by overcoming the ~0.09 eV barrier.

Recall that, in silicon, I^+ is unstable since the reaction $2I_{H_d}^+ \rightarrow I_T^{2+} + I_D^0$ is exothermic due to the negative-U property, provided that the defects have enough thermal energy to overcome the relevant energy barriers (0.3 eV for $I_{H_d}^0 \rightarrow I_D^0$). However, maybe this does not happen before 160 K, when the AA12 resonance disappears through a charge state change.

As mentioned before, it is unclear whether the charge state of the self-interstitials in the material is I^{2+} or I^0 since we do not know the position of the Fermi level after such a high dose of irradiation. Nevertheless, at 160 K, the paramagnetic (I^+) state is not detectable by EPR unless it is created by a burst of bandgap light (or possibly other excitation). Presumably, light converts I^{2+} or I^0 into I^+ and the light pulse is sufficiently short in duration to avoid long-range motion of the self-interstitial. The stability of the centers is consistent with the activation energy for diffusion of I^{2+}, but the neutral defect is expected to diffuse at much lower temperatures.

Hence, the properties of these two defect centers are consistent with the theory. Moreover, it is difficult to conceive a model of a larger defect, which would give rise to the unusual hyperfine interactions reported for AA12. Still, the possibility that it arises from a larger self-interstitial cluster is difficult to rule out.

4.5.2 *N*-Type Silicon

In contrast with *p*-type Si, early radiation experiments have shown that 1 MeV *e*-irradiation of *n*-type Si at cryogenic temperatures produces electron removal rates of less than 0.02 cm^{-2};[43] still, these can be increased by orders of magnitude if the radiation was performed above 60 K.[43,55] Consistently, it is known that the introduction rate for vacancies below 60 K in *n*-Si is also about two orders of magnitude lower than in *p*-Si.[17]

It seems that, in *n*-type Si, the primary radiation products are V^{2-} and I^0. Hence, there is no Coulomb repulsion. During irradiation, the self-interstitial is mobile through a Bourgoin–Corbett mechanism. Additionally, *V* can also be made mobile under irradiation conditions due to the electronic excitation of electron–hole pairs that recombine at the vacancy.[17] In the absence of Coulomb repulsion, it can annihilate the self-interstitials during the irradiation, reducing the survival rates. We do not know whether a similar diffusion process exists, or why should it be less effective, in *n*-Ge where, as we will consider in the next section, Frenkel pairs are much more stable.

The remaining self-interstitials that did not recombine with vacancies seem to be less mobile than in *p*-type silicon, maybe not moving before 140 K as it has been suggested.[18,45] They remain elusive to observation by EPR or DLTS.[18,22]

4.5.3 *N*-Type Germanium

Electrons with energies of 0.7–2 MeV have maximum transmitted energy of the order of the displacement energy (between 20 and 30 eV in Ge),[56] and therefore produce mainly single interstitial–vacancy pairs.* As in *n*-type Ge, the cross sections for the introduction of electrically active defects by ~1 MeV *e*-irradiation typically are as large as $\sigma \sim 10^{-23}$ cm^2,[56] the density of Frenkel pairs introduced can be comparable to the concentration of dopants.

According to theory, in *n*-Ge, the equilibrium charge states are V^{2-} and I^0. Thus, two electrons are trapped per Frenkel pair formed. Irradiation, therefore, strongly reduces the concentration of free charge carriers in the material, even converting *n*-type into *p*-type material,[59] and the introduction of electrically active defects can be quantified by monitoring the electrical properties of the material (Figure 4.6).

---

* For 2 MeV, some larger defects (likely related to divacancies and di-interstitials) start to be detected: see Reference 21, 57, 58.

The theory that the predominant irradiation products are V^{2-} and I^0 is consistent with experimental evidence that the Frenkel pairs are double acceptors.[60] DLTS measurements have assigned a double-acceptor level at $E_c - 0.14$ eV to the Frenkel pair (FP). This must correspond to a perturbed level of the self-interstitial $[V^{2-} - I^{(0/+)}]$.

The compensating centers introduced by bombardment with ~1 MeV electrons start to be observed immediately after irradiation at 4.2 K (although they seem to reach their equilibrium charge states only at ~20 K[60]), and their concentrations remain approximately stable in the dark up to about 65 K, when the dominant annealing stage is observed.[56] This annealing stage has been observed by electrical measurements[21,49,53,60,61] and energy release experiments[62] in both lightly doped and degenerate material, and it is generally accepted that it relates to the annihilation of Frenkel pairs[56]: The survival of frozen Frenkel pairs up to 65 K is unique to *n*-type Ge, and similar features are absent in electron or γ-irradiated *p*-type germanium, silicon, or diamond.[18,56] The main evidence for this assignment is the following:

1. The production and annealing of the defects appear to be independent of the chemical nature of the dopants used.[21]
2. The charge carrier traps removed at ~65 K account for 80–95% of the conductivity change in ~1 MeV irradiated samples.[60,61]
3. The cross-section formation of defects in *n*-Ge by irradiation with <2 MeV electrons is consistent with the theoretical curve, assuming a threshold displacement energy (E_d) between 20 and 30 MeV.[21,56]
4. The energy spectrum shows the introduction of a series of new defects for electron energies larger than 2 MeV, indicating that in samples submitted to irradiation with slower electrons predominate single displacements.[21,58]
5. The energy released during the annealing is consistent with the energy estimated for Frenkel pair annihilation.[61*]
6. The kinetics of this stage is also consistent with the diffusion-limited recombination of correlated pairs, with an activation energy of 0.14 ± 0.01 eV.[61,63,64]

The equilibrium charge state of the Frenkel pairs also supports the V^{2-}–I^0 model. Callcott and MacKay used short, infrequent high-field pulses to quantify the concentration of electrons trapped at the irradiation-induced compensation centers at ~20 K.[60] Both the magnitudes of the electron capture cross section for the higher-lying defect level and its dependence on the carrier energy (approximately quadratic on the electron temperature T_e) were found to be characteristic of a negative defect capturing a second electron. Consistently, the measured conductivity change with the defects full (doubly charged) was found to be approximately twice that observed with the defects empty, providing further evidence in support of a double-negative charge state.

Bourgoin also observed that in moderately doped material, lowering the Fermi level by a few meV slightly decreases the annealing temperature (determined by

* This stored energy, released when interstitials and vacancies annihilate by thermal annealing, is measured by carefully recording and comparing the temperature changes in a slowly heated sample before and after being submitted to irradiation.

isochronal annealing) of the ~65 K recombination stage.[64] This is to be expected since the lower the E_F, the higher the fraction of caged interstitials that can be ionized into I_H^+ (or even I_T^{2+}), leaving the Frenkel pair in the $I^+ - V^{2-}$ state and highly unstable due to the Coulomb attraction. Hence, the annealing rate is increased, and the annealing temperature is lowered. The electrical level suggested with this transformation was placed between 50 and 70 meV below E_c[64] and close to the calculated (0/+) level of the caged interstitial at about $E_c - 0.2$ eV, as well as to the level determined by the PAC trapping experiments (see Section 4.5.6). It is possible that the majority of the self-interstitials produced by irradiation will be in the caged interstitial configuration, maybe because the proximity of the vacancy favors the formation of I_H rather than I_D. An alternative scenario where only $I_D^0 - V^{2-}$ Frenkel pairs persist seems unlikely, as one would need to assume that the (0/+) level of I_D is shifted upwards due to the presence of the double-negative vacancy*; such a large shift of the electrical level would require a strong interaction between close components of the pair, difficult to conciliate with the thermal stability of the pair.

The Fermi level dependence of the Bourgoin example is also important because it allows us to exclude the fourfold coordinated defect,[65] the closest *I–V* pair generated by rotating a Si–Si unit, as a candidate to the dominant 65 K defect. Even though the fourfold coordinated defect is a double acceptor, it should be stabilized by lowering the Fermi level, which contradicts experiment.

The average distance between pair components $\langle d(I - V)\rangle$ depends on the sample and irradiation conditions, being smaller in more imperfect material, and increasing with the duration and temperature of irradiation. Three models for recombination have been widely discussed in the early literature (Figure 4.7).[19,42,5566–71] Today, the theoretical interpretation and the models of close pairs support the view that the distance between *V* and *I* is larger than the lattice parameter, and that they are separated by more than one energy barrier, as in the model for kinetics of the 65 K recombination stage in Ge developed by Zizine, Waite, and Whitehouse.[61,69,72] For ~1 MeV electron irradiation, the time dependence of the fraction of Frenkel pairs annealed during the 65 K stage is $a \propto (Dt)^{-1/2}$ in the short time limit or $1 - a \propto (Dt)^{-1/2}$ in the long time limit, as is characteristic of the diffusion-limited process, suggesting that the average *I–V* separation is much smaller than the average distance between FPs.

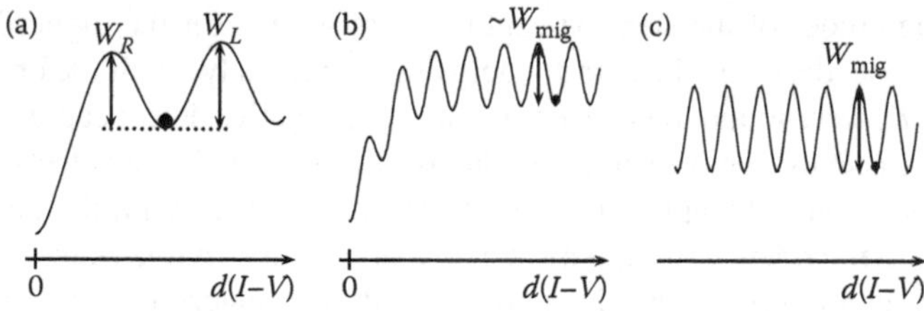

FIGURE 4.7 Models for the recombination of the Frenkel pairs: (a) close pair (Wertheim model), (b) correlated pair, and (c) separated pair.

* As happens to the (+/2+) level of interstitial iron close to an acceptor.[60]

For a random distribution of FPs, and assuming a density of the order of 10^{16} cm^{-3} (similar to the conditions of Callcott and MacKay's experiments[60]), this translates into an average distance between FPs of about 500 nm.

4.5.3.1 Deep-Level Transient Spectroscopy

In 1981, a transient spectroscopy study by Bourgoin et al. revealed the presence of several primary and secondary radiation defects produced by ≤2 MeV electron irradiation in *n*-type Ge at 4.2 K.[21] Two defect levels, M2 (E_c – 0.05 eV) and M3 (E_c – 0.12 eV), were then tentatively related to the self-interstitial. M2 and M3 annealed simultaneously at 200 K, consistent with the stability of the self-interstitial, as known today. However, little was reported about the characteristics of those centers, and for a long time, the results were not reproduced or supported by modern DLTS studies.

Very recently, however, Mesli et al. conducted *in situ* DLTS and Laplace-transform DLTS (LDLTS) of germanium irradiated with ≤3 MeV electrons, which opened up the possibility of monitoring the primary irradiation defects and their dynamics before they interacted with each other or with impurities.[22]

In *n*-type germanium, the Frenkel pairs were found to have a level of about 0.14 eV below the conduction band, although the recombination, starting well below 80 K, made the determination of the level rather difficult; the uncorrelated constituents, the vacancy, and the interstitial seemed to be stable up to 160 K. The presence of the self-interstitial was signaled by two coupled electron traps [the (0/+) and (+/2+) levels], placed at 0.08 and 0.24 eV below the conduction band (respectively), in excellent agreement with the theoretical predictions (placed at $E(+/2+) \simeq E_c - 0.1$ eV and $E(0/+) \simeq E_c - 0.2$ eV)[27] except for the relative energy order of the levels, which was thought to be a limitation of the theoretical method employed in Reference 27. This is also surprisingly close to the M2 and M3 levels found by the early DLTS study, and already attributed to self-interstitials.[21] The capacitance transients showed a clear increase of the average emission rate from the shallower level when measured at larger reverse bias, indicating a donor nature of the electron trap, also in resonance with the theory.

4.5.4 *p*-Type Germanium

Similar electrical measurements in *p*-Ge revealed a strikingly different scenario. It was found that the carrier concentration of heavily doped *p*-type Ge was little affected by ~1 MeV irradiation at 4.2, 78, or 300 K (Table 4.4).[50,56] Stored energy release experiments also reported no measurable energy release between 20 and 80 K, in clear contrast with *n*-Ge.[62*]

This can be explained by recalling that in germanium, V has no donor levels and I_D is electrically inert. Hence, it seems that radiation has left self-interstitials with

* A recent DLTS study has reported changes on the capacitance of n^+p junctions upon irradiation with MeV electrons at cryogenic temperatures.[22] The reason for the discrepancy of the results relative to the earlier studies is not clear, but may be related to a much higher sensitivity of the technique (Mesli A. 2008, private communication). Unfortunately, it was not possible to measure introduction rates (Mesli A. 2008, private communication).

the D configuration and any I formed at cage sites has presumably recombined with V. This may happen during irradiation when caged self-interstitials can migrate by a Bourgoin–Corbett mechanism, but any I caught at the D site is forced to remain neutral—and thus immobile at cryogenic temperatures. The barrier from I_D to I_H prevents I^0 from transforming to I_T^{2+}.

However, excitation with light- or higher-temperature anneals reveals the presence of hidden radiation defects. The main transformations below room temperature (RT) have been observed between 40 K and 70, 100, and 150 K, and between 200 K and 270 K[48,50,51]:

- *Illumination-induced donors and the 40–70 K stage*. If, following irradiation at 5 K, the sample was illuminated with less-than-bandgap light, the concentration of free holes was found to decay.[48] This decay process seemed to be decomposable into two exponential transients with shorter- and a longer-time constants, labeled "Exp. 1" and "Exp. 2," respectively.[50] The traps responsible for the conductivity changes were suggested to be donors with low capture cross section for holes, and both lifetimes seemed to be independent of the dopant used. The compensation centers revealed by illumination at 5 K remained stable up to temperatures between 40 and 70 K.[48] The 40–70 K annealing stage was also resolvable into two overlapping processes centered at 50 and 65 K, likely related to the Exp. 1 and Exp. 2 centers, respectively. Those donors are likely self-interstitials converted by illumination to the caged configuration, and the shallow donor level close to the conduction band explains the low capture cross section for holes.
- *Thermally induced reverse annealing between 100 and 150 K*: If the sample had not been submitted to illumination, or if the illumination-induced ionization had not been complete, the concentration of free holes suffered a decrease at about 100 K.[48] The compensating center, which becomes apparent at this temperature, has been called the "two-state" defect for it can be reversibly switched between two charge states up to about 200 K.
- *At about 200 K*, the two-state defect anneals out.

The two-state defect, observed between 100 and 200 K, was found to behave as a long-lived electron trap that can be cycled between two charge states, denoted "+" and "0" (by historical reasons), with a transition level at about 0.1 K or 0.2 eV below the conduction band edge.[31,48] The very low capture cross section for holes suggests that this was a donor or even a double donor level.[31,51] The stable state in p-Ge is "+" but a burst of ionizing radiation (with electrons or x-rays) leaves the defect in the "0" state. The "0" state is recovered in a scale of minutes above 120 K or by illumination with less-than-bandgap light. At about 200 K, the "two-state" donor defect disappears with an activation energy of 0.4 eV,[48] with consequent recovery of the concentration of free holes.[48]

The P1-EPR center, which has similar properties to the two-state defect, was observed by EPR on a spin-1/2 state, in p-type Ge irradiated with electrons at 77 K.[31] It is a Rhombic-I center (of the C_{2v} point symmetry group), with the twofold symmetry axis along $\langle 100 \rangle$. The paramagnetic state was correlated with the "0" state of the two-state defect, meaning that the latter must, therefore, have an odd number of electrons.[31] This means that the true charge state of the "0" state cannot be neutral as it was originally assumed.[48] The defect in the "0" state (Trueblood's "full trap") anneals at about 120 K with an activation energy of ~0.10 eV[31] as found for the transformation of two-state defects at 126 K.[48] However, the spin-0 state ("empty trap") is still present in the samples until about 220 K, when it anneals. The activation energies reported for the 200 K anneal process was estimated to be about 0.12 eV, within a factor of two.[31] The P1 EPR center has been observed in both doped and undoped Ge[31] and is therefore considered to be an intrinsic defect. Consistently, the two-state defect detected in the electrical measurements was also found to have roughly the same formation and annealing temperatures in B-, Al-, Ga-, In-, or Tl-doped samples (although the annealing rates at about 200 K show impurity-related effects).[48]

The electronic properties and thermal stability of the two-state defect are consistent with an assignment with the self-interstitial (Figure 4.8). We have argued that the barrier from I_D to I_H prevents I^0 from transforming to I_T^{2+}, explaining the necessity of the heating stage or illumination for the defects to be revealed.

The paramagnetic charge state disappears at around 120 K, as I assumes the equilibrium charge state I^{2+}.

The stable spin-0 state is still present until 220 K and can be filled with electrons, and subsequently observed by EPR, by means of white light or a short burst of electron irradiation.[31] Its excitation is governed by a single activation energy of 0.2 eV.[48] This is entirely supported by modern first-principles calculations. The two-state defect is now believed to be the self-interstitial, which at around 120 K, the positively charged interstitial loses an electron to the conduction band ($I_{\mathrm{H/T}}^{+}$ + 0.2eV $\rightarrow I_{\mathrm{T}}^{2+} + \mathrm{e}^-$). The ionization level close to the conduction band explains the higher capture cross section for electrons (10^{-15} cm^{-2}) than for holes ($<10^{-27}$ cm^{-2}) in the empty (spin-0) state of the defect.[51] Such capture cross section for holes is too small for an uncharged defect, and suggesting a positive or even double-positive defect with ionization level far from the valence band,[48] in agreement with the self-interstitial model.

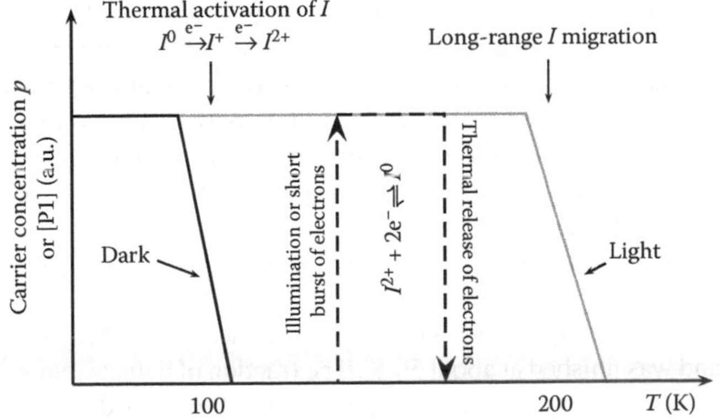

FIGURE 4.8 Self-interstitial model for the two-state defect observed in References 31, 48, 51.

A difficulty is the geometry of the defect. In the positive charge state, the equilibrium position of the caged self-interstitial lies between the H and T sites, but the reorientation barrier among the four H_d sites surrounding a T site is only 0.06 eV and comparable to the error of the calculation. The resultant effective symmetry is T_d. Since the electronic level for tetrahedral I^+ is degenerate, a further distortion can possibly arise from spin-orbit coupling or other effects not taken into account by our calculations. Thus, an assignment of the two-state defect to the self-interstitial is not inconsistent with the C_{2v} symmetry of the P1 spin-1/2 EPR center.[31] A difference to be noted between the conditions of the conductivity and EPR experiments, however, is the fact that much higher irradiation doses were used in Reference 31 to produce a density of P1 defects detectable by EPR. Nevertheless, the two-state defect seems to be a dominant center since it has been reported by different groups and observed in very different materials, including degenerate, moderately doped, and undoped p-Ge irradiated with electrons at both ~5 or ~77 K.[31,48,50,51]

The reported values for the activation energy of the 220 K annealing vary.[20,31,48] Flanagan finds the long-range anneal of I^+ or I^{2+} at 200 K to be characterized by a 0.4 eV activation energy and reports an effective frequency of $5 \times 106\ s^{-1}$,[48] characteristic of a diffusion process, suggesting trapping by the chemical acceptor in p-Ge. The calculated long-range migration barrier of the distorted I_H^+ is 0.31 eV (Table 4.2) and consistent with the 220 K anneal. Close to 200 K, neutral vacancies also become mobile,[20] and some will be captured by self-interstitials.

4.5.5 High-Purity Ge

Low-angle diffuse x-ray scattering was used by Ehrhart and Zillgen[52] to characterize the displacement fields for the distribution of intrinsic defects generated by high dose ($\phi t \approx 2 \times 10^{19}\ cm^{-2}$), 2.5 MeV irradiation in high-purity n-type germanium ($n = 1.2 \times 10^{14}\ cm^{-1}$). It was suggested that at 4 K, a large fraction of the defects is stabilized in the form of close Frenkel pairs, characterized by a nearly perfect cancelation of the long-range displacement fields of the interstitial and the vacancy. The introduction rate was measured to be 3 cm^{-1}, higher than that found by electrical methods, leading the authors to suggest that some of the defects were neutral and not detectable by conventional techniques. Although it is difficult to compare the sensitivity of such different methods, this interpretation is consistent with the present understanding that the self-interstitial is neutral in n-type Ge. Although the measured density of Frenkel pairs exceeded by about four orders of magnitude the concentration of dopants, no formation of agglomerates was observed. Thus, it was argued that the majority of the defects were in the isolated state.

Two main annealing stages were reported, at $T < 50$ K and $T > 100$ K:

- *In the $T < 50$ K range*: It was observed that annealing started already below 15 K and was finished at about 50 K. The fraction of defects removed during this low-temperature recovery stage was about 50% of the defects introduced by irradiation; it was argued that this stage was related to the recombination of close Frenkel pairs (correlated within 1–2 lattice constants).

- *From 100 K up to RT*: The annealed fraction remained approximately unaltered from 50 to 100 K. Then, from 100 K to RT, there was a second annealing stage, characterized by a smaller preexponential factor and tentatively related to the recombination of interstitial–vacancy pairs due to free migration of the interstitial and/or vacancy.

Similar experiments were performed in very lightly *p*-doped silicon irradiated with 2.5 MeV electrons.[46] Although the reported introduction rates are not higher than those expected from the comparison with electrical measurements (Table 4.4), also in that study, the estimated concentration of radiation-induced defects was about three orders of magnitude higher than that of chemical dopants, in both Czochralski- and floating-zone-Si, suggesting that most of the intrinsic defects were not trapped at impurities.[46]

4.5.6 From *n*-Type to *p*-Type Ge: Monitoring *I* Trapping with PAC Spectroscopy

Nuclear methods were used to study intrinsic defects trapped at or in the proximity of radioactive probes. Comparing two methods of Frenkel pair production, the trapping method and the neutrino recoil method, and using two methods of characterization (Mössbauer and PAC spectroscopy), a consistent picture of the formation of *V*-probe complexes was traced for a wide range of Fermi level energies across the bandgap. Also, another defect, trapped at In probes and detected by PAC spectroscopy, was suggested to be the positive self-interstitial.[20] This study provided an important piece of information about the electrical levels and migration energy of isolated *I* defects in germanium.

A defect giving rise to a quadrupole interaction $\nu_{Q2} = 415$ MHz also with axial symmetry and ⟨111⟩-oriented was identified with the self-interstitial. *Q*2 was observed in both *n*- and *p*-Ge, but not in heavily doped material. It was argued that it could not correspond to a probe in an interstitial position since it had not been produced by the neutrino recoil technique. Furthermore, it did not seem related to any impurity. If it was the self-interstitial, the trapping behavior could be explained by assuming a donor state close to the conduction band ($E_c - 0.04 \pm 0.02$ eV), and loss of trapping in heavily doped *p*-type by competitive trapping by the Ga dopants. This is likely to be one of the donor levels of the caged self-interstitial (placed at $E(+/2+) \simeq E_c - 0.1$ eV and $E(0/+) \simeq E_c - 0.2$ eV, respectively[27]). In fact, unlike the vacancy that was also observed in the same experiments, it seems that the strain interaction would make the capture of neutral self-interstitials by oversized In probes very inefficient, especially given the high density of donors present.

The findings of the PAC experiments were completely corroborated by theory. The first point of agreement is the absence of an acceptor level for the self-interstitial. As pointed out in Reference 73, while assuming the alternative picture of an acceptor level at $E_c - 0.04$ eV, the termination of the pairing in heavily doped *p*-type Ge ($p > 10^{16}$ cm^{-3}) could still be attributed to the competition between In probes and Ga centers. This leads to a problem in explaining why in *n*-Ge at Sb concentrations between 10^{17} and 10^{18} cm^{-3}—orders of magnitude higher than both the self-interstitial and probe concentration—pair formation occurs at the indium-111 (^{111}In) probes.

The $E_c - 0.04$ eV level is a donor level of the caged self-interstitial, either (0/+) or (+/+ +). According to theory, both lie in the top half of the bandgap.[27]

A curious result was that the long-range migration of both interstitials and vacancies occurred only at around 200 K. The migration energy of both defects was found to be the same and measured to be 0.5 ± 0.1 eV[74] or more recently 0.6 ± 0.1.[75] Some values reported for the ~200 K annealing stage by early experiments were slightly different.[20,31,48] According to theory, the calculated long-range migration barrier of the distorted I_H^+ is 0.3 eV, but the activation energy for migration of I_T^{2+} is very high (1.2 eV). It is thus likely that it ionized and migrated as I_H^+, with an effective migration energy that is Fermi-level dependent. An alternative explanation is that the measured activation energies are affected by the binding to the probe.

4.6 AGGREGATION OF SELF-INTERSTITIALS

4.6.1 Silicon

Multi-interstitial clusters can form either directly following irradiation with heavy particles or by diffusion and aggregation of smaller interstitial clusters. High-temperature annealing (600–800°C) results in a rapid dissolution of large self-interstitial clusters, which undergo Oswalt ripening, thereby releasing an oversaturation concentration of self-interstitials.[76] Such high concentration of self-interstitials triggers the diffusion of dopants, namely, B, a technologically detrimental phenomenon known as TED. The supersaturation concentration of the defects can be related to the respective formation energies, which determine the rates for capture and release of interstitials. Cowern et al.[76] used inverse modeling to show that self-interstitial clusters with 4 or 8 interstitials are exceptionally stable; then, for more than 10 self-interstitials, the formation energy per interstitial is approximately constant, and the defects resemble the largest rodlike "{113}" interstitial defects (Figure 4.9).

Unless they are large enough to be observed by microscopy imaging techniques, it is usually difficult to assign unambiguously an experimental defect signature to a complex intrinsic defect. By exclusion, centers that do not appear to be related to

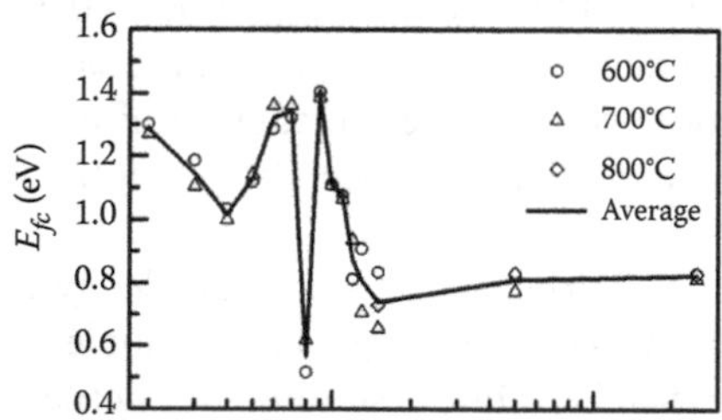

FIGURE 4.9 Experimental formation energies of interstitial clusters, estimated from the Oswalt ripening analysis. (From Cowern, N. E. B. et al. 1999. *Phys. Rev. Lett.* 82:1990.)

impurities or vacancies have been assigned to multi-interstitial defects. The assignments are usually based on their symmetry, electrical levels, and their stability or sequence of formation or annealing, that is, whether their appearance seems related to the aggregation or dissociation of other defects identified previously.

4.6.2 Di-Interstitial

Based on theoretical calculations, it is believed that I_2 is a rapidly diffusing species at room temperature.[77–79] Additionally, it is a multistable defect. Three of the most important configurations are shown in Figure 4.10.

Structure L is based on an early model for the P6 EPR center proposed by Lee et al.[80] However, this configuration is not stable. The two lowest-energy forms are K and C,[79,83] proposed by Kim et al.[84] and Coomer et al.,[6] respectively. Neither of the two lowest-energy structures has the monoclinic-II symmetry of the P6 center, and the L form actually relaxes to a structure with a {110} C_{1h} mirror plane.[6,82] This suggests that P6 is either a metastable version of I_2 or a larger interstitial complex. Most likely, I_2 has so escaped detection so far.

4.6.3 Tri-Interstitial

The tri-interstitial is also believed to be a fast-moving species.[77] Different configurations of I_3 are shown in Figure 4.11a through e. The most stable structure, I_3–IV, shown in Figure 4.11d, is one with C_2 symmetry, similar to an $S = 1$ paramagnetic center in diamond.[84,85]

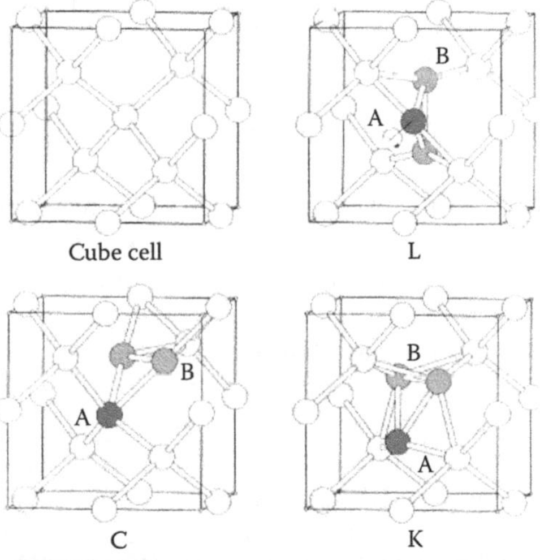

FIGURE 4.10 Di-interstitial defect models: from References 80, 81 (L), from Reference 82 (K), and from References 6, 79, 83. (Reprinted from *Nuclear Instruments Meth. Phys.* 186, Jones, R. et al., 10, Copyright 2002, with permission from Elsevier.)

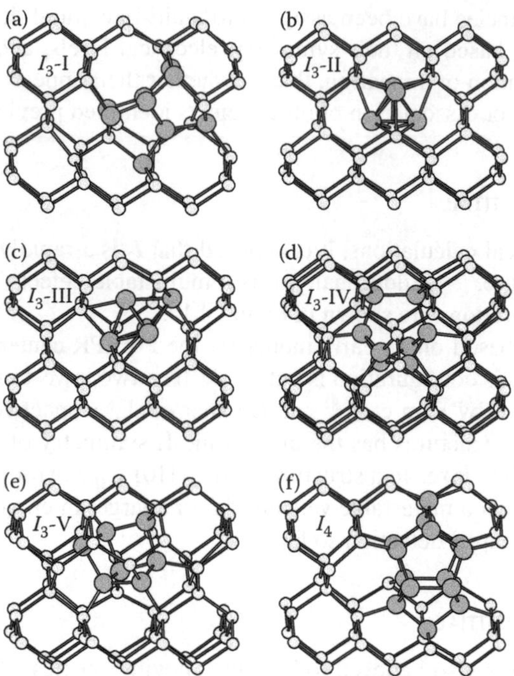

FIGURE 4.11 Models for the tri-interstitial and tetra-interstitial defects. (Reprinted with permission from Carvalho, A. et al. 2005. *Phys. Rev. B* 72:155208. Copyright 2005 by the American Physical Society.)

Most of the lowest-energy I_3 defects shown in Figure 4.11 have no levels in the gap,[86] and therefore are very difficult to detect. An exception is Coomer's I_3–I structure (Figure 4.11a). This consists of a {111}-habiting three-member ring of Si interstitials, centered near the tetrahedral interstitial site.[87] This structure has a deep donor level close to the valence band, and it has been linked to the *W* photoluminescence (PL) line.

The *W* PL line with zero-phonon lines at 1.0182 eV has been proposed as originating from interstitial aggregates and has been linked with a trigonal form of I_3, respectively.[87,88] The evidence for an interstitial origin to the *W*-line is quite strong. The defect is formed in most types of irradiated Si and has not then been connected with any impurity, and stress studies show that the symmetry is trigonal.[89] Moreover, phonon replicas of the ZPL show that the defect possesses a local vibrational mode (LVM) lying at 70.0 meV and above the Raman frequency. This mode shifts by –2.3 meV in ^{30}Si isotopically enriched crystals.[92] Such a shift is expected for a vibration involving Si atoms only, and the modes must arise from compressed Si–Si bonds in order to lie above the Raman edge. Vacancy defects are often reconstructed and lead to weak bonds, which could not possess a Si-related LVM.

The donor level of the I_3–I structure is consistent with the energy of the *W*-line. Moreover, it has three A_1 LVMs close to one reported for the *W*-center.[86] The shift of the donor level in SiGe is also consistent with I_3–I.[91] However, this structure is now believed to be only metastable, about 1.7 eV higher in energy than the lowest-energy structure.[77,86] A possibility is that the lowest-energy forms, which are very mobile, are not detectable, and only this metastable form gives rise to luminescence.

4.6.4 Tetra-Interstitial

At the temperature the *W*-line anneals out, another PL line, named "*X*," appears.[90] This line, at 1.0397 eV, has been assigned to the tetra-interstitial.

Examining the I_4 structure in Figure 4.11f,[92,93] it is easy to understand why four is a magic number in self-interstitial aggregation. This I_4 structure preserves the fourfold coordination of all atoms, and the bond lengths and angles are very close to bulk silicon. Thus, unlike I_2 and I_3, it is very stable and there are no other I_4 forms with comparable energy.

The *X*-line is one of the best characterized intrinsic defect centers. Stress-splitting studies suggested D_{2d} symmetry.[94] This unusual symmetry is the same as the I_4 defect. The donor state of I_4 is close to the valence band, consistent with the energy of the PL line.

Further, in natural Si, the *X*-line system has three LVMs at 66.2, 67.9, and 69.0 meV, and in ^{30}Si, they shift by −2.1, −2.3, and −2.1 meV, respectively, relative to the zero-phonon position. The existence of LVMs in this frequency range, just as in the *W*-line, suggests an interstitial origin to the *X*-line. I_4 has, in fact, several modes with similar frequencies.[86] Two of them have A_1 symmetry, and the third LVM observed in PL spectra seems to arise from the anharmonic coupling between two *E* modes whose frequencies are also very close to the experimental value.

The B3 EPR center has also been related to I_4^+[95–97]. The defect has an unusually high (D_{2d}) symmetry, and hyperfine interactions with shells of 2, 4, and 8 Si neighbors have been observed. This is exactly the coordination of Si shells for the I_4^+ defect. The ground state of the neutral I_4 defect has $S = 0$ and is then nonparamagnetic. Under illumination with light of energy around 1.03 eV, another EPR center, NL51, with $S = 1$ has been detected in proton-irradiated Si.[98,99] NL51 also has D_{2d} symmetry and its stability is similar to that of the B3 EPR center. This represents a triplet excited state and suggests that the *X*-line at 1.0398 eV results from the decay of the corresponding singlet excitonic state. The observation of an EPR signal due to I_4^+ demonstrates that the defect has a deep donor level. Early reports placed this at $E_v + 0.29$ eV,[100] but this location seems inconsistent with the energy of the *X*-line. A vibrational mode (582 cm^{-1}) detected by localized vibrational mode spectroscopy on neutron-irradiated silicon has also been attributed to a cluster with four or less interstitials,[101] most likely I_4.

4.7 GERMANIUM

In contrast with Si, where the isolated self-interstitial becomes easily trapped in impurities thereby frustrating detection, the multi-interstitial defects are well characterized; in germanium, where isolated self-interstitials have been detected, multi-interstitial clusters and complexes of self-interstitials with impurities are very difficult to nail down. One of the main reasons is the higher formation energy of the neutral self-interstitial, compared with the vacancy,[26] which renders the formation of interstitial defects more unlikely at high temperature.

Until recently, the only exception was the self-interstitial passivated by two hydrogen atoms (IH_2). Created by proton or deuteron implantation at 30 K and subsequent RT storage, this well-characterized defect has been observed by IR absorption spectroscopy and its identification was established with the aid of uniaxial stress measurements and *ab initio* calculations.[102]

It is thus reasonable to expect that trapped self-interstitials also exist in electron-irradiated material. According to the picture derived from the neutrino-recoil PAC spectroscopy experiments, when the self-interstitial becomes mobile at about 200–230 K in *p*-Ge, some interstitials recombine with the vacancies, but a fraction will escape recombination by forming more stable defect complexes.[103] That self-interstitials do exist in germanium at higher temperatures can also be inferred from the kinetics of boron diffusion.[104,105]

4.7.1 Di-Interstitial

The higher formation energy may be one of the reasons why small self-interstitial clusters also remain undetectable in Ge. According to Birner et al.,[106] the di-interstitial is multistable with at least four competing structures with close formation energies. All these are possibly donors with levels close to the valence band top. A defect with a level at E_v +0.138 eV, detected by DLTS in alpha-particle-irradiated *n*-type Ge, was tentatively assigned to a di-interstitial defect.[107,108] This defect, stable at room temperature, is thought to annihilate the divacancy.

In electron-irradiated oxygen-doped germanium, $\{I_2,O\}$ complexes have been detected at about 200 K, but not all of them are stable up to RT.[109]

4.7.2 Tri-Interstitial and Larger Defects

Modeling of tri-interstitial configurations in germanium has been more limited than in Si.[106] The *W* PL line is no longer found in $Si_{1-x}Ge_x$ alloys with $x > 5\%$.[93,110] Based on the assignment of the *X*- and *W*-lines to the I_4 and I_3 defects represented in Figure 4.11a and f, this is readily justified as the donor levels for both structures merge with the valence band with increasing Ge fraction.[91] As in Si, however, Coomer's I_3 (Figure 4.11a) is metastable.[106]

Similarly, the extended {311} planar interstitial aggregates that form above 700°C in Si after nonamortizing Si^+ implantation do not seem to form in $Si_{1-x}Ge_x$ with $x > 25\%$[111] or in Ge in the same conditions.[112] Rather, with increasing germanium content, the material seems to skip this stage of interstitial aggregation directly forming dislocation loops.[111] Nevertheless, the formation of extended defects in germanium appears to depend on the irradiation parameters, and extended interstitial-type defects in the {100} plane[113] as well as other planar defects[114,115] have been observed in deuteron- and hydrogen-implanted material. Obviously, the conditions determining the extent and the products of aggregation are still unclear. The involvement of morphs on the formation of extended defect structures is also still an open issue.

4.8 SUMMARY

Self-interstitials have attracted much research over the years, both for their compelling complexity and for their importance in explaining many of the defect-related phenomena at low and high temperature. The self-interstitial in silicon forms many complexes with impurities, and the reports on its observation on the isolated state have been very few. In contrast, in Ge, where Frenkel pairs can be stable up to 65 K in the dark, the self-interstitials have been characterized by a variety of techniques. One of the most intriguing findings during the early days of the study of irradiation damage in Si and Ge has been the asymmetry between *n*-Si and *p*-Si, and *n*-Ge and *p*-Ge. Many of these observations can now be understood on the basis of the different structures and ionization levels of the self-interstitial and vacancy in both materials.

Nevertheless, some mysteries remain unsolved. A unified understanding of how the behavior of the self-interstitial at low temperature relates to its behavior at higher temperature is one of the points where discussion remains open. Additionally, several intrinsic defects that are thought to be multi-interstitial complexes await a definite identification.

We conclude by remarking how the silicon self-interstitial is an illustrative example of a problem where theory and experiment have worked side by side to reveal the multiple dimensions of the microscopic reality.

REFERENCES

1. Rafferty, C. S., G. H. Gilmer, M. Jaraiz, D. Eaglesham, and H. J. Gossmann. 1996. *Appl. Phys. Lett.* 68:2395.
2. Srour, J. R. and C. J. Marshall. 2003. *IEEE Trans. Nuclear Sci.* 50:653.
3. Masri P., A. H. Harker, and A. M. Stoneham. 1983. *J. Phys. C: Solid State Phys.* 16:L613.
4. Car, R., P. J. Kelly, A. Oshiyama, and S. T. Pantelides. 1984. *Phys. Rev. Lett.* 20:1814.
5. Chadi, D. J. 1992. *Phys. Rev. B* 46:9400.
6. Coomer, J. 2000. Ph.D. Thesis, University of Exeter.
7. Bar-Yam, Y. and J. D. Joannopoulos. 1984 *Phys. Rev. B* 30:2216.
8. De Souza, M. M., C. K. Ngw, M. Shishkin, and E. M. S. Narayanan. 1999. *Phys. Rev. Lett.* 83:1799.
9. De Souza, M. M. and J. Goss. 2005. *Defects Diffusion Forum* 245–246:29.
10. Needs, R. J. 1999. *J. Phys.: Condens. Matter* 11:10437.
11. Lopez, G. M. and V. Fiorentini. 2004. *Phys. Rev. B* 69:155206.
12. Lee, W. C., S.-G. Lee, and K. J. Chang. 1998. *J. Phys: Condens. Matter* 10:995.
13. Eberlein, T. A. G., N. Pinho, R. Jones et al. 2001. *Physica B* 308–310:454.
14. Leung, W.-K., R. J. Needs, G. Rajagopal, S. Itoh, and S. Ihara. 1999. *Phys. Rev. Lett.* 83:2351.
15. Caliste, D., P. Pochet, T. Deutsch, and F. Lançon. 2007. *Phys. Rev. B* 75:125203.
16. Rinke, P., A. Janotti, M. Scheffler, and C. G. Van de Walle. 2009. *Phys. Rev. Lett.* 102:026402.
17. Watkins, G. D. 1992. The lattice vacancy in silicon, in *Deep Centers in Semiconductors*, ed. by S. T. Pantelides. New York: Gordon and Breach, 2nd edition, p. 177.
18. Watkins, G. D. 2000. *Mater. Sci. Semicond. Proc.* 3:227.
19. Bourgoin J. and M. Lannoo. 1983. *Point Defects in Semiconductors*. Berlin: Springer.
20. Haesslein, H., R. Sielemann, and C. Zistl. 1998. *Phys. Rev. Lett.* 80:2626.
21. Bourgoin, J. C., P. M. Mooney, and F. Poulin. 1981. *Inst. Phys. Conf. Series* 59:33.

22. Mesli, A., L. Dobaczewski, K. Bonde Nielsen, V. I. Kolkovski, M. C. Petersen, and A. Nylandsted Larsen. 2008. *Phys. Rev. B* 78:165202.
23. Alloyeau, D., B. Freitag, S. Dag, L. W. Wang, and C. Kisielowski. 2009. *Phys. Rev. B* 80:014114.
24. da Silva, A. R. J., A. Janotti, A. Fazzio, R. J. Baierle, and R. Mota. 2000. *Phys. Rev. B* 62:9903.
25. Moreira, M. D., R. H. Miwa, and P. Venezuela. 2004. *Phys. Rev. B* 70:115215.
26. Vanhellemont, J., P. Śpiewak, and K. Sueoka. 2007. *J. Appl. Phys.* 101:036103.
27. Carvalho, A., R. Jones, J. Goss et al. 2007. *Phys. Rev. Lett.* 99:175502.
28. Al-Mushadani, O. K. and R. J. Needs. 2003. *Phys. Rev. B* 68:235205.
29. Jones, R., A. Carvalho, J. P. Goss, and P. R. Briddon. 2009. *Mater. Sci. Eng. B* 159–160:112.
30. Mukashev, B. N., Kh. A. Abdullin, and Yu. V. Gorelkinskii. 1998. *Phys. Stat. Solidi (a)* 168:73.
31. Trueblood, D. L. 1967. *Phys. Rev.* 161:828.
32. Carvalho, A., R. Jones, C. Janke et al. 2008. *J. Phys. Condens. Matter* 20:135220.
33. Bourgoin, J. C. and J. W. Corbett. 1972. *Phys. Lett.* 38A:135.
34. Corbett, J. W. 1981. *Nucl. Instrum. Meth.* 182–183:457.
35. Spiewak, P., J. Vanhellemont, K. Sueoka, K. J. Kurzydlowski, and I. Romandic. 2008. *Mater. Sci. Semicond. Proc.* 11:328–331.
36. Cowern, N. E. B., S. Simdyankin, C. Ahn et al. 2013. *Phys. Rev. Lett.* 110:155501.
37. Seeger, A. and K. P. Chik. 1968. *Phys. Status Solidi* 29:455.
38. Watkins, G. D. 2008. *J. Appl. Phys.* 103:106106.
39. Caliste, D. and P. Pochet. 2006. *Phys. Rev. Lett.* 97:135901.
40. Rhodes, R. G. 1964. *Imperfections and Active Centres in Semiconductors*, International series of monographs on semiconductors Vol. 6. Oxford: Pergamon Press.
41. Vavilov, V. S. 1965. *Effects of Radiation on Semiconductors*. New York: Consultants Bureau.
42. MacKay, J. W. and E. E. Klontz. 1971. in *Irradiation Defects in Semiconductors*, Proceedings of the International Conference, Albany, USA, ed. by J. W. Corbett and G. D. Watkins. New York: Gordon and Breach Scientific Publishers.
43. Sivo L. L. and E. E. Klontz. 1969. *Phys. Rev.* 178:1264.
44. Watkins, G. D. 1963. *J. Phys. Soc. Japan* 18(Supplement II):22.
45. Watkins, G. D., J. R. Troxell, and A. P. Chatterjee. 1979. *Inst. Phys. Conf. Ser.* 46:16.
46. Zillgen, H. and P. Ehrhart. 1996. *Nucl. Instr. Meth. B* 127:27
47. Emtsev, V. V., P. Ehrhart, K. V. Emtsev, D. S. Poloskin, and U. Dedek. 2006. *Physica B* 376–377:173.
48. Flanagan, T. M. and E. E. Klontz. 1967. *Phys. Rev.* 167:789.
49. MacKay, J. W. and E. E. Klontz. 1959. *J. Appl. Phys.* 30:1269.
50. Whitehouse, J. E. 1966. *Phys. Rev.* 143:520.
51. Brown, W. L., W. M. Augustyniak, and T. R. Waite. 1959. *J. Appl. Phys.* 30:2158.
52. Ehrhart, P. and H. Zillgen. 1999. *J. Appl. Phys.* 85:3503.
53. Hyatt W. D. and J. S. Kehler. 1971. *Phys. Rev. B* 4:1903.
54. Abdullin, Kh. A., B. N. Mukashev, and Yu. V. Gorelkinskii. 1996. *Semicond. Sci. Technol.* 11:1696–1703.
55. Wertheim, G. K. 1959. *Phys. Rev.* 115:568.
56. Emtsev, V. V., T. V. Mashovets, and V. V. Mikhnovich. 1992. *Fiz. Tekh. Poluprovodn.* 26:20 1992 [1992. *Soviet. Phys. Semicond.* 26:12].
57. Mooney, P. M., F. Poulin, and J. C. Bourgoin. 1983. *Phys. Rev. B* 28:3372.
58. Christian Petersen M., C. E. Lindberg, K. Bonde Nielsen, A. Mesli, and A. Nylandsted Larsen. 2006. *Mater. Sci. Semicond. Proc.* 9:597.
59. Vook, F. L. 1965. *Phys. Rev.* 138:A1234.

60. Calcott, T. A. and J. W. Mackay. 1967. *Phys. Rev. B* 161:698.
61. Zizine, J. 1968. in *Radiation Effects in Semiconductors*, Proceedings of the Conference, Santa Fe, USA, ed. by F. L. Vook. New York: Plenum Press, p. 186.
62. Singh M. P. and J. W. MacKay. 1968. *Phys. Rev.* 175:985.
63. Meese, J. M. 1974. *Phys. Rev. B* 9:4373.
64. Bourgoin J. and F. Mollot. 1971. *Phys. Stat. Sol. (b)* 43:343.
65. Carvalho A., R. Jones, J. Goss et al. 2007. *Physica B* 401402:495.
66. Whan, R. E. 1968. in *Radiation Effects in Semiconductors*, Proceedings of the Conference, Santa Fe, USA, ed. by F. L. Vook, New York: Plenum Press, p. 195.
67. MacKay, J. M. 1967. in *Action des Rayonements sur les Composants a Semiconduteours*, Prooceedings of the International Colloquium, Toulouse, France, 1967 (Journées d'Electronique, France, 1967), p. A2–1
68. Fletcher, R. C. and W. L. Brown. 1193. *Phys. Rev.* 92:585.
69. Waite, T. R. 1957. *Phys. Rev.* 107:471.
70. Brucker, G. J. 1969. *Phys. Rev.* 183:712.
71. Stein, H. J. and F. L. Vook. 1967. *Phys. Rev.* 163:790.
72. Whitehouse, J. E. 1971. *J. Phys. Chem. Solids* 32:677.
73. Sielemann, R., H. Haesslein, and Ch. Zistl. 2001. *Physica B* 302–303:101.
74. Sielemann, R. 1998. *Nucl. Inst. Methods Phys. Res.* 146:329.
75. Sielemann, R., H. Haesslein, Ch. Zistl, M. Müller, L. Stadler, and V. V. Emtsev. 2001. *Physica B* 308–310:529.
76. Cowern, N. E. B., G. Mannino, P. A. Stolk, and F. Roozeboom. 1999. *Phys. Rev. Lett.* 82:1990.
77. Estreicher, S. K., M. Gharaibeh, P. A. Fedders, and P. Ordejón. 2001. *Phys. Rev. Lett.* 86:1247.
78. Du, Y. A., R. G. Hennig, and J. W. Wilkins. 2006. *Phys. Rev. B* 73:245203.
79. Eberlein, T. E. G., N. Pinho, R. Jones et al. 2001. *Physica B* 308:454.
80. Lee, Y.-H., N. N. Gerasimenko, and J. W. Corbett. 1976. *Phys. Rev. B* 14:4506.
81. Lee, Y. H. 1998. *Appl. Phys. Lett.* 73:1119.
82. Kim, J., F. Kirchoff, W. G. Aulbur, J. W. Wilkins, F. S. Khan, and G. Kresse. 1999. *Phys. Rev. Lett.* 83:1990.
83. Jones, R., T. A. G. Eberlein, N. Pinho et al. 2002. *Nuclear Instruments Meth. Phys.* 186:10.
84. Goss, J. P., B. J. Coomer, R. Jones et al. 2004. *Phys. Rev. B* 63:195208.
85. Twitchen, D. J., D. C. Hunt, M. E. Newton et al. 1999. *Physica B* 273–274:628.
86. Carvalho, A., R. Jones, J. Coutinho, V. J. B. Torres, and P. R. Briddon. 2005. *Phys. Rev. B* 72:155208.
87. Coomer, B. J., J. P. Goss, R. Jones, S. Oberg, and P. R. Briddon. 1999. *Physica B* 274:505.
88. Coomer, B. J., J. P. Goss, R. Jones, S. Oberg, and P. R. Briddon. 2001. *J. Phys.: Condens. Matter* 13:L1.
89. Davies, G. E., C. Lightowlers, and Z. E. Ciechanowska. 1987. *J. Phys. C: Solid State Phys.* 20:191.
90. Hayama, S., G. Davies, and K. M. Itoh. 2004. *J. Appl. Phys.* 96:1754.
91. Leitão, J. P., A. Carvalho, J. Coutinho et al. 2011. *Phys. Rev. B* 84:165211.
92. Humble, P. 1982. *Proc. R. Soc. Lond. A* 381:65.
93. Arai, N., S. Takeda, and M. Kohyama. 1997. *Phys. Rev. Lett.* 78:4265.
94 Ciechanowska, Z., G. Davies, and E. C. Lightowlers. 1984. *Solid State Commun.* 49:427.
95. Brower, K. L. 1976. *Phys. Rev. B* 14:872.
96. Mchedlidze T. and M. Suezawa. 2003. *J. Phys.: Condens. Matter* 15:3683.
97. Pierreux, D. and A. Stesmans. 2003. *Phys. Rev. B* 68:193208.
98. Mchedlidze, T. and M. Suezawa. 2003. *Physica B* 340–342:682.

99. Stallinga, P., T. Gregorkiewicz, and C. A. J. Ammerlaan. 1994. *Solid State Commun.* 90:401.
100. Mukashev, B. N., A. V. Spitsyn, N. Fukuoka, and H. Saito. 1982. *Japan. J. Appl. Phys.* 21:399.
101. Londos, C. A., G. Antonaras, and A. Chroneos. 2013. *J. Appl. Phys.* 114:043502.
102. Budde, M., B. Bech Nielsen, R. Jones, J. Goss, and S. Öberg. 1996. *Phys. Rev. B* 54:5485.
103. Sielemann, R., H. Haesslein, L. Wende, and Ch. Zistli. 1999. *Physica B* 273–274:565.
104. Janke, J., R. Jones, S. Öberg, and P. R. Briddon. 2008. *Phys. Rev. B* 77:075208 2008.
105. Uppal, S., A. F. W. Willoughby, J. M. Bonar et al. 2004. *J. Appl. Phys.* 96:1376.
106. Birner, S., J. P. Goss, R. Jones, and P. R. Briddon. 2000. Stockholm: ENDEASD Conference Proceedings.
107. Petersen, M. C. 2010. PhD thesis, Denmark: Aarus University.
108. Petersen J. W. and J. Nielsen. 1990. *Appl. Phys. Lett.* 56:1122.
109. Khirunenko, L. I., Yu.V. Pomozov, M. Sosnin et al. 2008. *Mater. Sci. Semicond. Proc.* 11:332.
110. Tan, J., G. Davies, S. Hayama, and A. N. Larsen. 2007. *Appl. Phys. Lett.* 90:041910.
111. Crosby, R. T., K. S. Jones, M. E. Law, A. Nylandsted Larsen, and J. Lundsgaard Hansen. 2003. *Mater. Sci. Semicond. Proc.* 6:205.
112. Hickey, D. P., Z. L. Bryan, K. S. Jones, and R. G. Elliman, E. E. Haller. 2008. *J. Vacuum Sci. Tech. B* 26:425.
113. Muto, S. and S. Takeda. 1995. *Phil. Mag. Lett.* 72:99.
114. Akatsu, T., K. K. Bourdelle, C. Richtarch, B. Faure, and F. Letertre. 2005. *Appl. Phys. Lett.* 86:181910.
115. David, M. L., F. Pailloux, D. Babonneau et al. 2007. *J. Appl. Phys.* 102:096101.
116. Chantre A., D. Bois. 1985. *Phys Rev B* 15:7979.
117. Lindberg, C. E., J. Lundsgaard Hansen, P. Bomholt et al. 2005. *Appl. Phys. Lett.* 87:172103.

5 Vacancies in Si and Ge

Eiji Kamiyama, Koji Sueoka, and Jan Vanhellemont

CONTENTS

5.1 Introduction ... 119
5.2 Calculation Methods for Vacancy Properties ... 120
5.2.1 *Ab Initio* ... 120
5.2.2 Molecular Dynamics ... 121
5.3 Vacancy Properties ... 122
5.3.1 Formation Energy ... 122
5.3.2 Vacancy Cluster Formation ... 124
5.4 Parameters Influencing Vacancy Properties ... 130
5.4.1 Internal and External Stress ... 130
5.4.2 Impurities Interacting with Vacancies ... 133
5.4.3 Surfaces, Interfaces, and Capping Layers ... 137
5.5 Impact of Vacancies on Material Properties and Vacancy Engineering: Two Case Studies ... 146
5.5.1 Oxygen Precipitation ... 146
5.5.2 Annealed Wafers ... 147
5.6 Conclusions and Outlook ... 154
References ... 154

5.1 INTRODUCTION

In this chapter, an overview is provided on recent progress in understanding the properties and impact of vacancies in silicon and germanium. Vacancies play an important role in a wide variety of defect formation processes in silicon and germanium, such as single crystal growth from a melt, wafer properties engineering, interstitial oxygen precipitate nucleation and growth, and dopant activation and diffusion.

It is not straightforward to assess the vacancy formation and migration energy through experiment as the vacancy interacts with nearly all impurities in the semiconductor lattice and, of course, also with its counterpart, the self-interstitial. Even more complicated is the fact that both intrinsic point defects can occur in different charge states making also it that the Fermi level will have an important impact. For this same reason, a very wide range of vacancy properties extracted from a plethora of experiments ranging from crystal growth, irradiation, dopant activation and diffusion, and metal diffusion can be found in the literature.

During the last decade, *ab initio* and molecular dynamics (MD) calculation techniques to calculate the properties of point defects in silicon and germanium have made tremendous progress and have become an indispensable tool for defect understanding and engineering.

The first part of the chapter provides an overview of recent results obtained by the authors with respect to the calculation of the impact of stress, doping and surfaces on vacancy properties, as well as the implications for single crystal growth from a melt. Also, the use of MD calculation to simulate the first stages of vacancy clustering in silicon and germanium is illustrated. The results indicate that *ab initio* calculation will play an important role in the development of more advanced crystal growth process as needed for the new family of 450-mm-diameter silicon crystals or for vacancy-poor, large-diameter germanium crystals.

The second part of the chapter presents two case studies in which vacancies play a key role, that is, interstitial oxygen precipitation in silicon and the development of annealed silicon wafers. Vacancies are indeed the origin of crystal-originated particles (COPs), which play a crucial role in the nucleation of interstitial oxygen precipitates and in the formation of the oxidation-induced stacking fault (OSF) ring and reactive ion etching (RIE) defects in a Si crystal. Ultra-high-temperature rapid thermal annealing (RTA) is the most advanced technique to produce annealed wafers (AWs) and is based on vacancy engineering and controlling the concentration of interstitial oxygen. The recent progress in AWs closely follows the trends in the development of electronic devices.

5.2 CALCULATION METHODS FOR VACANCY PROPERTIES

5.2.1 *Ab Initio*

There are several *ab initio* calculation methods depending on the basis of function and/or the approximation of electron exchange and correlation. One of the most successful approaches is the density functional theory (DFT), in which the Kohn–Sham equation is solved self-consistently using the local density approximation (LDA) or the generalized gradient approximation (GGA) for electron exchange and correlation. Since the maximum thermal equilibrium concentration of vacancies V in Si or Ge crystals is of the order of $1 \times 10^{15}/cm^3$, special care has to be taken to obtain reliable values for the vacancy properties. The technique most commonly used is (1) the supercell method with a three-dimensional periodic boundary condition, (2) plane waves for the basis of wave functions, and (3) the (ultra-soft) pseudopotential method to reduce the number of plane waves. Geometry optimization around the vacancy is very important to obtain a reliable value for the formation energy.

Figures 5.1a and b show the geometries of a single vacancy and a split vacancy. The split vacancy is a transition state of V in Si or Ge crystals. The Jahn–Teller distortion of V as shown in Figure 5.2 is obtained by using a supercell containing at least 128 atoms in the case of Si.

For neutral vacancies in Si, the calculated formation energies with Jahn–Teller distortion by GGA lie mostly between 3.2 and 3.7 eV.[1] By using more sophisticated functions such as sx-LDA or HSE06, the formation energy of the neutral V increases to around 4.3–4.4 eV.[2] For neutral vacancies in Ge, the formation energies obtained by GGA lie between 2.0 and 2.4 eV,[3] while the use of HSE06 yields about 2.9 eV.[4]

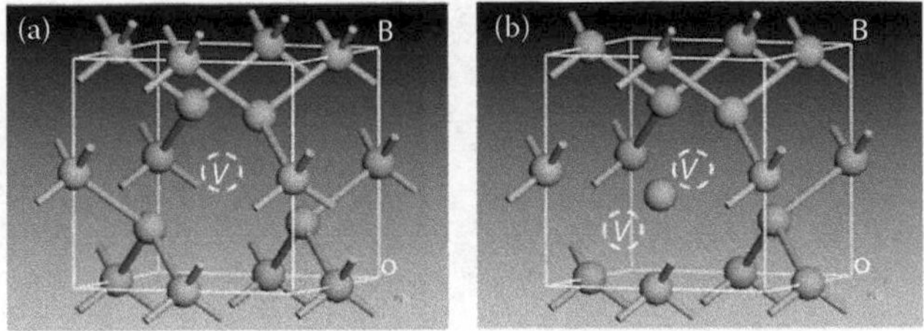

FIGURE 5.1 (a) Geometry of a vacancy and (b) geometry of a split vacancy. (From K. Sueoka, E. Kamiyama and H. Kariyazaki, *Journal of Applied Physics* **111**, 2012. With permission.)

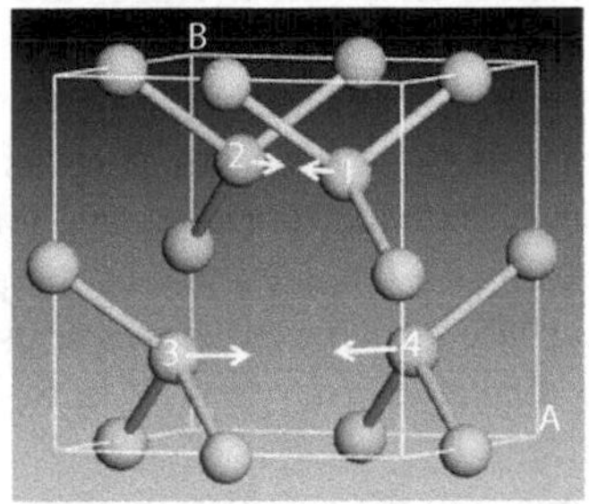

FIGURE 5.2 Crystal structure around the unrelaxed vacancy showing the first (1, 2, 3, and 4) and second (the other atoms) nearest-neighboring atoms of the vacancy. The arrows indicate the pairing of the first nearest neighbors by Jahn–Teller distortion. (From K. Sueoka, E. Kamiyama and H. Kariyazaki, *Journal of Applied Physics* **111**, 2012, 093529. With permission.)

The charge state of the vacancy in Si or Ge crystals can be evaluated by using "charged cell" with adding/removing one or two electrons. In this case, some kind of corrections, for example, the charge correction between image cells, should be considered. Neutral V is the most stable for intrinsic Si, while negatively charged V is the most stable for intrinsic Ge.[5]

5.2.2 Molecular Dynamics

MD is another extensively used calculation tool to evaluate vacancy properties and behavior in Si or Ge crystals. In the most common version, the trajectories of atoms in the simulation cell are determined by numerically solving the Newton's equations of motion. In classical MD, the forces between the atoms and the associated potential energy are determined by empirical potential functions. For crystals with the diamond structure such as Si and Ge, the Stillinger–Weber (SW) and Tersoff potentials are commonly used. The number of atoms in classical MD can be several orders of magnitude higher than that used in *ab initio* calculations as shown in Figure 5.3. Furthermore, it is the great merit of MD that it can deliver information of the dynamics of atoms for finite temperatures. The results of MD simulations can, therefore, be used to estimate macroscopic thermodynamic properties of a system based on the

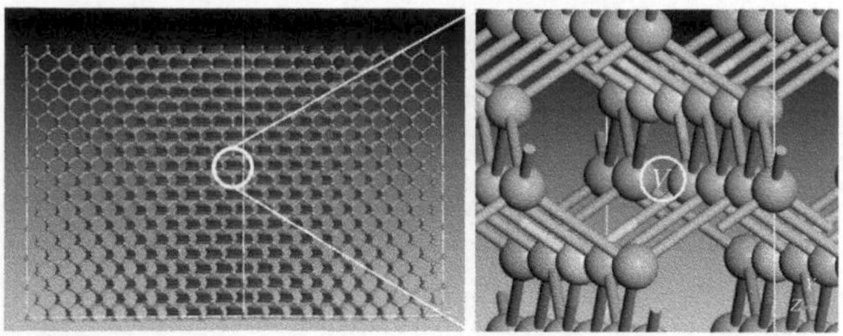

FIGURE 5.3 8000 Si atoms cell containing one vacancy.

ergodic hypothesis: the statistical ensemble averages are equal to the time averages of the system.

Since classical MD does not consider the electrons in the crystal, the configuration related to the electronic state of the system, such as the Jahn–Teller distortion around a vacancy, cannot be reproduced. Furthermore, it is obvious that the reliability of the MD results depending on the accuracy of the potential function. In spite of these limitations, classical MD is also used to derive useful information about the vacancy in Si and Ge. The formation energy of *V* in Si was estimated to be about 2.7 eV using the SW potential.[6]

Owing to the drastic decrease in the calculation costs and increase of calculation speed, *ab initio MD* will become a prominent tool to simulate vacancy dynamics in the near future.

5.3 VACANCY PROPERTIES

5.3.1 Formation Energy

By using *ab initio* or MD simulation, the formation energy of the vacancy E_f^V can be evaluated with the following equation:

$$E_f^V = E_{\text{tot}}[\text{Si}_{N-1} V_1] - \frac{N-1}{N} E_{\text{tot}}[\text{Si}_N]. \tag{5.1}$$

Here, $E_{\text{tot}}[\text{Si}_{N-1} V_1]$ is the total energy of the cell including one *V*. $E_{\text{tot}}[\text{Si}_N]$ is the total energy of the perfect cell without point defect. The migration energy E_m^V is obtained by the change of E_f^V at the transition state and E_f^V at the most stable state. Figure 5.4 shows a range of estimates for the neutral vacancy formation and migration energies computed at various levels of approximation for the total energy description until 2001.[7] The formation energies computed with *ab initio* and tight-binding are clustered between 3.5 and 4.0 eV with an average of 3.7 eV.

More recent *ab initio* simulations by using larger supercells yield formation energies between 3.2 and 3.7 eV using GGA.[1] More sophisticated functions, such as sx-LDA or HSE06, yield formation energies between 4.3 and 4.4 eV.[2] MD calculation results are somewhat more scattered and using the SW potential leads to a lower estimate of about 2.7 eV.[6] The migration energies computed with *ab initio* and MD

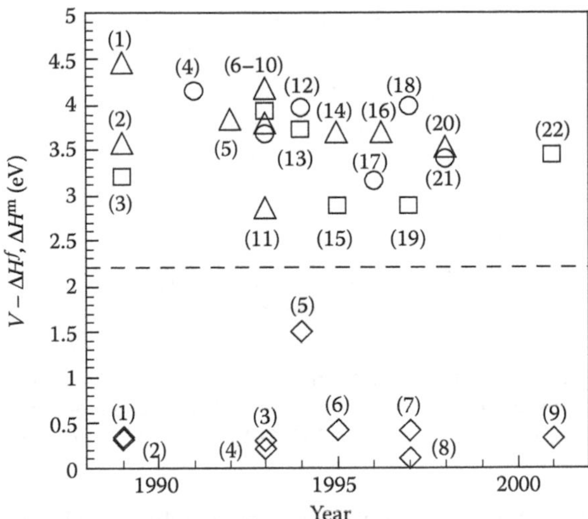

FIGURE 5.4 Formation (upper dataset) and migration energies (lower dataset) for the neutral V computed by atomic simulation. The numbers in brackets are references. (From T. Sinno, Thermophysical properties of intrinsic point defects in crystalline silicon, *Semiconductor Silicon 2002, ECS Proceedings Vol. 2002-2*, 2002, 212. Reproduced by permission of The Electrochemical Society.)

are in good agreement with values between 0.1 and 0.45 eV. Recent values using larger supercells are also in this range.

Figure 5.5 shows the dependence of the V formation energy on the hydrostatic pressure P.[8–10] It was found that E_f^V is almost constant for pressures between −1 and 1 GPa. The formation energies with 216-atom supercell are 3.543 eV for the Jahn–Teller distorted V (h-JT) and 3.791 eV for the split-V at $P = 0$. The migration energy of V is thus 0.248 eV at $P = 0$.

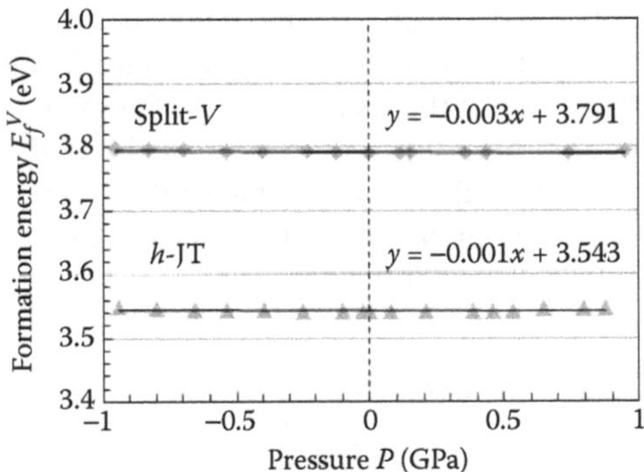

FIGURE 5.5 Calculated dependence of formation energy E_f^V of vacancies V on pressure P between −1 GPa (tensile) and 1 GPa (compressive). (From K. Sueoka, E. Kamiyama and J. Vanhellemont, *Journal of Crystal Growth* **363**, 2013, 97. With permission.)

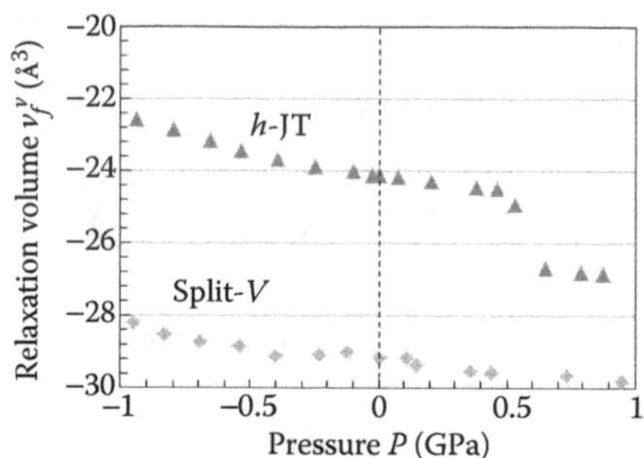

FIGURE 5.6 Calculated dependence of relaxation volume v_f^V of vacancies V on pressure P between −1 GPa (tensile) and 1 GPa (compressive). (From K. Sueoka, E. Kamiyama and J. Vanhellemont, *Journal of Crystal Growth* **363**, 2013, 97. With permission.)

The relaxation volume v_f^V is the volume change with respect to the perfect crystal by V formation. This value can be used to estimate the formation enthalpies of V under stress (pressure). The relaxation volumes V_f^V of Jahn–Teller distorted V (*h*-JT, ground state) and of split-V (transition state) decrease with increasing compressive P and increase with increasing tensile P as shown in Figure 5.6.[8–10]

It is interesting that the formation volume for the vacancy in Si (=atomic volume of Si + V relaxation volume) becomes negative under hydrostatic pressure. The relaxation volume is −24.105 Å^3 of *h*-JT and −28.062 Å^3 of split-V, at $P = 0$. Several negative values of formation volumes for V with Jahn–Teller distortion were reported as −20.8 Å^3,[11] −22.4 Å^3,[12] −22.1 Å^3,[13] and −27.1 Å^3.[14]

The charge state of the vacancy in Si crystals can be evaluated by using a "charged cell," by adding or removing one or two electrons. In this case, some corrections have to be performed, such as a charge correction between image cells. As an illustration, Figure 5.7 shows the GGA energies of the Si vacancy as a function of Fermi level.[15] The uncharged V is the most stable one for intrinsic Si, while V^{2+} (V^{2-}) is the most stable one in heavily doped *p* (*n*)-type Si.

For Ge, LDA and GGA fail to reproduce the semiconductor state as no bandgap is obtained. So, a more sophisticated approach is needed to be able to calculate the Ge bandgap and the charged state of the vacancy. Figure 5.8 shows the formation energy of the vacancy in Ge as a function of the Fermi level and the charge state calculated using the hybrid functional HSE06 and the LDA + U method.[4] The negatively charged V has the lowest formation energy in intrinsic and *n*-type Ge, while this is the case for the neutral V in *p*-type Ge.

5.3.2 Vacancy Cluster Formation

On the one hand, the properties of small vacancy clusters are difficult to determine, experimentally. On the other hand, it is also difficult to study these properties using *ab initio* and tight-binding atomistic calculations, due to large computational efforts required for these techniques, in the case of large atomistic domains. These limitations can be circumvented by classical MD calculations, based on an SW potential,

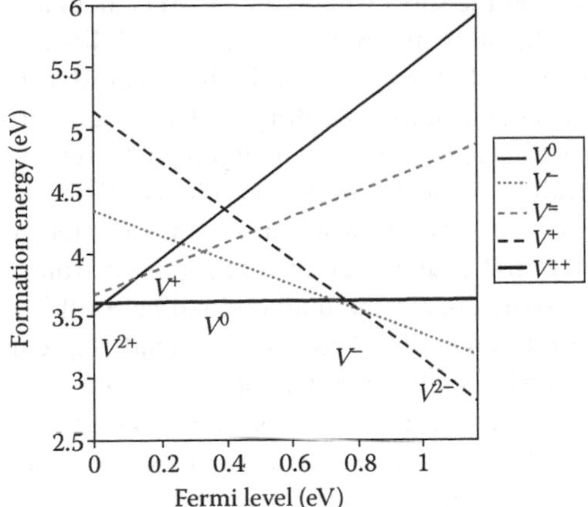

FIGURE 5.7 GGA energies of the Si vacancy as a function of the Fermi level. (From W. Windl, *ECS Transactions* **3**, 2006, 171. Reproduced by permission of The Electrochemical Society.)

which were used for a theoretical investigation of growth patterns of vacancy agglomerates V_i in Ge, for the range $1 \leq i \leq 35$.[16,17]

In order to avoid cluster dissociation, all MD calculations were performed for a constant temperature of 300 K, under the NVT-ensemble with periodic boundary conditions. First, the equilibrium enthalpy of the system was calculated for a perfect lattice consisting of 1728 atoms. These calculations are then repeated for the same structure with a number of vacancies. A sequence of 10 runs, each containing 70,000 steps (corresponding to 11 ps of real time), was performed for each cluster size and configuration. The results were subsequently averaged to minimize statistical uncertainty.

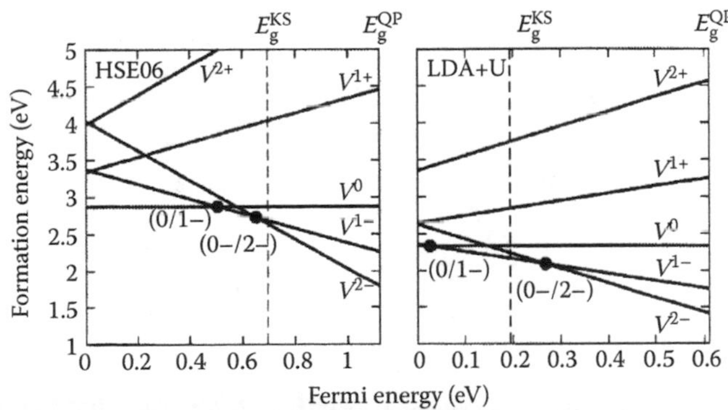

FIGURE 5.8 The vacancy formation energy in Ge as a function of the Fermi level and the charge state. (From P. Spiewak, J. Vanhellemont and K. Kurzydlowski, *Journal of Applied Physics* **110**, 2011, 063534. With permission.)

It should be noted that the short time scale of MD calculations is sufficient only for localized lattice relaxation and not for long-ranged diffusive rearrangement of cluster geometry. As a result, the choice of the initial cluster geometry is critically important for making conclusions regarding nucleation and growth pathways.[18,19]

Three cluster growth modes, shown in Figure 5.9, were investigated in order to find the energetically favorable mode. The choice of these growth modes was based on the results of the previous work on vacancy cluster growth in silicon single crystals.[18–22] In the first case, Ge atoms are removed from the six-member ring and 10-member cage structures in the diamond lattice. In this way, octahedral-like vacancy clusters are formed; called here as hexagonal ring clusters (HRC). In the second case, the successive shells on neighbors of a given Ge atom were taken out in order to form spherically shaped clusters (SPC). The third investigated cluster growth mode was the (111)-oriented stacking fault (SF), which consists of interconnected six-member vacancy ring arranged in a (111) plane.

The energetically favorable growth mode was determined by comparing the formation energies for clusters containing between 1 and 35 vacancies. The formation energy for a cluster that contains i vacancies can be calculated from

$$E^f(i) = E(N - i) - (N - i)/N\, E(N). \tag{5.2}$$

$E(N)$ is the energy of a perfect system with N atoms and $E(N - i)$ is the energy of a system with $N - i$ atoms (lattice with N sites and i vacancies). The results of these calculations shown in Figure 5.10 demonstrate that SF clusters have a higher

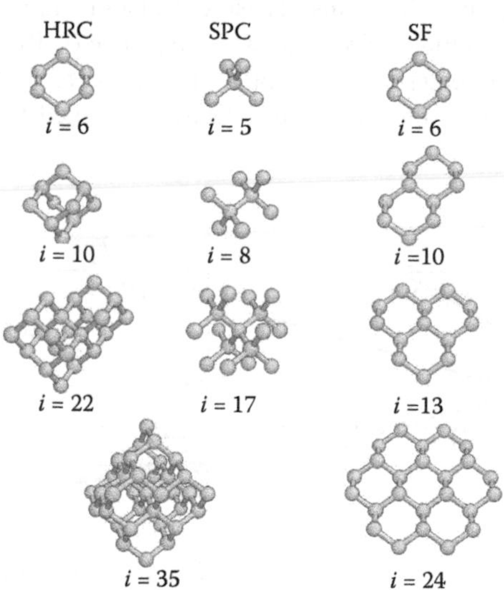

FIGURE 5.9 Configuration of vacancy cluster growth modes in germanium for the HRC-6, 10, 22, and 35, the SPC-5, 8, 17, and 35, and the SF-6, 10, 13, and 24 vacancy clusters. (From P. Spiewak et al. *Materials Research Society Symposia Proceedings* **994**, 2007, 0994-F03-08. Reprinted with permission.)

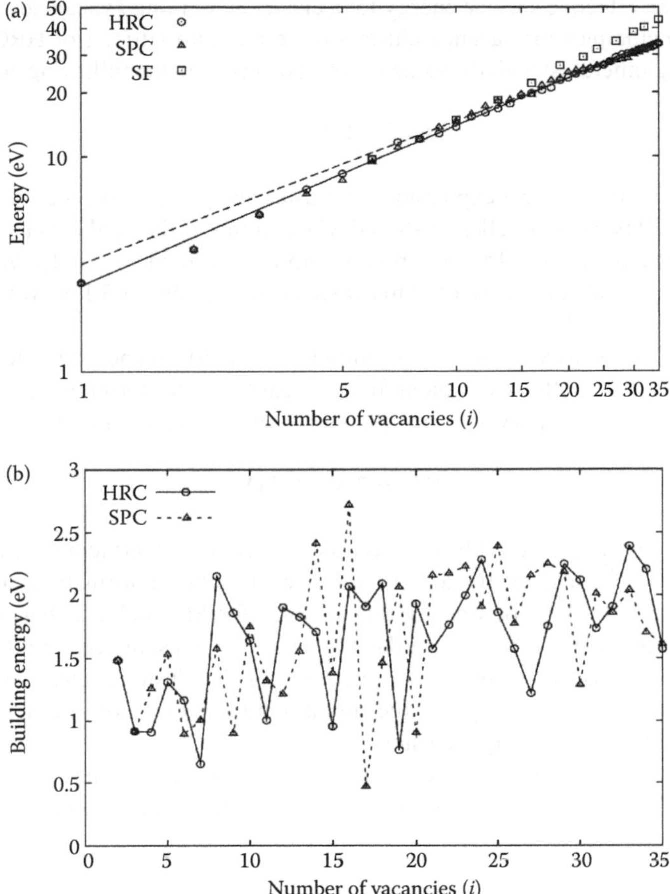

FIGURE 5.10 (a) Formation enthalpies of HRC, SPC, and SF vacancy clusters in single crystal germanium as a function of the number of vacancies i in the cluster—the dashed line shows the power law fit ($\sigma\, i^{2/3}$) and the solid line the correction (optimal fit) for the cluster size interval $1 \leq i \leq 19$ for the HRC and SPC clusters. (b) Binding energy of HRC and SPC vacancy clusters in germanium as a function of the cluster size i. (From P. Spiewak et al. *Materials Research Society Symposia Proceedings* **994**, 2007, 0994-F03-08. Reprinted with permission.)

formation energy than thc HRC and SPC ones, so that their formation is less likely. The formation energies for spherically shaped and hexagonal ring clusters seem to intersect one another, as in the case of silicon.[18] Such behavior suggests that the agglomeration mechanism for large vacancy clusters is rather complex and SPC and HRC clusters are competing.

Continuum models describing void formation and growth use a phenomenological expression for the vacancy cluster formation energy of the form

$$E^f(i) = \sigma\, i^{2/3}, \tag{5.3}$$

where σ is an effective surface energy for a cluster containing i vacancies. Fitting the formation enthalpies for vacancy clusters in germanium lattice (for HRC and SPC vacancy agglomerates) with the same power law results in the following equation[16,17]:

$$E^f(i) = 3.15\ i^{2/3}\ \text{eV}. \tag{5.4}$$

This phenomenological expression for cluster energy is most accurate for a cluster size $i > 20$ because smaller clusters deviate from the classical $i^{2/3}$ energy scaling behavior. The prefactor 3.15 corresponds with a surface energy of 1.1 J/m^2 (at 0 K) when assuming octahedral voids. This value is close to the 1.05 J/m^2 value that was obtained for silicon.[18]

For smaller vacancy clusters containing less than 20 vacancies, the MD calculations suggest that another expression for the vacancy cluster formation energy should be used. For the size interval $1 \le i \le 19$, the best fit is obtained for[16,17]

$$E^f(i) = 2.50\ i^{0.745}\ \text{eV}. \tag{5.5}$$

Equation 5.5 can be used to increase the accuracy of continuum models, which use phenomenological expression for the vacancy cluster formation energy. The exponent of 0.745 is in between 1 that is valid for SFs and 2/3 that is valid for spheres, (truncated) octahedra, or other clusters with constant aspect ratio. It could be explained by assuming that the aspect ratio of the cluster is changing from the one of a nearly plate-like shape for the smallest clusters to the one of more spherical or octahedral shapes for larger clusters.

The relative stability of vacancy clusters was investigated in terms of their binding energy. The binding energy for a cluster containing i vacancies can be calculated from

$$E^b(i) = E^f(i-1) + E^f(1) - E^f(i). \tag{5.6}$$

The results of such calculations, shown in Figure 5.10, demonstrate that the binding energy of vacancy clusters is a nonmonotonic function of the cluster size and cannot be expressed simply in term of nearest-neighbor interactions.[18,23] Similar to tight-binding calculations for silicon,[18] some of germanium HRC vacancy clusters ($i = 5, 8, 12, 16, 18, \ldots$) display higher stability (i.e., a higher $E^b(i)$) than clusters with an exact number of six-member rings ($i = 6, 10, 14, 18, \ldots$). Bongiorno et al.[20] proposed a two-step explanation for such behavior in silicon. The first step is to select those aggregates obtained by extracting atoms from complete six-member rings. A structure formed in this way minimizes the number of dangling bonds created at the cluster boundary and, in turn, minimizes the internal surface energy along the <111> or equivalent direction (it is well known that the (111) surface in the diamond lattice corresponds to the minimal surface energy). The second step addresses the relaxation features of the internal surface. The rearrangement of the dangling bonds favors in some cases a cluster, which does not strictly fulfill the previous topological conditions. This is the case for cluster with sizes $i = 5, 8, 12, 16$ in which the structural rearrangement modifies the number of dangling bonds with respect to the unrelaxed structure. In any case, the aggregation of a new vacancy to a stable cluster always gives rise to a less stable structure, as is the case for $i = 7, 11, 15, 19$. A similar relaxation process is observed for SPC clusters, which contain 5, 8, 10, 14, 16, 19 vacancies. In this case, the (100) internal surfaces are relaxed as well.

The binding energy of 1.48 eV for the germanium divacancy obtained in this work from MD calculations is in good agreement with the value of 1.5 eV obtained by Janke et al.[24] who used DFT in a cluster method. It is higher than the 1.1 eV obtained by Sueoka and Vanhellemont[25] who employed supercell DFT LDA/GGA calculations.

However, Sueoka and Vanhellemont[25] used a different equation for calculating the binding energy. They calculated binding energy of i isolated vacancies through the reaction $iV \rightarrow V_i$ up to $i = 4$, while Equation 5.6 determines binding energy of isolated single vacancy and $i - 1$ vacancy cluster through the reaction $V + V_{i-1} \rightarrow V_i$. In order to compare MD results obtained here with Sueoka and Vanhellemont estimates, the binding energy of isolated i vacancies was calculated, according to the following equation:

$$E^b_{\text{isolated}}(i) = iE(N-1) - [E(N-i) + (i-1)E(N)], \tag{5.7}$$

where $E(N)$ is the enthalpy of a perfect lattice with N atoms and $E(N - i)$ is the enthalpy of a system with $N - i$ atoms (lattice with N sites and i vacancies). The results of these calculations are shown in Figure 5.10 for a cluster size up to 35 vacancies.

Table 5.1 shows that the binding energies calculated for the smaller clusters are higher than the values reported by Sueoka and Vanhellemont.[25] One reason for this difference might be the small supercell of 64 Ge atoms in a perfect lattice that was used in the DFT calculations.

The choice of a sufficiently large supercell is important for accurate binding energy estimation because of the lattice relaxation around a vacancy and vacancy clusters, as has been shown previously for silicon.[26,27] Although the relaxation intensity decays rapidly away from the vacancy, it decreases more slowly along the zigzag chain of atoms in the <110> directions. The same lattice relaxation as in silicon, which is significant up to the fifth shell of the nearest-neighbor atoms around the vacancy, was observed for germanium.[28]

MD simulations were also performed for a large vacancy cluster. A perfect germanium lattice site containing 46,656 atoms was used to create a spherical

TABLE 5.1
Binding Energy (eV) of i Isolated Ge Vacancies after the Reaction $iV \rightarrow V_i$

$2V \rightarrow V_2$	$3V \rightarrow V_3$	$4V \rightarrow V_4$	Reference
1.48	2.40	3.65	16
1.10	1.97	2.70	25

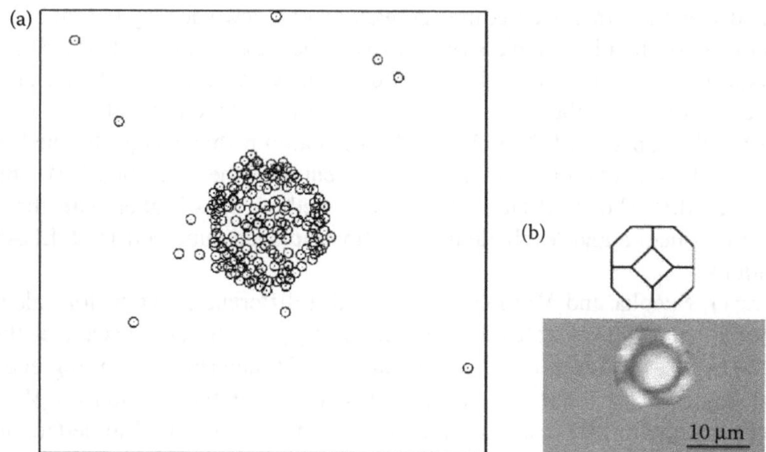

FIGURE 5.11 (a) Atoms in the vicinity of the 314 vacancies cluster in germanium, as viewed along the <001> direction, with the square displacement exceeding 22 pm. The frame indicates the edges of the computational domain.[17] (b) Optical micrograph of a "crystal-originated particle" or COP on a polished germanium wafer. The COP schematically shown on top originates from a truncated octahedral void in the bulk of the as-grown Cz germanium crystal.

vacancy cluster containing 314 vacancies. An SW potential with the parameters of Wang and Stroud[29] was used in a 2 ns MD annealing run at 1100 K, allowing the cluster to relax and take the shape with the lowest total energy. Similar to the previous simulations, the NVT-ensemble with periodic boundary conditions was employed. The result of the simulation of this very large cluster is shown in Figure 5.11 and suggests that the vacancy cluster morphology converges to a (truncated) octahedral shape, which is bounded by the low-energy (111) and (100) planes. This is in agreement with the experimentally observed surface pit shape on polished germanium wafers.[30,31]

5.4 PARAMETERS INFLUENCING VACANCY PROPERTIES

5.4.1 Internal and External Stress

Si crystals (wafers) in mass production are always under some form of stresses. For example, during crystal growth, thermal stress is generated in Si crystals in particular near the melt/solid interface. This stress is mainly compressive and should be considered as internal stress (the surface of the crystal is assumed to be stress-free). Therefore, the point defects (vacancy V and self-interstitial I) during crystal growth also experience internal stress. To estimate the impact of external and internal stresses on the V parameters, the differences in the formation volumes of V between both types of stress should be considered.[8–10] As an example, the cases of isotropic external stress (=hydrostatic pressure) and isotropic internal stress are discussed. We consider the formation of V in a perfect Si crystal (without any sinks or sources of vacancies) with a finite size (surrounded by stress-free surfaces) under thermal equilibrium. To create *V*, one moves an atom from the interior crystal site to a crystal surface site. This results in a volume increase of Ω_{Si} (=atomic volume) at the surface. Therefore, under hydrostatic pressure *P*, one needs the work of $P(\Omega_{Si}+v_f^V)$ with v_f^V, the relaxation volume, to form a vacancy. In a Si crystal under internal stress σ, on the other hand, we should not include the contribution of $\sigma\Omega_{Si}$ as the surface is free of external normal stress. That is, for the assumption of internal stress σ, one only needs the work of σv_f^V to form a vacancy.

The formation enthalpy H_f^V of *V* under hydrostatic pressure *P* and under internal stress σ is obtained from

$$H_f^V(\sigma) = E_f^V(\sigma) + \sigma \times v_f^V(\sigma) \tag{5.8}$$

and

$$H_f^V(\sigma) = E_f^V(\sigma) + \sigma \times v_f^V(\sigma), \tag{5.9}$$

respectively. Here, E_f^V is the formation energy of vacancy. For the self-interstitial *I*, similar equations can be used to obtain the formation enthalpies under hydrostatic pressure and under internal stress.[8–10] The migration enthalpy H_m is obtained by the change of H_f at the transition state and H_f at the most stable state.

The formation enthalpies of Jahn–Teller distorted *V* (*h*-JT, most stable) and split-*V* decrease with increasing pressure *P* as shown in Figure 5.12 (a). The calculations lead to H_f^V of neutral *V* and H_m^V of neutral *V* dependencies on *P* (–3 GPa < *P* < 3 GPa) given by

$$H_f^V = 3.543 - 0.024 \times P^2 - 0.009 \times P(\text{eV}) \tag{5.10}$$

and

$$H_m^V = 0.249 + 0.005 \times P^2 - 0.030 \times P(\text{eV}) \tag{5.11}$$

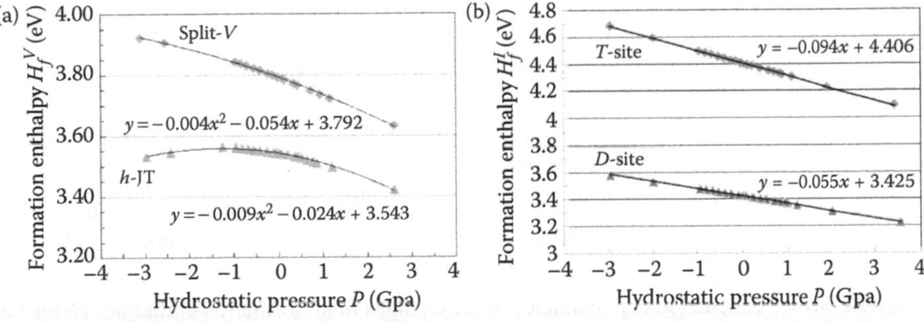

FIGURE 5.12 Dependence of formation enthalpy (a) H_f^V for *h*-JT (most stable) and split-*V* (transition state), and (b) formation enthalpy H_f^I for *D*-site and *T*-site on the hydrostatic pressure *P*[9].

with P given in GPa. The formation enthalpies of I decrease linearly with increasing pressure P as shown in Figure 5.12 (b). The calculations show that the formation enthalpy H_f^I of the neutral I at the D-site and the migration enthalpy H_m^I of the neutral I are given by

$$H_f^I = 3.425 - 0.055 \times P(\text{eV}) \tag{5.12}$$

and

$$H_m^I = 0.981 - 0.039 \times P(\text{eV}) \tag{5.13}$$

with P given in GPa. This result indicates that hydrostatic pressure leads to an increase of the equilibrium concentration and diffusion of self-interstitials.

These results indicate that hydrostatic pressure leads to an increase of the equilibrium concentration and diffusion of vacancies but this increase is considerably smaller than that of self-interstitials. Taking into account recombination between vacancies and self-interstitials, hydrostatic pressure thus makes a Si crystal more interstitial-rich. The obtained results are consistent with the experimental observations on dopant diffusion in Si under hydrostatic pressure. That is, the hydrostatic pressure enhances the diffusivity of B through an interstitial-based mechanism while it retards the diffusivity of Sb through a vacancy-based mechanism in Si.

Figure 5.13 (a) shows the change of formation enthalpies H_f for I and V due to internal stress σ (-1 GPa $< \sigma <$ 1 GPa). It is clear that H_f^I increases while the H_f^V decreases with increasing σ. Figure 5.13 (b) shows the change of migration enthalpies H_m for I and V due to internal stress σ.

The calculations lead to the dependencies of H_f^V, H_m^V, and H_f^I, H_m^I on σ for perfect Si given by

$$H_f^V = 3.543 - 0.154 \times \sigma(\text{eV}) \tag{5.14}$$

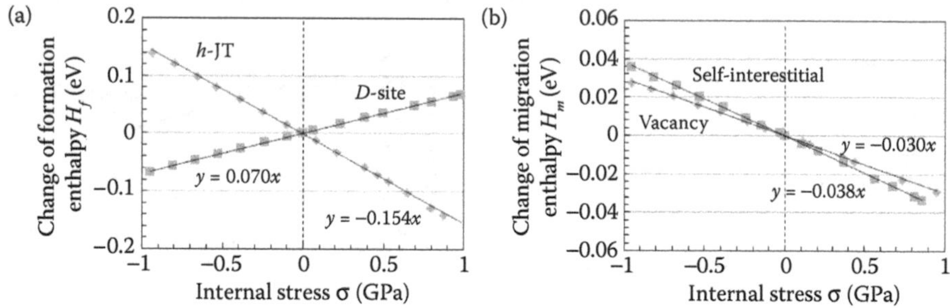

FIGURE 5.13 Calculated dependencies of the changes in (a) formation enthalpies H_f for I at D-site (most stable), V of h-JT (most stable), and (b) migration enthalpies H_m for I, on internal stress σ between -1 GPa (tensile) and 1 GPa (compressive) in a perfect Si crystal. (From K. Sueoka, E. Kamiyama and J. Vanhellemont, *Journal of Crystal Growth* **363**, 2013, 97. With permission.)

$$H_m^V = 0.248 - 0.030 \times \sigma(\text{eV}) \tag{5.15}$$

and

$$H_f^I = 3.425 + 0.070 \times \sigma(\text{eV}) \tag{5.16}$$

$$H_m^I = 0.981 - 0.038 \times \sigma(\text{eV}) \tag{5.17}$$

with σ given in GPa.

These results indicate that compressive internal stress leads to an increase of the equilibrium concentration and diffusion of V and to a decrease of the equilibrium concentration of I. By the recombination between V and I, compressive internal stress thus makes Si crystals more V-rich.

It is noticeable that the impact of hydrostatic pressure and compressive internal stress on point defects is opposite. That is, hydrostatic pressure makes the Si crystal more I-rich, while internal compressive stress makes the Si crystal more V-rich. Here, we briefly comment on the previous studies under biaxial (external) compressive stress. The experimental results revealed that biaxial stress retards the diffusivity of B through an interstitial-based mechanism while it enhances the diffusivity of Sb through a vacancy mechanism.[32] These results are opposite to those under hydrostatic pressure, and as shown above, the difference can be qualitatively explained by taking into account the contribution of $P\Omega_{\text{Si}}$ to the enthalpies or not.

The impact of thermal stress σ_{th} on intrinsic point defect behavior during the growth of large-diameter single crystal Si can be evaluated based on these results. The impact of compressive thermal stress, which is internal stress, makes the growing Si crystal more vacancy-rich. Figure 5.14 indicates the importance of taking into account the impact of thermal stress σ_{th} on the critical (v/G), with v the pulling rate and G the temperature gradient, which defines the pulling condition for defect-free crystals.[8–10] Since the thermal stress increases with increasing crystal diameter, it is clear that the larger thermal stress makes it a real challenge to develop defect-free 450-mm-diameter Si crystals.

5.4.2 Impurities Interacting with Vacancies

It is well known that neutral and electrically active impurities interact with intrinsic point defects in a Si single crystal growing from a melt.[33] For the neutral impurities (N, C, Ge, O, and Sn), N and C atoms suppress V-type point defect clusters, Ge atoms have only a limited impact on V concentration, O and Sn atoms make the crystal more V-rich. For the electrically active dopants (B, P, As, and Sb), B atoms make the crystal more I-rich while the n-type dopants make the crystal more V-rich.

Several mechanisms to explain the impact of dopants have already been identified,[34–37] that is, (1) local strain (stress) related to dopant atom size, (2) lattice strain (stress) related to dopant atom size, and (3) change in the Fermi level. These

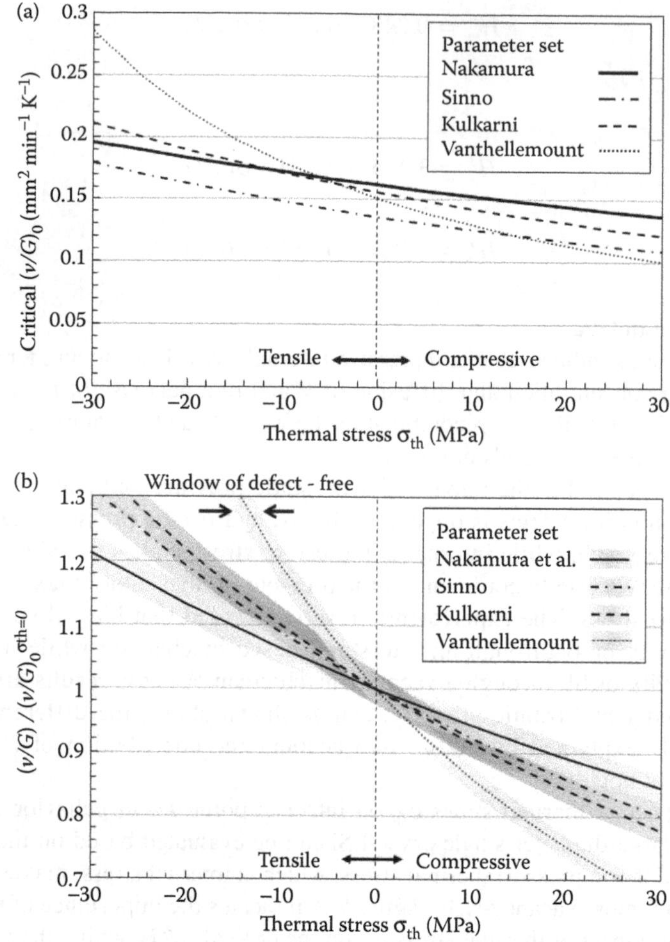

FIGURE 5.14 (a) Dependency of the critical (*v/G*) on thermal stress σ_{th} between −30 MPa (tensile) and 30 MPa (compressive), and (b) dependence of the critical (*v/G*) normalized with respect to the critical (*v/G*) at the stress-free value of $(v/G)_0\sigma_{th} = 0$, and the window for defect-free Si on the thermal stress σ_{th}. The windows for defect-free Si in (b) are shown by the shaded bands. (From K. Sueoka, E. Kamiyama and J. Vanhellemont, *Journal of Crystal Growth* **363**, 2013, 97. With permission.)

mechanisms, however, do not allow to explain the experimental results for *n*-type dopants. Here, we show the results of a theoretical study showing that the impact of dopants is mainly due to a change in the *V* and *I* formation energies around dopant atoms.

The *V* formation energy within a sphere with 6 Å radius around the dopant atom is calculated as follows.[33] A dopant atom is introduced at the center of a perfect 216-atom supercell and a vacancy is placed at the 1st–5th closest neighbor positions from the dopant atom. Figure 5.15a shows the sites of *V* from the dopant atom in relation

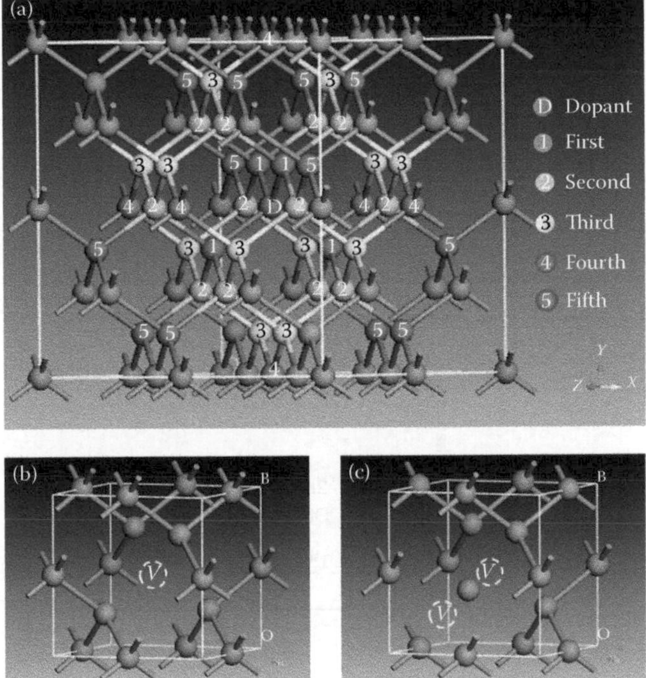

FIGURE 5.15 (a) Sites of a vacancy closest to a dopant atom in a 64-atom supercell. Geometries of a vacancy with (b) Jahn–Teller distortion (V_{JT}) and T_d symmetry (V_{Td}), and (c) split vacancy (split-V), in a conventional cubic unit cell of Si. (From K. Sueoka, E. Kamiyama and J. Vanhellemont, *Journal of Applied Physics* **114**, 2013, 153510. With permission.)

to a 64-atom supercell. Figure 5.15b shows the geometry of a vacancy and Figure 5.15c shows that of a split vacancy. The formation energies of V are calculated with Jahn–Teller distortion (V_{JT}), T_d symmetry (V_{Td}), and split vacancy (split-V) in the nondoped cells. The formation energy $E_V^{f,\text{non}}$ of Jahn–Teller distortion (V_{JT}) in perfect Si is 3.578 eV for neutral vacancies V and is about 0.214 eV lower than that for the T_d symmetry. The activation energy of V (V_{JT} is the ground state and split-V is the transition state) is 0.267 eV.

Figure 5.16a through c plots the dependencies of the change in the vacancy formation energy from that for Jahn–Teller distortion in perfect crystal Si on the distance from *p*-type, neutral, and *n*-type dopants. The dotted lines from the 1st to the 5th in these figures indicate the distance from the dopant before the cell size and ionic coordinates are relaxed.

It can be confirmed that $E_V^{f,\text{dope}}$ at the 1st site differs for the same types of dopants. Note that the formation energy of vacancies with larger dopant atoms is less than that with smaller dopant atoms. Since the electrical state is almost the same for the same types of dopants, this result is mainly due to the difference in local strain. Furthermore, $E_V^{f,\text{dope}}$ of V at and far from the 2nd sites are close for neutral dopants without changing the electrical state, and close to that of V in perfect Si. This

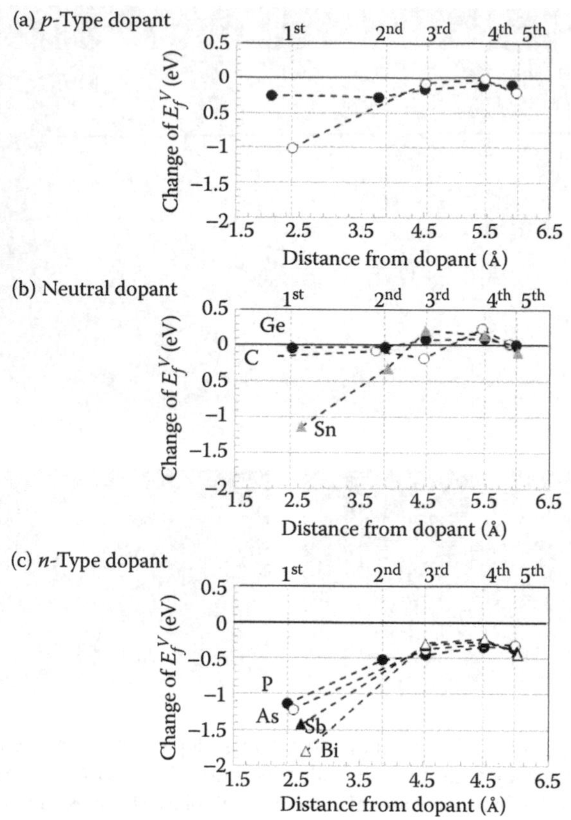

FIGURE 5.16 Dependence of the change of vacancy formation energy from that for Jahn–Teller distortion in a perfect Si crystal on the distance from a (a) *p*-type dopant atom, (b) neutral dopant atom, and (c) *n*-type dopant atom. The open circles for Sn, Sb, and Bi indicate split vacancies. The dotted lines from 1st to 5th indicate the distance from the dopant atom before cell size and ionic coordinates are relaxed. (From K. Sueoka, E. Kamiyama and J. Vanhellemont, *Journal of Applied Physics* **114**, 2013, 153510. With permission.)

indicates that local strain effects are only important at the 1st site from the dopant atom. The type and magnitude of local strain differ for the *n*-type dopants P, As, Sb, and Bi. However, starting from the 2nd site, $E_V^{f,\mathrm{dope}}$ is nearly the same for all *n*-type dopants and is about 0.3–0.4 eV less than that in perfect Si. This illustrates that not local strain but the electrical state around *n*-type dopants mainly determines the *V* formation energy.

Figure 5.17 shows the atomic configurations of the most stable vacancies around these dopant atoms. The displacement of Si atoms around the vacancies depends on the dopant element and can be understood from the visualizations.

Similar calculations were performed for self-interstitials.[33] Figure 5.18 plots the binding energies (energy reductions) of point defects with nine dopant atoms. Several reported values are also provided in this figure. Both the present and previous

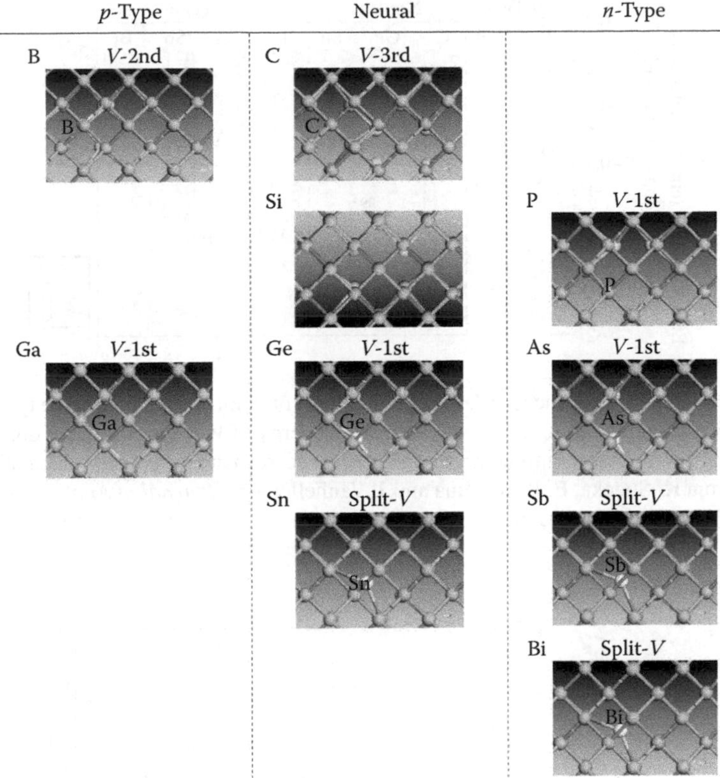

FIGURE 5.17 Atomic configurations of most stable vacancies around p-type, neutral, and n-type dopant atoms. Light gray spheres indicate dopant atoms and green spheres indicate Si atoms forming vacancies. (From K. Sueoka, E. Kamiyama and J. Vanhellemont, *Journal of Applied Physics* **114**, 2013, 153510. With permission.)

calculations clearly demonstrate that (1) I is more stable than V around p-type dopants while V is more stable than I around n-type dopants, and (2) V becomes more stable around larger dopants while I becomes more stable around smaller dopants.

The thermal equilibrium concentration of total V and I as a function of quenching temperature T for different dopants was obtained. Figure 5.19 plots the calculated dependence of $C_V^{\mathrm{eq,tot}}(T_m) - C_I^{\mathrm{eq,tot}}(T_m)$ on dopant types and concentrations at the melting temperature. On the basis of the Voronkov model, the plots in Figure 5.19 can very well explain the experimental results of the impact of dopant concentration on the intrinsic point defect concentrations during Si crystal growth.

5.4.3 Surfaces, Interfaces, and Capping Layers

All real crystals are finite and thus have free surfaces. This means that diffusing intrinsic point defects in Si or Ge bulk crystals will at a certain point in time reach the crystal surface if it diffuses through a bulk with few dislocations or grain boundaries as is the case in Si or Ge single crystals or wafers. This is why the internal stress has

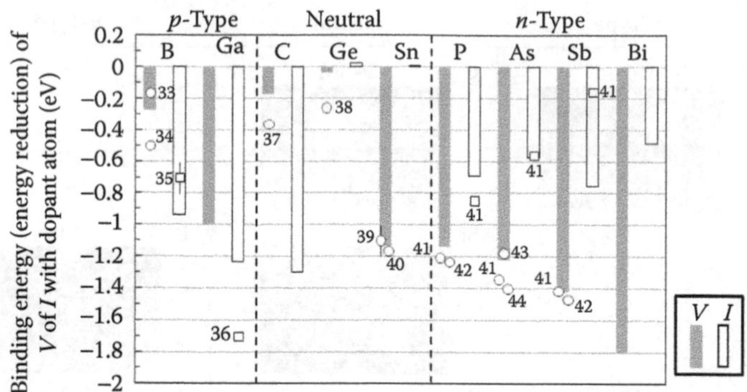

FIGURE 5.18 Binding energies (energy reductions) of point defects with three types of dopant atoms obtained by using the lowest formation energy of *V* or *I* around each dopant. Open circles and open squares with numbers (references) correspond to reported calculations for *V* and *I*. (From K. Sueoka, E. Kamiyama and J. Vanhellemont, *Journal of Applied Physics* **114**, 2013, 153510. With permission.)

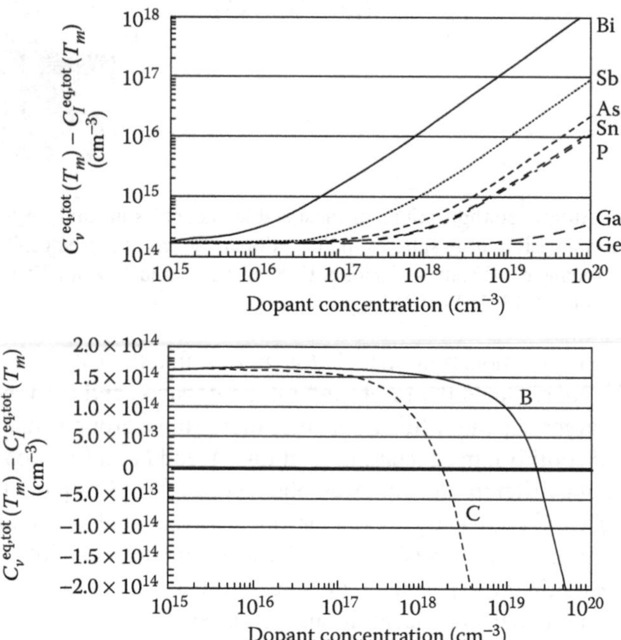

FIGURE 5.19 Calculated dependence of $C_V^{eq,tot}(T_m) - C_I^{eq,tot}(T_m)$ at Si melting temperature on dopant concentration. (Adapted from K. Sueoka, E. Kamiyama and J. Vanhellemont, *Journal of Applied Physics* **114**, 2013; Parameters proposed by K. Nakamura, T. Saishoji and J. Tomioka, Simulation of the point defect diffusion and growth condition for defect-free Cz silicon crystal, *Semiconductor Silicon 2002, PV 2002-2*, The Electrochemical Society Proceedings Series, 2002, p. 554.)

to be considered in a Si crystal as mentioned in the previous section. When intrinsic point defects arrive at the surface, they will interact with the surface because there are dangling bonds, surface-induced charges, and free space for the relaxation of atom positions. Owing to this, the formation energy of an intrinsic point defect at or just below the surfaces is different from that in the bulk and it will depend on the distance from the surface. Based on these atomistic schemes, the boundary conditions in the classical continuous models used to calculate the concentrations of intrinsic point defects can be justified. In this section, the formation energies of intrinsic point defects near interfaces and capping layers will also be briefly discussed.

The (001) Si and Ge surfaces were investigated in detail[39] while other surface orientations, such as (011) and (111), show similar phenomena, qualitatively. It is well accepted that the c(4 × 2) dimer structure is the most stable state in the (001) surface for both Si and Ge (001) clean surfaces,[40,41] although some unclear points remain such as the observation of a p(2 × 1) structure below 20 K in Si.[42] Plate models with dimer structures on both surfaces of the plate and a vacuum slab, as shown in Figure 5.20, are used to examine the effect of a clean Si or Ge surface with c(4 × 2) structure. The

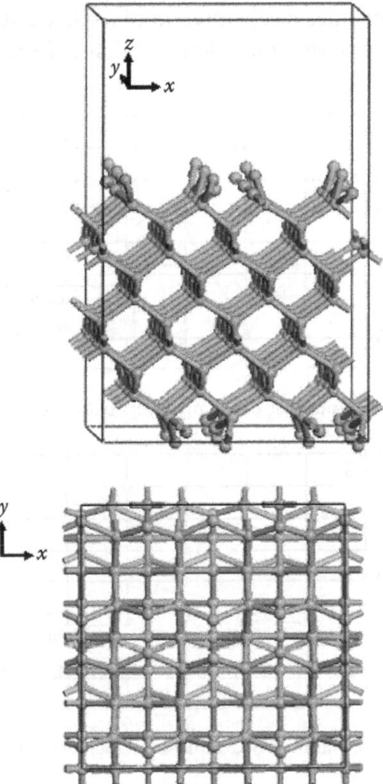

FIGURE 5.20 A plate model with c(4 × 2) dimer structures on both surfaces of the plate and a vacuum, (top) bird's eye view, (bottom) planar view. (From E. Kamiyama, K. Sueoka and J. Vanhellemont, *ECS Journal of Solid State Science and Technology* **2**, 2013, 104. Reproduced by permission of The Electrochemical Society.)

energy bands of dangling bonds at the surface before the formation of dimers cross the Fermi level, and the system is therefore unstable. The formation of dimers leads to the splitting of the energy band crossing the Fermi level into two bands of bonding. These two bands are related to the bonding and antibonding states, lying below and above the Fermi level, respectively, and the system becomes stable again.[43]

The intrinsic point defect formation energies were calculated by using the chemical potential of a Si or Ge atom in an usual 64-atom cell with no vacuum slab and using a three-dimensional boundary condition. Figure 5.21 shows the formation energy dependencies on *d*, the number of atomic layers below the plate surface. When a vacancy or a self-interstitial moves inside from the surface, the interaction with the surface through the overlap of wave functions, the transfer of electric charges from the surface, and the relaxation of atom positions by the deformation of surface dimer structures due to the introduction of the point defect become less effective, and the formation energy must be close to the bulk value. The position dependencies for Si and Ge are similar for two aspects: (i) the formation energies of both Si and Ge self-interstitials are lower than that of a vacancy at the top surface where self-interstitials can thus form easier than vacancies or an isolated atom can move from a melt and becomes a self-interstitial as discussed further, (ii) the formation energies' dependence of the intrinsic point defect formation energy on the plate

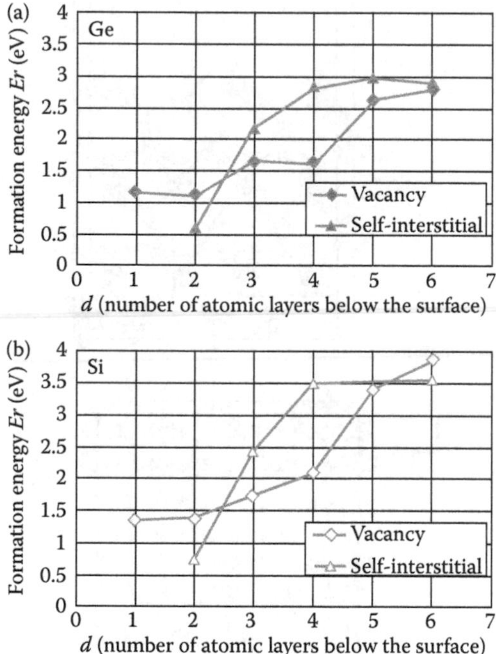

FIGURE 5.21 Intrinsic point defect formation energy dependencies on the depth below the plate surface expressed by d, the number of atomic layers below the surface for the 12-layer plate model shown in Figure 5.20. (From E. Kamiyama, K. Sueoka and J. Vanhellemont, *ECS Journal of Solid State Science and Technology* **2**, 2013, 104. Reproduced by permission of The Electrochemical Society.)

thickness expressed by N, the number of atomic layers in the plate is shown in Figure 5.22, in which the point defect was set at the center of the film. The thickness dependencies of both intrinsic point defect formation energies are similar for Si and Ge, and also similar for a vacancy and a self-interstitial. It is, however, interesting to note that the ordering of the formation energies of a vacancy and a self- of both Si and Ge self-interstitials increase drastically as they move inside and become higher than that of the vacancy at the third layer and finally saturate at the fifth layer. Opposite to self-interstitials, the formation energies of both Si and Ge vacancies increase slowly, whereby the Si vacancy has a higher formation energy than the self-interstitial starting from the sixth layer.

As for the view of the relation between the nanostructures and the bulk crystals, the interstitial is opposite when comparing Si and Ge. This is due to the lower formation energy of a vacancy in Ge than in Si because chemical bonds in Ge are weaker than those in Si. Therefore, this trend will continue until a bulk crystal is obtained ($1/N = 0$). The maximum values obtained for the 12 layer models in Figure 5.22 (except for the Ge self-interstitial) can be considered to be close to the bulk values. Opposite to this, the values for the Ge self-interstitial did not yet show saturation even for the 16-layer model. This is probably because the effect of the surface-induced charge reaches so deep inside the film, even down to the 15th layer from the surface[44] as discussed further. Therefore, the formation energy of the Ge self-interstitial may still be underestimated compared to the bulk value.

Regarding the charge transfer between a vacancy and a surface, the changes in the charges of the atoms from the first to the fourth layers against the vacancy position in a Si crystal are shown in Figure 5.23.[45] A vacancy is introduced in the (a) first layer, (b) second layer, …, (h) ninth layer, and (i) nowhere. These figures are top views, as viewed from the surface, and the projections of the original vacancy positions

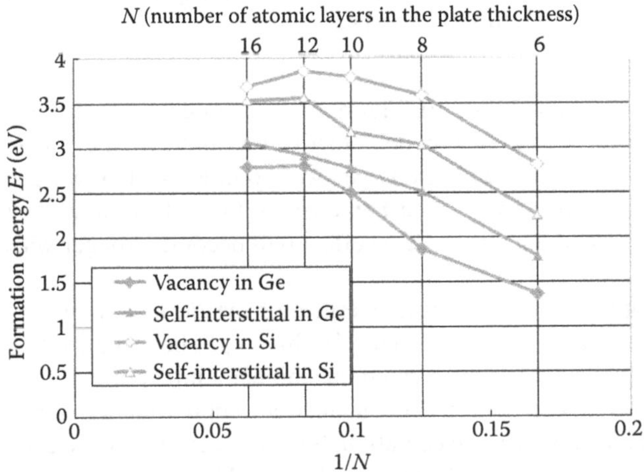

FIGURE 5.22 Formation energy as a function of the thickness of the plate, expressed by N, the number of atomic layers in the plate, used in the model. (From E. Kamiyama, K. Sueoka and J. Vanhellemont, *ECS Journal of Solid State Science and Technology* **2**, 2013, 104. Reproduced by permission of The Electrochemical Society.)

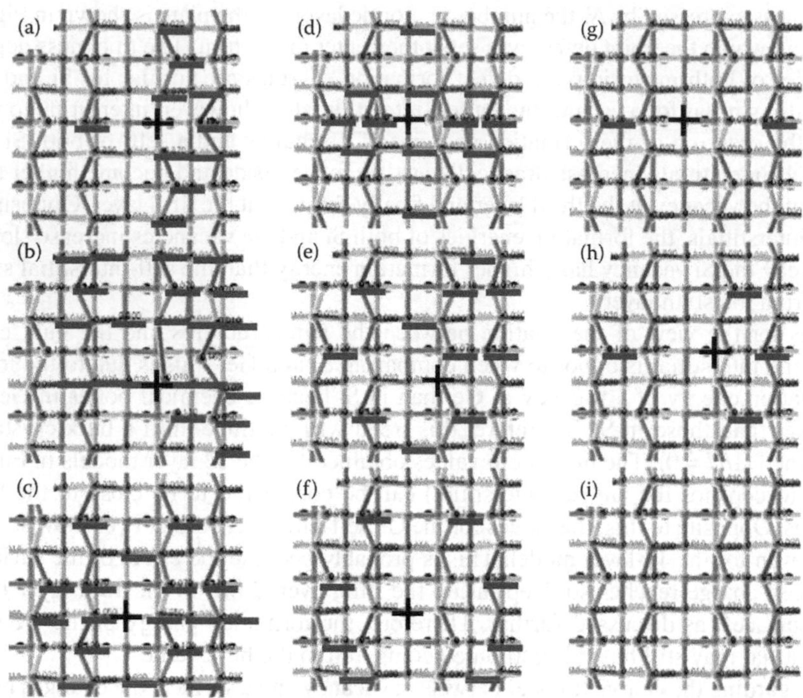

FIGURE 5.23 The changes of charges in atoms from first to fourth layers against vacancy introducing position in (a) first, (b) second, (c) third, (d) fourth, (e) fifth, (f) sixth, (g) eighth, and (h) ninth layers, and (i) nowhere. These figures are top views, seen from the surface, and the projections of the original vacancy positions are marked with crossing lines. The different numbers of charges from (i) are underlined. (From E. Kamiyama and K. Sueoka, *Journal of Applied Physics* **111**, 2012, 013521. With permission.)

are marked with crossed lines. In Figure 5.23a through h, the different numbers of charges from the "no vacancy model," that is, Figure 5.23i, are underlined. In these figures, some of the numbers of charges were omitted if they were always zero or less meaningful. Apparently, even the vacancy located in the ninth layer affected these charges around the dimers. This effect will become stronger when the vacancy moves to the surface.

Another topic for the formation energies of the intrinsic point defects near the surface is crystal growth from a melt.[39,46] This might be calculated not by *ab initio* but a method such as MD, which can treat a higher temperature. The basic equations of the *ab initio* calculations indeed include an approximation supposing that the temperature is 0 K. *Ab initio* calculation, however, can supply the ground states that dominate the system one is calculating. Classical MD can treat the melt–solid interface at the melting point directly,[47] but in general it is not easy to extract the physical meanings from the calculated results. At the stage of Figures 5.21 and 5.22, only the interaction of point defects with the single crystal surface in contact with vacuum is considered. For self-interstitials, the possible paths of a Si atom from the vacuum into the surface region of the Si single crystal are investigated. For the different possible paths, both the stability in each site as a function of depth below the surface and the barrier height for point defect diffusion are calculated.

Figure 5.21 illustrates the lower formation energy of the vacancy and the self-interstitial at the surface. Hence, the total concentration of intrinsic point defects near the surface is expected to be higher than in the bulk. The microscopic mechanism that lowers the formation energy of a vacancy is the decrease of dangling bonds by moving and rearranging of atoms and their bonds. At the free surface in contact with vacuum, the atoms can move easily even at $T = 0$ K, compared with atoms in a bulk. A similar phenomenon will most probably also occur at the melt–solid interface. Once a vacancy is formed in the first layer, it can easily transfer to the second layer as there is no diffusion barrier or extremely low barrier between the two layers. In some cases, a vacancy in the first layer does not transfer to the second layer but annihilates with an interstitial coming from the melt or from the bulk of the crystal. A vacancy in the second atomic layer will be transported to the bulk automatically when the crystal growth continues and extra Si atoms supplied from the melt form an additional crystal layer at the top surface (which becomes a new first layer); that is, the region just below the surface tends to have a higher concentration of a vacancy than the equilibrium concentration inside bulk crystals at a given temperature. Many vacancies, which were transported to go up the energy slope or released in the bulk by crystal growth, diffuse back to the surface due to the formation energy slope in Figure 5.21 or annihilate by recombination with a self-interstitial. According to the same mechanism, a self-interstitial transported to go up the energy slope or released in the bulk by crystal growth will also diffuse back to the surface or annihilate by recombination. This supports a macroscopic model whereby the recombination of a vacancy and a self-interstitial is a much more important process in the bulk of the crystal than close to the surface and that the near surface layers thus act as a reservoir/source of intrinsic point defects.[48]

Although the melt–solid interface is most probably quite different from the crystal surface on which the results so far were obtained, one can assume that boundary conditions for the point defect concentrations at the interface for crystal growth from the melt in continuum simulations can be set at fixed values, such as "thermal equilibrium values at the melting point." When discussing "thermal equilibrium" values of intrinsic point defect concentrations, one should therefore take into account the contribution from the melt–solid interface.

It should be noted that Ge surface has a unique character for positively charged particles such as positrons or self-interstitials, diffusing from the inside to the surface. This is due to the different distribution of the surface-induced charges as predicted by *ab initio* calculations[44] and illustrated in Figure 5.24. This difference can be observed using Doppler broadening spectroscopy (DBS) of positron annihilation spectroscopy (PAS).[49] PAS is a powerful tool for the study of vacancy-type defects in semiconductors,[50,51] for example, in irradiated Si and Ge. Positron trapping at vacancies can be observed by the narrowing of the Doppler-broadened 511 keV annihilation γ-ray peak. By varying the incident positron energy, one can probe different sample depths in the submicron range. This method is called Doppler broadening spectroscopy.

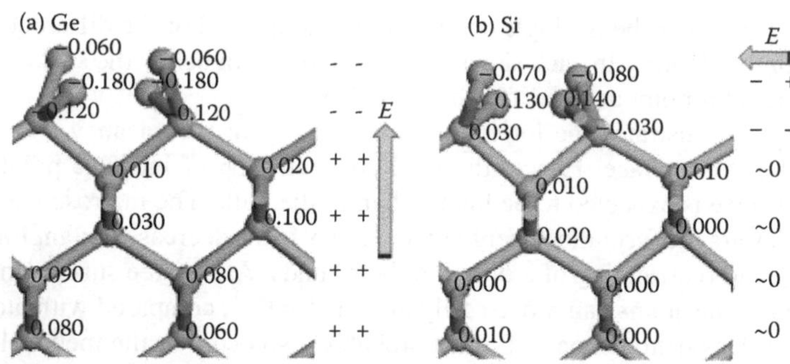

FIGURE 5.24 Comparison of surface-induced charges between Si and Ge from Mulliken analysis using *ab initio* calculations with ultrasoft potentials. Electric field directions in the dipole of the surface-induced charges are indicated by arrows whereby the length of the arrows is an indication of the relative field strength without being quantitative. (From E. Kamiyama, K. Sueoka et al. *ECS Solid State Letters* **2**, 2013, 89. Reproduced by permission of The Electrochemical Society.)

The so-called S-parameter, extracted from DBS, is closely related with positron annihilation by the valence electrons and is sensitive to open volume defects. An increase of the S-parameter can be simply interpreted as an increase of the number of vacancy-type defects. The S-parameter obtained for an incident positron energy lower than 2 keV, however, in general differs from the one obtained for positron energies above 15 keV. Incident positrons with lower energy will indeed mainly interact with the near surface area of the sample, and will be sensitive for thin native oxide layers, surface contamination, surface roughness, surface defects, and so on. The S-parameter decreases, in general, toward the surface for most materials due to positron diffusion toward and annihilation at the surface.[52]

The DBS spectra were measured by single DB and characterized using the standard S-parameter, defined as the ratio between the number of counts in the central region of the 511 keV annihilation peak (511 ± 0.76 keV) and the total number of counts in the peak region. The spectra were measured as a function of incident positron energy ranging from 0 to 20 keV. The obtained results for the Si/Ge substrates are shown in Figure 5.25 revealing a saturation of the S-parameter for positron energies above about 12 keV. As a reference for the Si bulk and at the same time the near surface of Ge, a Ge-on-Insulator (GOI) wafer with an undoped Ge top layer produced by the Ge condensation technique is also included. It is clear that the S-parameter of the bulk Si sample decreases toward the surface (corresponding with a lower positron energy).

The average depth range of positrons[53,54] is about 550 nm when the incident positron energy is 12 keV. The S-parameter of the GOI sample coincides with that of bulk Ge when the incident positron energy is below 5 keV and with that of Si when the incident positron energy is 15 keV or higher as illustrated in Figure 5.25. This confirms the result obtained for the Ge and Si substrates and guarantees the reproducibility and reliability of the PAS measurements on both Ge and Si samples for the whole energy range. There is no clear increase of the standard deviation of the

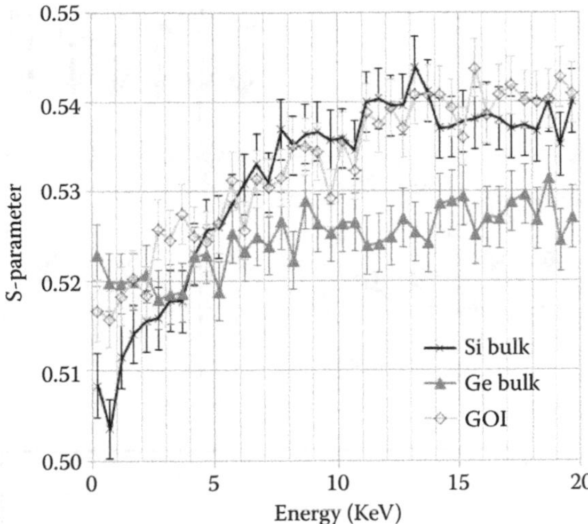

FIGURE 5.25 The dependence of the S-parameter of the bulk Si and Ge samples and the reference GOI sample on the incident positron energy and thus on the depth below the crystal surface. (From E. Kamiyama et al. *ECS Solid State Letters* **2**, 2013, 89. Reproduced by permission of The Electrochemical Society.)

S-parameters in a particular sample or for a lower positron energy. The observed continuous increase of S-parameter for energies above 5 keV in case of the GOI sample is due to the Si substrate as it is very similar to that of the bulk Si sample. A unique characteristic of Ge has been reported with respect to the behavior of Ge self-interstitials near the single crystal surface. In most materials, the surface acts as a sink or a source of self-interstitials. It has, however, been reported that Ge self-interstitials are reflected by the Ge surface.[55,56] Both Si and Ge self-interstitial atoms usually have a positive charge and are with that respect similar to positrons.

Calculation of the formation energies of point defects near the interface of two materials is actually complicated. This is because the two materials have different lattice constants, and there is generally no information about the structure at the interface. As an illustration, the interface between alpha-quartz and silicon is shown in Figure 5.26. In this figure, the formation energies of the intrinsic point defects only change very near to the interface. This is a useful result for considering the effects of capping layers. A short range effect of the presence of foreign atoms at the surface, on the formation energy has been reported for Frenkel pairs at the Si (001) surface,[57] although more layers would be needed to compare with the results in Figure 5.26. The absolute values of formation energies below the second layer calculated by *ab initio* are meaningless as discussed in Section 5.3.1.

As a result, the formation energies of intrinsic point defects below the capping layers are mainly determined by the external stress introduced by the capping layers, and this impact is illustrated by "external stress" (=hydrostatic pressure) in Figure 5.12 in Section 5.4.1.[8] These results indicate that hydrostatic pressure leads to a

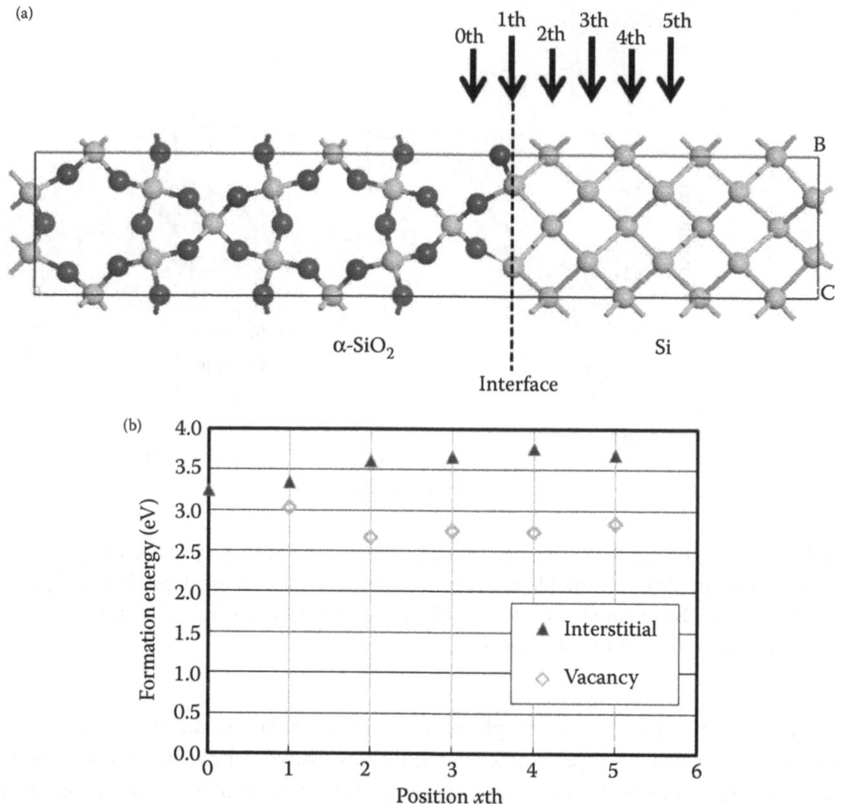

FIGURE 5.26 (a) Interface between alpha-quartz and silicon. (b) Calculated formation energies for a vacancy or self-interstitial Si at each position. In this model, self-interstitial Si is set at a tetragonal site.

slight increase of the equilibrium concentration and diffusion of vacancies, but this increase is, however, considerably smaller than that of self-interstitials. By the recombination between vacancies and self-interstitials, the hydrostatic pressure makes Si crystals more interstitial-rich. The results we obtained are consistent with the experimental results where hydrostatic pressure enhances the diffusivity of B through an interstitial-based mechanism while it retards the diffusivity of Sb through a vacancy mechanism[58] in Si by capping a Si–Ge layer.

5.5 IMPACT OF VACANCIES ON MATERIAL PROPERTIES AND VACANCY ENGINEERING: TWO CASE STUDIES

5.5.1 Oxygen Precipitation

Oxygen precipitation in Czochralski (Cz)-grown silicon crystals has been studied for nearly half a century due to its impact on material and device properties.[59] The importance of oxygen precipitation relates to the internal gettering (IG) of contamination metals, wafer strength (both the dislocation pinning and wafer softening), and so on. Therefore, the control of oxygen precipitation in target device processes is very important in mass production, and much effort was taken to understand the mechanism of oxygen precipitation.

The oxygen precipitation behavior in Cz-Si crystals has a close relationship with intrinsic point defects such as self-interstitials (I) and vacancies (V). The following equation is widely accepted for the precipitation reaction:

$$(1 + y)\ \mathrm{Si} + 2\mathrm{O}_i + xV = \mathrm{SiO}_2 + yI + \text{stress}, \sigma, \tag{5.18}$$

where x is the fraction of V consumed and y a fraction of I emitted. This reaction indicates that V enhances oxygen precipitation.[60]

RTA is a widely used technique to increase the vacancy concentration resulting in enhanced oxygen precipitation taking into account (1) RTA ambient, (2) RTA temperature, and (3) RTA cooling rate. The relationship of thermal equilibrium concentration C^* and that of the diffusion constant D between V and I are $C_V^* > C_I^*$ and $D_I > D_V$ in the temperature range for enhancement of oxygen precipitation with RTA. In Ar ambient, the concentration of V and I in the depth of the wafers becomes constant values of thermal equilibrium by maintaining the RTA temperature. During the wafer cooling, most of I are out-diffused and V remain in Si wafers when the cooling rate is rather high. In the case of N_2 (O_2) ambient, vacancies (self-interstitials) are injected from Si/nitride (oxide) interface. Therefore, the vacancy concentration is higher in N_2 ambient and lower in O_2 ambient in comparison with Ar ambient.

Figure 5.27 (upper) shows the depth profile of the precipitate density with RTA in Ar at 1280°C for 60 s, after two-step annealing at 800°C for 4 h + 1000°C for 16 h.[61] It was found that the oxygen precipitation enhanced with increasing cooling rate. Precipitation enhancement was observed in cases of RTA at ≥1240°C with cooling rate of ≥25 K/s in Ar. Figure 5.27 (lower) shows the depth profile of precipitate density with the same annealing and RTA in N_2 ambient as in Figure 5.27 (upper). It is obvious that the precipitate density was higher than that in Ar under the same RTA condition. The M-like profile appeared clearly with a decrease in the cooling rate. In the case of RTA in O_2 ambient, oxygen precipitation was drastically suppressed.

Figure 5.28 shows the oxide precipitate density as a function of simulated $C_V - C_I$. It was found that the density of nucleated precipitates at 800°C increases rapidly when it is higher than approximately $6 \times 10^{12}\ \mathrm{cm}^{-3}$.[61]

5.5.2 Annealed Wafers

Tying up with the development of electronic devices, the Si wafer manufacturing process has been remarkably developed comparing with Ge. It includes a combination of many different technologies, such as gettering, defect- and flatness-control, and all these processing steps interact with each other. The currently used wafer types for the processing of electronic devices are categorized into polished wafers (PWs), epitaxial wafers (EWs), and AWs, depending on the manufacturing method. Since AWs, which are the subject of this section, are obtained by annealing PWs, first the history of a Si crystal growth development and related gettering technologies is briefly summarized.

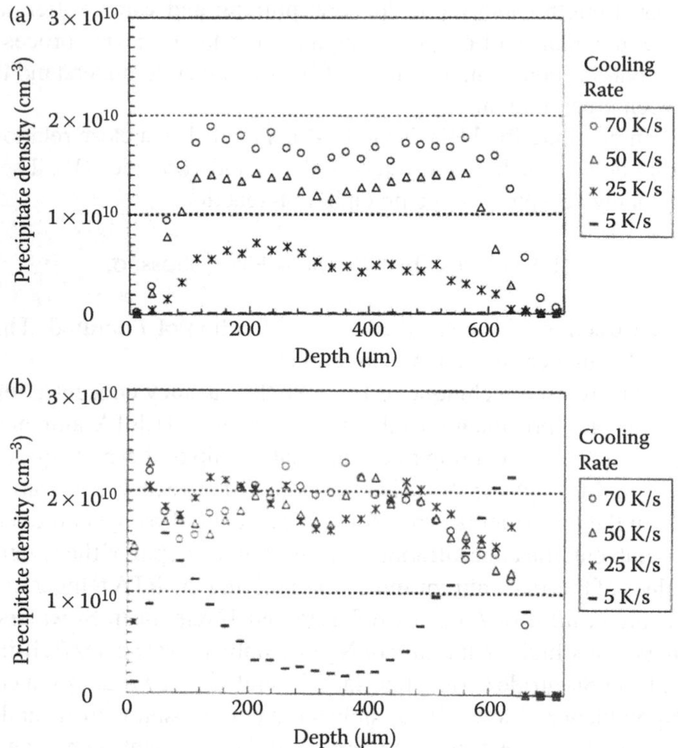

FIGURE 5.27 Depth profile of the precipitate density after two-step annealing at 1000°C for 16 h followed by 800°C for 4 h in a case of RTA in (a) Ar, and in (b) N_2, at 1280°C for 60 s. (From M. Akatsuka et al. *Japanese Journal of Applied Physics* **40**, 2001, 3055. Copyright 2001. The Japan Society of Applied Physics. With permission.)

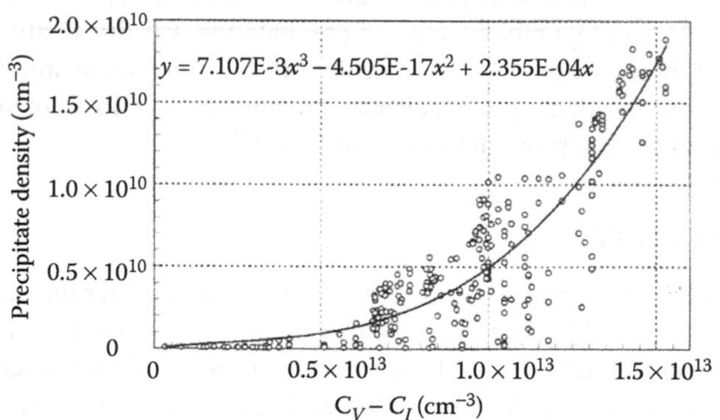

FIGURE 5.28 Relationship between $C_V - C_I$ and density of precipitates nucleated at 800°C. (From M. Akatsuka et al. *Japanese Journal of Applied Physics* **40**, 2001, 3055. Copyright 2001. The Japan Society of Applied Physics. With permission.)

After the discovery of COPs,[62] the so-called "pure-Si," "perfect Si," or "defects-free Si" crystals have been developed by reducing the as-grown defects according to the theory proposed by Voronkov more than 30 years ago,[63] and finally have been commercially manufactured.[64] At that time, "defect-free Si" PWs have been widely accepted for device manufacturing makers due to the cost merit since the primary wafers for the device use are manufactured only by slicing and polishing processes after crystal growth without any additional treatment to remove the defects present in Cz-grown crystals as is the case for EWs or AWs. "Defect-free Si," however, turned out not to be a real defect-free crystal when using, for example, the newly developed evaluation method based on RIE. This method is superior for detecting very small defects in the crystals. The defects that are observed by using RIE, the so-called RIE defects, are present in as-grown crystals even after the optimized growth conditions based on the Voronkov criterion to have no voids and no dislocation loops. These RIE defects are small oxygen precipitates formed during the cooling part of a crystal growth process, and they can cause degradation of the electronic properties of devices.[65] Currently, PWs, EWs, and AWs are for that reason equally considered as candidates as substrate for state-of-the-art devices and there is no straightforward reason to select one of them for all types of devices although each device manufacturer tends to choose wafers according to its own past records and experience.

As gettering technologies, in the past, extrinsic gettering (EG), by either sandblast or polysilicon back sealing, was used extensively. These techniques cannot be used for double-side PWs and meanwhile, DZ-IG (denuded zone–internal gettering) wafers[66,67] have been developed by adjusting the bulk microdefect (BMD) density by controlling the supersaturation of oxygen atoms in the wafer in the late 1980s. Nitrogen or carbon atoms have been introduced as a dopant to increase the controllability of nucleation centers for BMDs. Voids inside the wafers play a role as BMD nucleation centers too and interact strongly with nitrogen atoms that tend to change the shape of the voids. After the year 2000, nitrogen-doped IG wafers have been widely accepted by the device makers since double-side polishing became a requirement in order to achieve a higher wafer flatness. In the conventional DZ-IG technique, the annealing temperature to grow oxide precipitates inside wafers is in the range between 1000°C and 1300°C. This technology inherently focused on PWs with no apparent grown-in defects, the so-called perfect silicon. This technique requires a better control of the interstitial oxygen concentration in the grown crystal to obtain uniformity of the gettering ability within a wafer. An alternative approach is to use high-temperature annealing to erase the thermal history of Cz crystal growth, which is the subject of this section. Such anneal is performed at temperatures close to the melting point of Si, and has totally different effects than the oxygen precipitation anneals at temperatures in the range between 1000°C and 1300°C mentioned above.

The first attempt to change by annealing the wafer properties and to partly erase the thermal history of grown Cz crystals is the DZ-IG process mentioned earlier. At the early times of DZ-IG, however, there were still problems in the layer of the wafer in which devices were formed and DZ-IG only showed a limited or no improvement compared with a simple Cz crystal PW and in some cases even degradation of the performance of devices was observed. Improvement of the crystalline quality of Cz crystal PW by using a high-temperature hydrogen annealing was reported.[68]

Subsequently, high-temperature wafer annealing using inert gas, typically Ar, also showed a similar effect.[69] Annealing both in hydrogen and in inert atmosphere has the following beneficial effects: (i) both gases remove the surface oxide film of the wafers, (ii) dissolved oxygen atoms diffuse out of the wafer and due to that their concentration near the wafer surface decreases, (iii) the inner oxide films of the voids near the surface of the wafer, in which the oxygen atom concentration has decreased, are removed, (iv) voids themselves dissolve and transform into mobile vacancies, (v) these mobile vacancies migrate toward the wafer surfaces and annihilate. Hereby, another option to annihilate voids in (iv) and (v) is to inject Si self-interstitials by wafer surface oxidation, thus filling the voids by absorbing the self-interstitials. Thus, voids near the wafer surfaces of Cz crystals can be efficiently removed by both types of high-temperature annealing. Since Si surfaces are both a sink and a source for intrinsic point defects as mentioned in Section 5.4.3, the reaction in (v) is possible. A key point in annihilating voids is the promotion of reaction (ii), in which temperature and anneal time are the main factors and ramping up to 1200°C is necessary. This series of reactions that make the voids shrink can only proceed when mobile vacancies migrate toward the wafer surfaces and annihilate there.

This technology has specific advantages because of the resulting high gate oxide integrity due to the annihilation of voids while at the same time keeping the gettering ability of the bulk of the wafer due to the remaining high oxygen concentration inside the wafer.[70] The detailed morphology of the voids was studied by transmission electron microscopy (TEM), which revealed that after hydrogen annealing at 1200°C for 1 h the voids became polyhedral.[71] In these studies, however, only knowledge was obtained on the annihilation of voids with sizes of 100 nm and more for which one needed anneals at 1200°C for several hours. This was due to the fact that no suitable commercial furnaces were available for semiconductor use allowing annealing temperatures higher than 1300°C. In those days, before the discovery of RIE defects, the target in erasing the thermal history of grown Cz crystals was mainly the removal of large voids in the near surface area. However, the lowering of oxygen concentration near the wafer surface in steps (i) and (ii) was unavoidable due to the required long-time thermal anneals. This resulted in lower wafer strength in the vicinity of surface, which was in conflict with new requirements arising from the development of new processing steps for advanced devices.

In order to improve the total bulk strength of AWs dramatically, an increase of the density of small-sized BMDs with a uniform distribution within the wafers was applied.[72] This strategy successfully reduced wafer deformation by suppressing the propagation of dislocations using BMDs. It should be noted that BMDs are also effective in capturing contamination introduced during wafer processing. The correlations between misalignment during the photolithography in a device manufacturing process and BMDs, and the optimization of the formation of BMDs and the control process successfully demonstrate the effectiveness of the engineered wafers. The correlation between misalignment and BMDs was also confirmed by Hirano in the case of flash memories.[73] It was also found that BMDs suppress slip in addition and cause a high degree of gettering. On the basis of these facts, it is believed that wafers with high oxygen content generate a larger number of BMDs, which prevent slip in wafers, thereby reducing deformation, misalignment, and defocus during photolithography.

Figure 5.29 shows the dependence of the density of oxygen precipitates on the ramping-up and ramping-down processes in a batch-type furnace.[74] The density of oxygen precipitates depends strongly on the ramping-up rate but it is independent of the ramping-down rate. A lower ramping-up rate increases the density of the precipitates in hydrogen-annealed Si wafers. Few oxygen precipitates are found in the bulk region when the ramping-up rate is 30°C/min or higher at temperatures in the range between 900°C and 1200°C, as shown in Figure 5.30. This is true regardless of the oxygen concentration in the wafers. This is probably because the radii of the oxide precipitate nuclei do not exceed the critical radius for nucleation during the rapid ramping-up process. In the case of a low ramping-up rate, oxygen precipitates grow and their radii increase beyond the critical radius during the ramping-up process, showing that the nucleation of oxygen precipitates in the bulk region can be controlled by adjusting the ramping-up rate during hydrogen annealing.

The effect of oxygen precipitation in annealed Si wafers on the plastic strain induced during rapid thermal processing (RTP) has been studied, with focus on the density of oxide precipitates. It was found that the strain decreased with an increase in the precipitate density. Furthermore, the strain was drastically suppressed when the density of oxygen precipitates (average size = 80 nm) was greater than 10^{10} cm^{-3}. This effect is called "precipitation hardening" and it is due to the blocking of the movement of dislocations during RTP. The dependence of the strain on the density of oxide precipitates is shown in Figure 5.31. The strain decreased markedly throughout the periphery of the wafer with an increase in the density of oxide precipitates. The solid line in Figure 5.31 represents the Orowan stress, which is due to

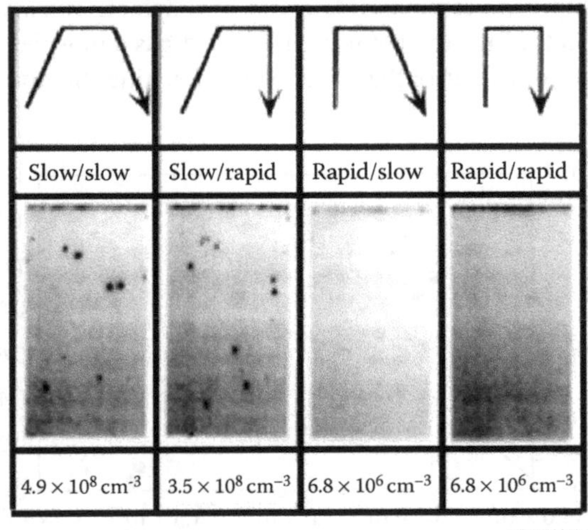

FIGURE 5.29 IR tomography images of oxygen precipitates in hydrogen-annealed Si wafers with [Oi]= 1.6×10^{18} cm^{-3}. (From K. Izunome, *Proceedings of the 6th International Symposium on Advanced Science and Technology of Silicon Materials (JSPS Si Symposium)*, Nov. 19–23, 2012, Kona, Hawaii, USA, pp. 9–13. With permission.)

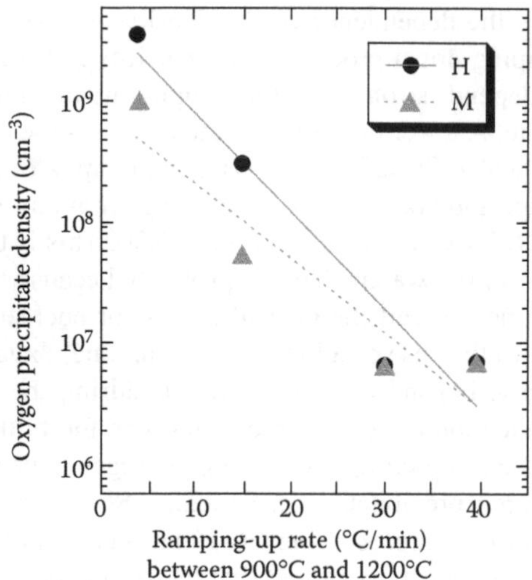

FIGURE 5.30 Effect of the ramping-up rate on oxygen precipitate density in hydrogen-annealed Si wafers with [Oi] = 1.6×10^{18} atoms/cc (M), and $1.72–1.74 \times 10^{18}$ cm^{-3} (H), as observed by IR tomography. (From K. Izunome, *Proceedings of the 6th International Symposium on Advanced Science and Technology of Silicon Materials (JSPS Si Symposium)*, Nov. 19–23, 2012, Kona, Hawaii, USA, pp. 9–13. With permission.)

"precipitation hardening." The oxide precipitates partially suppress the movement of dislocations during RTP.[75]

It was also attempted to remove BMDs and void defects quickly and completely by using RTP at an ultra-high temperature in an inert gas atmosphere or with oxygen gas, which can be used to control the injection of interstitial Si atoms and interstitial

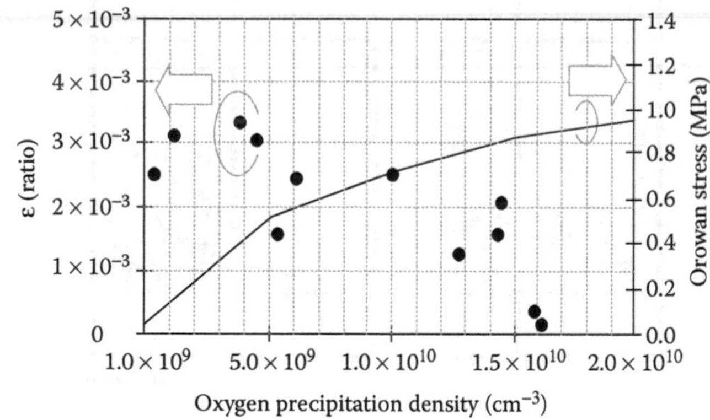

FIGURE 5.31 Dependence of the plastic strain and calculated Orowan stress on the density of oxygen precipitates during RTP. (From K. Izunome, *Proceedings of the 6th International Symposium on Advanced Science and Technology of Silicon Materials (JSPS Si Symposium)*, Nov. 19–23, 2012, Kona, Hawaii, USA, pp. 9–13. With permission.)

oxygen atoms into the substrate. A 300-mm-diameter Cz-Si single-crystalline wafer was heated to a maximum temperature of 1350°C, in an Ar atmosphere and then in an O_2 atmosphere. As a result of combining these two types of heat-treatment ambients, not only BMDs but also void defects near the surface were eliminated almost completely, as shown in Figure 5.32.[76]

Recently, the concept of erasing the thermal history of grown Cz crystals has been further developed. The nuclei leading to OSF ring formation can be annihilated using a thermal process at a temperature higher than 1300°C.[77] The relation between the RIE defects and the nuclei that lead to OSF ring formation is still unclear since the OSF ring is not observed directly but needs appropriate oxidation conditions. They must, however, be somehow related because they are observed after similar crystal growth conditions. Finally, it should be pointed out that the key points of this new technology are RTA above 1300°C and close to the Cz crystal growth temperature. It can be a great advantage that one can tune the intrinsic point defect and oxygen atom concentrations in a wafer by choosing the appropriate RTA treatment. This way one can not only annihilate voids, but also the nuclei that lead to OSF ring formation, RIE defects, and at the same time strengthen the near surface part of the wafer by introducing oxygen atoms into the surface. For the use of this RTA technology in the actual mass production, however, one needs sufficiently low levels of metal contamination and dopant atoms and a major challenge is also preventing slip formation during the RTA.

As mentioned above, Si surfaces act as sinks and sources both for vacancies and self-interstitials, which allow to improve the quality of Cz crystal wafers by using the appropriate anneals. Ge surfaces, however, behave differently in particular with respect to interstitials.[44] Therefore, a different anneal process might be required for Ge wafers. To the knowledge of the authors, there are no published results on the development of annealed Ge wafers.

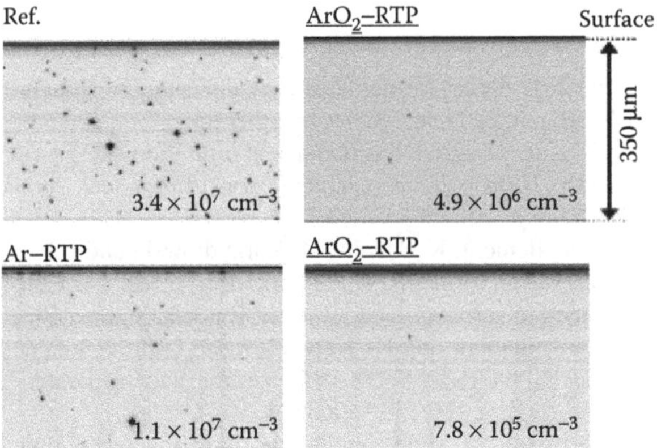

FIGURE 5.32 IR tomography images of void defects formed during various RTPs. (From K. Izunome, *Proceedings of the 6th International Symposium on Advanced Science and Technology of Silicon Materials (JSPS Si Symposium)*, Nov. 19–23, 2012, Kona, Hawaii, USA, pp. 9–13. With permission.)

5.6 CONCLUSIONS AND OUTLOOK

During the last decade, tremendous progress has been made in understanding the properties and behavior of the vacancy in single crystals Si and Ge. This is to a large extent due to the maturing of *ab initio* calculation techniques and the increase of computing power enabling to calculate, for example, the impact of Fermi level and of external and internal stress on intrinsic point defect parameters. Based on these results, the impact on important processes such as defect formation during crystal pulling or device processing can be predicted and dedicated experiments designed to check the results of the *ab initio* calculations. Very recently, the first experimental evidence was presented of the important impact of thermal stress on the conditions for the so-called perfect silicon growth as was predicted earlier based on *ab initio* calculations.[78] This result is very important for the development of the future 450-mm-diameter Si crystals and wafers.

The *ab initio* results also indicate that much more experimental and theoretical work is needed to obtain more reliable values of the formation and migration energies of the intrinsic point defects in stress-free Si as well as of the associated preexponential factors.[79] Hereby, besides crystal pulling data, one should also use other experimental data, such as the (stress-dependent) self-diffusion coefficient D_{SD}, that can be written as an expression of $C_I D_I$ and $C_V D_V$. The experimental stress-dependent D_{SD} values could then be fitted assuming the same stress dependence of the intrinsic point defect parameters as used for the interpretation of the crystal pulling experiments. This approach would probably allow to obtain more reliable values for the stress-free formation and migration energies and prefactors of both point defects and, of course, the stress dependence of these parameters.

It can, therefore, be expected that in the coming years, a number of dedicated crystal pulling experiments or experiments under high pressure will be performed to confirm and quantify further the various impacts of stress, doping, and alloying on point and extended defect behavior in Si, Ge, and their alloys.

REFERENCES

1. B. Puchala, Modeling defect mediated dopant diffusion in silicon, PhD Thesis, The University of Michigan, 2009, and references therein.
2. P. Spiewak and K. Kurzydlowski, Formation and migration energies of the vacancy in Si calculated using the HSE06 range-separated hybrid functional, *Physical Review B* **88**, 2013, 195204.
3. P. Spiewak, J. Vanhellemont, K. Sueoka, K. Kurzydlowski and I. Romandic, First principles calculations of the formation energy and deep levels associated with the neutral and charged vacancy in germanium, *Journal of Applied Physics*, **103**, 2008, 086103.
4. P. Spiewak, J. Vanhellemont and K. Kurzydlowski, Improved calculation of vacancy properties in Ge using the Heyd-Scuseria-Ernzerhof range-separated hybrid functional, *Journal of Applied Physics* **110**, 2011, 063534.
5. K. Sueoka, P. Spiewak and J. Vanhellemont, *Ab initio* analysis of point defects in plane-stressed Si or Ge crystals, *ECS Transactions* **11**, 2007, 375.
6. F. Stillinger and T. Weber, Computer simulation of local order in condensed phases of silicon, *Physical Review B* **31**, 1985, 5262.
7. T. Sinno, Thermophysical properties of intrinsic point defects in crystalline silicon, *Semiconductor Silicon 2002, ECS Proceedings Vol. 2002-2*, 2002, 212.

8. K. Sueoka, E. Kamiyama and H. Kariyazaki, A study on density functional theory of the effect of pressure on the formation and migration enthalpies of intrinsic point defects in growing single crystal Si, *Journal of Applied Physics* **111**, 2012, 093529.
9. K. Sueoka, E. Kamiyama, H. Kariyazaki and J. Vanhellemont, DFT study of the effect of hydrostatic pressure on formation and migration enthalpies of intrinsic point defects in single crystal Si, *Physica Status Solidi C* **9**, 2012, 1947.
10. K. Sueoka, E. Kamiyama and J. Vanhellemont, Theoretical study of the impact of stress on the behavior of intrinsic point defects in large-diameter defect-free Si crystals, *Journal of Crystal Growth* **363**, 2013, 97.
11. S. Centoni, B. Sadigh, G. Gilmer, T. Lenosky, T. Diaz de la Rubia and C. Musgrave, First-principles calculation of intrinsic defect formation volumes in silicon, *Physical Review B* **72**, 2005, 195206.
12. W. Windl, M. Daw, N. Carlson and M. Laudon, Multiscale modeling of stress-mediated diffusion in silicon—Volume tensors, *Materials Research Society Symposium Proceedings* **677**, 2001, AA9.4.1.
13. A. Antonelli, E. Kaxiras and D. Chadi, Vacancy in silicon revisited: Structure and pressure effects, *Physical Review B* **81**, 1998, 2088.
14. O. Sugino and A. Oshiyama, Microscopic mechanism of atomic diffusion in Si under pressure, *Physical Review B* **46**, 1992, 12335.
15. W. Windl, *Ab-initio* calculations of the energetics and kinetics of defects and impurities in silicon, *ECS Transactions* **3**, 2006, 171.
16. P. Spiewak, K. J. Kurzydlowski, J. Vanhellemont, P. Wabinski, K. Mlynarczyk and I. Romandic, Simulation of vacancy cluster formation and binding energies in single crystal Germanium, *Materials Research Society Symposia Proceedings* **994**, 2007, 0994-F03-08.
17. P. Spiewak, M. Muzyk, K. J. Kurzydlowski, J. Vanhellemont, K. Mlynarczyk, P. Wabinski and I. Romandic, Molecular dynamics simulation of intrinsic point defects in germanium, *Journal of Crystal Growth* **303**, 2007, 12.
18. M. Prasad and T. Sinno, Atomistic-to-continuum description of vacancy cluster properties in crystalline silicon, *Applied Physics Letters* **80**, 2002, 1951.
19. M. Prasad and T. Sinno, Internally consistent approach for modeling solid-state aggregation. I. Atomistic calculations of vacancy clustering in silicon, *Physical Review B* **68**, 2003, 045206; ibidem, Internally consistent approach for modeling solid-state aggregation. II. Mean-field representation of atomistic processes, *Physical Review B* **68**, 2003, 045207.
20. A. Bongiorno, L. Colombo and T. Diaz De la Rubia, Structural and binding properties of vacancy clusters in silicon, *Europhysics Letters* **43**, 1998, 695.
21. A. van Veen, H. Schut, A. Rivera and A. V. Fedorov, Growth of vacancy clusters during post-irradiation annealing of ion implanted silicon, in *Ion-Solid Interactions for Materials Modification and Processing, Volume 396 of MRS Symposia Proceedings*, D. B. Poker, D. Ila, Y. -S. Cheng, L. R. Harriott and T. W. Sigmon, eds, Materials Research Society, Pittsburgh, 1996, p. 155.
22. G. H. Gilmer, T. Diaz de la Rubia, D. M. Stock and M. Jaraiz, Diffusion and interactions of point-defects in silicon—Molecular-dynamics simulations, *Nuclear Instruments and Methods in Physics Research Section B: Beam Interactions with Materials and Atoms* **102**, 1995, 247.
23. D. J. Chadi and K. J. Chang, Magic numbers for vacancy aggregation in crystalline Si, *Physical Review B* **38**, 1988, 1523.
24. C. Janke, R. Jones, S. Oberg and P. R. Briddon, Supercell and cluster density functional calculations of the thermal stability of the divacancy in germanium, *Physical Review B* **75**, 2007, 195208.
25. K. Sueoka and J. Vanhellemont, *Ab initio* studies of intrinsic point defects, interstitial oxygen and vacancy or oxygen clustering in germanium crystals, *Materials Science in Semiconductor Processing* **9**, 2006, 494.

26. C. Z. Wang, C. T. Chan and K. M. Ho, Tight-binding molecular-dynamics study of defects in silicon, *Physical Review Letters* **66**, 1991, 189.
27. S. Öğüt, H. Kim and J. R. Chelikowsky, *Ab initio* cluster calculations for vacancies in bulk Si, *Physical Review B* **56**, 1997, 11353.
28. Antônio J. R. da Silva, R. Baierle, R. Mota and A. Fazzio, Native defects in germanium, *Physica B: Condensed Matter* **302–303**, 2001, 364.
29. Z. Q. Wang and D. Stroud, Monte Carlo studies of liquid semiconductor surfaces: Si and Ge, *Physical Review B* **38**, 1988, 1384.
30. J. Vanhellemont and E. Simoen, Brother silicon, sister germanium, *Journal of The Electrochemical Society* **154**, 2007, H572.
31. J. Vanhellemont, P. Spiewak, K. Sueoka and I. Romandic, Intrinsic (point) defects in silicon and germanium, similar but so different, in *Proceedings of the Forum on the Science and Technology of Silicon Materials 2007*, November 12–14, 2007, H. Yamada-Kaneta and H. Ono, eds, Shinkousoku Printing Inc., Niigata, Japan, p. 113.
32. M. Aziz, Stress effects on defects and dopant diffusion in Si, *Materials Science in Semiconductor Processes* **4**, 2001, 397, and references therein.
33. K. Sueoka, E. Kamiyama and J. Vanhellemont, Density functional theory study on the impact of heavy doping on Si intrinsic point defect properties and implications for single crystal growth from a melt, *Journal of Applied Physics* **114**, 2013, 153510, and references therein.
34. K. Tanahashi, H. Harada, A. Koukitsu and N. Inoue, Modeling of point defect behavior by the stress due to impurity doping in growing silicon, *Journal of Crystal Growth* **225**, 2001, 294.
35. T. Abe, Generation and annihilation of point defects by doping impurities during FZ silicon crystal growth, *Journal of Crystal Growth* **334**, 2011, 4.
36. J. Vanhellemont, E. Kamiyama and K. Sueoka, Silicon single crystal growth from a melt: On the impact of dopants on the *v/G* criterion, *ECS Journal of Solid State Science and Technology* **2**, 2013, P166.
37. K. Nakamura, R. Suewaka, T. Saishoji and J. Tomioka, The effect of impurities on the grown-in defects in CZ-Si crystals, (B, C, N, O, Sb, As, P), *Proceedings of the Forum on the Science and Technology of Silicon Materials* 2003, Hayama, Japan, p.161, and references therein.
38. K. Nakamura, T. Saishoji and J. Tomioka, Simulation of the point defect diffusion and growth condition for defect free Cz silicon crystal, *Semiconductor Silicon 2002, PV 2002-2*, The Electrochemical Society Proceedings Series, 2002, Philadelphia, p. 554.
39. E. Kamiyama, K. Sueoka and J. Vanhellemont, Formation energy of intrinsic point defects in Si and Ge and implications for Ge crystal growth, *ECS Journal of Solid State Science and Technology* **2**, 2013, 104.
40. K. Inoue, Y. Morikawa, K. Terakura and M. Nakayama, Order-disorder phase transition on the Si (001) surface: Critical role of dimer defects, *Physical Review B* **49**, 1994, 14774.
41. S. D. Kevan, Surface states and reconstruction on Ge (001), *Physical Review B* **32**, 1985, 2344.
42. T. Uda, H. Shigekawa, Y. Sugawara, S. Mizuno, H. Tochihara, Y. Yamashita, J. Yoshinobu, K. Nakatsuji, H. Kawai and F. Komori, Ground state of the Si(001) surface revisited—Is seeing believing?, *Progress in Surface Science* **76**, 2004, 147.
43. D. Chen and J. J. Boland, Chemisorption-induced disruption of surface electronic structure: Hydrogen adsorption on the Si(100) – 2 × 1 surface, *Physical Review B* **65**, 2002, 165336.
44. E. Kamiyama, K. Sueoka and J. Vanhellemont, Surface-induced charge at a Ge (100) dimer surface and its interaction with vacancies and self-interstitials, *Journal of Applied Physics* **113**, 2013, 093503.

45. E. Kamiyama and K. Sueoka, First principles analysis on interaction between vacancy near surface and dimer structure of silicon crystal, *Journal of Applied Physics* **111**, 2012, 013521.
46. E. Kamiyama K. Sueoka and J. Vanhellemont, *Ab initio* study of vacancy and self-interstitial properties near single crystal silicon surfaces, *Journal of Applied Physics* **111**, 2012, 083507.
47. T. Motooka, K. Nishihira, R. Oshima, H. Nishizawa and F. Hori, Atomic diffusion at solid/liquid interface of silicon: Transition layer and defect formation, *Physical Review B*, **65**, 2002, 081304.
48. E. Kamiyama, K. Sueoka and J. Vanhellemont, *Ab initio* analysis of a vacancy and a self-interstitial near single crystal silicon surfaces: Implications for intrinsic point defect incorporation during crystal growth from a melt, *Physica Status Solidi A* **209**, 2012, 1880.
49. E. Kamiyama, K. Sueoka, K. Izunome, K. Kashima, O. Nakatsuka, N. Taoka, S. Zaima and J. Vanhellemont, Doppler broadening spectroscopy of positron annihilation near Ge and Si (001) single crystal surfaces, *ECS Solid State Letters* **2**, 2013, 89.
50. R. Krause-Rehberg and H. S. Leipner, *Positron Annihilation in Semiconductors, Defect Studies*, Springer Series in Solid-State Sciences 127, Springer, Berlin, 1999.
51. http://positronannihilation.net/index_files/Page364.htm.
52. J. Slotte, M. Rummukainen, F. Tuomisto, V. P. Markevich, A. R. Peaker, C. Jeynes and R. M. Gwilliam, Evolution of vacancy-related defects upon annealing of ion-implanted germanium, *Physical Review B* **78**, 2008, 085202.
53. S. Valkealahti and R. M. Nieminen, Monte-Carlo calculations of keV electron and positron slowing down in solids, *Applied Physics A* **32**, 1983, 95.
54. P. Asoka-Kumar, K. G. Lynn and D. O. Welch, Characterization of defects in Si and SiO_2–Si using positrons, *Journal of Applied Physics* **76**, 1994, 4935.
55. H. Bracht, S. Schneider, J. N. Klug, C. Y. Liao, J. Lundsgaard Hansen, E. E. Haller, A. Nylandsted Larsen, D. Bougeard, M. Posselt and C. Wündisch, Interstitial-mediated diffusion in germanium under proton irradiation, *Physical Review Letters* **103**, 2009, 255501.
56. H. Bracht, S. Schneider and R. Kube, Diffusion and doping issues in germanium, *Microelectronic Engineering* **88**, 2011, 452.
57. S. Fetah, A. Chikouche, A. Dkhissi, G. Landa and P. Pochet, Stability of Frenkel pairs in Si(1 0 0) surface in the presence of germanium and oxygen atoms, *Microelectronic Engineering* **88**, 2011, 503.
58. Y. Zhao, M. Aziz, H. Gossmann, S. Mitha and D. Schiferl, Activation volume for boron diffusion in silicon and implications for strained films, *Applied Physics Letters*, **74**, 1999, 31.
59. K. Sueoka, Oxygen precipitation in lightly and heavily doped Czochralski Silicon, *9th International Symposium on High Purity Silicon, The Electrochem. Soc. Proceedings* **3**, 2006, 71–87.
60. K. Sueoka, N. Ikeda, T. Yamamoto and S. Kobayashi, Morphology and growth process of thermally induced oxide precipitates in Czochralski silicon, *Journal of Applied Physics* **74**, 1993, 5437.
61. M. Akatsuka, M. Okui, N. Morimoto and K. Sueoka, Effect of rapid thermal annealing on oxygen precipitation behavior in silicon wafers, *Japanese Journal of Applied Physics* **40**, 2001, 3055.
62. J. Ryuta, E. Morita, T. Tanaka and Y. Shimanuki, Crystal-originated singularities on Si wafer surface after SC1 cleaning, *Japanese Journal of Applied Physics* **29**, 1990, L1947.
63. V. V. Voronkov, The mechanism of swirl defects formation in silicon, *Journal of Crystal Growth* **59**, 1982, 625.
64. J. G. Park, G. S. Lee, J. M. Park, S. M. Chon and H. K. Chung, Effect of crystal defects such as COPs, large dislocations, and OSF ring on device integrity degradation, *Proceedings*

of the Third International Symposium on Defects in Silicon, Electrochemical Society Series, **99**, 1999, 324.

65. K. Nakashima, Y. Watanabe, T. Yoshida and Y. Mitsushima, A method to detect oxygen precipitates in silicon wafers by highly selective reactive ion etching, *Journal of The Electrochemical Society* **147**, 2000, 4294.
66. T. Y. Tan, E. E. Gardner and W. K. Tice, Intrinsic gettering by oxide precipitate induced dislocations in Czochralski Si, *Applied Physics Letters* **30**, 1977, 175.
67. K. Nagasawa, Y. Matsushita and S. Kishino, A new intrinsic gettering technique using microdefects in Czochralski silicon crystal: A new double preannealing technique, *Applied Physics Letters* **37**, 1980, 622.
68. Y. Matsushita, M. Wakatsuki and Y. Saito, Improvement of silicon surface quality by H_2 anneal, *Extended Abstracts of the 18th (1986 International) Conference on Solid State Devices and Materials*, Business Center for Academic Societies Japan, Tokyo, 1986, p. 529.
69. N. Adachi, T. Hisatomi, M. Sano and H. Tsuya, Reduction of grown-in defects by high temperature annealing, *Journal of Electrochemical Society*, **147**, 2000, 350.
70. Y. Matsushita, S. Samata, M. Miyashita and H. Kubota, Improvement of thin oxide quality by hydrogen annealed wafer, *Proc. 1995 Int. Electron Devices Meeting Technical Digest*, IEEE, Piscataway, 1994, pp. 321.
71. H. Fujimori, H. Matsushita, I. Oose and T. Okabe, Depth effect of the morphology change induced by hydrogen annealing of grown-in defects in silicon, *Journal of The Electrochemical Society* **147**, 2000, 3508.
72. K. Izunome, Advanced silicon wafers for leading-edge semiconductor devices, *Proceedings of the 6th International Symposium on Advanced Science and Technology of Silicon Materials (JSPS Si Symposium)*, Nov. 19–23, 2012, Kona, Hawaii, USA, pp. 9–13.
73. Y. Hirano, K. Yamazaki, F. Inoue, K. Imaoka, K. Tanahashi and H. Yamada-Kaneta, Impact of defects in silicon substrate on flash memory characteristics, *Journal of The Electrochemical Society* **154**, 2007, H1027.
74. K. Izunome, H. Shirai, K. Kashima, J. Yoshikawa and A. Hojo, Oxygen precipitation in Czochralski-grown silicon wafers during hydrogen annealing, *Applied Physics Letters* **68**, 1996, 49.
75. K. Araki, H. Sudo, T. Aoki, T. Senda, H. Isogai, H. Tsubota, M. Miyashita, et al., Effect of oxygen precipitation in nitrogen-doped annealed silicon wafers on thermal strain induced by rapid thermal processing, *Japanese Journal of Applied Physics* **49**, 2010, 080205.
76. T. Senda, K. Araki, H. Saito, S. Maeda, K. Kashima and K. Izunome, Annihilation behaviour of void defects in Czochralski silicon single crystal using rapid thermal process with ultra high temperature, *Proceedings of the 6th International Symposium Advanced Science and Technology of Silicon Materials (JSPS Si Symposium)*, Kona, Hawaii, Nov. 19–23, 2012, pp. 55–58.
77. K. Araki, S. Maeda, T. Senda, H. Sudo, H. Saito and K. Izunome, Impact of rapid thermal oxidation at ultrahigh-temperatures on oxygen precipitation behavior in Czochralski-silicon crystals, *ECS Journal of Solid State Science and Technology* 2, 2013, P66.
78. K. Nakamura, R. Suewaka and B. Ko, Experimental study of the impact of stress on the point defect incorporation during silicon growth, *ECS Solid State Letters* 3(3), 2014, N5.
79. J. Vanhellemont, E. Kamiyama and K. Sueoka, Comment on experimental study of the impact of stress on the point defect incorporation during silicon growth, *ECS Solid State Letters*, **3**(N5), 2014; *ECS Solid State Letters*, **3**(5), 2014, X3.

6 Self- and Dopant Diffusion in Silicon, Germanium, and Their Alloys

Hartmut Bracht

CONTENTS

6.1 Introduction ... 159
6.2 Diffusion in Silicon ... 161
6.2.1 Self-Diffusion ... 162
6.2.2 Dopant Diffusion ... 169
6.2.2.1 Mechanisms of Boron Diffusion ... 172
6.2.2.2 Mechanisms of Phosphorus Diffusion ... 173
6.2.2.3 Mechanisms of Arsenic Diffusion ... 175
6.2.3 Comparison between Self- and Dopant Diffusion ... 178
6.2.4 Nonequilibrium Diffusion ... 178
6.3 Diffusion in Germanium ... 182
6.3.1 Self-Diffusion ... 184
6.3.2 Dopant Diffusion in Germanium ... 187
6.3.2.1 Donor Diffusion ... 188
6.3.2.2 Acceptor Diffusion ... 190
6.3.2.3 Formation of Dopant–Defect Complexes ... 192
6.3.3 Nonequilibrium Diffusion ... 193
6.4 Diffusion in Silicon–Germanium Alloys ... 197
6.4.1 Self-Diffusion ... 197
6.4.2 Dopant Diffusion ... 202
6.5 Conclusion and Outlook ... 204
Acknowledgments ... 206
References ... 206

6.1 INTRODUCTION

Over the last six decades, our daily life has been revolutionized by the invention of solid-state electronic devices. The key factor for this development was the preparation of high-purity semiconductors as well as the ability to control the impurity level, that is, the incorporation of defects on the atomic scale. Currently, nanoelectronic devices increasingly consist of layered materials with defect densities that limit the integrity of their structure. Further miniaturization is intimately connected with an improved understanding of the properties and interaction of atomic defects to effectively control the technological processing steps to fabricate advanced functional devices. Defect control still remains a challenge for progress in semiconductor devices. Although atomistic calculations are increasingly used to predict defect types, defect interactions, their stability, mobility, and electronic properties, the experimental relevance of the theoretical results needs to be verified as the theoretical results are mainly representative for $T = 0$ K. In this respect, diffusion studies are of pivotal significance as they provide valuable information about the atomic mechanisms of mass transport, the type of point defects involved, and their thermodynamic properties. The diffusion behavior reflects defect interactions and defect properties at elevated temperatures that, in particular, are relevant for the fabrication of semiconductor devices. This chapter treats the diffusion of self- and dopant atoms in silicon, germanium, and their alloys. The defect interactions and defect properties deduced from diffusion studies are summarized and compared to theoretical calculations.

Numerous diffusion experiments with group-IV semiconductors have been performed over the past few decades [1]. These studies mainly comprise self- and dopant diffusion experiments that, according to common practice, were performed separately. More recently, the availability of highly enriched silicon (Si) and germanium (Ge) isotopes and the development of epitaxial deposition systems enabled the growth of high-purity epitaxial layers with homogeneous chemical but alternating isotopic composition. Such isotopically controlled semiconductor heterostructures grown, for example, by means of chemical vapor deposition (CVD) and molecular beam epitaxy (MBE) are very advantageous for studying self- and dopant diffusion [2–12]. Previous self-diffusion studies utilizing radioactive isotopes of the host material are limited at the time of diffusion due to the short radioactive half-life of ^{31}Si (2.6 h). In contrast, no severe limitation exists when self-diffusion is investigated with isotope heterostructures. Combined with state-of-the-art depth profiling methods such as secondary ion mass spectrometry (SIMS) [2–4,12], neutron reflectometry (NR) [8,12,13], and atom-probe tomography (APT) [14–17], even small diffusion lengths of a few nanometers can be determined. This allows extending diffusion experiments to a wider temperature range so as to identify possible changes in the dominant mechanisms of diffusion [12]. Moreover, self- and dopant diffusion can be performed simultaneously with isotope heterostructures [6,7,9–11]. These experiments reveal the impact of dopant diffusion and doping on self-diffusion and provide direct access to the charge states of the native point defects [6,7,9–11]. Self-diffusion studies with semiconductors of natural abundance aiming at similar results are extremely laborious because experiments with samples of different background doping level have to be performed. Compared to separate self- and dopant-diffusion experiments, the simultaneous diffusion of self- and dopant atoms also yields more detailed insights on the mechanisms of self- and dopant diffusion [7,9]. Last but not least, isotopically controlled semiconductor heterostructures are highly valuable for studying atomic transport processes under nonequilibrium conditions realized, for example, by irradiation [18–20]. Apart from diffusion studies, isotopically controlled semiconductors are also advantageous for spectroscopic investigations in identifying the actual constituents of, for example, deep luminescence defect centers in silicon [21,22]. Finally, isotope heterostructures are well suited to investigate ion-beam-induced self-atom mixing [23–26] and the impact of isotope doping and modulation on heat transport [27–31].

In the following sections, self- and dopant diffusion in Si, Ge, and their alloys are discussed. Results gained from diffusion studies under thermal equilibrium conditions utilizing isotopically modulated semiconductor heterostructures are compared to self- and dopant-diffusion experiments performed separately. Moreover, nonequilibrium self-diffusion experiments investigated with isotope structures of Si and Ge are presented that complement our understanding on the mechanisms of self- and dopant diffusion, the type of native point defects involved, and their interaction with dopants. All this information is required to develop efficient doping strategies that help to improve the fabrication of group-IV-based nanoelectronic devices.

6.2 DIFFUSION IN SILICON

The vast majority of experiments on diffusion in Si concerns the diffusion of foreign atoms [1]. Foreign atom diffusion can be classified into those that are dissolved on mainly (i) interstitial, (ii) both interstitial and substitutional, and (iii) mainly substitutional lattice sites. Mainly, interstitial dissolved foreign atoms, such as hydrogen (H), iron (Fe), nickel (Ni), and copper (Cu), reveal the highest diffusion coefficients in Si as displayed in Figure 6.1.

The diffusion of these elements proceeds by atomic jumps from one interstice to another, that is, via an interstitial diffusion mechanism (see Figure 6.2a). Hybrid elements such as platinum (Pt), gold (Au), zinc (Zn), sulfur (S), and carbon (C) occupy both interstitial and substitutional sites. The diffusion of hybrid elements is described by an interstitial–substitutional exchange, namely, the kick-out and the dissociative mechanisms (see Figure 6.2b and c). The respective diffusion coefficients are lower than those of interstitial foreign atoms but higher than the diffusion of mainly substitutional dissolved atoms (see Figure 6.1). The group-III elements, boron (B), aluminum (Al), and gallium (Ga), and the group-V elements, phosphorus (P), arsenic (As), and antimony (Sb), mainly occupy substitutional sites and give rise to hole (p-type) and electron (n-type) conductivity in Si. These dopants diffuse via native point defects, that is, vacancies (V) and self-interstitials (I) (see Figure 6.2) that approach the substitutional atom and form mobile dopant–defect pairs (see Figure 6.2d and e). As demonstrated in Figure 6.1, the n- and p-type dopants diffuse faster than the Si self-atoms. It has been noted that the diffusion coefficients of all foreign atoms in Si are either higher or very close to the self-diffusion coefficient. This is specific for Si. In Ge, for example, B diffuses several orders of magnitude slower than Ge self-atoms (see Section 6.3). The diffusion coefficient D_A illustrated in Figure 6.1, where A represents self- and foreign atoms, describes the diffusion for electronic intrinsic and thermal equilibrium conditions, that is, the Fermi level is located in the middle of the band gap and the concentrations of V and I are in thermal equilibrium.

In Sections 6.2.1 and 6.2.2, the individual contributions to self- and dopant diffusion deduced from diffusion experiments with Si isotope heterostructures are

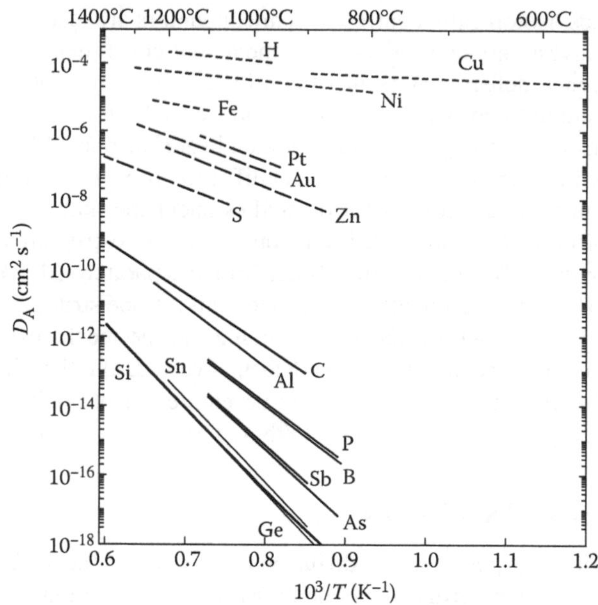

FIGURE 6.1 Temperature dependence of the diffusion coefficients of foreign atoms in Si in comparison to self-diffusion. The diffusion data of mainly interstitial dissolved elements is indicated by the short-dashed lines (H [32]; Cu [33]; Ni [34]; Fe [35]). Long-dashed lines illustrate diffusion data of hybrid elements (Pt [36]; Au [37]; Zn [38]; S [39]). These elements are dissolved on substitutional and interstitial sites but mainly diffuse in an interstitial configuration via the kick-out and dissociative mechanisms (see Figure 6.2). The solid lines represent diffusion data of group-III and group-V dopants (B [7,40]; Al [41]; P [7,42,43]; As [7,44]; Sb [45]). The diffusion of the isovalent elements (lower solid lines: Ge [46]; Sn [47]) is very similar in magnitude to Si self-diffusion (thick solid line: [12]).

discussed and the individual contributions to D_A due to the different mechanisms of diffusion are specified.

6.2.1 Self-Diffusion

Self-diffusion is the most fundamental process of matter transport and is mediated by native point defects. In principle, a direct exchange of lattice atoms that does not involve any native point defects could also contribute to self-diffusion [48]. No experimental evidence has been found for this contribution to self-diffusion so far. Compared to self-diffusion in close-packed metals, which is mainly mediated by V, both V and I contribute to self-diffusion in Si under thermal equilibrium. The presence of both V and I in Si becomes evident in the diffusion behavior of hybrid and dopant elements and by the type of microscopic defects that evolve during Si single crystal growth. Considering contributions of V and I to Si self-diffusion, the self-diffusion coefficient D_{Si} for thermal equilibrium conditions is given by (see, e.g., Reference 6).

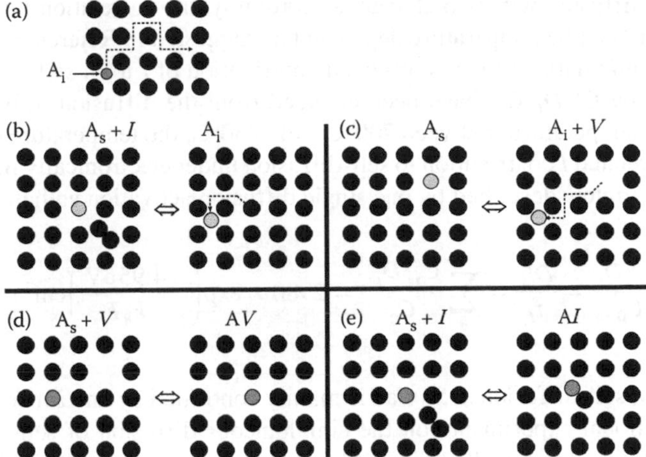

FIGURE 6.2 Schematic 2D representation of (a) the direct interstitial and the indirect (b) kick-out, (c) dissociative, (d) vacancy, and (e) interstitialcy diffusion mechanisms of an element A in a solid. A_i, A_s, V, and I denote interstitially and substitutionally dissolved foreign atoms, vacancies, and self-interstitials, respectively. AV and AI are defect pairs of the corresponding defects.

$$D_{\mathrm{Si}} = D_{\mathrm{Si}}^{V} + D_{\mathrm{Si}}^{I} = \sum_k f_{V^k} \frac{C_{V^k}^{\mathrm{eq}} D_{V^k}}{C_0} + \sum_u f_{I^u} \frac{C_{I^u}^{\mathrm{eq}} D_{I^u}}{C_0} \tag{6.1}$$

with the Si atom density $C_0 = 5 \times 10^{22}$ cm^{-3}. The first (second) term considers the contribution of single V^k (I^u) with the charge state k (u) to self-diffusion via the vacancy (interstitialcy) mechanism. These mechanisms are schematically illustrated in Figure 6.2d and e. The site exchange of a tagged Si atom requires a native point defect next to it and a diffusion jump of the defect. The probability of finding a native defect next to a tagged Si atom depends on its concentration $C_{V^k,I^u}^{\mathrm{eq}}$ established under thermal equilibrium. This concentration is determined by the formation energy of the defect. The diffusion D_{V^k,I^u} of the defect is described by the migration energy barrier between initial and final states. The nonrandom diffusion of tagged Si atoms compared to the random diffusion of V^k and I^u is considered by the correlation factor f_{V^k,I^u}. Although the correlation factor f_{V^k} for self-diffusion via V^k equals 0.5 for all charge states [49], f_{I^u} likely depends on the charge state of I^u as with the charge state also the preferred structure of I can change. Theory predicts I configurations with self-atoms that occupy tetrahedral and hexagonal interstices or form split interstitials, that is, dumbbell interstitials, where two self-atoms share one lattice site. Different structures of I cause different migration energies, migration paths, and correlation factors. For the dumbbell configuration, a correlation factor f_I of 0.59 is calculated by means of molecular dynamics simulations [50], whereas a statistical treatment of diffusion via tetrahedral interstices yields 0.73 [51,52]. In the case where the preferred configuration of I changes with temperature, the dominant contribution of I^u to self-diffusion will also change. Accordingly, the correlation factor for self-diffusion via I can be temperature dependent as suggested in Reference 52.

Experimental data on the uncorrelated contribution of I to Si self-diffusion, that is described by $C_I^{eq} D_I / C_0$, have been deduced from the diffusion of hybrid atoms [53–59]. For temperatures between 700°C and 1250°C, the temperature dependence of the uncorrelated I contribution to self-diffusion under electronically intrinsic conditions is accurately described by one single diffusion activation enthalpy [55].

$$\frac{C_I^{eq} D_I}{C_0} = \frac{D_{Si}^I}{f_I} = \sum_u \frac{C_{I^u}^{eq} D_{I^u}}{C_0} = 2980 \times \exp\left(-\frac{4.95\,\text{eV}}{k_B T}\right) \text{cm}^2\,\text{s}^{-1} \tag{6.2}$$

Obviously, self-diffusion in Si via I is mainly controlled by one I-mediated mechanism. Additional experiments on the simultaneous diffusion of self- and dopant atoms in isotopically controlled Si multilayer structures revealed the individual contributions of charged I^u to the total I contribution to self-diffusion [6]. These contributions given by

$$\frac{C_{I^0}^{eq} D_{I^0}}{C_0} = 2732 \exp\left(-\frac{4.96\,\text{eV}}{k_B T}\right) \text{cm}^2\,\text{s}^{-1} \tag{6.3}$$

$$\frac{C_{I^+}^{eq} D_{I^+}}{C_0} = 69.6 \exp\left(-\frac{4.82\,\text{eV}}{k_B T}\right) \text{cm}^2\,\text{s}^{-1} \tag{6.4}$$

$$\frac{C_{I^{2+}}^{eq} D_{I^{2+}}}{C_0} = 469 \exp\left(-\frac{5.02\,\text{eV}}{k_B T}\right) \text{cm}^2\,\text{s}^{-1} \tag{6.5}$$

are illustrated in Figure 6.3 and demonstrate that neutral I^0 mainly control I-mediated self-diffusion. The comparison to the Si self-diffusion coefficient D_{Si} obtained from direct self-diffusion studies [2] supports a value of about 0.6 for the correlation factor of self-diffusion via I. This correlation factor is in accordance with theoretical predictions of I migration via dumbbells (split interstitials) [50]. Hence, the current experimental and theoretical results indicate that I-mediated self-diffusion in undoped Si is mainly controlled by neutral I^0, which according to theoretical calculations are likely dumbbells, that is, split interstitials. With increasing acceptor doping, the contribution of singly and doubly positively charged I to self-diffusion increases. Finally, under extrinsic p-type doping, that is, the position of the Fermi level is close to the valence band, positively charged I dominate the I-mediated self-diffusion in Si.

The uncorrelated contribution of V to Si self-diffusion given by $C_V^{eq} D_V / C_0$ was also determined from the diffusion behavior of hybrid atoms. The supersaturation of V established during out-diffusion of Zn from homogeneously Zn-doped Si crystals leads to a V-controlled Zn diffusion. Accordingly, the Zn out-diffusion behavior and

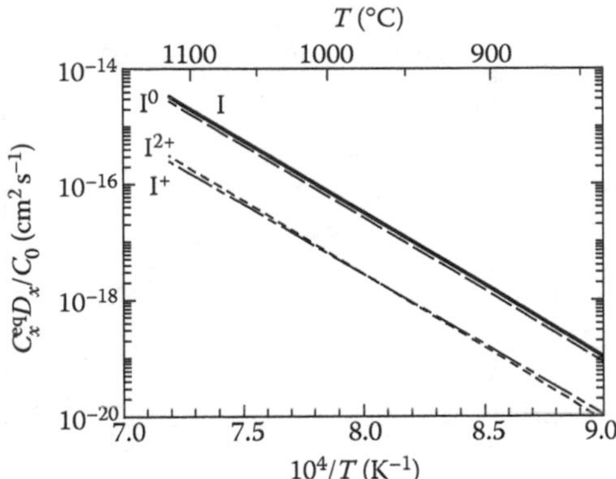

FIGURE 6.3 Temperature dependence of the transport coefficients $C_{I^u}^{eq}D_{I^u}/C_0$ of I^0 (long-dashed line), I^+ (long–short-dashed line), and I^{2+} (short-dashed line) deduced from modeling the simultaneous diffusion of self- and dopant atoms in Si isotope multilayer structures [7]. The dashed lines are reproduced by Equations 6.3 through 6.5. The thick solid line represents the sum of all contributions that adds up to Equation 6.2.

thus the corresponding Zn profiles bear information about the uncorrelated V contribution to self-diffusion. However, microscopic defects induced by Zn out-diffusion have hindered an accurate determination of the uncorrelated V contribution to self-diffusion [60]. For this reason, the V contribution to self-diffusion is more accurately deduced by means of Equation 6.1 from the directly measured Si self-diffusion coefficient D_{Si} [2] and the I contribution to self-diffusion (see Equation 6.2). Assuming $f_V = f_{V^k} = 0.5$ [49] and $f_I = f_{I^u} = 0.56$ [50] for all charge states, the uncorrelated V contribution to Si self-diffusion for intrinsic doping conditions is given below [7] (Equation 18 in Reference 7 shows a preexponential factor that is a factor of two to small. Erroneously, the temperature dependence of D_{Si}^V is given and not of D_{Si}^V/f_V with $f_V = 0.5$):

$$\frac{C_V^{eq}D_V}{C_0} = \frac{D_{Si}^V}{f_V} = 86 \times \exp\left(-\frac{4.56\,\text{eV}}{k_B T}\right)\text{cm}^2\,\text{s}^{-1} \tag{6.6}$$

The V contribution is displayed in Figure 6.4 along with the individual uncorrelated contributions of charged V^k to Si self-diffusion. The contributions of the various charge states k of V^k were deduced from the simultaneous diffusion of self- and dopant atoms in Si [6] and are described by

$$\frac{C_{V^0}^{eq}D_{V^0}}{C_0} = 13.3\exp\left(-\frac{4.42\,\text{eV}}{k_B T}\right)\text{cm}^2\,\text{s}^{-1} \tag{6.7}$$

$$\frac{C_{V^-}^{eq}D_{V^-}}{C_0} = 38.9\exp\left(-\frac{4.63\,\text{eV}}{k_B T}\right)\text{cm}^2\,\text{s}^{-1} \tag{6.8}$$

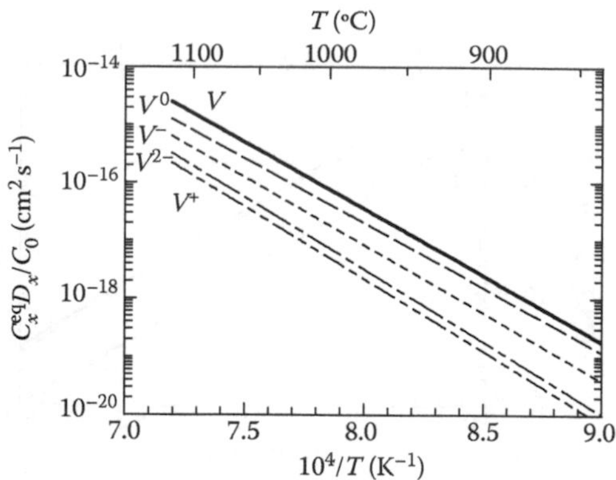

FIGURE 6.4 Temperature dependence of the transport coefficients $C^{eq}_{V^u}D_{V^u}/C_0$ of V^0 (long-dashed line), V^- (short-dashed line), V^{2-} (long–short-dashed line), and V^+ (long–short–short-dashed line) deduced from modeling the simultaneous diffusion of self- and dopant atoms in Si isotope multilayer structures [7]. The dashed lines are reproduced by Equations 6.7 through 6.10 and add up to the total V contribution to self-diffusion given by Equation 6.6.

$$\frac{C^{eq}_{V^{2-}}D_{V^{2-}}}{C_0} = 240\exp\left(-\frac{4.93\,\text{eV}}{k_B T}\right)\text{cm}^2\,\text{s}^{-1} \tag{6.9}$$

$$\frac{C^{eq}_{V^{+}}D_{V^{+}}}{C_0} = 150\exp\left(-\frac{4.92\,\text{eV}}{k_B T}\right)\text{cm}^2\,\text{s}^{-1} \tag{6.10}$$

The experiments demonstrate that V-mediated self-diffusion in Si is mainly controlled by neutral V.

The total contributions of I and V to Si self-diffusion and their sum according to Equation 6.1 are displayed in Figure 6.5 in comparison to the direct measured Si self-diffusion coefficient D_{Si} [2,5,12]. For temperatures between 850°C and 1400°C, the individual contributions of I and V given by Equations 6.2 and 6.6 add up accurately to D_{Si} (see black dashed line in Figure 6.5). The most recent self-diffusion data obtained for temperatures between 730°C and 850°C clearly exceed the sum of D^I_{Si} and D^V_{Si} extrapolated to lower temperatures. Combining the earlier and most recent self-diffusion data deduced from self-diffusion in undoped isotope heterostructures, a clear bowing is evident in the temperature dependence of D_{Si}. Assuming that self-diffusion is controlled by V and I with a diffusion activation enthalpy of 4.95 eV for I-mediated self-diffusion as suggested by Equation 6.2, an activation enthalpy of self-diffusion via V of 3.5–3.6 eV is obtained [5,12]. The corresponding V contribution to self-diffusion not only differs from the results of the Zn out-diffusion study but is also in conflict with the activation enthalpy of V-mediated dopant diffusion in Si [12]. An overall consistent interpretation of Si

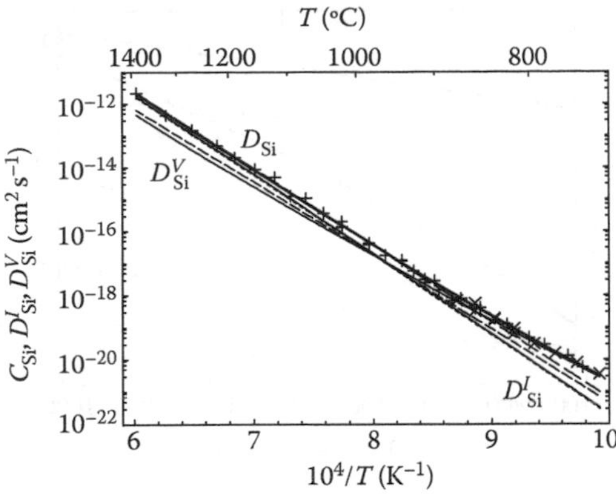

FIGURE 6.5 Silicon self-diffusion coefficients D_{Si} deduced from self-diffusion experiments under thermal equilibrium conditions. (+): data from SIMS analyses of self-diffusion profiles in isotopically controlled $^{\mathrm{nat}}\mathrm{Si}/^{28}\mathrm{Si}/^{\mathrm{nat}}\mathrm{Si}$ sandwich structures [2,12]. (×): self-diffusion data obtained from Raman shift measurements of annealed $^{28}\mathrm{Si}/^{30}\mathrm{Si}$ isotope superlattices [5]. A clear upward bowing is observed in the logarithmic representation of D_{Si} versus the inverse temperature $1/T$. The temperature dependence is accurately described with the I contribution D^{I}_{Si} to self-diffusion deduced from Zn diffusion in Si [55] (see thin solid line and Equation 6.2 with $f_I = 0.6$ [12]). The lower thin solid line (for $T > 1000°\mathrm{C}$) shows the V contribution D^{V}_{Si} to self-diffusion taking into account temperature-dependent thermodynamic properties of V displayed in Figure 6.6. The upper thick solid line represents the total self-diffusion coefficient $D_{\mathrm{Si}} = D^{I}_{\mathrm{Si}} + D^{V}_{\mathrm{Si}}$. The dashed lines show the I (short dashed) and V (long dashed) contribution to D_{Si} according to Equations 6.2 and 6.6 with $f_I = 0.56$ and $f_V = 0.5$, respectively. The thick-dashed lines show the corresponding sum of D^{I}_{Si} and D^{V}_{Si} that equals D_{Si} for 1400–850°C but deviates for lower temperatures.

self-diffusion is achieved assuming temperature-dependent thermodynamic properties of V [12]. The resulting contribution D^{V}_{Si} to self-diffusion is shown in Figure 6.5 as a thin solid line.

The temperature dependence of the formation and migration enthalpy (entropy) of V is displayed in Figure 6.6. The formation (migration) enthalpy of V increases from 3.2 eV (0.5 eV) to 3.9 eV (0.8 eV) for temperatures between 700°C and 1400°C. In the same temperature range, the activation enthalpy Q_V of V-mediated self-diffusion increases from 3.7 to 4.7 eV. The temperature dependence yields Q_V values close to and higher than 4 eV for temperatures $\gtrapprox$ 900°C that are in good agreement with V-mediated dopant diffusion in Si (see Section 6.2.2).

A possible temperature dependence of defect properties was already proposed by Seeger and Chik [61] to explain the high activation enthalpy and entropy of Si self-diffusion. The native defects mediating self-diffusion were proposed to be extended or spread-out defects. This concept of spread-out defects could not be verified because former self-diffusion experiments were restricted to a much narrower temperature range than recent self-diffusion studies performed with isotopically

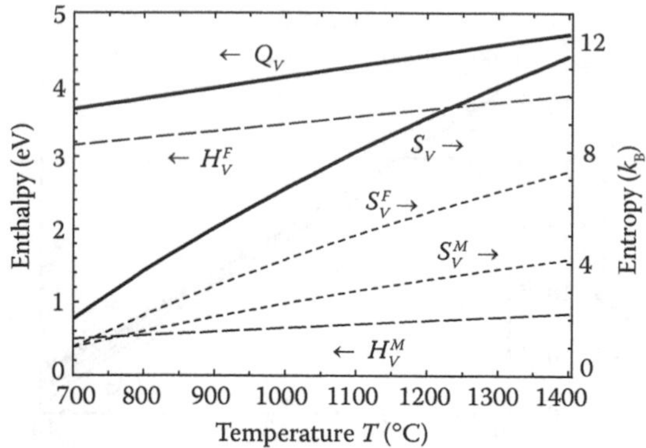

FIGURE 6.6 Temperature dependence of the thermodynamic properties of V in Si deduced from the temperature dependence of Si self-diffusion. The activation enthalpy (entropy) Q_V (S_V) of self-diffusion via V, which is given by the sum of the enthalpy (entropy) of V formation $H_V^F (S_V^F)$ and migration $H_V^M (S_V^M)$, is referred to the left (right) y-axis.

controlled heterostructures. As far as the present self-diffusion data on Si and the results on I and V contributions are concerned, only the thermodynamic properties of V seem to be temperature dependent. Available experimental results on the I contribution to self-diffusion do not indicate temperature-dependent I properties.

Although the concept of temperature-dependent defect properties is very intriguing and provides a consistent interpretation of self- and dopant diffusion in Si (see Reference 12), an independent verification is still necessary. The diffusion of foreign atoms such as Sb and tin (Sn) in Si, whose migration is mainly mediated by V [47], could verify the concept, provided the experiments are performed over a wide range of temperatures to identify deviations from a singly Arrhenius-type temperature dependence.

Recent molecular dynamics (MD) calculations seem to support the concept of extended defects in both Si and Ge [62]. The calculations suggest that extended defects exist in Si and Ge. However, temperature-dependent I properties are not supported by available experimental data (see, e.g., Reference 12). Moreover, the type of I does not seem to change significantly with temperature, that is, the neutral self-interstitial I^0 clearly dominates in Si under intrinsic conditions (see Figure 6.3) for temperatures between 850°C and 1400°C. Considering Si vacancies, V-mediated self-diffusion is described with V properties that change with temperature. The type of V still remains unsolved but considering Figure 6.4 the contribution of V^- to Si self-diffusion is not far-off the contribution of V^0.

Currently, the type and properties of V in Si are still a matter of debate. In the following section, the diffusion behavior of the donor dopants phosphorus (P) and arsenic (As) and of the acceptor dopant (B) in Si is discussed. The diffusion of these dopants in isotopically controlled heterostructures significantly contributed to our understanding of the mechanisms of dopant diffusion and the charge states of the defects involved.

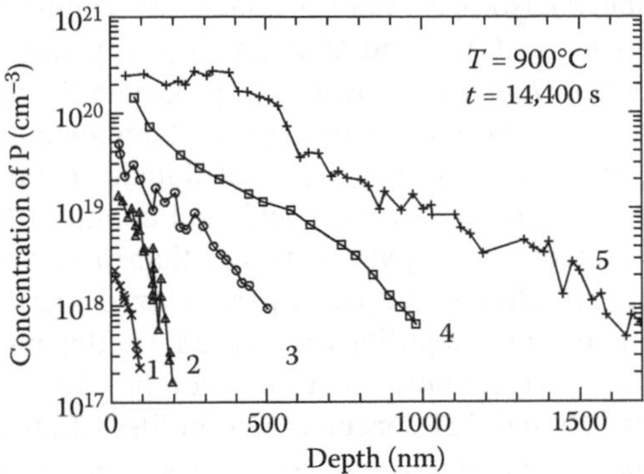

FIGURE 6.7 Concentration profiles of electrically active phosphorus (P) in silicon measured after diffusion annealing at the temperature and time indicated in the figure [63,64]. The profiles 1–5 demonstrate the impact of increasing *n*-type doping level on the P profile shape.

6.2.2 Dopant Diffusion

Numerous experiments on dopant diffusion in Si were performed over the past few decades to identify the atomic mechanisms of diffusion. The studies were motivated by the demand to control the diffusion and activation of dopants in the fabrication of Si-based electronic devices. Compared to self-diffusion, dopant diffusion in semiconductors can be more complex since doping affects the position of the Fermi level and thus the formation of charged defects. Moreover, dopant diffusion can establish native point defect concentrations that deviate from thermal equilibrium. This complexity of dopant diffusion becomes, for example, evident by P diffusion in Si. Figure 6.7 illustrates concentration profiles of electrically active P measured by Yoshida et al. [63,64] after annealing at 900°C for 14,400 s. The profiles demonstrate impressively the impact of the P doping level on the shape of the P diffusion profile.

Diffusion research aims not only to describe but also to predict the distribution of dopants in semiconductors. Considering the mechanisms of diffusion illustrated in Figure 6.2, an unambiguous identification of the relevant dopant diffusion reactions that comprises the charge states of the point defects involved is challenging. In particular, dopant diffusion profiles are not necessarily sensitive to the type and the properties of the native point defects involved. This only holds, for example, for in-diffusion when the formation of the substitutional dopant A_s via the vacancy and interstitialcy mechanisms (see Figure 6.2d and e) is controlled by the removal of V and I, respectively. This so-called native-defect-controlled diffusion mode holds for the diffusion of B and P in Si under sufficiently high doping levels (see, e.g., References 6, 7). However, the diffusion of As in Si does not lead to V and I concentrations that deviate from thermal equilibrium. This mode of dopant diffusion is called foreign atom controlled [6,7]. In order to identify the mechanisms of dopant diffusion and the type and properties of the native point defects involved, experiments on the impact of dopant diffusion on self-diffusion are very valuable. Such experiments could first be performed when enriched stable Si isotopes were available. In conjunction with epitaxial deposition systems such as CVD and MBE, high-purity single crystalline layers of isotopically enriched Si can be grown on appropriate Si substrates. Structures consisting of alternating layers of Si of natural abundance and highly enriched ^{28}Si are very beneficial to study the impact of dopant diffusion and doping on self-diffusion. The interference between self- and dopant diffusion yields more insight on the mechanisms of dopant diffusion than previous diffusion studies with natural Si (see, e.g., Reference 1). In particular, deviations in native defect concentrations from thermal equilibrium caused by dopant diffusion become directly evident in the corresponding self-diffusion profile.

Figure 6.8 shows ^{28}Si and ^{30}Si concentration profiles of an MBE grown Si isotope multilayer structure measured with SIMS. The isotope structure consists of five bilayers of ^{28}Si(120 nm)/^{nat}Si with a 200-nm-thick natural Si layer on top. After the growth of the single-crystalline isotope structure, a 250-nm-thick amorphous natural Si layer was deposited by means of low-temperature MBE. The dopant that is intended to be diffused into the isotope structure was implanted in the top amorphous layer. An As implant located within the amorphous layer is illustrated in Figure 6.8. Similar isotope structures were implanted with B and P to investigate the interference between self- and dopant diffusion [7].

Typical dopant profiles of B, P, and As and the corresponding self-atom profiles measured with SIMS after diffusion annealing at about 1000°C are illustrated in Figure 6.9a through c, respectively. The inhomogeneous self-atom profile clearly reveals that dopant diffusion affects self-diffusion. All three dopant profiles and the corresponding self-atom profiles were considered for modeling the experimental diffusion profiles at 1000°C [7]. Moreover, additional profiles recorded for other temperatures were taken into account to refine the mechanisms of dopant diffusion in Si. Finally, dopant diffusion profiles given in the literature were described with the same diffusion model to verify its consistency [7]. An example is shown in Figure 6.10. The P profiles of Figure 6.7 are accurately described with the refined P diffusion model described below.

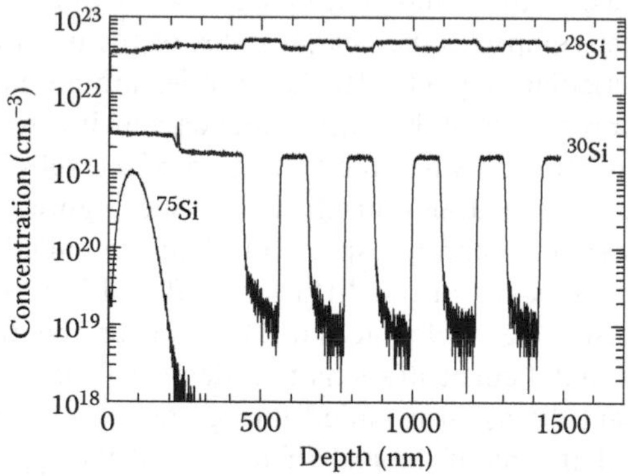

FIGURE 6.8 SIMS concentration profiles of ^{28}Si, ^{30}Si, and ^{75}As of an As-implanted Si isotope multilayer structure [7].

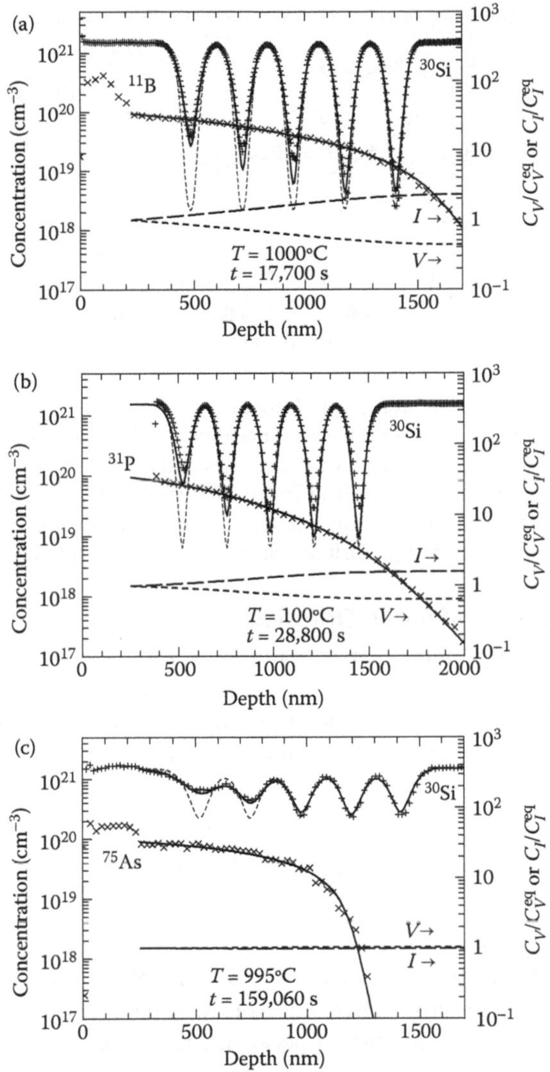

FIGURE 6.9 SIMS concentration profiles (symbols) of ^{30}Si and (a) ^{11}B, (b) ^{31}P, and (c) ^{75}As after annealing the respective dopant-implanted isotope multilayer structure at the temperatures and times indicated. The solid lines in (a–c) are best fits to the experimental dopant and self-atom profiles obtained on the basis of reactions (6.11), (6.15), and (6.22). The lower dashed lines show the corresponding concentrations $C_{V,I}$ of V (short-dashed line) and I (long-dashed line) normalized to their thermal equilibrium concentration $C_{V,I}^{eq}$ (see right ordinate). Diffusion of B and P into Si leads to a super- and undersaturation of I and V, respectively, whereas As diffusion does not lead to significant deviations from the local equilibrium concentration of I and V. The upper thin-dashed lines in (a), (b), and (c) shows the ^{30}Si profiles that are expected for self-diffusion under electronic intrinsic and thermal equilibrium conditions. Before SIMS profiling of the P-diffused Si isotope structure, the top amorphous layer was removed by chemical–mechanical polishing [7].

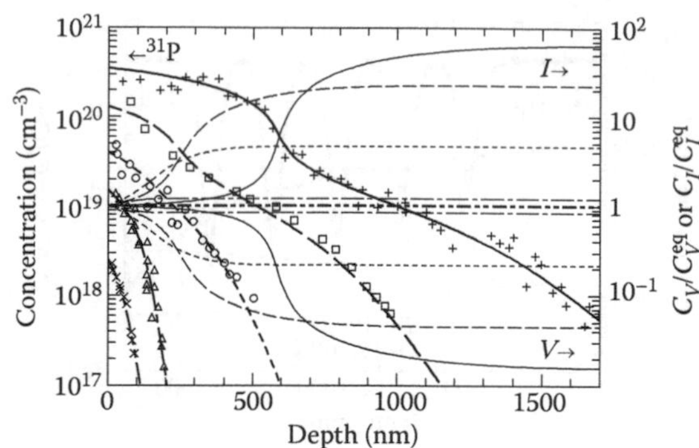

FIGURE 6.10 Concentration profiles of electrically active phosphorus (P) in silicon measured by Yoshida et al. [63,64] after diffusion annealing at 900°C for 14,400 s. With increasing doping level, the penetration depth of P increases and the profile shape changes. This P diffusion behavior is accurately reproduced with the P diffusion model described by reactions (6.15). The corresponding V and I profiles obtained from modeling P diffusion and normalized to their respective thermal equilibrium concentration are referred to the right y-axis.

6.2.2.1 Mechanisms of Boron Diffusion

Boron diffusion profiles in the isotope multilayers and the corresponding self-atom profiles (see Figure 6.9a) are accurately reproduced on the basis of the kick-out and dissociative mechanisms, which, respectively, are schematically illustrated in Figure 6.2b and c, and considered by the following reactions [7]:

$$\begin{aligned} B_i^0 &\rightleftharpoons B_s^- + I^u + (1-u)h \\ B_i^0 + V^k &\rightleftharpoons B_s^- + (1+k)h \end{aligned} \tag{6.11}$$

with neutral (I^0), singly and doubly positively (I^+, I^{2+}) charged self-interstitials, and neutral (V^0), singly positive (V^+), and singly and doubly negative (V^-, V^{2-}) vacancies. h denotes a hole to satisfy charge neutrality of the reactions. The dissociative mechanism assures local equilibrium between I and V after sufficiently long diffusion times. Reactions (6.11) not only describe the simultaneous diffusion of B and Si in isotope multilayers [7] but also B diffusion profiles given in the literature [65]. Moreover, the enhanced [66,67] and retarded [68,69] diffusion of B in Si due to I and V injection, respectively, is consistently explained on the basis of reactions (6.11) [7]. These experiments demonstrate that B mainly diffuses via B interstitials B_i or BI pairs rather than via BV pairs [70]. Reactions (6.11) also predict a linear dependence of B diffusion on the acceptor concentration that is in accord with isoconcentration diffusion studies reported by Miyake [71]. The observed dependence of B diffusion on the acceptor concentration shows that the charge difference between B_s and B_i must be one [6]. Accordingly, B_i is a neutral defect since B_s, as single acceptor, is certainly singly negatively charged. Successful modeling of B diffusion in Si on the basis of reactions (6.11) reveals that mainly the mobile point defect B_i^0 contributes to B diffusion. Accordingly, the B diffusion coefficient D_B deduced from modeling experimental B profiles on the basis of reactions (6.11) is described by (see, e.g., Reference 6)

$$D_B = \frac{C^{eq}_{B_i^0} D_{B_i^0}}{C^{eq}_{B_s^-}} \tag{6.12}$$

where $C^{eq}_{B_s^-}$ equals the equilibrium concentration of substitutional B_s under the established experimental conditions. Equation 6.12 describes the physical meaning of the B diffusion coefficient. The linear dependence of D_B on free carrier concentration is a consequence of the doping dependence of $C^{eq}_{B_s^-}$. With increasing n-type (p-type) doping, the formation of the single acceptor dopant B_s^- is favored (hindered) due to the so-called Fermi-level effect [72]. Note that the concentration of neutral defects is independent of the doping level [72]. D_B displayed in Figure 6.1 represents the B diffusion coefficient for electronically intrinsic conditions, that is, the free carrier concentration equals the intrinsic carrier concentration n_{in}. The temperature dependence of D_B is reproduced by

$$D_B(n_{in}) = 0.87 \exp\left(-\frac{3.46\,\text{eV}}{k_B T}\right) \text{cm}^2\,\text{s}^{-1} \tag{6.13}$$

and was determined from modeling the simultaneous self- and B diffusion in the Si isotope multilayers [7] and from fitting B diffusion profiles given in the literature [65]. For any other n-type or p-type doping level, the B diffusion coefficient follows from

$$D_B = D_B(n_{in})\left(\frac{n_{in}}{n}\right) = D_B(n_{in})\left(\frac{p}{n_{in}}\right) \tag{6.14}$$

with $np = (n_{in})^2$ (see, e.g., Reference 6).

6.2.2.2 Mechanisms of Phosphorus Diffusion

The P diffusion profiles in the isotope multilayers and the corresponding self-atom profiles (see Figure 6.9b) as well as P profiles measured by Yoshida et al. [63,64] (see Figure 6.10) are accurately reproduced on the basis of the kick-out, dissociative, vacancy, and dopant–defect pair-assisted recombination mechanisms, which, respectively, are described by the following reactions [7]:

$$\begin{aligned} P_i^q &\rightleftharpoons P_s^+ + I^u + (1+u-q)e^- \\ P_i^q + V^k &\rightleftharpoons P_s^+ + (1-k-q)e^- \\ PV^- &\rightleftharpoons P_s^+ + V^k + (2+k)e^- \\ PV^- + I^u &\rightleftharpoons P_s^+ + (2-u)e^- \end{aligned} \tag{6.15}$$

The interstitial phosphorus P_i^q with the charge state q is assumed to exist as neutral and singly positively charged defect. For V^k and I^u, the same charge states k and u than those in reactions (6.11) are assumed, that is, $V \in \{V^+,V^0,V^-,V^{2-}\}$ and $I \in \{I^{2+},I^+,I^0\}$. e^- denotes an electron that satisfies charge neutrality of the reactions. The dissociative and the dopant–defect pair-assisted recombination mechanisms assure local equilibrium after sufficiently long diffusion times. Reactions (6.15) accurately describe the simultaneous diffusion of P and Si in isotope multilayers [7] and P diffusion profiles given in the literature [63,64]. Moreover, reactions (6.15) are in accord with the retarded and less-retarded P diffusion in intrinsic [68,69] and extrinsic P-doped [69] Si, respectively, observed under vacancy injection realized by Si surface nitridation. From the degree of retardation, it was concluded that the diffusion of P is mainly mediated by PI pairs or interstitial phosphorus P_i [73] under electronic intrinsic conditions. For extrinsic P doping, the retardation of P diffusion under nitridation decreases [69]. This indicates an increased contribution of vacancies to P diffusion that is considered by the vacancy mechanism. The P diffusivity was found to be proportional to the square of the P concentration for doping levels exceeding 10^{20} cm^{-3} [43]. This observation points to singly negatively charged dopant–vacancy pairs PV^- as the mobile foreign-atom-related defect involved in the vacancy mechanism (see reaction (6.15)) [74]. For lower doping levels, where mobile interstitial-related foreign atom defects dominate P diffusion, isoconcentration studies of Makris and Masters [42] showed that the P diffusivity increases approximately linear with the free electron concentration n. Accordingly, the charge difference between P_s^- and the PI pair or the interstitial P_i must be one. Hence, mobile P_i^0 and mobile PV^- defects contribute to P diffusion in Si [6,74]. Accurate modeling of the interference between P and self-diffusion requires considering both neutral P_i^0 and positively charged P_i^+ interstitial P [7]. The mobile P_i^+ defect contributes to the extended tail observed in the P diffusion profile. This extended tail is not entirely caused by the supersaturation of I established during P diffusion. The supersaturation required to explain the extended tail of P profiles would predict an enhanced self-diffusion in the tail region that is not confirmed by simultaneous P and self-diffusion experiments [7]. This shows that the P diffusion coefficient D_P consists of three different contributions [6,7], that is,

$$D_P = D^*_{P_i^0} + D^*_{P_i^+} + D^*_{PV^-} = \frac{C^{eq}_{P_i^0} D_{P_i^0}}{C^{eq}_{P_s^+}} + \frac{C^{eq}_{P_i^+} D_{P_i^+}}{C^{eq}_{P_s^+}} + \frac{C^{eq}_{PV^-} D_{PV^-}}{C^{eq}_{P_s^+}} \tag{6.16}$$

whose significance changes with the doping level. The right-hand side of Equation 6.16 gives the physical meaning of the individual contributions to P diffusion in Si [6]. The first term $C^{eq}_{P_i^0} D_{P_i^0}/C^{eq}_{P_s^+}$ with a charge difference of one between P_i^0 and P_s^+ predicts a linear dependence of D_P on the free electron concentration n. On the other hand, the charge difference of zero between P_i^+ and P_s^+ leads to a contribution of P diffusion that is independent of doping. The last term describes a contribution to P diffusion that depends on n^2. The total P diffusion coefficient D_P displayed in Figure 6.1 is representative for electronic intrinsic conditions and reproduced by

$$D_P(n_{in}) = 0.75\exp\left(-\frac{3.42\,\text{eV}}{k_BT}\right)\text{cm}^2\,\text{s}^{-1} \tag{6.17}$$

In order to calculate D_P for any doping level, the individual contributions $D^*_{P_i^0}$, $D^*_{P_i^+}$, and $D^*_{PV^-}$ to the P diffusion coefficient must be known. These contributions were determined from modeling the simultaneous self- and P diffusion and ordinary P diffusion profiles given in the literature [7].

$$D^*_{P_i^0}(n_{in}) = 2.53\exp\left(-\frac{3.68\,\text{eV}}{k_BT}\right)\text{cm}^2\,\text{s}^{-1} \tag{6.18}$$

$$D^*_{P_i^+}(n_{in}) = 0.57\exp\left(-\frac{3.43\,\text{eV}}{k_BT}\right)\text{cm}^2\,\text{s}^{-1} \tag{6.19}$$

$$D^*_{PV^-}(n_{in}) = 41.4\exp\left(-\frac{4.44\,\text{eV}}{k_BT}\right)\text{cm}^2\,\text{s}^{-1} \tag{6.20}$$

Equations 6.18 and 6.19 represent the individual contributions to P diffusion for intrinsic conditions. For any other doping level n, the P diffusion coefficient follows from

$$D_P(n) = D^*_{P_i^0}(n_{in})\left(\frac{n}{n_{in}}\right) + D^*_{P_i^+}(n_{in}) + D^*_{PV^-}(n_{in})\left(\frac{n}{n_{in}}\right)^2 \tag{6.21}$$

A graphical representation of the doping dependence of P diffusion at 1000°C is illustrated in Figure 6.11.

6.2.2.3 Mechanisms of Arsenic Diffusion

The As diffusion profiles in the isotope multilayers and the corresponding self-atom profiles (see Figure 6.9c) as well as As profiles measured by Murota et al. [75,76] (see Reference 7) are accurately reproduced on the basis of the vacancy and interstitialcy mechanisms illustrated in Figure 6.2d and e and the dopant–defect pair-assisted mechanisms of I–V recombination, which, respectively, are described by the following reactions [7]:

$$\begin{aligned}
\text{As}V^j &\rightleftharpoons \text{As}_s^+ + V^k + (1+k-j)\text{e}^- \\
\text{As}I^0 &\rightleftharpoons \text{As}_s^+ + I^u + (1+u)\text{e}^- \\
\text{As}V^j + I^u &\rightleftharpoons \text{As}_s^+ + (1-j-u)\text{e}^- \\
\text{As}I^0 + V^k &\rightleftharpoons \text{As}_s^+ + (1-k)\text{e}^-
\end{aligned} \tag{6.22}$$

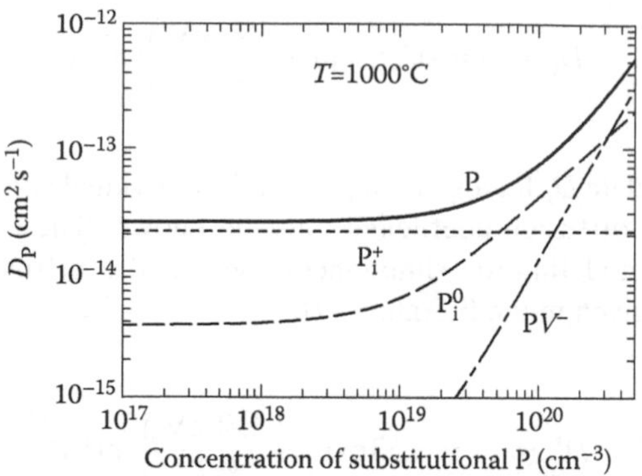

FIGURE 6.11 Dependence of the P diffusion coefficient in Si on the concentration of substitutional P_s for 1000°C. The thick solid line represents the total P diffusion coefficient D_P according to Equation 6.21. The dashed lines show the individual contributions of the neutral P_i^0 and positively P_i^+ charged interstitial P_i and of the negatively charged phosphorus–vacancy pair PV^- to D_P.

For V^k and I^u, the same charge states k and u than those in reactions (6.11) and (6.15) are considered and neutral and singly negatively charged dopant–vacancy pairs AsV^0 and AsV^- are assumed. The diffusion reaction system (6.22) considered for As diffusion in Si is consistent with the experimentally observed As diffusion behavior. The diffusion of As in Si under intrinsic conditions is known to be enhanced under both V and I injection [69,77]. This indicates that the interstitialcy or kick-out mechanism and the vacancy mechanism contribute to As diffusion. Isoconcentration studies revealed an As diffusion coefficient that is almost proportional to the As concentration [44]. This and the fact that As diffusion is similarly enhanced under both V and I injection justify to assume equal contributions of neutral AsV^0 and AsI^0 pairs to As diffusion. Besides these neutral defect pairs, singly negatively charged AsV^- pairs are also considered in reactions (6.22). The AsV^- contribution to As diffusion predicts an As diffusivity that is proportional to n^2. This is seemingly in conflict with the results of isoconcentration studies, but better fits to the experimental As profiles are obtained than in the case where only neutral defect pairs are assumed. Actually, the apparent almost linear concentration dependence of As diffusion consists of a linear and quadratic dependence [7]. In total, three individual defect pairs contribute to the As diffusion coefficient D_{As}, that is,

$$D_{As} = D^*_{AsV^0} + D^*_{AsI^0} + D^*_{AsV^-}$$

$$= \frac{C^{eq}_{AsV^0} D_{AsV^0}}{C^{eq}_{As_s^+}} + \frac{C^{eq}_{AsI^0} D_{AsI^0}}{C^{eq}_{As_s^+}} + \frac{C^{eq}_{AsV^-} D_{AsV^-}}{C^{eq}_{As_s^+}} \tag{6.23}$$

whose significance again depends on the doping level. The total As diffusion coefficient D_{As} displayed in Figure 6.1 is representative for electronic intrinsic conditions and reproduced by

$$D_{\mathrm{As}}(n_{\mathrm{in}}) = 0.47\exp\left(-\frac{4.20\,\mathrm{eV}}{k_{\mathrm{B}}T}\right)\mathrm{cm}^2\,\mathrm{s}^{-1} \tag{6.24}$$

Extensive modeling of the interference between As and self-diffusion and of As profiles given in the literature provide information about the individual contributions to As diffusion [7]. For electronic intrinsic conditions, these contributions are

$$D^*_{\mathrm{As}V^0}(n_{\mathrm{in}}) = D^*_{\mathrm{As}I^0}(n_{\mathrm{in}}) = 11\exp\left(-\frac{4.12\,\mathrm{eV}}{k_{\mathrm{B}}T}\right)\mathrm{cm}^2\,\mathrm{s}^{-1} \tag{6.25}$$

$$D^*_{\mathrm{As}V^-}(n_{\mathrm{in}}) = 1.32\times10^4\exp\left(-\frac{5.14\,\mathrm{eV}}{k_{\mathrm{B}}T}\right)\mathrm{cm}^2\,\mathrm{s}^{-1} \tag{6.26}$$

The As diffusion coefficient D_{As} for any other doping level n is given by

$$D_{\mathrm{As}}(n) = D^*_{\mathrm{As}V^0}(n_{\mathrm{in}})\left(\frac{n}{n_{\mathrm{in}}}\right) + D^*_{\mathrm{As}I^0}(n_{\mathrm{in}})\left(\frac{n}{n_{\mathrm{in}}}\right) + D^*_{\mathrm{As}V^-}(n_{\mathrm{in}})\left(\frac{n}{n_{\mathrm{in}}}\right)^2 \tag{6.27}$$

and illustrated as function of the substitutional $\mathrm{As_s}$ concentration for 950°C in Figure 6.12.

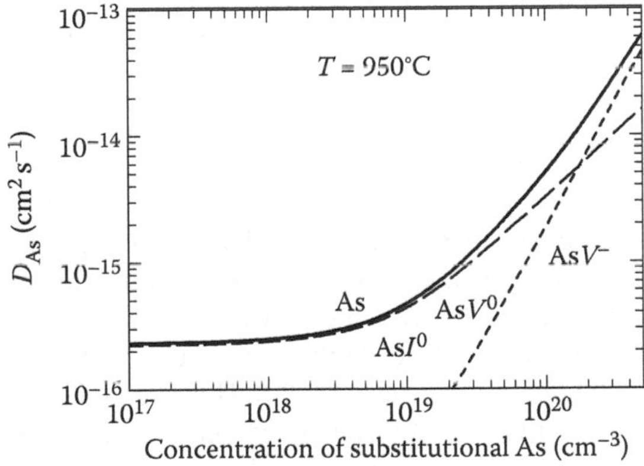

FIGURE 6.12 Dependence of the As diffusion coefficient in Si on the concentration of substitutional $\mathrm{As_s}$ for 950°C. The thick solid line represents the total As diffusion coefficient D_{As} according to Equation 6.27. The dashed lines show the individual contributions of the neutral As V^0 and As I^0 pairs and of the negatively charged AsV^- pair to D_{As}. The equal contributions of AsV^0 and AsI^0 to As diffusion are in accord with experimental results on As diffusion in Si that show a similar enhancement of As diffusion under both V and I injection [69,77].

6.2.3 Comparison between Self- and Dopant Diffusion

The experiments performed on self- and dopant diffusion in Si clearly indicate the significance of both *V* and *I* in the atomic transport. The strong correlation of the diffusion of hybrid metals such as Au, Pt, and Zn [53–59] with Si self-diffusion [2,5,12] supports an activation enthalpy of self-diffusion via *I* of about 5 eV. Such high values for the enthalpy of *I*-mediated self-diffusion in Si are also predicted by theoretical calculations [78–80]. Hence, a general agreement seems to exist on *I*-mediated diffusion in Si. Moreover, total energy calculations of the formation enthalpy of *V* and *I* predict donor and acceptor states for both type of defects [81,82]. The experiments on simultaneous self- and dopant diffusion [7] confirm the existence of *V* acceptor states in the upper half of the Si band gap and *V* donor states in the lower half [7]. Moreover, a reversed level ordering for the donor levels introduced by self-interstitials is suggested [7,83] that was already predicted theoretically by Car et al. [81]. Altogether the type and charge states of the native point defects deduced from modeling dopant diffusion in Si isotope multilayers are in good agreement with theoretical predictions. In contrast, the present situation of *V*-mediated self- and dopant diffusion is more complex. The temperature dependence of self-diffusion clearly reveals a deviation from a single Arrhenius equation (see Section 6.2.1) that on first sight indicates an activation enthalpy of about 3.6 eV for *V*-mediated self-diffusion [5,12]. However, this value is at variance with the activation enthalpy of *V*-mediated dopant diffusion that clearly exceeds 3.6 eV as, for example, demonstrated by Equations 6.20, 6.25, and 6.26. Other examples are discussed in the literature [7,12,84]. A consistent interpretation of *V*-mediated self- and dopant diffusion is achieved, assuming temperature-dependent *V* formation and migration enthalpies. Although, this concept is convincing, an independent verification is still lacking. The key for a consistent interpretation of *V*-mediated atomic transport in Si remains the property of *V*, in particular, at elevated temperatures. The *V* properties at cryogenic temperatures are well characterized by Watkins et al. [85–87].

6.2.4 Nonequilibrium Diffusion

Additional insight on the properties of native point defects in Si can be obtained by studying self-diffusion under irradiation. Diffusion experiments of this type could first be performed by means of isotopically controlled Si heterostructures [18] because conventional self-diffusion studies with radioactive Si isotopes are hampered due to the short lifetime. Prior to that, the impact of irradiation on diffusion in Si was mainly studied with dopants [88–90]. However, the complexity of dopant diffusion in Si under conventional diffusion conditions (see Section 6.2.2) and the likely occurrence of additional dopant–defect interactions under irradiation hinder an unambiguous characterization of *V* properties.

Self-diffusion under irradiation is characterized by a nonthermal formation of *V* and *I* in equal numbers. This differs to self-diffusion under thermal equilibrium with homogeneous concentrations of *V* and *I*, whose equilibrium concentrations will certainly be different. Depending on the irradiation conditions, the concentration of Frenkel defects, that is, of *I* and *V* pairs, exceeds the concentrations of

V and I formed under thermal equilibrium. As a consequence, self-diffusion in Si under irradiation is expected to be enhanced. Figure 6.13 demonstrates a ^{30}Si self-diffusion profile measured with SIMS after 2 MeV proton irradiation of a Si isotope structure at 832°C for 4 h. In comparison to the ^{30}Si profile of the as-grown ^{nat}Si(80 nm)/^{28}Si(280 nm)/^{nat}Si(100 nm) isotope structure, the annealed structure shows a clear diffusion-induced intermixing at both ^{nat}Si/^{28}Si interfaces.

First, the intermixing at the deep ^{28}Si/^{nat}Si interface is considered. The intermixing can be accurately described by the solution of Fick's law for self-diffusion across a buried interface [2] with a composition-independent self-diffusion coefficient. Radiation-enhanced self-diffusion (RESD) data obtained in this way for samples irradiated at constant temperature and proton flux and several times t reveal a $\sqrt{t}$ dependence of the diffusion length. The flux density dependence of self-diffusion was found to vary as $\sqrt{\Phi}$ [18]. These results reveal a steady state of native defect concentrations that is established during irradiation [91]. Data for RESD corresponding to $\Phi = 1.0$ μA and the intermixing of the ^{28}Si/^{nat}Si interface at a distance of 360 nm are shown in Figure 6.14 in comparison to self-diffusion under thermal equilibrium [2,12]. The temperature dependence of RESD is described by the Arrhenius equation $D = D_0 \exp(-Q/k_B T)$ with the Boltzmann constant k_B and $Q = (0.92 \pm 0.26)$eV and ln $(D_0/\text{cm}^2\ \text{s}^{-1}) = -27.5 \pm 2.7$ [18]. As demonstrated in the literature [91,92], the activation enthalpy Q of RESD equals 0.5 H_X^m of the migration enthalpy H_X^m of the native defect $X \in \{V,I\}$, which controls self-diffusion under irradiation. This migration enthalpy was assigned to the single vacancy and, accordingly, $H_V^m = (1.8 \pm 0.5)$ eV was proposed [18].

A more comprehensive analysis of RESD is based on numerical solutions of the underlying diffusion equation system. It is evident from Figure 6.13 that the intermixing at the topmost interface is weaker than at the deeper ^{nat}Si/^{28}Si interface. Since

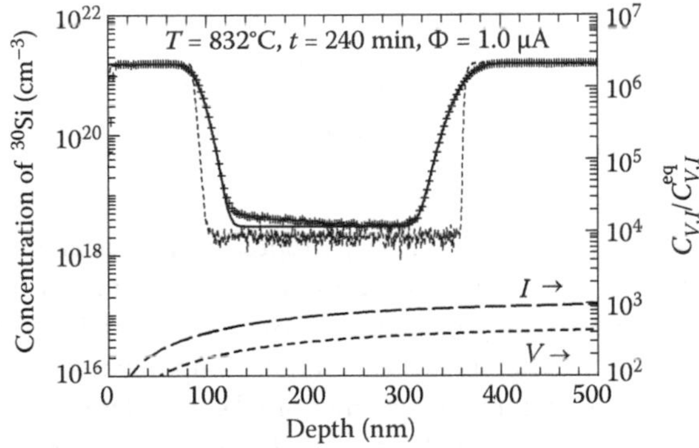

FIGURE 6.13 Concentration-depth profiles of ^{30}Si in a ^{nat}Si/^{28}Si/^{nat}Si isotope heterostructure measured with SIMS before (top short-dashed line) and after annealing (+) at 832°C for 240 min and concurrent 2 MeV proton irradiation with a proton flux of $\Phi = 1.0$ μA. The upper solid line represents a best fit to the measured radiation-enhanced self-diffusion profile. The lower long- and short-dashed lines show the corresponding calculated supersaturations S_I and S_V, respectively [18].

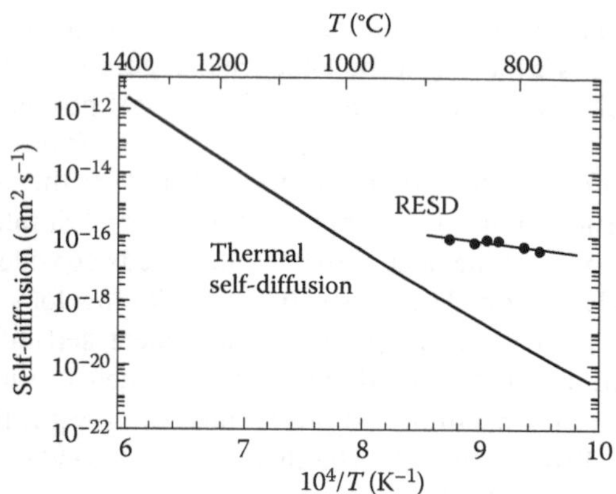

FIGURE 6.14 Temperature dependence of radiation-enhanced self-diffusion (RESD: symbol and upper solid line [18]) in Si in comparison to Si self-diffusion under thermal equilibrium conditions (lower solid line [2,12]). The activation enthalpy of RESD equals $0.5H_X^m = (1.8 \pm 0.5)\text{eV}$ of the native defect $X \in \{V,I\}$ mediating RESD.

self-diffusion is directly mediated by native point defects, this depth dependence in the intermixing reflects an inhomogeneous distribution of I and V. The self-diffusion profile obtained under proton irradiation is accurately described assuming diffusion equations for V and I that consider the formation and annihilation of native defects during irradiation [18]. Moreover, Fick's law of self-diffusion with a depth-dependent self-diffusion coefficient given by

$$\begin{aligned} D_{\text{Si}}(x,t) &= f_V \frac{C_V(x,t)D_V}{C_0} + f_I \frac{C_I(x,t)D_I}{C_0} \\ &= S_V f_V \frac{C_V^{\text{eq}} D_V}{C_0} + S_I f_I \frac{C_I^{\text{eq}} D_I}{C_0} \\ &= S_V D_{\text{Si}}^V + S_I D_{\text{Si}}^I \end{aligned} \tag{6.28}$$

was considered, where $S_{V,I} = C_{V,I}(x,t)/C_{V,I}^{\text{eq}}$ represent the local V and I supersaturations established by irradiation. $f_{V,I}$ are the correlation factors for self-diffusion via the vacancy and interstitialcy mechanisms (see Section 6.2.1). It is evident from Equation 6.28 that self-diffusion under irradiation compared to the equilibrium case, which is expressed by the individual contributions $D_{\text{Si}}^{V,I}$, is determined by $S_{V,I}$. Figure 6.13 shows the supersaturations of V and I deduced from modeling the radiation-enhanced ^{30}Si diffusion profile. A gradient in the concentration of V and I is established during irradiation due to the interplay between formation and annihilation of V and I in the bulk and their annihilation at the surface. Considering data of $D_{\text{Si}}^{V,I}$ deduced from metal and equilibrium self-diffusion studies, self-diffusion under irradiation mainly provides information on $S_{V,I}$. Assuming realistic numbers for the formation and annihilation of V and I and for the thermal equilibrium concentrations

C_I^{eq} of I from Zn diffusion studies [55], the equilibrium concentrations C_V^{eq} of V were determined [18].

Figure 6.15 illustrates the temperature dependence of C_V^{eq} deduced from RESD [18] in comparison to lower [93] and upper bounds [94] of C_V^{eq} data reported in the literature. The dashed line in Figure 6.15 indicates a definitive upper bound for C_I^{eq} [55]. The temperature dependence of $C_V^{eq}(=\exp(S_V^f/k_B)\exp(-H_V^f/k_BT))$ yields a vacancy formation enthalpy of $H_V^f = (2.1 \pm 0.7)\text{eV}$ and entropy of $S_V^f = (-0.8 \pm 6.8)k_B$. Irrespective of the limited accuracy of H_V^f and S_V^f, the data for C_V^{eq} exceeds the results obtained from modeling of point defect dynamics during crystal growth [93,95].

The sum $H_V^m + H_V^f$ deduced for the vacancy formation and migration enthalpy is about 3.9 eV fully consistent with the temperature dependence of the activation enthalpy $Q_V(T)$ illustrated in Figure 6.6. $Q_V(T)$ varies from 3.8 to 4.0 eV for temperatures between 800°C and 900°C where most of the RESD experiments were conducted so far.

Although present experiments on self-diffusion under equilibrium and nonequilibrium conditions are consistently explained with temperature-dependent thermodynamic properties of the vacancy, additional independent evidences are required to verify this interpretation since other concepts are also reported in the literature to unravel the contradictory results on the migration enthalpy of vacancies at low and high temperatures. For example, Caliste and Pochet [96,97] show that the observed temperature dependence for the V migration enthalpy can be explained by the existence of three diffusion regimes for divacancies. Their numerical studies of self-diffusion via mono- and divacancies combine structure calculations by *ab initio* and Monte Carlo methods. The calculations demonstrate that the effective diffusion behavior of V via transient formation of divacancies can be non-Arrhenian depending on the temperature range and the V concentration. Based on this concept,

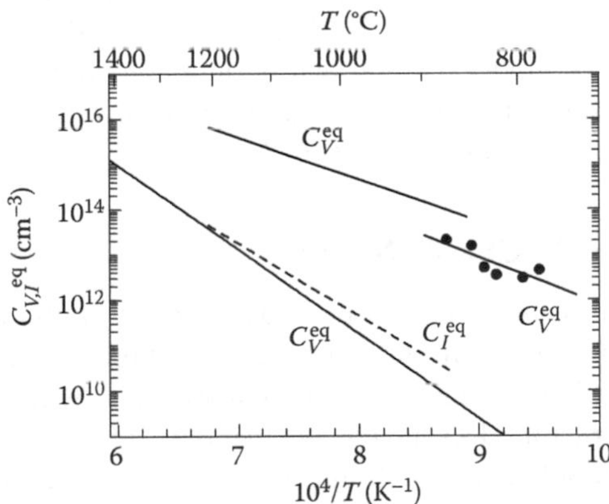

FIGURE 6.15 Temperature dependence of the thermal equilibrium concentrations C_I^{eq} (symbols, solid line) deduced from modeling RESD in comparison to data of $C_{V,I}^{eq}$ (solid lines) reported in the literature. The upper [94] and lower [93] solid lines form the boundary of C_V^{eq} data reported in the literature. The definitive upper bound for C_I^{eq} (dashed line: [55]) was considered for RESD modeling.

the authors provide a plausible interpretation of the high migration enthalpy of (1.8 ± 0.5) eV deduced from self-diffusion experiments under proton irradiation [18], that is, the migration enthalpy assigned to a monovacancy rather reflects an effective migration energy due to the transient formation of divacancies. The formation of divacancies is plausible since proton irradiation realizes V concentrations in Si that exceed those under thermal equilibrium by several orders of magnitude. However, the concept of effective diffusion suggested by Caliste and Pochet [96] is not applicable for Si self-diffusion under thermal equilibrium and thus not suitable to explain the temperature dependence of the activation enthalpy $Q_V(T)$ of self-diffusion via V. Experimental estimates of the V concentration in Si yield concentrations in the range of 10^{15} cm^{-3} at the melting point [98,99]. Metal diffusion experiments provide an upper bound of about 2×10^{14} cm^{-3} for the V concentration at 870°C [55] with even lower values for lower temperatures. The calculations of Pochet and Caliste [96,97] assume V concentrations of 10^{16} cm^{-3} that clearly exceed the thermal equilibrium concentration of V. Considering two orders of magnitude lower concentrations, the intermediate diffusion regime, where diffusion of monovacancies is affected by the transient formation of divacancies, shifts to temperatures beyond those accessible by equilibrium self-diffusion studies. This questions an effective self-diffusion via mono- and divacancies under thermal equilibrium. However, Voronkov and Falster [95] and more recently also Cowern et al. [62] suggest that V and I in Si and also in Ge exist in two structural forms, a localized one and an extended one, of strongly differing diffusivities. This interpretation also results in a temperature dependence of Q_V but in contrast to the interpretation of Kube et al. [12], where the form of the defect changes gradually with temperature, different forms of the same defect exist simultaneously. Which concept actually holds for Si has to be seen in the future.

In Section 6.3, experimental results on self- and dopant diffusion in Ge are discussed. Compared to Si, self- and dopant diffusion in Ge is mainly mediated by V. No evidence on the contribution of I to self-diffusion has been determined. Accordingly, atomic transport in Ge under thermal equilibrium conditions is less complex than in Si and thus also suited to verify the concept of localized and extended forms of vacancies that was also proposed for Ge [61,62].

6.3 DIFFUSION IN GERMANIUM

The first bipolar transistor developed by J. Bardeen, W. Brattain, and W. Shockley was made of elemental Ge [100]. However, the interest in Ge strongly decreased with the development of Si-based field effect transistors (FET) with a metal-oxide-semiconductor (MOS) structure. Despite the fact that the electron and hole mobilities are higher in Ge than in Si [101], the extremely stable high-ohmic Si oxide (SiO_2) compared to the less stable Ge oxide (GeO_2) made Si more attractive as the base material for today's microelectronics. Over the past decade, Ge as material for microelectronic applications has received renewed attention [102–104]. This renewed interest arises from their promising applications in Si-based integrated-circuit technology. With state-of-the-art epitaxial deposition techniques, the former obstacles that strongly limited the use of Ge as the base material for electronic devices no longer exist. Using Ge or SiGe epitaxial layers instead of Si, one can take advantage of the higher carrier mobility in Ge-rich layers [101].

Successful integration of Ge in future nanoelectronic devices requires detailed control on the diffusion of *n*- and *p*-type dopants and their electrical activation. In the case of Si, the understanding of the existence and properties of both vacancies (V) and self-interstitials (I) and their contributions to self- and dopant diffusion (see Section 6.2) has led to advanced doping strategies that are realized by today's fabrication of Si-based nanoelectronic devices. Compared to Si, much less research has been conducted until about 2005 on the mechanisms of dopant diffusion in Ge and, in general, on the interaction between native and foreign atom defects.

Figure 6.16 shows the temperature dependence of the diffusion coefficient of various foreign atoms in Ge in comparison to self-diffusion [105]. The solid lines indicate the diffusivities of mainly substitutionally dissolved elements B [106], Si [107], Al [108], P [109], Zn [110], As [111,112], and Sb [113]. The diffusion of the hybrid elements Cu [114,115] (Equation 4 in Reference 115 should read $D^{\text{eff}}_{\text{Cu(1)}} = 7.8 \times 10^{-5} \exp(-(0.084\text{eV}/k_B T))\text{cm}^2\,\text{s}^{-1}$), Ag [114], Au [114], and Ni [116], which are dissolved both on interstitial and substitutional sites, is indicated by the long-dashed lines. The data represent the interstitial foreign-atom-controlled mode of diffusion via the dissociative mechanism (see Figure 6.2c). The upper thin-dashed line in Figure 6.16 shows the diffusivity of interstitial Cu_i that was deduced in Reference 115 from the interstitial-controlled Cu diffusion coefficient and the solubility of Cu on interstitial and substitutional lattice sites. Cu_i is the fastest-diffusing

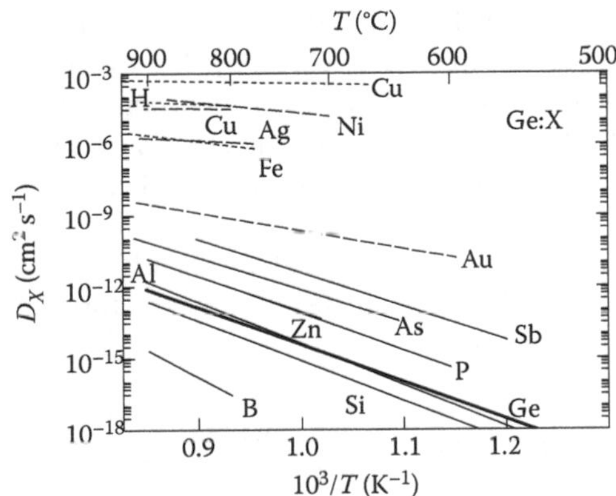

FIGURE 6.16 Temperature dependence of the diffusion coefficients of foreign atoms in Ge (thin lines) in comparison to self-diffusion (thick line) [8,105]. The diffusion data are representative for electronically intrinsic conditions. Solid lines represent data of the elements B [106], Al [108], Si [107], Zn [110], P [109], As [111], and Sb [113]. The elements are mainly dissolved on substitutional lattice sites. Long-dashed lines illustrate diffusion data of the hybrid elements Cu [115,116], Ag [114], Au [114], and Ni [116]. These elements are dissolved on both substitutional and interstitial sites and diffuse in an interstitial configuration via the dissociative mechanism (see Figure 6.2c). The short-dashed lines indicate the diffusion of mainly interstitial-dissolved elements H [32] and Fe [117]. The upper short-dashed line shows the diffusivity deduced for interstitial Cu_i [115].

species among the other interstitial foreign atoms such as H [32] and Fe [117]. The data for B [106], Al [108], Si [107], Zn [110], and As [111] represent results of diffusion studies utilizing SIMS [106–108,111] and the spreading resistance technique [110] as profiling methods. The diffusivities of Sb [113] and P [109] shown in Figure 6.16 were determined by means of incremental sheet resistivity measurements. More recent data on P and Sb diffusion in Ge is given in Section 6.3.2. It is evident from Figure 6.16 that the diffusivity of B and Si is slower than Ge self-diffusion. Note that in the case of Si, foreign atom diffusion always exceeds self-diffusion [118] (see Figure 6.1). This indicates that both V and I contribute almost the same extent to Si self-diffusion [2,119] whereas in the case of Ge, V dominates self- and dopant diffusion [9,105,111] (see Section 6.3.2). The lower diffusion of Si and B compared to Ge self-diffusion points to a repulsive interaction between V and B. Such a repulsive interaction is supported by theoretical calculations [120].

Over the past decade, some studies on self- and dopant diffusion under thermal equilibrium [8–10,111,112,121,122] and nonequilibrium conditions [19,123–126] have been performed to identify the atomic mechanisms of mass transport in Ge and the type of defects involved. In Sections 6.3.1 and 6.3.2, experimental studies are reviewed that significantly contribute to our understanding on diffusion, doping, and defect reactions in Ge.

6.3.1 Self-Diffusion

The first studies on self-diffusion in Ge date back more than 50 years [127,128]. The impact of doping and hydrostatic pressure on self-diffusion provided the first evidence on the acceptor nature of V in Ge [105]. The excellent agreement between Ge self-diffusion and the V contribution to self-diffusion deduced from Cu diffusion in dislocation-free Ge has proved the dominance of V in Ge [114,115,129]. Recently, experiments on self-diffusion in Ge could be extended to lower temperatures by means of isotopically controlled Ge multilayer structures in conjunction with NR measurements [8]. Considering the former Ge self-diffusion data of Werner et al. [105] and the results from the NR study [8], the temperature dependence of Ge self-diffusion D_{Ge} is best described by

$$D_{\mathrm{Ge}} = 25.4 \times \exp\left(-\frac{3.13\,\mathrm{eV}}{k_{\mathrm{B}}T}\right)\mathrm{cm}^2\,\mathrm{s}^{-1} \tag{6.29}$$

for temperatures between 429°C and 904°C [8].

In contrast to Si self-diffusion, self-diffusion in Ge is accurately described by one single diffusion activation enthalpy [8]. This reveals that mainly one type of native point defect, the vacancy, determines self-diffusion in Ge. A contribution of I to self-diffusion is not evident. Accordingly, self-diffusion in Ge is mainly controlled by one single vacancy form with constant, that is, temperature-independent, vacancy formation, and migration enthalpies (entropies). The concept of extended V first proposed by Seeger and Chik [61] and recently adapted by Cowern et al. [62] for diffusion in Si and Ge is not applicable for V in Ge.

Investigations of Ge self-diffusion under n-type doping indicate that V are doubly negatively charged. This was independently verified by Brotzmann et al. [9] and Naganawa et al. [10] by dopant diffusion in isotopically controlled Ge multilayers. The impact of p-type doping on Ge self-diffusion was recently investigated with homogeneously B-doped Ge isotope structures [11]. Figure 6.17 illustrates the temperature dependence of self-diffusion in p-type compared to undoped Ge. p-type doping clearly retards self-diffusion compared to electronically intrinsic conditions. The doping dependence of self-diffusion suggests two V-related acceptor levels in the band gap of Ge [11]. The first acceptor level is located 0.28 eV above the valence band maximum (VBM) and the second level at 0.14 eV above VBM. This level ordering is inverted, that is, the single acceptor state lies above the double acceptor state. This is indicated in the inset of Figure 6.17. As a consequence of the level ordering, doubly negatively charged V^{2-} prevails under n-type doping and neutral V^0 under p-type doping. Singly negatively charged V^- does not dominate and control self-diffusion under any doping level [11]. This is demonstrated in Figure 6.18 by the individual contributions of neutral, singly, and doubly negatively charged V to the total Ge self-diffusion coefficient at 700°C that was deduced from the doping dependence of self-diffusion. Accordingly, the total Ge self-diffusion coefficient for thermal equilibrium conditions mainly consists of three individual contributions:

$$\begin{aligned} D_{\mathrm{Ge}} &= D_{\mathrm{Ge}}^{V^0} + D_{\mathrm{Ge}}^{V^-} + D_{\mathrm{Ge}}^{V^{2-}} \\ &= f_{V^0}\frac{C_{V^0}^{\mathrm{eq}} D_{V^0}}{C_0} + f_{V^-}\frac{C_{V^-}^{\mathrm{eq}} D_{V^-}}{C_0} + f_{V^{2-}}\frac{C_{V^{2-}}^{\mathrm{eq}} D_{V^{2-}}}{C_0} \end{aligned} \tag{6.30}$$

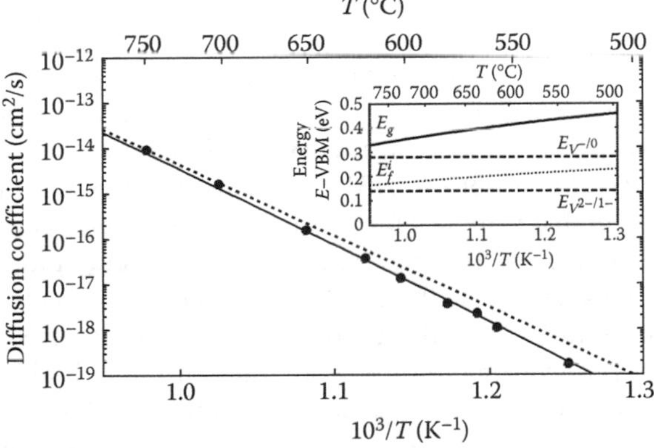

FIGURE 6.17 Self-diffusion in Ge as a function of the inverse temperature for electronically intrinsic (dotted line: [8]) and p-type doping conditions (•, solid line: [11]). The inset illustrates the positions of the V-related energy levels (dashed lines) deduced from the doping dependence of self-diffusion. Expressions given by Vanhellemont and Simoen [187] were considered for the temperature dependence of the band gap $E_g\,(T)$ (solid line in the inset) and the position of the intrinsic Fermi level $E_f^i(\sim 0.5E_g(T))$ (dotted line in the inset).

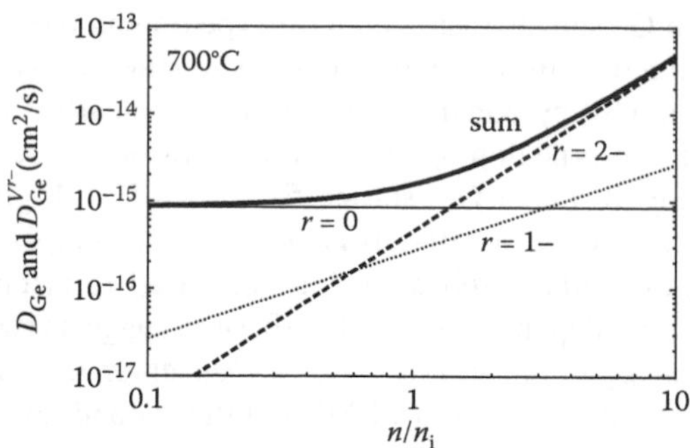

FIGURE 6.18 Individual contributions of neutral ($r = 0$: thin solid line), singly ($r = 1$: dotted line), and doubly ($r = 2$: dashed line) negatively charged V to Ge self-diffusion at 700°C. The total Ge self-diffusion coefficient is given by the sum of the individual contributions and shown by the thick solid line. The singly negatively charged V^- does not dominate self-diffusion under any doping level.

where $f_V^{r-} = 0.5$ and $C_0 = 4.4 \times 10^{22}$ cm^{-3} represent the correlation factor for self-diffusion via vacancies [49] and the Ge atom density, respectively. As illustrated in Figure 6.18, the relative contributions of V^- and V^{2-} to Ge self-diffusion change with the doping level, whereas the contribution of V^0 remains constant.

The energy level scheme of V shown in the inset of Figure 6.17 is consistent with recent results of deep-level transient spectroscopy (DLTS) studies on defects introduced in Ge by low-temperature electron irradiation [130]. Present theoretical calculations do not predict an inverse-level ordering of the first and second acceptor states [131]. But an inverse ordering cannot be excluded [132] because theoretical results are representative for 0 K and it remains unclear how the level positions depend on temperature.

Diffusion studies under equilibrium conditions all indicate that self-interstitials do not significantly contribute to the atomic transport in Ge. Theoretical calculations based on density functional theory (DFT) confirm the dominance of V over I under thermal equilibrium [131,133]. This is expressed by the formation energy of the defect that is higher for I than for V. However, I can be formed non-thermally, for example, by electron irradiation. Haesslein et al. [134] identified a donor level at 0.04 eV below the conduction band of Ge in electron-irradiated Ge by means of perturbed angular correlation (PAC) spectroscopy. They attributed this level to I. According to their work, the self-interstitial introduces donor and the vacancy acceptor states. The acceptor nature of V is consistent with the doping dependence of self-diffusion. The donor nature of I is consistent with the diffusion behavior in Ge under nonequilibrium conditions that is discussed in Section 6.3.3.

In the next section, the characteristics of dopant diffusion in Ge are summarized. Compared to dopant diffusion in Si, dopant diffusion in Ge is fully described on the basis of the vacancy mechanism. Additional defect reactions that lead to deactivation of substitutional dopants operate at high doping levels.

6.3.2 Dopant Diffusion in Germanium

Figure 6.19 illustrates the diffusion coefficients of n-type (P [112], As [112], Sb [112]) and p-type (Al [108], Ga [121], In [122]) dopants in comparison to Ge self-diffusion [8,105]. The diffusion of the n-type dopants clearly exceeds the diffusion of the p-type dopants. Moreover, donor diffusion exceeds Ge self-diffusion and increases from P to Sb [112]. The diffusion of the p-type dopants is very similar to self-diffusion and in the case of B [106] even significantly lower. The following equations describe the temperature dependence of the dopant diffusion coefficient D_A with $A \in \{B, Al, Ga, In, P, As, Sb\}$ illustrated in Figure 6.19. The temperatures given in brackets show the T range of the respective experimental study [106,108,112,121,122].

$$D_B = 1.97 \times 10^5 \times \exp\left(-\frac{4.65\,\text{eV}}{k_B T}\right) \text{cm}^2\,\text{s}^{-1}, \quad T \in [800°\text{C} - 900°\text{C}] \tag{6.31}$$

$$D_{Al} = 1.0 \times 10^3 \times \exp\left(-\frac{3.45\,\text{eV}}{k_B T}\right) \text{cm}^2\,\text{s}^{-1}, \quad T \in [554°\text{C} - 905°\text{C}] \tag{6.32}$$

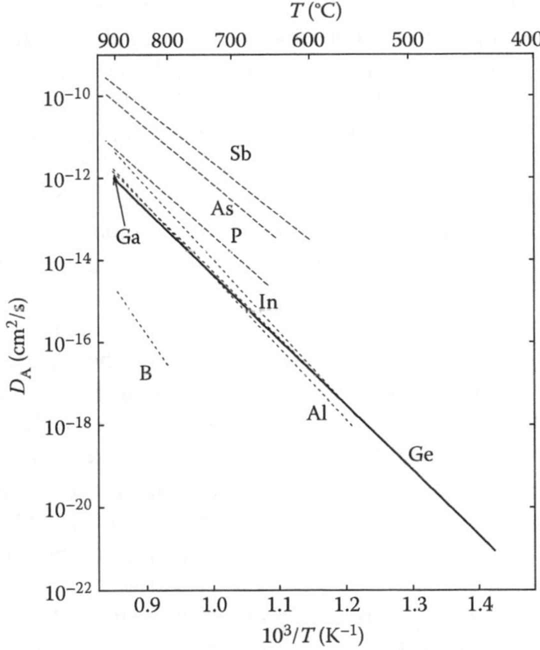

FIGURE 6.19 Diffusion coefficients of the n-type dopants (dashed lines): phosphorus (P) [112], arsenic (As) [112], and antimony (Sb) [112] in Ge compared to Ge self-diffusion (solid line) [8] and to the p-type dopants (dotted lines): boron (B) [106], aluminum (Al) [108], gallium (Ga) [121], and indium (In) [122]. Diffusion data are representative for dopant diffusion under electronically intrinsic and equilibrium conditions. Each line spans the range of the respective experimental data.

$$D_{\mathrm{Ga}} = 80 \times \exp\left(-\frac{3.21\,\mathrm{eV}}{k_{\mathrm{B}}T}\right)\mathrm{cm}^2\,\mathrm{s}^{-1}, \quad T \in [575°\mathrm{C} - 910°\mathrm{C}] \tag{6.33}$$

$$D_{\mathrm{In}} = 5.12 \times 10^3 \times \exp\left(-\frac{3.51\,\mathrm{eV}}{k_{\mathrm{B}}T}\right)\mathrm{cm}^2\,\mathrm{s}^{-1}, \quad T \in [550°\mathrm{C} - 900°\mathrm{C}] \tag{6.34}$$

$$D_{\mathrm{P}} = 9.1 \times \exp\left(-\frac{2.85\,\mathrm{eV}}{k_{\mathrm{B}}T}\right)\mathrm{cm}^2\,\mathrm{s}^{-1}, \quad T \in [650°\mathrm{C} - 920°\mathrm{C}] \tag{6.35}$$

$$D_{\mathrm{As}} = 32 \times \exp\left(-\frac{2.71\,\mathrm{eV}}{k_{\mathrm{B}}T}\right)\mathrm{cm}^2\,\mathrm{s}^{-1}, \quad T \in [640°\mathrm{C} - 920°\mathrm{C}] \tag{6.36}$$

$$D_{\mathrm{Sb}} = 16.7 \times \exp\left(-\frac{2.55\,\mathrm{eV}}{k_{\mathrm{B}}T}\right)\mathrm{cm}^2\,\mathrm{s}^{-1}, \quad T \in [600°\mathrm{C} - 920°\mathrm{C}] \tag{6.37}$$

Equation 6.31 shows that the slow diffusion of B is associated with a high diffusion activation enthalpy of 4.65 eV that exceeds the activation enthalpy of self-diffusion (see Equation 6.29) by more than 1 eV. This indicates that B atoms are not likely associated with V. This is confirmed by DFT investigations that reveal that the mobile BV pair is unstable [120,135]. The high diffusion activation enthalpy of B rather indicates a diffusion of B via I. An I-mediated diffusion is supported by the DFT study of Janke et al. [135] and experimentally by the enhanced diffusion of B under an I supersaturation established by irradiation [20,123,124,126,136] (see Section 6.3.3). Compared to B, the diffusion of the other p-type dopants is fully consistent with the vacancy mechanism. The slightly higher diffusion activation enthalpy of Al [108], Ga [121], and In [122] compared to self-diffusion [8] reflects an interaction of these p-type dopants to V that is less attractive than in the case of the n-type dopants P, As, and Sb [9]. The difference in the diffusion behavior of p- and n-type dopants in Ge is due to Coulomb interactions between the substitutional dopant and V. The acceptor and donor dopant are, respectively, singly negatively and singly positively charged on substitutional lattice site. However, the vacancy is neutral and doubly negatively charged under p- and n-type doping, respectively. Accordingly, an attractive Coulomb interaction exists between substitutional donors and V. This facilitates the formation of mobile dopant–vacancy pairs and thus the diffusion of the donor atoms.

In the case of Ge, mainly dopant–V pairs exist in thermal equilibrium because the formation of I is energetically suppressed due to its high formation energy. Information about the charge states of dopant–V pairs in Ge is gained from the shape of the dopant diffusion profiles and the doping dependence of diffusion [7].

6.3.2.1 Donor Diffusion

The doping dependence of donor diffusion is illustrated in Figure 6.20 for As diffusion in Ge. Different As vapor pressures were realized in the diffusion experiment in

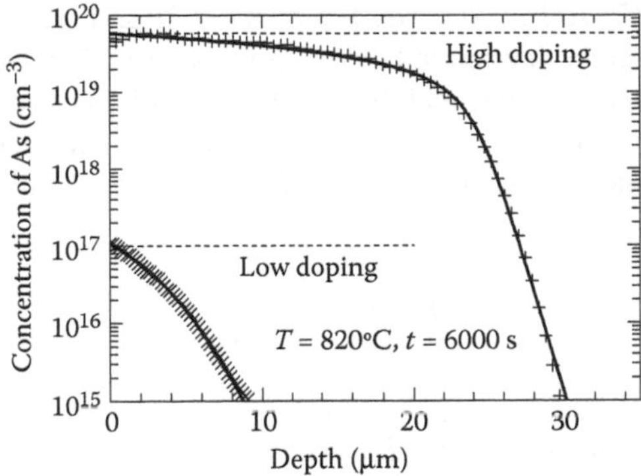

FIGURE 6.20 Concentration profiles of As measured by means of the spreading resistance technique after diffusion annealing at 820°C for 6000s [112]. The profiles illustrate the strong doping dependence of As diffusion due to different As surface concentrations.

order to establish low and high doping levels of As at the Ge surface [112]. It is evident from Figure 6.20 that the doping level strongly affects both the shape and the penetration depth of the As profile. The lower profile shown in Figure 6.20 reflects a doping level of about 10^{17} cm^{-3}, which is below the intrinsic carrier concentration n_i at the diffusion temperature. In this case, the dopant profile is described by a concentration-independent intrinsic As diffusion coefficient [112]. The upper profile reflects a doping level with a free carrier concentration n, which exceeds n_i, that is, extrinsic doping conditions are realized. In this case, a dopant diffusion profile develops that reveals a concentration-dependent dopant diffusion coefficient. A very similar diffusion behavior is observed for P and Sb in Ge [112]. The strong doping dependence of the diffusion coefficient D_A of the donor atom A with A ∈ {P, As, Sb} is accurately described by

$$D_A(n) = D_A^*(n_i)\left(\frac{n}{n_i}\right)^2 \tag{6.38}$$

where D_A (n_i) is the dopant diffusion coefficient for intrinsic doping conditions. The dependence of D_A (n) on the square of the free carrier concentration shows that the charge difference between the substitutional donor and the dopant–vacancy pair must be two. This dependence unambiguously demonstrates that the mobile dopant–vacancy pairs involved in the vacancy mechanism of P, As, and Sb diffusion are singly negatively charged [112]. Considering the dominance of V^{2-} under n-type doping (see Figure 6.18), donor diffusion in Ge is accurately described by the vacancy mechanism with the following defects and charge states involved [112]

$$AV^- \leftrightarrow A_s^+ + V^{2-} \tag{6.39}$$

Compared to Si, donor diffusion in Ge is by far less complex. Although P and As diffusion in Si are each described by three different individual contributions due to

P_i^0, P_i^+, PV^- and AsV^0, AsI^0, AsV^- (see Equations 6.21 and 6.27), respectively, donor diffusion in Ge is mainly determined by one contribution. Donor diffusion in Ge does not lead to deviations from the local thermal V equilibrium concentration [9,111,112]. This reveals that the diffusion proceeds in the foreign-atom-controlled mode [6]. Accordingly, the donor diffusion coefficient D_A with A ∈ {P, As, Sb} is described by

$$D_A = \frac{C^{eq}_{AV^-} D_{AV^-}}{C^{eq}_{A_s^+}} \tag{6.40}$$

The strong doping dependence of D_A expressed by Equation 6.38 is a consequence of the impact of doping on the thermal equilibrium concentration $C^{eq}_{AV^-}$ and $C^{eq}_{A_s^+}$ of AV^- and A_s^+, respectively. $C^{eq}_{AV^-}$ increases and $C^{eq}_{A_s^+}$ decreases with n. This leads to the dependence of D_A on n^2. In Si, n-type dopant diffusion mainly increases with n. Only for higher doping levels, a stronger doping dependence of donor diffusion becomes evident (see Section 6.2.2). This is due to the formation of both neutral and singly negatively charged AV pairs in Si.

The doping dependence of donor diffusion in Ge is generally supported by numerous experimental studies. Vainonen-Ahlgren et al. [137] studied As diffusion from a GaAs overlayer. Their experiments clearly demonstrate a strong doping dependence of As diffusion [111]. Further experiments on donor diffusion in ion-implanted Ge also consistently support the doping dependence of diffusion [103,138–142]. However, implantation damage gives rise to enhanced donor diffusion, in particular, at short diffusion times [141]. Accordingly, diffusion data deduced from implanted and annealed samples are often not representative for equilibrium diffusion conditions. In this respect, P diffusion data reported by Chui et al. [138] that clearly exceed the equilibrium intrinsic diffusion coefficient of P more likely reflect a transient enhanced diffusion due to implantation damage than an equilibrium P diffusion.

6.3.2.2 Acceptor Diffusion

The diffusion of p-type dopants such as Ga and In in Ge does not reveal any significant doping dependence [121,122]. This is illustrated by the In profiles shown in Figure 6.21. Detailed analysis of In diffusion in Ge reveals an In diffusivity for extrinsic conditions that is twice as high as for intrinsic conditions [122]. The overall In diffusion behavior in Ge demonstrates that the InV pair must be singly negatively charged and therewith possesses the same charge state as the In acceptor. Data reported in the literature on Ga diffusion in Ge also indicate a concentration-independent diffusion [121]. Accordingly, the GaV pair is likely also singly negatively charged. Presumably, Al in Ge behaves similar [108]. Overall, p-type dopants A ∈ {Al,Ga,In} in Ge diffuse via the vacancy mechanism with the following defects and charge states involved:

$$(AV)^- \rightleftharpoons A_s^- + V^0 \tag{6.41}$$

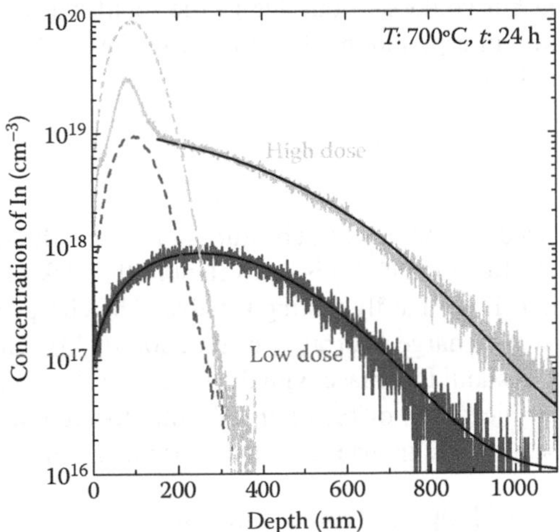

FIGURE 6.21 Indium concentration profiles (lower and upper solid lines) measured with SIMS after diffusion annealing of the low- and high-dose In-implanted Ge samples at the temperature and time indicated. The low- and high-doses as-implanted In profiles are illustrated by the lower and upper dashed lines. The implantation was performed with 350 keV In ions and doses of 1.1×10^{14} cm^{-2} and 1.1×10^{15} cm^{-2}. The black solid lines are simulations of In diffusion performed on the basis of reaction (6.41).

This reaction also considers the charge state of V that prevails under p-type doping (see Section 6.3.1). According to reaction (6.41), the acceptor diffusion coefficient D_A with A ∈ {Al, Ga, In} is given by

$$D_{\mathrm{A}} = \frac{C^{\mathrm{eq}}_{\mathrm{A}V^-} D_{\mathrm{A}V^-}}{C^{\mathrm{eq}}_{\mathrm{A}_s^-}} \tag{6.42}$$

The doping dependence of AV^- and A_s^- cancels between each other. The factor of two difference between intrinsic and extrinsic In diffusion only becomes evident when the differential equations representing the extrinsic diffusion of acceptors via the vacancy mechanism are considered in detail [6,122].

The diffusion of the acceptor dopant B in Ge behaves very different from that in Al, Ga, and In. Its diffusivity is several orders of magnitude lower than self-diffusion (see Figure 6.16) and exhibits a diffusion activation enthalpy of 4.65 eV [106] (see Equation 6.31) that clearly exceeds the activation enthalpy of self-diffusion (see Equation 6.29). This can be explained with a repulsive elastic interaction between B and V as confirmed by theoretical calculations [120,143]. However, the slow diffusion of B can also reflect the contribution of the interstitialcy or kick-out mechanism to dopant diffusion. This contribution is usually covered by the dominance of the vacancy mechanism but could step out due to the repulsive interaction between B and V. Following this interpretation, B diffusion in Ge could be mediated by the interstitialcy mechanism

$$(BI)^+ \rightleftharpoons B_s^- + I^{2+} \tag{6.43}$$

that considers the acceptor nature of substitutional B and the donor nature of Ge self-interstitials [134]. Reaction (6.43) is equivalent to the kick-out reaction since the BI pair and interstitial B are hardly distinguishable. The charge state of the BI pair is assumed to fulfill the charge neutrality of reaction (6.43). The reaction predicts that the diffusion of B should increase with the square of the acceptor concentration [6]. This could be verified by B diffusion in Ge with different acceptor background doping levels. To the author's knowledge, such experiments were not yet performed.

6.3.2.3 Formation of Dopant–Defect Complexes

Dopant–defect complexes such as A_2X and AX_2 can be formed when a dopant–defect pair AX gets close to a substitutional dopant A_s or to an isolated native defect X with $X \in \{V, I\}$. This is expressed by the reactions

$$AX + A_s \rightleftharpoons A_2X \tag{6.44}$$

$$AX + X \rightleftharpoons AX_2 \tag{6.45}$$

It is evident from Equation 6.39 that the formation of A_2V complexes is very favorable for n-type dopants in Ge due to the attractive Coulomb interaction between AV^- and A_s^+. Indeed, a significant deactivation of n-type dopants becomes evident in the diffusion behavior of P and As at high concentrations [9]. This is accurately described assuming the formation of neutral A_2V complexes. AV_2 complexes are not favored under equilibrium diffusion conditions as a Coulomb barrier exists between AV^- and V^{2-}.

Considering p-type dopants, the formation of A_2V complexes is not very likely due to the Coulomb repulsion between AV^- and A_s^- (see Equation 6.41). Instead, reaction (6.45) can cause a successive deactivation of p-type dopants since the formation of both AV^- and AV_2 is not hindered by a Coulomb barrier (see Equation 6.41). Hence, deactivation of Ga in Ge [144] very likely proceeds via reaction (6.45). Finally, B in Ge is considered. As already mentioned, B diffuses exceptionally slowly [106] and is likely mediated by I [20,124,126,136]. In the case where the concentration of I increases in B-doped Ge, the concentration of BI^+ is also expected to increase. With increasing concentration of BI^+, a deactivation of substitutional B^- is favored like in Si [136]. Under conventional annealing conditions, the concentration of I is too low to initiate B complex formation. This situation changes when I are formed athermally, for example, by ion implantation and/or irradiation as will be seen in Section 6.3.3.

Dopant–defect complexes can also form with other foreign atoms intentionally or unintentionally introduced in Ge. The impact of codoping on the diffusion of n-type dopants has been evidenced, for example, for carbon [9,145]. Carbon is an isovalent impurity and possesses a very low solubility in Ge ([C] $\leq 2.5 \times 10^{14}$ cm^{-3}) [146]. However, carbon concentrations well above the solubility limit can be introduced by means of epitaxial layer growth. The impact of carbon on As diffusion is demonstrated by Ge samples unintentionally doped with carbon during Ge epitaxy [9]. Arsenic aggregation is observed within the Ge layers doped with carbon to concentrations of about 10^{20} cm^{-3}. The dopant aggregation fully correlates with the carbon profile and is consistently described by means of [9]

$$\mathrm{As}V^{-} + \mathrm{C}_{\mathrm{s}}^{0} \leftrightarrow \mathrm{C}V\mathrm{As}^{0} + \mathrm{e}^{-} \tag{6.46}$$

Reaction (6.46) describes the trapping of negatively charged AsV^- pairs by neutral substitutionally dissolved carbon. In the course of this reaction, CVAs complexes are formed. The stability of carbon–vacancy–dopant CVA complexes is confirmed by DFT calculations [147]. The formation of CVAs complexes or more general of CVA complexes with A $\in$ {P,As,Sb} leads to a retardation of donor diffusion [9]. A retarded diffusion of donors by carbon is very beneficial for the fabrication of ultra-shallow dopant profiles [131,148]. However, donor deactivation due to the formation of A_2V complexes is not suppressed. In fact, the additional formation of neutral CVA complexes further limits the activation of donors.

Other strategies to engineer the effective diffusion and activation of dopants concern codoping with large isovalent foreign atoms [149], double donor doping [150–152], and codoping with fluorine [153]. These and additional strategies of defect engineering in Ge are discussed in detail in recent reviews [131,148] and will not be repeated here.

In summary, self- and dopant diffusion in Ge under conditions close to thermal equilibrium are consistently described on the basis of the vacancy mechanism. The charge states of the vacancies involved are characterized and the doping dependence of diffusion and the interactions of the point defects that lead to dopant deactivation are clarified.

In the following section, self- and dopant diffusion in Ge under nonequilibrium conditions realized by irradiation are treated. It is demonstrated that diffusion in Ge under irradiation behaves very unusual in comparison to Si.

6.3.3 Nonequilibrium Diffusion

The effect of proton irradiation on self- and dopant diffusion in Ge was recently studied by means of isotopically controlled Ge and natural Ge samples doped with P, As, and B [19,20,154]. Analysis of self- and dopant diffusion in Ge under irradiation revealed an unexpected diffusion behavior. This is demonstrated by Figures 6.22 through 6.24 that, respectively, indicate experimental results on self-diffusion, B diffusion, and As diffusion under concurrent annealing and irradiation. Although the diffusion of self- and B atoms is clearly enhanced compared to thermal equilibrium conditions [20], the diffusion of n-type dopants such as P and As remains unaffected [154], that is, equals the diffusion under thermal equilibrium conditions.

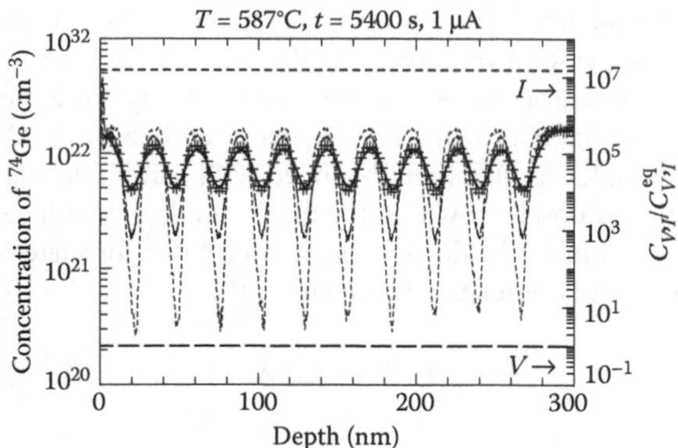

FIGURE 6.22 Concentration profile of ^{74}Ge (+: only every 9th data point is shown for clarity) measured with SIMS after annealing a $(^{nat}Ge/^{70}Ge)_{10}$ isotope multilayer structure at 587°C for 5400s and concurrent irradiation with 2.5 MeV protons at a flux of 1 μA. The solid line represents the best fit to the experimental ^{74}Ge profile based on the model for Ge self-diffusion under irradiation proposed by Schneider et al. [20]. The calculated concentrations of V and I (lower and upper dashed lines) are normalized to their thermal equilibrium values and referred to the right axis. The distributions of both V and I are homogeneous with V concentrations close to thermal equilibrium and I concentrations in high supersaturation. The thin short-dashed profile reflects the ^{74}Ge profiles of the as-grown isotope structure. The long-dashed line represents the Ge profile beneath a covered part of the Ge sample. This part was not irradiated and, accordingly, reflects self-diffusion under thermal equilibrium conditions.

In Reference 19, a retarded donor diffusion under irradiation is reported, which is at variance with the results shown in Figure 6.24. The seemingly retarded donor diffusion under irradiation in Reference 19 is suggested by an enhanced out-diffusion of the dopant to the sample surface. This out-diffusion was not sufficiently suppressed by a SiO_2 layer deposited on the Ge surface and leads to a dopant dose loss of about 70% of the total implanted dose. Experiments with P-implanted samples capped with silicon nitride and As-implanted samples capped with SiO_2 do not reveal a retarded diffusion under irradiation [154]. Interestingly, donor diffusion in Ge under irradiation fully equals donor diffusion under thermal equilibrium conditions. On the contrary, self- and B diffusion shown in Figures 6.22 and 6.23 are clearly enhanced and, interestingly, do not reveal any depth dependence that is expected in the case the Ge surface acts as efficient sink for native defects. This strongly differs from the behavior of self-diffusion in Si under irradiation shown in Figure 6.13 and discussed in Section 6.2.4. The RESD profile in Si reveals a distinct depth dependence that reflects an inhomogeneous distribution of V and I [18]. Accordingly, compared to Si, the unusual behavior of self- and dopant diffusion in Ge under irradiation must be due to Ge-specific properties that differ from Si.

In order to uncover the origin of the unusual diffusion behavior in Ge under irradiation, each specific experimental observation and its consequence are described. First, the absence of any depth-dependent broadening in self- and dopant diffusion shows that the distribution of the native defects must be homogeneous. Second, the equilibrium diffusion of V-mediated donor diffusion under irradiation demonstrates

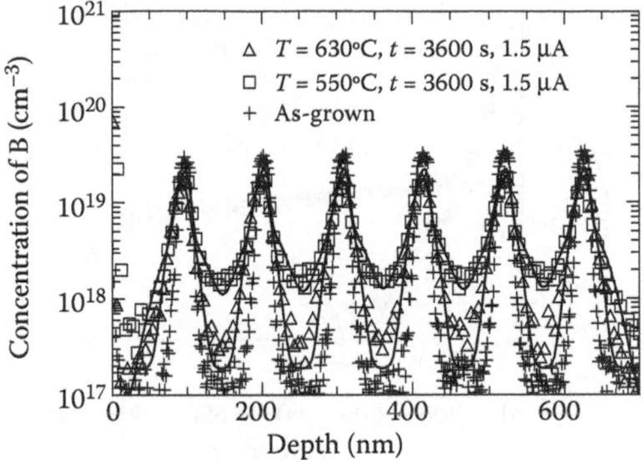

FIGURE 6.23 Concentration profiles of B (symbols: only every 6th data point is shown) in Ge measured with SIMS after concurrent annealing and irradiation with 2.5 MeV protons (Δ, □) and experimental conditions as indicated in the figure. The distribution of B in the as-grown structure (+) is shown for comparison. The B profiles reveal a stronger diffusional broadening at low compared to high temperatures. The solid lines are theoretical B profiles calculated on the basis of the B diffusion model described by Schneider et al. [20]. The model considers contributions of substitutional B_s, B*I* pairs, and immobile B clusters to the total B concentration.

that the concentration of *V* in Ge under irradiation must be close to thermal equilibrium. Third, the enhanced and homogeneous broadening of B profiles under irradiation indicates that B diffusion is mainly mediated by *I*, whose concentration is homogeneous and exceeds the thermal equilibrium value by several orders of magnitude. These characteristic aspects of self- and dopant diffusion in Ge under concurrent annealing and irradiation are all consistently explained with a Ge surface that acts as efficient sink for *V* but as barrier for *I* [20,154]. This Ge surface property leads to a strong imbalance between *I* and *V* during irradiation. This imbalance is characterized by homogeneous *V* and *I* distributions, *V* concentrations close to thermal equilibrium, and *I* concentrations that exceed the thermal equilibrium concentration by several orders of magnitude.

The equilibrium diffusion of P and As in Ge under irradiation is a quite unusual behavior since from Si we know that dopant diffusion is significantly enhanced under irradiation [88–90]. This directly shows that the concentration of *V* must be in thermal equilibrium even under irradiation and that donor diffusion via dopant–*I* pairs is not favorable even under *I* supersaturation. The negligible dopant diffusion via A*I* pairs is explained by Coulomb repulsion between the positively charged donor A_s^+ and the singly or doubly positively charged *I* [134,148]. This repulsion hinders the formation of mobile A*I* pairs. But is there in the end any difference between equilibrium and nonequilibrium diffusion of donors in Ge? The answer is yes. The striking difference is that concurrent annealing and irradiation suppresses the deactivation of donors as demonstrated in Figure 6.24 [154]. A reaction must be operative that hampers the formation of A_2V complexes. This is best described by means of a competitive inhibition reaction. Such reactions are known from biochemical enzymatic reactions. In the specific case, *I* act as inhibitor for the formation of A_2V.

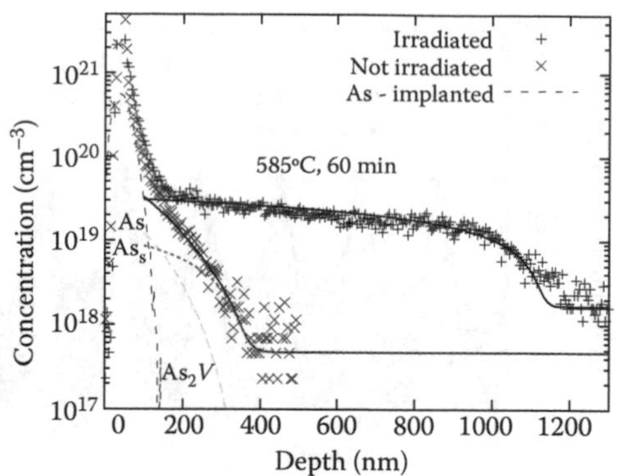

FIGURE 6.24 Concentration profiles of As in Ge measured with SIMS after As implantation (thin-dashed line) and additional annealing (symbols) at the temperature and time indicated. The diffusion behavior of As under irradiation (+) and nonirradiation (×) is very different as demonstrated by the differences in profile shape and penetration depth. Arsenic diffusion is accurately described on the basis of reactions (6.39), (6.44), and the inhibitor reaction described in the text. The position of the amorphous/crystalline (a/c) interface formed by As implantation at about 100 nm from the surface is taken as onset of the diffusion profile. Arsenic within about 100 nm from the surface likely exists in clusters that are more complex than As_2V. These clusters are considered to dissolve during annealing and serve as constant source for dopant diffusion. Below the a/c interface, the As profile after concurrent annealing and irradiation (+) is accurately described by the vacancy mechanism (6.39) with full activation of the dopant that reflects the contribution of the inhibitor reaction. The dopant profile after annealing without irradiation reveals the formation of A_2V complexes and a corresponding lower level of activation. The profiles of the electrical active substitutional As_s and inactive As_2V complexes deduced from the simulation of As diffusion without irradiation are indicated by the two lower dashed lines.

The inhibitor reaction $A_s^+ + I^{2+} \leftrightarrow AI^{3+}$ lowers the rate constant for the formation of the A_2V complex via reaction (6.44). The suppression increases with increasing supersaturation of I. Physically, the inhibitor reaction describes the tendency of the substitutional donor A_s^+ to form an AI pair with I^{2+} that can be favored due to elastic interactions. However, the repulsive Coulomb interaction finally hinders the formation of AI.

The limited efficiency of the Ge surface to annihilate I is the key for the explanation of the unusual behavior of self- and dopant diffusion in Ge under concurrent annealing and irradiation. This Ge surface property is supported by the detection of I in Ge by means of high-resolution transmission electron microscopy [155]. The identification of I would be hardly possible in the case the Ge surface is a perfect sink.

It is evident from Figure 6.23 that I strongly affect the diffusion of B under irradiation. Recently, Cowern et al. [62] used B diffusion measurements to probe the nature of I in Ge. They report on two distinct self-interstitial forms, that is, one form with a low and another with high activation entropy. This was concluded from the analysis of B diffusion based on the g/λ approach [156]. However, the applicability of this approach for the analysis of B diffusion in Ge under nonequilibrium conditions is questionable since several assumptions considered in this approach are not fulfilled. A more rigorous analysis of B diffusion in Ge, which is based on the full system of differential diffusion equations [20], does not confirm the existence of two forms of I defects in Ge. Instead, the analysis reveals that the concentration of mobile BI pairs strongly exceed the concentration of substitutional B_s under irradiation. The formation entropy of the BI pair was determined to 30 k_B [20], whereas for the migration entropy of I, a value of about 4 k_B was deduced. Although the total activation entropy, that is, sum of formation and migration entropy, of the BI pair and the isolated I are not accessible, the analysis confirms the presence of two distinct forms of I, but the more extended form is a BI pair rather than another form of a self-interstitial. This distinction between BI and I is difficult in the g/λ approach.

6.4 DIFFUSION IN SILICON–GERMANIUM ALLOYS

Since several years, SiGe alloys have entered the fabrication of metal oxide semiconductor field effect transistor (MOSFET) to meet the requirements for the continuous downscaling of Si-based electronic devices [157–159]. By adding Ge to Si, the lattice constant increases with respect to Si. Thin epitaxial layers of Si on SiGe experience tensile strain. Strained Si layers are advantageous for the channel region of MOSFET structures because the mobility of free carriers is enhanced [160,161]. In order to maintain the strain in Si/SiGe structures during device processing, the Si and Ge interdiffusion across the Si/SiGe interface must be controlled. However, the interdiffusion is not only affected by the local alloy composition but also by the local strain due to the lattice mismatch between Si and SiGe [162,163]. In order to unravel the different aspects that affect the interdiffusion of Si/SiGe and Si/Ge structures, experiments on the diffusion of Si and Ge in unstrained $Si_{1-x}Ge_x$ alloys without any chemical gradient have been performed. These experiments provide valuable information about the tracer diffusion coefficients of Si and Ge in SiGe that are important quantities in the Darken equation [164] that describes the Si–Ge interdiffusion coefficient and its concentration dependence. In this section, the self-diffusion of Si and Ge in unstrained SiGe alloys is reviewed. Moreover, the diffusion behavior of dopants in unstrained SiGe is treated. In order to limit the content of this section, only experimental and theoretical results on self- and dopant diffusion are presented that span almost the whole composition range of SiGe.

6.4.1 SELF-DIFFUSION

Several studies on the diffusion of Ge in unstrained $Si_{1-x}Ge_x$ are reported in the literature [46,165–171]. However, only a few studies concern the diffusion of Si in SiGe since experiments with the short-lived radioactive ^{31}Si isotope are very challenging [169,170]. The availability of highly enriched stable isotopes of Si and Ge enabled the MBE growth of SiGe isotope structures that are highly suitable for studying simultaneously the diffusion of Si and Ge in relaxed SiGe over the whole composition range [46,171]. Figure 6.25 shows concentration profiles of Ge and Si isotopes of as-grown $^{na}Si_{1-x}{}^{nat}Ge_x/{}^{28}Si_{1-x}{}^{70}Ge_x/{}^{nat}Si_{1-x}{}^{nat}Ge_x$ isotope heterostructures with $x = 0.45$ (see Figure 6.25a) and $x = 0.70$ (see Figure 6.25c). Additional $^{nat}Si_{1-x}{}^{nat}Ge_x/{}^{28}Si_{1-x}{}^{70}Ge_x/$

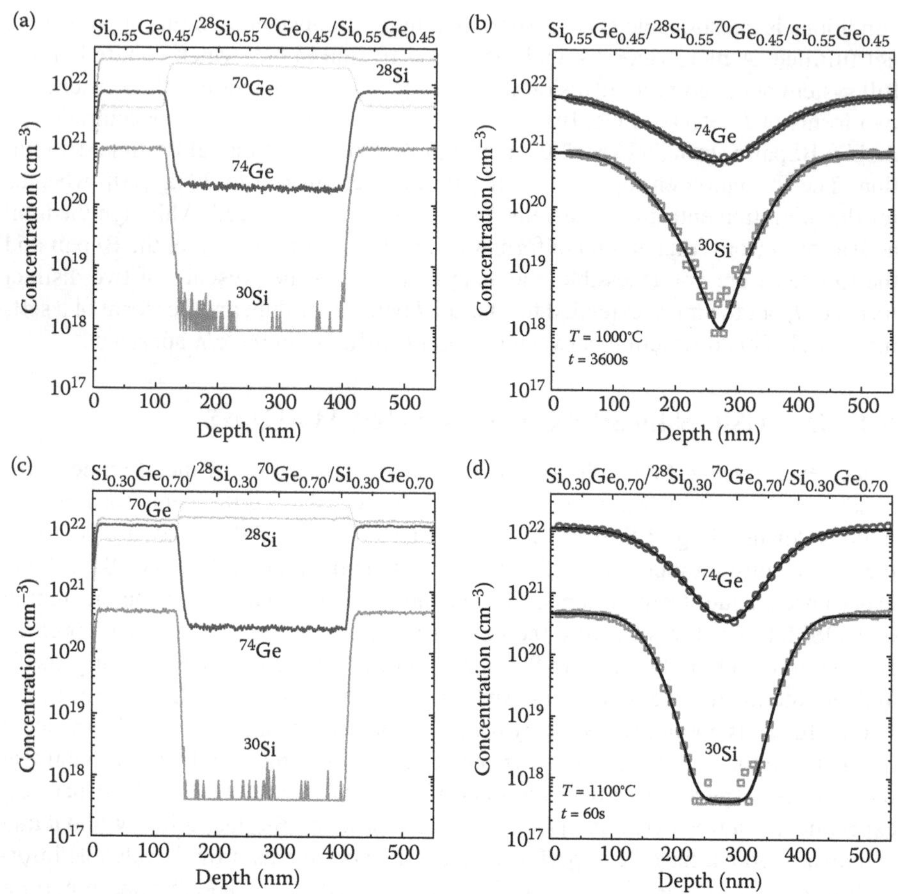

FIGURE 6.25 SIMS concentration profiles of ^{30}Si and ^{74}Ge in epitaxial $^{nat}Si_{1-x}{}^{nat}Ge_x$/$^{28}Si_{1-x}{}^{70}Ge_x$/$^{nat}Si_{1-x}{}^{nat}Ge_x$ isotope structures with $x = 0.45$ and 0.70 before ((a) and (c)) and after annealing ((b) and (d)) for temperatures and times indicated in the figure. The solid lines in (b) and (d) are best fits based on the solution of Fick's second law of self-diffusion. Only every ninth (fourth) data point is shown for Ge (Si) in (b) and (d) for clarity. (Reprinted with permission from R. Kube et al. Composition dependence of Si and Ge diffusion in relaxed $Si_{1-x}Ge_x$ alloys. *J. Appl. Phys.* **107**, 073520, 2010. Copyright 2010, American Institute of Physics.)

$^{nat}Si_{1-x}{}^{nat}Ge_x$ heterostructures with $x = 0.05$ and 0.25 were prepared by MBE and utilized for self-diffusion experiments [171]. After annealing, the diffusion profiles of Si and Ge were measured with SIMS. Typical Si and Ge concentration profiles measured with SIMS after diffusion annealing are shown in Figure 6.25b and d. Additional profiles for other SiGe alloys are illustrated in Reference 171.

The temperature dependences of the Si and Ge diffusion coefficients obtained from the analysis of the self-diffusion profiles in $Si_{1-x}Ge_x$ with $x \in \{0.05, 0.25, 0.45, 0.70\}$ are shown in Figure 6.26 in comparison to self-diffusion in Si [2], self-diffusion in

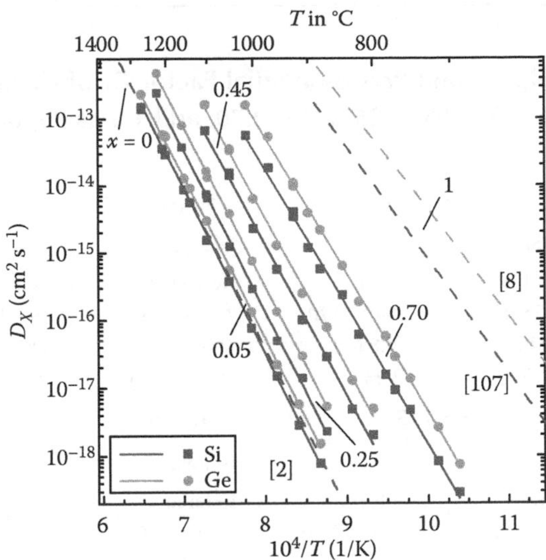

FIGURE 6.26 Temperature dependence of the diffusion coefficient of ^{30}Si (filled squares) and ^{74}Ge (filled circles) in $Si_{1-x}Ge_x$. For comparison, the lower, second upper, and upper dashed lines show the temperature dependence of Si self-diffusion [2,119], Si diffusion in Ge [107], and Ge self-diffusion [8], respectively. (Reprinted with permission from R. Kube et al. Composition dependence of Si and Ge diffusion in relaxed $Si_{1-x}Ge_x$ alloys. *J. Appl. Phys.* **107**, 073520, 2010. Copyright 2010, American Institute of Physics.)

Ge [8], and to Si diffusion in Ge [107]. The corresponding Si and Ge self-diffusion data for all investigated SiGe alloy compositions are listed in Reference 46. The temperature dependence of Si and Ge diffusion reveals that with increasing Ge content the difference in the diffusion of Si and Ge increases, that is, the diffusion of Ge becomes increasingly faster with increasing Ge content compared to Si diffusion. This trend is explained with an enhanced contribution of V to Ge diffusion in SiGe with increasing Ge content [171]. The experimentally observed Ge diffusion behavior confirms recent theoretical results [172–174] that the vacancy in SiGe is more likely coordinated by Ge atoms.

Within the temperature range considered in the experiments, the diffusion of Si and Ge follows the Arrhenius expression $D = D_0 \exp(-(Q/k_BT))$ with an activation enthalpy Q and preexponential factor D_0 listed in Table 6.1 for each particular alloy. Within the accuracy of the experimental results (about 0.05 eV), the activation enthalpy of Si diffusion equals the diffusion activation enthalpy of Ge but tends to be slightly higher than that of Ge. The preexponential factor D_0 of Ge diffusion in $Si_{1-x}Ge_x$ for $0 \leq x \leq 0.70$ is higher than for Si diffusion and reflects the faster diffusion of Ge in SiGe compared to Si.

The dependence of the activation enthalpy of Si and Ge diffusion on the Ge content is illustrated in Figure 6.27 in comparison to earlier results on Ge diffusion in SiGe [165–170]. The self-diffusion study with $Si_{1-x}Ge_x$ isotope heterostructures shows that the diffusion activation enthalpy of Si and Ge decreases with increasing

TABLE 6.1
Activation Enthalpy Q and Preexponential Factor D_0 of Ge and Si Diffusion in $Si_{1-x}Ge_x$ with $x = 0$, 0.05, 0.25, 0.45, 0.70, and 1.0 as Reported in the Literature

$Si_{1-x}Ge_x$	$x = 0$		$Si_{1-x}Ge_x$	$x = 0.05$	
Isotope	^{30}Si	^{74}Ge	Isotope	^{30}Si	^{74}Ge
Reference	[2,119]	[46]	Reference	[46]	[46]
T range (°C)	855–1388	880–1270	T range (°C)	880–1250	880–1250
Q (eV)	4.76	4.83	Q (eV)	4.82	4.77
D_0 (cm^2 s^{-1})	560	923	D_0 (cm^2 s^{-1})	795	915
$Si_{1-x}Ge_x$	$x = 0.25$		$Si_{1-x}Ge_x$	$x = 0.45$	
Isotope	^{30}Si	^{74}Ge	Isotope	^{31}Si	^{71}Ge
Reference	[46]	[46]	Reference	[46]	[46]
T range (°C)	870–1225	870–1225	T range (°C)	800–1105	800–1105
Q (eV)	4.77	4.71	Q (eV)	4.44	4.42
D_0 (cm^2 s^{-1})	2233	3022	D_0 (cm^2 s^{-1})	1075	2074
$Si_{1-x}Ge_x$	$x = 0.70$		$Si_{1-x}Ge_x$	$x = 1.0$	
Isotope	^{31}Si	^{71}Ge	Isotope	^{28}Si	^{71}Ge
Reference	[46]	[46]	Reference	[107]	[8]
T range (°C)	690–1015	690–1015	T range (°C)	550–900	429–904
Q (eV)	4.05	4.05	Q (eV)	3.32	3.13
D_0 (cm^2 s^{-1})	366	1162	D_0 (cm^2 s^{-1})	42	25.4

Source: H. Bracht, E.E. Haller, and R. Clark-Phelps, *Phys. Rev. Lett.* **81**, 393, 1998; E. Hüger et al. *Appl. Phys. Lett.* **93**, 162104, 2008; R. Kube et al. *J. Appl. Phys.* **107**, 073520, 2010; H.H. Silvestri et al. *Semicond. Sci. Technol.* **21**, 758, 2006; H. Bracht, *Physica B* **376–377**, 11, 2006.

Ge concentration. The activation enthalpy $Q(x)$ as a function of the alloy composition x indicates a distinct upward bowing both for Si and Ge. Accordingly, $Q(x)$ deviates from Vegard's law $Q(x) = (1 - x)Q(0) + xQ(1)$ that describes $Q(x)$ with a linear interrelation of the activation enthalpy $Q(0)$ in pure Si and $Q(1)$ in pure Ge. The deviation from Vegard's law is described with a quadratic correction term according to

$$Q(x) = (1 - x)Q(0) + xQ(1) + x(1 - x)\Theta \tag{6.47}$$

where Θ is the bowing parameter. Taking into account Equation 6.47, the composition dependence of the activation enthalpy of self-diffusion in $Si_{1-x}Ge_x$ is best described with a bowing parameter $\Theta = (1.54 \pm 0.15)$ eV and (1.63 ± 0.12) eV for Si and Ge, respectively [46].

The earlier work of McVay and DuCharme [165–167] utilized SiGe polycrystals for their diffusion experiments with the radioactive isotope ^{71}Ge. They report a strong decrease of the activation enthalpy of Ge diffusion from 4.65 to 3.1 eV for

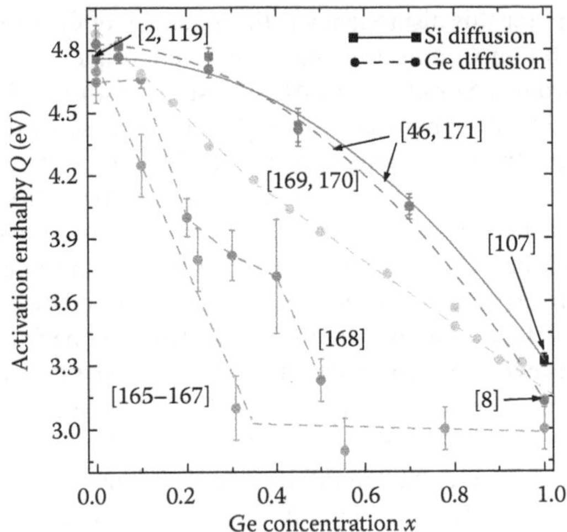

FIGURE 6.27 Activation enthalpy of Si diffusion (filled squares, solid line) and Ge diffusion (filled circles, dashed lines) in $Si_{1-x}Ge_x$ versus the Ge content x in comparison to Si self-diffusion [2,119], Ge self-diffusion [8], and Si diffusion in Ge [107]. The composition dependence comprises Si and Ge diffusion in $Si_{1-x}Ge_x$ reported by Kube et al. [46,171] and Ge diffusion in $Si_{1-x}Ge_x$ reported by McVay and DuCharme [165–167], Zangenberg et al. [168], and Strohm et al. [169,170]. (Reprinted with permission from R. Kube et al. Composition dependence of Si and Ge diffusion in relaxed $Si_{1-x}Ge_x$ alloys. *J. Appl. Phys.* **107**, 073520, 2010. Copyright 2010, American Institute of Physics.)

SiGe alloy compositions in the range of $x = 0$ and $x = 0.3$ and an almost constant activation enthalpy of 3.0 eV for higher Ge contents. Based on these results, McVay and DuCharme concluded that I dominate Ge diffusion in SiGe for $x \leq 0.3$ and V for $x > 0.3$. A more recent work of Zangenberg et al. [168] on Ge self-diffusion in relaxed epitaxial $Si_{1-x}Ge_x$ layers with x between 0 and 0.5 suggests that V mediate Ge diffusion in SiGe for $x \approx 0.5$. Even more recently, radiotracer diffusion experiments of Strohm et al. [169,170] conducted with single crystalline $Si_{1-x}Ge_x$ samples indicate an almost linear composition dependence of the activation enthalpy of Ge diffusion with a break at about $x = 0.35$. Finally, the most recent work of Kube et al. reveals an upward bowing of the $Q(x)$ with increasing Ge content x. The experiments of all these groups show different trends in the composition dependence of self-diffusion in SiGe. In particular, Strohm et al. [170] report that the diffusion of Si in $Si_{1-x}Ge_x$ is equal to Ge diffusion even for $x = 1.0$. This is, however, at variance with the experiments on Si diffusion in Ge reported by Silvestri et al. [107]. This study clearly demonstrates a slower diffusion of Si in Ge compared to Ge self-diffusion. The higher activation enthalpy of Si diffusion in Ge compared to Ge self-diffusion indicates a repulsive interaction between Si and V. This repulsive interaction is confirmed by atomistic calculations of Chroneos et al. [133].

The composition dependence suggested by the self-diffusion study with SiGe isotope heterostructures [46,171] is clearly at variance with the results reported by McVay and DuCharme [165–167], Zangenberg et al. [168], and Strohm et al. [169,170].

One possible reason for this discrepancy is the tendency of Ge to evaporate from the surface during high-temperature treatments. As a consequence, the near surface of the SiGe alloy becomes Si-rich. The diffusion study with the SiGe isotope structures is not affected by a near-surface loss of Ge since the self-diffusion coefficients are determined from the broadening of buried SiGe isotope interfaces [46]. On the contrary, in conventional radiotracer diffusion studies, the radioactive isotopes are dropped onto the surface or implanted with low energy. In particular, Strohm et al. [169,170] implanted the radioactive isotopes of ^{71}Ge and ^{31}Si to depths of at most 38 and 56 nm, respectively. The diffusion profiles were obtained by ion beam sputtering and a technique to detect the activity of the radioactive isotope from specific sections of the sample. However, in radiotracer studies, only the radioactive isotopes and not the matrix composition are detected. Very likely, the high-temperature diffusion experiments conducted by Strohm et al. suffer from a depletion of Ge close to the surface. Consequently, the extracted diffusion coefficients are systematically lower compared to the actual diffusivity in bulk SiGe. At lower temperatures, the depletion of Ge is less pronounced and accordingly the bulk self-diffusion coefficient for the respective SiGe alloy is obtained. Overall, this suggests an activation enthalpy of self-diffusion for a particular alloy composition that is lower than the values determined by means of the SiGe isotope structures. This interpretation is supported by the fact that the Ge diffusion data reported by Strohm et al. [169,170] for 25 at.% Ge and 45 at.% equal the results of Kube et al. [46,171] for low temperatures but are systematically lower at higher temperatures.

In contrast to elemental Si and Ge, no experimental results on the individual contributions of V and I to either Ge or Si diffusion in SiGe are available. Moreover, information about the charge state of the dominant native point defects mediating self-diffusion in SiGe are lacking. Recent theoretical calculations of Castrillo et al. [175,176] attempt to fill this gap of knowledge. The authors developed a physical-based model to describe self- and interdiffusion in SiGe for relaxed and strained conditions. The model parameters involved in their atomistic kinetic Monte Carlo approach are calibrated based on available experimental results on diffusion in SiGe under relaxed and strained conditions. In good agreement with our understanding on self-diffusion in Si and Ge (see Sections 6.2.1 and 6.3.1), Castrillo et al. consider that neutral V^0 and I^0 dominate in Si and V^{2-} in Ge under intrinsic conditions. With increasing Ge content, the relative contribution of charged I to self-diffusion decreases. This simulates the increasing (decreasing) significance of the vacancy (interstitialcy) mechanism for atomic transport in SiGe with increasing Ge content. Experiments on the impact of doping on self-diffusion in SiGe alloys would be highly desirable to refine the atomistic model for simulation of diffusion processes in SiGe.

6.4.2 Dopant Diffusion

Numerous experimental studies on dopant diffusion in SiGe have already been reported over the past two decades [1]. However, only a few systematic studies on dopant diffusion that cover the full SiGe composition range are reported. In particular, diffusion in relaxed SiGe structures is very important in investigating the mechanism of dopant diffusion and its dependence on the alloy composition. Based on these results, the impact of strain on dopant diffusion can be evaluated more consistently.

First, the results on the composition dependence of Sb diffusion are reported by Larsen et al. [177]. They investigated with SIMS the diffusional broadening of an Sb-doped SiGe layer sandwiched between undoped SiGe for temperatures between 850°C and 1028°C. Relaxed $Si_{1-x}Ge_x$ structures with x between 0.0 and 0.5 were grown by means of MBE for the diffusion study. The composition dependence of the Sb diffusion activation enthalpy is shown in Figure 6.28 in comparison to the diffusion activation enthalpy of Si and Ge. The activation enthalpy considered by Larsen et al. for Sb diffusion in pure Ge was taken from Sharma [178]. The value of about 2.5 eV is consistent with Q_{Sb} given by Equation 6.37. It is evident that the activation enthalpy of Sb diffusion shows a similar upward bowing as self-diffusion in SiGe. Assuming Equation 6.47, the upward bowing of Q_{Sb} is best described by $\Theta = (1.50 \pm 0.10)$ eV.

Laitinen et al. [179] have measured the diffusion of implanted radioactive ^{72}As and ^{73}As in relaxed $Si_{1-x}Ge_x$ within the full composition range. The activation enthalpy of As diffusion illustrated in Figure 6.28 also reveals an upward bowing. The upward bowing is described with $\Theta = (1.47 \pm 0.10)$ eV [46] and thus very similar to the composition dependence of Q_{Sb}. It is noted that the activation enthalpy of As diffusion reported by Laitinen et al. [179] for pure Si and Ge deviates by 0.4 and 0.3 eV from the results given by Equations 6.24 and 6.36, respectively. Recent experiments on

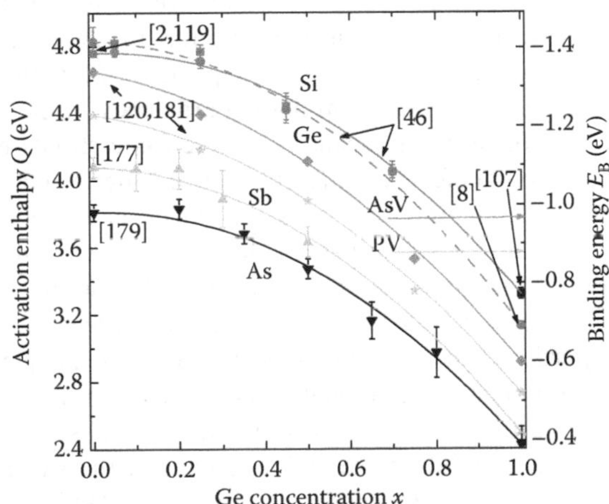

FIGURE 6.28 Upward bowing behavior of the activation enthalpy of Si diffusion (squares, straight line: [46]), Ge diffusion (circles, dashed line: [46]), Sb diffusion (triangles—up: [177], straight line: [46]), and As diffusion (filled triangles—down: [179], straight line: [46]) in $Si_{1-x}Ge_x$ as a function of the Ge content x. In comparison, the composition dependence of the binding energy of As*V* pairs (rhombs, straight line) and P*V* pairs (asterisks, straight line) [120,181] is shown. The negative binding energy implies that the defect cluster is more stable than its constituent point defect components [182]. (Reprinted with permission from R. Kube et al. Composition dependence of Si and Ge diffusion in relaxed $Si_{1-x}Ge_x$ alloys. *J. Appl. Phys.* **107**, 073520, 2010. Copyright 2010, American Institute of Physics.)

As diffusion in $Si_{1-x}Ge_x$ isotope heterostructures with $x = 0.05$ support a diffusion activation enthalpy of $Q_{As} = 4.23$ eV [180] that is close to the value of 4.20 eV for As diffusion in Si (see Equation 6.24). The origin for the apparent systematic underestimation of the Q_{As} values reported by Laitinen et al. remains unclear but could also be related to a preferred Ge depletion at high temperatures.

The similar bowing parameter for self- and dopant diffusion suggests the same origin of the nonlinear behavior. This could be related to the formation and migration enthalpy of native defects that mediate both self- and dopant diffusion. Unfortunately, information about the composition dependence of these thermodynamic properties of native defects in SiGe are not available. However, it is very likely that with the nonlinear behavior of the bond lengths and the lattice parameter of SiGe [181], the formation and migration enthalpy exhibit a deviation from Vegard's law. The impact of the host lattice on defect properties has been recently confirmed by electronic structure calculations. The calculations demonstrate that the nonlinear behavior of the stability of donor–vacancy pairs (*E* centers) does not depend on the donor atom but on the host lattice. The calculated nonlinear stability of the *E* center in SiGe is also illustrated in Figure 6.28. A bowing parameter of $\Theta = -0.54$ eV was calculated for both the P- and As-related *E* centers [181].

6.5 CONCLUSION AND OUTLOOK

Considering the diffusion behavior of self- and dopant atoms in Si and Ge, it follows from Sections 6.2 and 6.3 that strong differences exist. This not only holds for diffusion under thermal equilibrium but also for diffusion under nonequilibrium conditions. These differences are the consequences of the different properties of the native point defects and their annihilation efficiency in the bulk and at free surfaces or interfaces. In Si, both vacancies and self-interstitials exist in thermal equilibrium and both point defects are mainly neutral under electronically intrinsic conditions and annihilate readily at surfaces, interfaces, or extended defects in the bulk. On the contrary, vacancies dominate in Ge under thermal equilibrium, that is, self-interstitials are not relevant for atomic transport under thermal equilibrium, and the vacancy is doubly negatively charged even under intrinsic conditions. These differences in the properties of native point defects in Si and Ge explain the disparity in the dopant diffusion behavior. Although dopant diffusion in Si is very complex due to the interaction of the dopant atoms with both vacancies and self-interstitials, dopant diffusion in Ge is mainly explained by means of the vacancy mechanism. Differences in the magnitude of dopant diffusion in Si and Ge and, in particular, on the doping dependence of diffusion are related to the different charge states of the mobile dopant–defect pairs.

Under nonequilibrium conditions realized, for example, by irradiation, atomic transport in Ge behaves very unusual compared to Si. Although self- and dopant diffusion in Si are enhanced under irradiation, the diffusion behavior in Ge under irradiation reveals no significant radiation enhancement of donor diffusion via vacancies but a strong enhancement of boron and only a slight enhancement of self-diffusion. This, on first sight, atypical diffusion behavior is a consequence of the opposite charge states of vacancies and self-interstitials. Coulomb repulsion between donors and positively charged self-interstitials hinders the formation of mobile dopant–defect pairs. Accordingly, donor diffusion via the interstitialcy mechanism is suppressed. Moreover, self-interstitials formed in the bulk by irradiation are not annihilated at free Ge surfaces but rather reflected, whereas the vacancies formed by irradiation in equal numbers are readily annihilated at surfaces. This strong disparity in the annihilation efficiency is the key for the unusual diffusion behavior in Ge under irradiation.

The opposite charge states of vacancies V and self-interstitials I in Ge are likely also the origin of their efficient recombination. Implantation damage and I-related defect clusters dissolve readily and recrystallization of amorphized Ge proceeds rapidly even at 400°C [183–185,187]. This suggests an I–V recombination that is more efficient than in the case of Si, where implantation damage is less effectively removed by means of postimplantation annealing.

Considering the SiGe alloy system, the self- and dopant diffusion behavior should change gradually from the pure Si to the pure Ge case. Available data on self-diffusion in SiGe do not reveal a consistent picture of the composition dependence of self-diffusion. However, most recent self-diffusion results gained from studies with isotopically controlled SiGe heterostructures are considered to be reliable because the experiments are definitely not affected by Ge depletion of near-surface regions. Current experiments on the composition dependence of self- and dopant diffusion reveal a similar upward bowing of the diffusion activation enthalpy. This reflects a decreasing contribution of the I-mediated diffusion compared to the V-mediated diffusion; however, experiments that directly quantify the individual contributions are still lacking. This could be faced with dopant diffusion in isotopically controlled SiGe structures. Such studies would provide valuable information about the type and charge state of the native point defects that dominate in SiGe for a specific alloy composition.

Nowadays, electronic devices have already reached nanosized dimensions. The properties of dopants and self-atoms determined for bulk materials are not necessarily applicable to nanoscale Si and Ge structures because in small dimensions the thermodynamics and kinetics of point defects can be different from that in the bulk. Surface and interface states can affect the behavior of point defects in the whole nanostructure due to the high surface/interface-to-volume ratio. An intriguing example is the diffusion behavior in Ge under irradiation that is likely affected by surface states that cause a Coulomb repulsion of positively charged self-interstitials. Further miniaturization of electronic devices will thus be intimately connected with an improved understanding of the properties and interaction of atomic defects in nanosized volumes. This interaction will concern the interaction of defects in the volume and the interaction of defects with surfaces and interfaces. Although atomistic calculations are increasingly used to predict defect types, defect interactions, their stability, mobility, and electronic properties, the relevance of theoretical results needs to be verified experimentally. Characterization of atomic defects and their interaction with other point defects, surfaces, and interfaces in nanosized materials is very challenging. Not only experiments under well-defined conditions with 2D/or –3D nanostructures must be performed but also appropriate analyses techniques are necessary to resolve nanosized 2D/3D structures and their properties.

ACKNOWLEDGMENTS

The author acknowledges the contributions of bachelor, master, and PhD students and close collaborations with Professor Arne Nylandsted Larsen (University of Aarhus, Denmark) and Professor Eugene E. Haller (University of California at Berkeley and Lawrence Berkeley National Laboratory) over the past few years. Most of the work was funded by the Deutsche Forschungsgemeinschaft (DFG) as well as an individual grant within the Heisenberg program of the DFG.

REFERENCES

1. N.A. Stolwijk and H. Bracht, *Diffusion in Silicon, Germanium and Their Alloys*, Landolt-Börnstein New Series, Vol. III/33, Subvolume A, Springer, New York, 1998.
2. H. Bracht, E.E. Haller, and R. Clark-Phelps, Silicon self-diffusion in isotope heterostructures. *Phys. Rev. Lett.* **81**, 393, 1998.
3. A. Ural, P.B. Griffin, and J.D. Plummer, Self-diffusion in Si: Similarity between the properties of native point defects. *Phys. Rev. Lett.* **83**, 3454, 1999
4. S.R. Aid, T. Sakaguchi, K. Toyonaga, Y. Nakabayashi, S. Matumoto, M. Sakuraba, Y. Shimamune, Y. Hashiba, J. Murota, K. Wada, and T. Abe, Silicon self-diffusivity using isotopically pure $_{30}$Si epitaxial layers. *Mater. Sci. Eng. B* **114–115**, 330, 2004.
5. Y. Shimizu, M. Uematsu, and K.M. Itoh, Experimental evidence of the vacancy-mediated silicon self-diffusion in single crystalline silicon. *Phys. Rev. Lett.* **98**, 095901, 2007.
6. H. Bracht, Self- and foreign-atom diffusion in semiconductor isotope heterostructures. I. Continuum theoretical calculations. *Phys. Rev. B* **75**, 035210, 2007.
7. H. Bracht, H.H. Silvestri, I.D. Sharp, and E.E. Haller, Self- and foreign-atom diffusion in semiconductor isotope heterostructures. II. Experimental results for silicon. *Phys. Rev. B* **75**, 035211, 2007.
8. E. Hüger, U. Tietze, D. Lott, H. Bracht, D. Bougeard, E.E. Haller, and H. Schmidt, Self-diffusion in germanium isotope multilayers at low temperatures. *Appl. Phys. Lett.* **93**, 162104, 2008.
9. S. Brotzmann, H. Bracht, J. Lundsgaard Hansen, A. Nylandsted Larsen, E. Simoen, E.E. Haller, J.S. Christensen and P. Werner, Diffusion and defect reactions between donors, C, and vacancies in Ge. I. Experimental results. *Phys. Rev. B* **77**, 235207, 2008.
10. M. Naganawa, Y. Shimizu, M. Uematsu, K.M. Itoh, K. Sawano, Y. Shiraki, and E.E. Haller, Charge states of vacancies in germanium investigated by simultaneous observation of germanium self-diffusion and arsenic diffusion. *Appl. Phys. Lett.* **93**, 191905, 2008.
11. T. Südkamp, H. Bracht, G. Impellizzeri, J. Lundsgaard Hansen, A. Nylandsted Larsen, and E.E. Haller, Doping dependence of self-diffusion in germanium and the charge states of vacancies. *Appl. Phys. Lett.* **102**, 242103, 2013.
12. R. Kube, H. Bracht, E. Hüger, H. Schmidt, J. Lundsgaard Hansen, A. Nylandsted Larsen, J.W. Ager III, E.E. Haller, T. Geue, and J. Stahn, Contributions of vacancies and self-interstitials to self-diffusion in silicon under thermal equilibrium and nonequilibrium conditions. *Phys. Rev. B* **88**, 085206, 2013.
13. E. Hüger, R. Kube, H. Bracht, J. Stahn, T. Geue, and H. Schmidt, A neutron reflectometry study on silicon self-diffusion at 900°C. *Phys. Status Solidi B* **249**, 2108, 2012.
14. Y. Shimizu, Y. Kawamura, M. Uematsu, K.M. Itoh, M. Tomita, M. Sasaki, H. Uchida, and M. Takahashi, Atom probe microscopy of three-dimensional distribution of silicon isotopes in Si_{28}/Si_{30} isotope superlattices with sub-nanometer spatial resolution. *J. Appl. Phys.* **106**, 076102, 2009.

15. O. Moutanabbir, D. Isheim, D.N. Seidman, Y. Kawamura, and K.M. Itoh, Ultraviolet-laser atom-probe tomographic three-dimensional atom-by-atom mapping of isotopically modulated Si nanoscopic layers. *Appl. Phys. Lett.* **98**, 013111, 2011.
16. Y. Shimizu, Y. Kawamura, M. Uematsu, M. Tomita, T. Kinno, N. Okada, M. Kato et al. Depth and lateral resolution of laser-assisted atom probe microscopy of silicon revealed by isotopic heterostructures. *J. Appl. Phys.* **109**, 036102, 2011.
17. Y. Shimizu, H. Takamizawa, Y. Kawamura, M. Uematsu, T. Toyama, K. Inoue, E.E. Haller, K.M. Itoh, and Y. Nagai, Atomic-scale characterization of germanium isotopic multilayers by atom probe tomography. *J. Appl. Phys.* **113**, 026101, 2013.
18. H. Bracht, J. Fage Pedersen, N. Zangenberg, A. Nylandsted Larsen, E.E. Haller, G. Lulli, and M. Posselt, Radiation enhanced silicon self-diffusion and the silicon vacancy at high temperatures. *Phys. Rev. Lett.* **91**, 245502, 2003.
19. H. Bracht, S. Schneider, J.N. Klug, C.Y. Liao, J. Lundsgaard Hansen, E.E. Haller, A. Nylandsted Larsen, D. Bougeard, M. Posselt, and C. Wündisch, Interstitial-mediated diffusion in germanium under proton irradiation. *Phys. Rev. Lett.* **103**, 255501, 2009.
20. S. Schneider, H. Bracht, J.N. Klug, J. Lundsgaard Hansen, A. Nylandsted Larsen, D. Bougeard, and E.E. Haller, Radiation-enhanced self- and boron diffusion in germanium. *Phys. Rev. B* **87**, 115202, 2013.
21. M.L.W. Thewalt, Spectroscopy of excitons and shallow impurities in isotopically enriched silicon—electronic properties beyond the virtual crystal approximation. *Solid State Commun.* **133**, 715, 2005.
22. M. Steger, A. Yang, T. Sekiguchi, K. Saeedi, M.L.W. Thewalt, M.O. Henry, K. Johnston et al. Pohl, Photoluminescence of deep defects involving transition metals in Si: New insights from highly enriched $_{28}$Si. *J. Appl. Phys.* **110**, 081301, 2011.
23. R.S. Averback, D. Peak, and L.J. Thompson, Ion-beam mixing in purc and in immiscible copper bilayer systems. *Appl. Phys. A* **39**, 59, 1986.
24. Y. Shimizu, M. Uematsu, K.M. Itoh, A. Takano, K. Sawano, and Y. Shiraki, Quantitative evaluation of silicon displacement induced by arsenic implantation using silicon isotope superlattices. *Appl. Phys.* Express **1**, 021401, 2008.
25. Y. Kawamura, Y. Shimizu, H. Oshikawa, M. Uematsu, E.E. Haller, and K.M. Itoh, Quantitative evaluation of germanium displacement induced by arsenic implantation using germanium isotope superlattices. *Phys. B* **404**, 4546, 2009.
26. H. Bracht, M. Radek, R. Kube, S. Knebel, M. Posselt, B. Schmidt, E.E. Haller, and D. Bougeard, Ion-beam mixing in crystalline and amorphous germanium isotope multilayers. *J. Appl. Phys.* **110**, 093502, 2011.
27. T. Ruf, R.W. Henn, M. Asen-Palmer, E. Gmelin, M. Cardona, H.-J. Pohl, G.G. Devyatych, and P.G. Sennikov, Thermal conductivity of isotopically enriched silicon. *Solid State Commun.* **115**, 243, 2000 and **127**, 257, 2003.
28. N. Yang, G. Zhang, and B. Li, Ultralow thermal conductivity of isotope-doped silicon nanowires. *Nanoletters* **8**, 276, 2008.
29. M. Kazan, G. Guisbiers, S. Pereira, M.R. Correia, P. Masri, A. Bruyant, S. Volz, and P. Royer, Thermal conductivity of silicon bulk and nanowires: Effects of isotopic composition, phonon confinement, and surface roughness. *J. Appl. Phys.* **107**, 083503, 2010.
30. H. Bracht, N. Wehmeier, S. Eon, A. Plech, D. Issenmann, J. Lundsgaard Hansen, A. Nylandsted Larsen, J.W. Ager III, and E.E. Haller, Reduced thermal conductivity of isotopically modulated silicon multilayer structures. *Appl. Phys. Lett.* **101**, 064103, 2012.
31. H. Bracht, S. Eon, R. Frieling, A. Plech, D. Issenmann, D. Wolf, J. Lundsgaard Hansen, A. Nylandsted Larsen, J.W. Ager III, and E.E. Haller, Thermal conductivity of isotopically controlled silicon nanostructures. *New J. Phys.* **16**, 015021, 2014.
32. A. Van Wieringen and N. Warmoltz, On the permeation of hydrogen and helium in single crystal silicon and germanium at elevated temperatures. *Physica* **22**, 849, 1956.

33. A.A. Istratov, Ch. Flink, H. Hieslmair, E.R. Weber, and T. Heiser, Intrinsic diffusion coefficient of interstitial copper in silicon. *Phys. Rev. Lett.* **81**, 1243, 1998.
34. M.K. Bakhadyrkhanov, S. Zainabidinov, and A. Khamidov, Some characteristics of diffusion and electrotransport of nickel in silicon. *Sov. Phys. Semicond.* **14**, 243, 1980.
35. J.D. Struthers, Solubility and diffusivity of gold, iron, and copper in silicon. *J. Appl. Phys.* **27**, 1560, 1956, and **28**, 516, 1957.
36. W. Lerch, N.A. Stolwijk, H. Mehrer, and Ch. Poisson, Diffusion of platinum into dislocated and non-dislocated silicon. *Semicond. Sci. Technol.* **10**, 1257, 1995.
37. B. Kühn, Diffusion of Gold in Silizium—Isokonzentrationsmessungen und Untersuchungen des Einflusses von Oberflächenschichten. Doctoral thesis, University of Stuttgart, 1991.
38. H. Bracht, N.A. Stolwijk, I. Yonenaga, and H. Mehrer, Interstitial–substitutional diffusion kinetics and dislocation-induced trapping of zinc in plastically deformed silicon. *Phys. Stat. Sol.* (a) **137**, 499, 1993.
39. F. Rollert, N.A. Stolwijk, and H. Mehrer, Diffusion of sulphur-35 into silicon using an elemental vapor source. *Appl. Phys. Lett.* **63**, 506, 1993.
40. D.A. Antoniadis, A.G. Gonzales, and R.W. Dutton, Boron in near-intrinsic <100> and <111> silicon under inert and oxidizing ambients—diffusion and segregation. *J. Electrochem. Soc.* **125**, 813, 1978.
41. H. Mitlehner and H.-J. Schulze, Current developments in high-power thyristors. *EPE J.* **4**, 36, 1994.
42. J.S. Makris and B.J. Masters, Phosphorus isoconcentration diffusion studies in silicon. *J. Electrochem. Soc.* **120**, 1252, 1973.
43. R.B. Fair and J.C.C. Tsai, A quantitative model for the diffusion of phosphorus in silicon and the emitter dip effect. *J. Electrochem. Soc.* **124**, 1107, 1977.
44. B.J. Masters and J.M. Fairfield, Arsenic isoconcentration diffusion studies in silicon. *J. Appl. Phys.* **40**, 2390, 1969.
45. A. Nylandsted Larsen and P. Kringhøj, Diffusion of Sb in relaxed $Si_{1?x}Ge_x$. *Appl. Phys. Lett.* **68**, 2684, 1996.
46. R. Kube, H. Bracht, J. Lundsgaard Hansen, A. Nylandsted Larsen, E.E. Haller, S. Paul, and W. Lerch, Composition dependence of Si and Ge diffusion in relaxed $Si_{1?x}Ge_x$ alloys. *J. Appl. Phys.* **107**, 073520, 2010.
47. P. Kringhøj and A. Nylandsted Larsen, Anomalous diffusion of tin in silicon. *Phys. Rev. B* **56**, 6396, 1997.
48. K.C. Pandey, Diffusion without vacancies or interstitials: A new concerted exchange mechanism. *Phys. Rev. Lett.* **57**, 2287, 1986.
49. K. Compaan and Y. Haven, Correlation factors for diffusion in solids. *Trans. Faraday Soc.* **52**, 786, 1956.
50. M. Posselt, F. Gao, and H. Bracht, Correlation between self-diffusion in Si and the migration mechanisms of vacancies and self-interstitials: An atomistic study. *Phys. Rev. B* **78**, 035208, 2008.
51. K. Compaan and Y. Haven, Correlation factors for diffusion in solids. Part 2.—Indirect interstitial mechanism. *Trans. Faraday Soc.* **54**, 1498, 1958.
52. R. Chen and S.T. Dunham, Correlation factors for interstitial-mediated self-diffusion in the diamond lattice: Kinetic lattice Monte Carlo approach. *Phys. Rev. B* **83**, 134124, 2011.
53. N.A. Stolwijk, B. Schuster, and J. Hölzl, Diffusion of gold in silicon studied by means of neutron-activation analysis and spreading-resistance measurements. *Appl. Phys. A* **33**, 133, 1984.
54. N.A. Stolwijk, J. Hölzl, W. Frank, E.R. Weber, and H. Mehrer, Diffusion of gold in dislocation-free or highly dislocated silicon measured by the spreading-resistance technique. *Appl. Phys. A* **39**, 37, 1986.

55. H. Bracht, N.A. Stolwijk, and H. Mehrer, Properties of intrinsic point defects in silicon determined by zinc diffusion experiments under nonequilibrium conditions. *Phys. Rev. B* **52**, 16542, 1995.
56. S. Mantovani, F. Nava, S. Nobili, and G. Ottaviani, In-diffusion of Pt in Si from the PtSi/Si interface. *Phys. Rev. B* **33**, 5536, 1986.
57. J. Hauber, W. Frank, and N.A. Stolwijk, Diffusion and solubility of platinum in silicon. *Mater. Sci. Forum* **38–41**, 707, 1989.
58. F. Morehead, N.A. Stolwijk, W. Meyberg, and U. Gösele, Self-interstitial and vacancy contributions to silicon self-diffusion determined from the diffusion of gold in silicon. *Appl. Phys. Lett.* **42**, 690, 1983.
59. W.R. Wilcox and T.J. LaChapelle, Mechanism of gold diffusion into silicon. *J. Appl. Phys.* **35**, 240, 1964.
60. A. Giese, H. Bracht, N.A. Stolwijk, and D. Baither, Microscopic defects in silicon induced by zinc out-diffusion. *Mater. Sci. Engineer.* **B71**, 160, 2000.
61. A. Seeger and K.P. Chik, Diffusion Mechanisms and Point Defects in Silicon and Germanium. *Phys. Stat. Sol.* **29**, 455, 1968.
62. N.E.B. Cowern, S. Simdyankin, C. Ahn, N.S. Bennett, J.P. Goss, J.-M. Hartmann, A. Pakfar, S. Hamm, J. Valentin, E. Napolitani, D. De Salvador, E. Bruno, and S. Mirabella, Extended Point Defects in Crystalline Materials: Ge and Si. *Phys. Rev. Lett.* **110**, 155501, 2013.
63. M. Yoshida, E. Arai, H. Nakamura, and Y. Terunuma, Excess vacancy generation mechanism at phosphorus diffusion into silicon. *J. Appl. Phys.* **45**, 1498, 1974.
64. M. Yoshida, Numerical solution of phosphorus diffusion equation in silicon. *Jpn. J. Appl. Phys.* **18**, 479, 1979.
65. W.A. Orr Arienzo, R. Glang, R.F. Lever, R.K. Lewis, and F.F. Morehead, Boron diffusion in silicon at high concentrations. *J. Appl. Phys.* **63**, 116, 1988.
66. S. Matsumoto, Y. Ishikawa, and T. Niimi, Oxidation enhanced and concentration dependent diffusions of dopants in silicon. *J. Appl. Phys.* **54**, 5049, 1983.
67. Y. Ishikawa, I. Nakamichi, S. Matsumoto, and T. Niimi, The effect of thermal oxidation of silicon on boron diffusion in extrinsic conditions. *Jpn. J. Appl. Phys.* **26**, 1602, 1987.
68. S. Mizuo, T. Kusaka, A. Shintani, M. Nanba, and H. Higuchi, Effect of Si and SiO2 thermal nitridation on impurity diffusion and oxidation induced stacking fault size in Si. *J. Appl. Phys.* **54**, 3860, 1983.
69. P. Fahey, G. Barbuscia, M. Moslehi, and R.W. Dutton, Kinetics of thermal nitridation processes in the study of dopant diffusion mechanisms in silicon. *Appl. Phys. Lett.* **46**, 784, 1985.
70. H.-J. Gossmann, T.E. Haynes, P.A. Stolk, D.C. Jacobson, G.H. Gilmer, J.M. Poate, H.S. Luftman, T.K. Mogi, and M.O. Thompson, The interstitial fraction of diffusivity of common dopants in Si. *Appl. Phys. Lett.* **71**, 3862, 1997.
71. M. Miyake, Oxidation-enhanced diffusion of ion-implanted boron in silicon in extrinsic conditions. *J. Appl. Phys.* **57**, 1861, 1985.
72. W. Shockley and J.L. Moll, Solubility of flaws in heavily-doped semiconductors. *Phys. Rev.* **119**, 1480, 1960.
73. T.Y. Tan and U. Gösele, Point defects, diffusion processes, and swirl defect formation in silicon. *Appl. Phys. A* **37**, 1, 1985.
74. M. Uematsu, Simulation of boron, phosphorus, and arsenic diffusion in silicon based on an integrated diffusion model, and the anomalous phosphorus diffusion mechanism. *J. Appl. Phys.* **82**, 2228, 1997.
75. J. Murota, E. Arai, K. Kobasyashi, and K. Kudo, Arsenic diffusion in silicon from doped polycrystalline silicon. *Jpn. J. Appl. Phys.* **17**, 457, 1978.
76. J. Murota, E. Arai, K. Kobasyashi, and K. Kudo, Relationship between total arsenic and electrically active arsenic concentrations in silicon produced by the diffusion process. *J. Appl. Phys.* **50**, 804, 1979.

77. P.M. Fahey, P.B. Griffin, and J.D. Plummer, Point defects and dopant diffusion in silicon. *Rev. Mod. Phys.* **61**, 1989, 289.
78. M. Tang, L. Colombo, J. Zhu, and T. Diaz de la Rubia, Intrinsic point defects in crystalline silicon: Tight-binding molecular dynamics studies of self-diffusion, interstitial-vacancy recombination, and formation volumes. *Phys. Rev. B* **55**, 14279, 1997.
79. W.-K. Leung, R. J. Needs, and G. Rajagopal, S. Itoh, and S. Ihara, Calculations of silicon self-interstitial defects. *Phys. Rev. Lett.* **83**, 2351, 1999.
80. F. Bruneval, Range-separated approach to the RPA correlation applied to the van der Waals bond and to diffusion of defects. *Phys. Rev. Lett.* **108**, 256403, 2012.
81. R. Car, P.J. Kelly, A. Oshiyama, and S.T. Pantelides, Microscopic theory of atomic diffusion mechanisms in silicon. *Phys. Rev. Lett.* **52**, 1814, 1984.
82. G.M. Lopez and V. Fiorentini, Structure, energetics, and extrinsic levels of small self-interstitial clusters in silicon. *Phys. Rev. B* **69**, 155206, 2004.
83. Y. Shimizu, M. Uematsu, K.M. Itoh, A. Takano, K. Sawano, and Y. Shiraki, Behaviors of neutral and charged silicon self-interstitials during transient enhanced diffusion in silicon investigated by isotope superlattices. *J. Appl. Phys.* **105**, 013504, 2009.
84. H. Bracht and A. Chroneos, The vacancy in silicon: A critical evaluation of experimental and theoretical results. *J. Appl. Phys.* **104**, 076108, 2008.
85. G.D. Watkins, An EPR study of the lattice vacancy in silicon. *J. Phys. Soc. Jpn.* **18**, Suppl. II, 22, 1963.
86. G.D. Watkins, J.R. Troxell, and A.P. Chatterjee, Vacancies and interstitials in silicon. *Inst. Phys. Conf. Ser.* **46**, 18, 1979.
87. G.D. Watkins, The vacancy in silicon: Identical diffusion properties at cryogenic and elevated temperatures. *J. Appl. Phys.* **103**, 106106, 2008.
88. B.J. Masters and E.F. Gorey, Proton-enhanced diffusion and vacancy migration in silicon. *J. Appl. Phys.* **49**, 2717, 1978.
89. P. Lévêque, A. Yu. Kuznetsov, J.S. Christensen, B.G. Svensson, and A. Nylandsted Larsen, Irradiation enhanced diffusion of boron in delta-doped silicon. *J. Appl. Phys.* **89**, 5400, 2001.
90. P. Lévêque, J.S. Christensen, A. Yu. Kuznetsov, B.G. Svensson, and A. Nylandsted Larsen, Influence of boron on radiation enhanced diffusion of antimony in delta-doped silicon. J. *Appl. Phys.* **91**, 4073, 2002.
91. R. Sizmann, The effect of radiation upon diffusion in metals. *J. Nucl. Mater.* **69–70**, 386, 1978.
92. U. Gösele, W. Frank, and A. Seeger, Interpretation of experiments on radiation-enhanced diffusion in silicon. *Inst. Phys. Conf. Ser.* **46**, 538, 1979.
93. T. Sinno, T. Frewen, E. Dornberger, R. Hölzl, and Chr. Hoess, Parameterization of transient models of defect dynamics in Czochralski silicion crystal growth. *Mater. Res. Soc. Symp. Proc.* **700**, S8.3.1, 2001.
94. A. Giese, Charakterisierung von Leerstelleneigenschaften in Silizium und Germanium mittels der interstitiell-substitutionellen Fremddiffusion. Ph.D. thesis, University of Muenster, 2000.
95. V.V. Voronkov and R. Falster, The diffusivity of the vacancy in silicon: Is it fast or slow?. *Mater. Sci. Semicond. Process.* **15**, 697, 2012.
96. D. Caliste and P. Pochet, Vacancy-assisted diffusion in silicon: A three-temperature-regime model. *Phys. Rev. Lett.* **97**, 135901, 2006.
97. P. Pochet and D. Caliste, Point defect diffusion in Si and SiGe revisited through atomistic simulations. *Mater. Sci. Semicond. Process.* **15**, 675, 2012.
98. V.V. Voronkov, The mechanism of swirl defects formation in silicon. *J. Cryst. Growth* **59**, 625, 1982.
99. V. Voronkov and R. Falster, Vacancy-type microdefect formation in Czochralski silicon. *J. Cryst. Growth* **194**, 76, 1998.

100. J. Bardeen and W.H. Brattain, Physical principles involved in transistor action. *Phys. Rev.* **75**, 1208, 1949.
101. S.M. Sze, *Physics of Semiconductor Devices*, John Wiley and Sons, New York, 2001.
102. C. Claeys and E. Simoen (eds), *Germanium-Based Technologies—From Materials to Devices*, Elsevier, Amsterdam, 2007.
103. D.P. Brunco, B. De Jaeger, G. Eneman, J. Mitard, G. Hellings, A. Satta, V. Terzieva et al. Germanium MOSFET devices: Advances in materials understanding, process development, and electrical performance. *J. Electrochem. Soc.* **155**, H552, 2008.
104. E. Simoen, J. Mitard, G. Hellings, G. Eneman, B. De Jaeger, L. Witters, B. Vincent et al. Challenges and opportunities in advanced Ge pMOSFETs. *Mater. Sci. Semicond. Process.* **15**, 588, 2012.
105. M. Werner, H. Mehrer, and H.D. Hochheimer, Effect of hydrostatic pressure, temperature, and doping on self-diffusion in germanium. *Phys. Rev. B* **32**, 3930, 1985.
106. S. Uppal, A.F.W. Willoughby, J.M. Bonar, N.E.B. Cowern, T. Grasby, R.J.H. Morris, and M.G. Dowsett, Diffusion of boron in germanium at 800–900°C. *J. Appl. Phys.* **96**, 1376, 2004.
107. H.H. Silvestri, H. Bracht, J. Lundsgaard Hansen, A. Nylandsted Larsen, and E.E. Haller, Diffusion of silicon in crystalline germanium. *Semicond. Sci. Technol.* **21**, 758, 2006.
108. P. Dorner, W. Gust, A. Lodding, H. Odelius, and B. Predel, SIMS Untersuchungen zur Volumendiffusion von Al in Ge. *Acta Metall.* **30**, 941, 1982.
109. S. Matsumoto and T. Niimi, Concentration dependence of a diffusion coefficient at phosphorus diffusion in germanium. *J. Electrochem. Soc.* **125**, 1307, 1978.
110. A. Giese, H. Bracht, N.A. Stolwijk, and H. Mehrer, Diffusion of nickel and zinc in germanium. *Defect Diffusion Forum* **143–147**, 1059, 1997.
111. H. Bracht and S. Brotzmann, Atomic transport in germanium and the mechanism of arsenic diffusion. *Mater. Sci. Semicond. Process.* **9**, 471, 2006.
112. S. Brotzmann and H. Bracht, Intrinsic and extrinsic diffusion of phosphorus, arsenic, and antimony in germanium. *J. Appl. Phys.* **103**, 033508, 2008.
113. G.N. Wills, Solid state diffusion of antimony in germanium, from the vapour phase, in a vacuum furnace. *Solid State Electron.* **10**, 1, 1967.
114. H. Bracht, N.A. Stolwijk, and H. Mehrer, Diffusion and solubility of copper, silver, and gold in germanium. *Phys. Rev. B* **43**, 14465, 1991.
115. H. Bracht, Copper related diffusion phenomena in germanium and silicon. *Mater. Sci. Semicond. Process.* **7**, 113, 2004.
116. F. van der Maesen and J.A. Brenkman, The solid solubility and the diffusion of nickel in germanium. *Phillips Res. Rep.* **9**, 225, 1954.
117. B.I. Boltaks, *Diffusion in Semiconductors*, Infosearch Ltd., London, 1963, p. 183.
118. H. Bracht, Diffusion mechanisms and intrinsic point-defect properties in silicon. *MRS Bull.* **25**, 22, 2000.
119. H. Bracht, Diffusion mediated by doping and radiation-induced point defects. *Physica B* **376–377**, 11, 2006.
120. A. Chroneos, B. P. Uberuaga, and R. W. Grimes, Carbon, dopant, and vacancy interactions in germanium. *J. Appl. Phys.* **102**, 083707, 2007.
121. I. Riihimäki, A. Virtanen, S. Rinta-Anttila, P. Pusa, J. Räisänen, and The ISOLDE Collaboration, Vacancy-impurity complexes and diffusion of Ga and Sn in intrinsic and pdoped germanium. *Appl. Phys. Lett.* **91**, 091922, 2007.
122. R. Kube, H. Bracht, A. Chroneos, M. Posselt, and B. Schmidt, Intrinsic and extrinsic diffusion of indium in germanium. *J. Appl. Phys.* **106**, 063534, 2009.
123. E. Bruno, S. Mirabella, G. Scapellato, G. Impellizzeri, A. Terrasi, F. Priolo, E. Napolitani, D. De Salvador, M. Mastromatteo, and A. Carnera, Mechanism of B diffusion in crystalline Ge under proton irradiation. *Phys. Rev. B* **80**, 033204, 2009.

124. E. Napolitani, G. Bisognin, E. Bruno, M. Mastromatteo, G.G. Scapellato, S. Boninelli, D. De Salvador, S. Mirabella, C. Spinella, A. Carnera, and F. Priolo, Transient enhanced diffusion of B mediated by self-interstitials in preamorphized Ge. *Appl. Phys. Lett.* **96**, 201906, 2010.
125. H. Bracht, S. Schneider, and R. Kube, Diffusion and doping issues in germanium. *Microelectron. Eng.* **88**, 452, 2011.
126. G.G. Scapellato, E. Bruno, A.J. Smith, E. Napolitani, D. De Salvador, S. Mirabella, M. Mastromatteo, A. Carnera, R. Gwilliam, and F. Priolo, Role of self-interstitials on B diffusion in Ge. *Nucl. Instrum. Methods Phys. Res. B* **282**, 8, 2012.
127. H. Letaw Jr., W.M. Portnoy, and L. Slifkin, Self-diffusion in germanium. *Phys. Rev.* **102**, 636, 1956.
128. M.W. Valenta and C. Ramasastry, Effect of heavy doping on the self-diffusion of germanium. *Phys. Rev.* **106**, 73, 1957.
129. N.A. Stolwijk, W. Frank, J. Hölzl, S.J. Pearton, and E.E. Haller, Diffusion and solubility of copper in germanium. *J. Appl. Phys.* **57**, 5211, 1985.
130. A. Mesli, L. Dobaczewski, K. Bonde Nielsen, Vl. Kolkovsky, M. Christian Petersen, and A. Nylandsted Larsen, Low-temperature irradiation-induced defects in germanium: In situ analysis. *Phys. Rev. B* **78**, 165202, 2008.
131. A. Chroneos and H. Bracht, Diffusion of n-type dopants in germanium. *Appl. Phys. Rev.* **1**, 011301, 2014.
132. J. Coutinho, R. Jones, V.J.B. Torres, M. Barroso, S. Öberg, and P.R. Briddon, Electronic structure and Jahn–Teller instabilities in a single vacancy in Ge. *J. Phys.: Condens. Matter* **17**, L521, 2005.
133. A. Chroneos, H. Bracht, R. W. Grimes, and B.P. Uberuaga, Vacancy-mediated dopant diffusion activation enthalpies for germanium. *Appl. Phys. Lett.* **92**, 172103, 2008.
134. H. Haesslein, R. Sielemann, and C. Zistl, Vacancies and self-interstitials in germanium observed by perturbed angular correlation spectroscopy. *Phys. Rev. Lett.* **80**, 2626, 1998.
135. C. Janke, R. Jones, S. Öberg, and P.R. Briddon, Ab initio investigation of boron diffusion paths in germanium. *J. Mater. Sci.: Mater. Electron.* **18**, 775, 2007.
136. S. Mirabella, D. De Salvador, E. Napolitani, E. Bruno, and F. Priolo, Mechanisms of boron diffusion in silicon and germanium. *J. Appl. Phys.* **113**, 031101, 2013.
137. E. Vainonen-Ahlgren, T. Ahlgren, J. Likonen, S. Lehto, J. Keinonen, W. Li, and J. Haapamaa, Identification of vacancy charge states in diffusion of arsenic in germanium. *Appl. Phys. Lett.* **77**, 690, 2000.
138. C.O. Chui, K. Gopalakrishnan, P.B. Griffin, J.D. Plummer, and K.C. Saraswat, Activation and diffusion studies of ion-implanted p and n dopants in germanium. *Appl. Phys. Lett.* **83**, 3275, 2003.
139. P. Tsouroutas, D. Tsoukalas, I. Zergioti, N. Cherkashin, and A. Claverie, *J.* Modeling and experiments on diffusion and activation of phosphorus in germanium. *Appl. Phys.* **105**, 094910, 2009.
140. T. Canneaux, D. Mathiot, J.-P. Ponpon, and Y. Leroy, Modeling of phosphorus diffusion in Ge accounting for a cubic dependence of the diffusivity with the electron concentration. *Thin Solid Films* **518**, 2394, 2010.
141. M.S. Carroll and R. Koudelka, Accurate modelling of average phosphorus diffusivities in germanium after long thermal anneals: Evidence of implant damage enhanced diffusivities. *Semicond. Sci. Technol.* **22**, S164, 2007.
142. Y. Cai, R. Camacho-Aguilera, J.T. Bessette, L.C. Kimerling, and J. Michel, High phosphorous doped germanium: Dopant diffusion and modeling. *J. Appl. Phys.* **112**, 034509, 2012.
143. A. Chroneos, Dopant-vacancy cluster formation in germanium. *J. Appl. Phys.* **107**, 076102, 2010.

144. G. Impellizzeri, S. Mirabella, A. Irrera, M.G. Grimaldi, and E. Napolitani, Ga-implantation in Ge: Electrical activation and clustering. *J. Appl. Phys.* **106**, 013518, 2009.
145. G. Luo, C.C. Cheng, C.Y. Huang, S.L. Hsu, C.H. Chien, W.X. Ni, and C.Y. Chang, Suppressing phosphorus diffusion in germanium by carbon incorporation. *Electron. Lett.* **41**, 1354, 2005.
146. E.E. Haller, W.L. Hansen, P. Luke, R. McMurray, and B. Jarrett, Carbon in high-purity Germanium. *IEEE Trans. Nucl. Sci.* **29**, 745, 1982.
147. A. Chroneos, R.W. Grimes, B.P. Uberuaga, and H. Bracht, Diffusion and defect reactions between donors, C, and vacancies in Ge. II. Atomistic calculations of related complexes. *Phys. Rev. B* **77**, 235208, 2008.
148. H. Bracht, Defect engineering in germanium. *Phys. Status Solidi A* **211**, 109, 2014.
149. H.A. Tahini, A. Chroneos, R.W. Grimes, U. Schwingenschlögl, and H. Bracht, Point defect engineering strategies to retard phosphorous diffusion in germanium. *Phys. Chem. Chem. Phys.* **15**, 367, 2013.
150. J. Kim, S.W. Bedell, and D.K. Sadana, Improved germanium n + /p junction diodes formed by coimplantation of antimony and phosphorus. *Appl. Phys. Lett.* **98**, 082112, 2011.
151. H.A. Tahini, A. Chroneos, R.W. Grimes, and U. Schwingenschlögl, Co-doping with antimony to control phosphorous diffusion in germanium. *J. Appl. Phys.* **113**, 073704, 2013.
152. A. Chroneos, R. W. Grimes, H. Bracht, and B. P. Uberuaga, Engineering the free vacancy and active donor concentrations in phosphorus and arsenic double donor-doped germanium. *J. Appl. Phys.* **104**, 113724, 2008.
153. G. Impellizzeri, S. Boninelli, F. Priolo, E. Napolitani, C. Spinella, A. Chroneos, and H. Bracht, Fluorine effect on As diffusion in Ge. *J. Appl. Phys.* **109**, 113527, 2011.
154. S. Schneider and H. Bracht, Suppression of donor-vacancy clusters in germanium by concurrent annealing and irradiation. *Appl. Phys. Lett.* **98**, 014101, 2011.
155. D. Alloyeau, B. Freitag, S. Dag, L.W. Wang, and C. Kisielowski, Atomic-resolution three-dimensional imaging of germanium self-interstitials near a surface: Aberration-corrected transmission electron microscopy. *Phys. Rev. B* **80**, 014114, 2009.
156. N.E.B. Cowern, K.T.F. Janssen, G.F.A. van de Walle, and D.J. Gravesteijn, Impurity diffusion via an intermediate species: The B-Si system. *Phys. Rev. Lett.* **65**, 2434, 1990.
157. A. Pruijmboom, C.E. Timmering, J.M.L. van Rooij-Mulder, D.J. Gravesteijn, W.B. de Boer, W.J. Kersten, J.W. Slotboom, C.J. Vriezema, and R. de Kruif, Heterojunction bipolar transistors with $Si_{1?x}Ge_x$ base. *Micr. El. Ing.* **19**, 427, 1992.
158. L.J. Schowalter, Heteroepitaxy and Strain: Applications to Electronic and Optoelectronic Materials. *MRS Bull. Mater. Res. Soc.* **21**, 18, 1996.
159. D.L. Harame, S.J. Koester, G. Freeman, P. Cottrel, K. Rim, G. Dehlinger, D. Ahlgren et al. The revolution in SiGe: Impact on device electronics. *Appl. Surf. Sci.* **224**, 9, 2004.
160. E.E. Haller, Germanium: From its discovery to SiGe devices. *Mater. Sci. Semicond. Process.* **9**, 408, 2006.
161. M.L. Lee, E.A. Fitzgerald, M.T. Bulsara, M.T. Currie, and A. Lochtefeld, Strained Si, SiGe, and Ge channels for high-mobility metal-oxide semiconductor field-effect transistors. *J. Appl. Phys.* **97**, 011101, 2005.
162. S.M. Prokes, O.J. Glembocki, and D.J. Godbey, Stress and its effect on the interdiffusion in $Si_{1?x}Ge_x$/Si superlattices. *Appl. Phys. Lett.* **60**, 1087, 1992.
163. G. Xia, O.O. Olubuyide, J.L. Hoyt, and M. Canonico, Strain dependence of Si–Ge interdiffusion in epitaxial Si/$Si_{1-y}Ge_y$/Si heterostructures on relaxed $Si_{1-x}Ge_x$ substrates. *Appl. Phys. Lett.* **88**, 013507, 2006.

164. L.S. Darken, Diffusion, mobility and their interrelation through free energy in binary metallic systems. *Trans A.I.M.E.* **175**, 184, 1948.
165. G.L. McVay and A.R. DuCharme, The diffusion of germanium in silicon. *J. Appl. Phys.* **44**, 1409, 1973.
166. G.L. McVay and A.R. DuCharme, Diffusion of Ge in SiGe alloys. *Phys. Rev. B* **9**, 627, 1974.
167. G.E. Pike, W.J. Camp, C.H. Seager, and G.L. McVay, Percolative aspects of diffusion in binary alloys. *Phys. Rev. B* **10**, 4909, 1974.
168. N.R. Zangenberg, J.L. Hansen, J. Fage-Pederson, and A.N. Larsen, Ge self-diffusion in epitaxial $Si_{1?x}Ge_x$ layers. *Phys. Rev. Lett.* **87**, 125901, 2001.
169. A. Strohm, T. Voss, W. Frank, J. Räisänen, and M. Dietrich, Self-diffusion of 71Ge in Si-Ge. *Physica B* **308–310**, 542, 2001.
170. A. Strohm, T. Voss, W. Frank, P. Laitinen, and J. Räisänen, Self-diffusion of $_{71}$Ge and $_{31}$Si in Si-Ge alloys. *Z. Metallk.* **93**, 737, 2002.
171. R. Kube, H. Bracht, J. Lundsgaard Hansen, A. Nylandsted Larsen, E.E. Haller, W. Lerch, and S. Paul, Simultaneous diffusion of Si and Ge in isotopically controlled $Si_{1-x}Ge_x$ heterostructures. *Mater. Sci. Semicond. Process* **11**, 378, 2008.
172. P. Venezuela, G.M. Dalpian, A.J.R. da Silva, and A. Fazzio, Vacancy-mediated diffusion in disordered alloys: Ge self-diffusion in $Si_{1-x}Ge_x$. *Phys. Rev. B* **65**, 193306, 2002.
173. P. Ramanarayanan, K. Cho, and B.M. Clemens, Effect of composition on vacancy mediated diffusion in random binary alloys: First principles study of the $Si_{1?x}Ge_x$ system. *J. Appl. Phys.* **94**, 174, 2003.
174. M. Haran, J.A. Catherwood, and P. Clancy, Effects of Ge content on the diffusion of group-V dopants in SiGe alloys. *Appl. Phys. Lett.* **88**, 173502, 2006.
175. P. Castrillo, M. Jaraiz, R. Pinacho, and J. E. Rubio, Atomistic modeling of defect diffusion and interdiffusion in SiGe heterostructures. *Thin Solid Films* **518**, 2448, 2010.
176. P. Castrillo, R. Pinacho, M. Jaraiz, and J. E. Rubio, Physical modeling and implementation scheme of native defect diffusion and interdiffusion in SiGe heterostructures for atomistic process simulation. *J. Appl. Phys.* **109**, 103502, 2011.
177. A. N. Larsen and P. Kringhoj, Diffusion of Sb in relaxed $Si_{1?x}Ge_x$. *Appl. Phys. Lett.* **68**, 2684, 1996.
178. B. L. Sharma, Diffusion in silicon and germanium. *Defect Diffusion Forum* **70–71**, 1, 1990.
179. P. Laitinen, I. Riihimäki, J. Räisänen, and the ISOLDE Collaboration, Arsenic diffusion in relaxed Si1-xGex. *Phys. Rev. B* **68**, 155209, 2003.
180. C. Y-T. Liao, Diffusion in SiGe and Ge. PhD thesis, University of California at Berkeley, 2010.
181. A. Chroneos, H. Bracht, C. Jiang, B. P. Uberuaga, and R. W. Grimes, Nonlinear stability of E centers in $Si_{1?x}Ge_x$: Electronic structure calculations. *Phys. Rev. B* **78**, 195201, 2008.
182. A. Chroneos, R. W. Grimes, B. P. Uberuaga, S. Brotzmann, and H. Bracht, Vacancy-arsenic clusters in germanium. *Appl. Phys. Lett.* **91**, 192106, 2007.
183. A. Satta, E. Simoen, T. Janssens, T. Clarysse, B. De Jaeger, A. Benedetti, I. Hoflijk, B. Brijs, M. Meuris, and W. Vandervorst, Shallow junction ion implantation in Ge and associated defect control. *J. Electrochem. Soc.* **153**, G229, 2006.
184. S. Satta, T. Janssens, T. Clarysse, E. Simoen, M. Meuris, A. Benedetti, I. Hoflijk, B. De Jaeger, C. Demeurisseand, and W. Vandervorst, P implantation doping of Ge: Diffusion, activation, and recrystallization. *J. Vac. Sci. Technol. B* **24**, 494, 2006.
185. M. Posselt, B. Schmidt, W. Anwand, R. Grötzschel, V. Heera, A. Mücklich, H. Hortenbach et al. P implantation into preamorphized germanium and subsequent annealing: Solid phase epitaxial regrowth, P diffusion, and activation. *J. Vac. Sci. Technol. B* **26**, 430, 2008.

186. B. C. Johnson, P. Gortmaker, and J. C. McCallum, Intrinsic and dopant-enhanced solid-phase epitaxy in amorphous germanium. *Phys. Rev. B* **77**, 214109, 2008.
187. S. Koffel, N. Cherkashin, F. Houdellier, M. J. Hytch, G. Benassayag, P. Scheiblin, and A. Claverie, End of range defects in Ge. *J. Appl. Phys.* **105**, 126110, 2009.
188. J. Vanhellemont and E. Simoen, On the diffusion and activation of n-type dopants in Ge. *Mater. Sci. Semicond. Process.* **15**, 642, 2012.

7 Hydrogen in Si and Ge

Stefan K. Estreicher, Michael Stavola, and Jörg Weber

CONTENTS

7.1 Introduction 217
7.2 Interstitial Hydrogen 219
7.2.1 Diffusivity and Solubility of H in Si and Ge 219
7.2.2 Configurations and Electrical Activity of H in Defect-Free Si and Ge 220
7.2.2.1 Silicon 220
7.2.2.2 Germanium 223
7.2.2.3 Open Questions: Diffusion Properties of H^0 and H^- 224
7.2.3 Vibrational Properties 225
7.3 Hydrogen Pairs 226
7.3.1 Hydrogen Molecules 226
7.3.2 Open Question: Formation of H_2 229
7.3.3 H_2^* in Si and Ge 229
7.4 Interaction of H with Impurities and Defects 230
7.4.1 Passivation of Shallow Impurities 230
7.4.2 Hydrogen Interactions with Native Defects 233
7.4.3 Hydrogen Platelets and the Smart-Cut® Process 233
7.4.4 Hydrogenation of Deep Levels 235
7.4.5 Activation of Impurities by Hydrogen 236
7.5 Contrasting the Properties of H in High-Purity, O-Rich, and C-Rich Si Materials 237
7.6 Vibrational Lifetimes 240
7.7 Hydrogen in the Photovoltaic Industry 242
7.8 Conclusions 244
Acknowledgments 244
References 245

7.1 INTRODUCTION

This review of hydrogen begins with a survey of the fundamental properties of H in Si and Ge: diffusivity, solubility, equilibrium sites, electrical activity, and vibrational properties. Hydrogen is best known for its ability to change the electrical properties in semiconductors by forming covalent bonds with defects or with host atoms adjacent to them. This affects the position of the electrically active levels in the gap, sometimes leading to passivation. We also discuss the behavior of H in different types of Si: high-purity, O-rich, and C-rich Si materials. We review the properties of the two common H pairs, the interstitial H_2 molecules and the H_2^* complex, especially in its $H_2^*(C)$ form. The lifetimes of high-frequency H-related vibrational modes are reviewed. We also discuss two industrial applications of H, the "smart cut" process used to fabricate silicon-on-insulator (SOI) layers and the hydrogenation of multicrystalline silicon (mc-Si) material for photovoltaic applications.

The story of hydrogen in Si and Ge, its heavy isotope deuterium, and its light pseudo-isotope muonium, began more than six decades ago. Various aspects of this story have been reviewed by several authors.[1–9] The first measurements of the solubility and diffusivity of hydrogen in Si near its melting point date back to 1956,[10] but the number of experimental and theoretical studies substantially grew only after 1983, when it was realized that H easily passivates nearly all shallow acceptors, thus considerably decreasing the electrical conductivity of *p*-type Si.[11,12] Numerous other H-related interactions in Si have since then been uncovered experimentally and described theoretically.

In the case of Ge, the solubility and diffusivity were measured in 1960[13] but the first systematic studies of H-related complexes began in 1978 when it was realized that hydrogen traps at and activates several normally electrically inactive impurities.[14] However, much more is known about H and H interactions in Si than in Ge or SiGe alloys.

Today, H also plays a crucial role in various technologically important processes, in particular the "smart-cut" process,[15] which allows the production of high-purity SOI layers, and the hydrogenation[16] of bulk mc-Si for the large-scale production of solar cells. Hydrogen is also easily introduced unintentionally into Si devices during a range of processing steps, as H is present in most of the ambients and chemicals involved, or is an impurity in metals used for contacts.

Despite all that is known, hydrogen continues to be fascinating for both basic and applied research. For example, trace amounts of H in O-rich Si crystals greatly enhance the diffusivity of interstitial oxygen in the range 300–450°C.[17,18] This catalytic effect is poorly understood. Another example involves the hydrogenation of *p*-type mc-Si solar cells, a process that substantially increases their efficiency.[16] Yet, it is not known which passivated defects are primarily responsible for this increase. Furthermore, numerous H-related infrared active modes have been observed and identified, but the lifetimes of the various vibrational excitations vary by orders of magnitude[19,20] and sometimes exhibit surprisingly large isotope effects associated with H[21] or its Si[22] neighbors.

In this chapter, we focus on the key points for those topics that have already been reviewed and discuss recent developments and open questions. We begin with a summary of the properties of isolated H and in high-purity Si and Ge. Then, we survey H pairs and the interactions of H with impurities and defects. We discuss the reservoirs of H in high-purity, O-rich, and C-rich Si materials. We then comment on the lifetimes of H-related vibrational excitations, and conclude with a brief discussion of H in industrial application.

7.2 INTERSTITIAL HYDROGEN

7.2.1 Diffusivity and Solubility of H in Si and Ge

The diffusivity (D) and solubility (S)[6,23] of hydrogen in Si were first published in 1956.[10] The time dependence of the permeation of H through thin-walled cylinders of Si was measured from 1090°C to 1200°C and yielded (Figure 7.1)

$$D = 9.4 \times 10^{-3}\ \exp(-0.48\,\text{eV/kT})\ \text{cm}^2/\text{s};\ S = 2.4 \times 10^{21}\ \exp(-1.88\,\text{eV/kT})\ \text{cm}^{-3} \tag{7.1}$$

Subsequent experiments[24,25] have refined the value of the solubility of H in Si, and today's most accurate value[26] is $S = 9.1 \times 10^{21}\ \exp(-1.80\text{eV/kT})\ \text{cm}^{-3}$. This corresponds to about $4 \times 10^{16}\ \text{cm}^{-3}$ near the melting point of Si but the solubility drops to very low values at a few hundred degrees Celsius. Thus, an efficient method to introduce substantial concentrations of H into Si is to anneal the sample in an H_2 ambient around 1200–1300°C and then to rapidly quench to room temperature.[27,28] H becomes trapped in the form of H_2 molecules in high-purity Si or at C-related defects in C-rich materials. Relatively high concentrations of H can then be released from these traps by thermal anneals at a few hundred degrees Celsius.

The activation energy for the diffusivity (Equation 7.1) has been confirmed in several studies. For example, the AA9 electron paramagnetic resonance (EPR) spectrum of neutral bond-centered hydrogen (H^0) (observed under illumination) anneals out in the dark near 200 K with the same activation energy.[29] Furthermore, uniaxial stress EPR experiments (under illumination) at 135 K have shown that H reorients in the dark from bond-centered (BC) to BC site with again the same activation energy.[30] Thus, the 0.48 eV activation energy in Equation 7.1 corresponds to the BC-to-BC diffusion of H^+_{BC} and is consistent from 1473 to 135 K.[23,31]

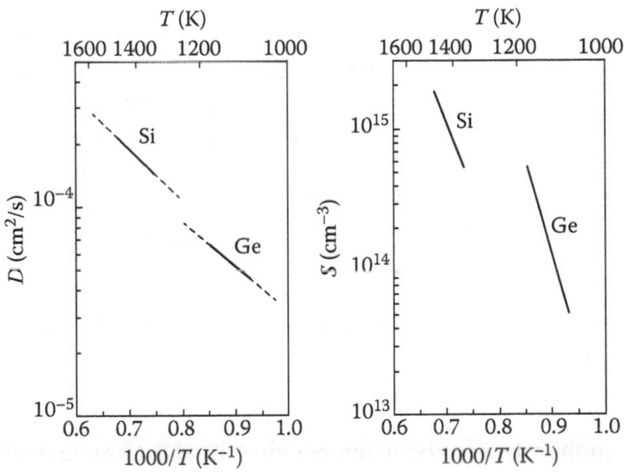

FIGURE 7.1 Diffusivity and solubility of H in Si[10] and Ge[13].

Permeation experiments[13] have also been performed in Ge over the temperature range 800–910°C to yield (Figure 7.1)

$$D = 2.7 \times 10^{-3}\exp(-0.38\,\mathrm{eV/kT})\ \mathrm{cm^2/s} \text{ and } S = 3.2 \times 10^{24}\exp(-2.3\,\mathrm{eV/kT})\ \mathrm{cm^{-3}} \tag{7.2}$$

As will be discussed below, it is likely that the diffusing species in Ge is H^-. These results are in reasonable agreement with subsequent measurements made at elevated temperatures. For example, Ge crystals were grown in a hydrogen ambient spiked with tritium whose decay was then detected in the grown material.[32] Hydrogen concentrations approximately consistent with Equation 7.2 were found.

The results in Equations 7.1 and 7.2 are sometimes extrapolated to temperatures as low as room temperature. However, the effective diffusivity and solubility of H that are observed in many experiments performed at reduced temperatures are affected dramatically by the interaction of H with other defects or with other H atoms. Equations 7.1 and 7.2 would predict a diffusivity near 10^{-7} cm²/s at 200°C for both Si and Ge and, therefore, a penetration depth of H near 200 μm for a 1 h hydrogenation treatment. In most practical circumstances, where H is introduced at high concentrations, the effective diffusivity of H is limited by trapping, an idea that was introduced to explain hydrogen in-diffusion profiles observed for Ge.[33]

If extrapolated to 200°C, the solubilities of H given by Equations 7.1 and 7.2 for H are negligible. In practice, however, if Si is treated for roughly 1 h in a hydrogen plasma at 200°C, a H concentration of ~10^{18} cm^{-3} is found in a layer ~1 μm thick.[34] The behavior of H in *n*-type Ge appears to be similar to its behavior in Si. However, for *p*-type Ge, experimental results suggest the existence of a surface barrier that leads to a high concentration of hydrogen very near the surface. These results agree with effusion measurements on plasma treated, highly doped, *p*-type Ge.[35] Out-diffusion of hydrogen was found already at 200°C, a temperature much lower than in Si (350–400°C).

For doped Si treated in hydrogen plasma, the H concentration near the surface mimics the doping concentration and can easily be 10^{17} cm^{-3} or more. Figure 7.2 shows a secondary ion mass spectrometry (SIMS) depth profile of H introduced into *p*-type Si from a hydrogen plasma.[34] The effective solubility is determined by the formation of hydrogen aggregates (such as H_2 molecules or platelets) or by the interaction of H with impurities and defects. Thus, H turns out to be a more important impurity in Si and Ge than Equations 7.1 and 7.2 suggest.

7.2.2 Configurations and Electrical Activity of H in Defect-Free Si and Ge

7.2.2.1 Silicon

Isolated H is mobile below room temperature in most semiconductors, making experimental studies of its properties challenging. One method to produce isolated H is by the implantation of protons into samples held at cryogenic temperature so that H diffusion is frozen. Isolated H in Si has been studied by electron paramagnetic

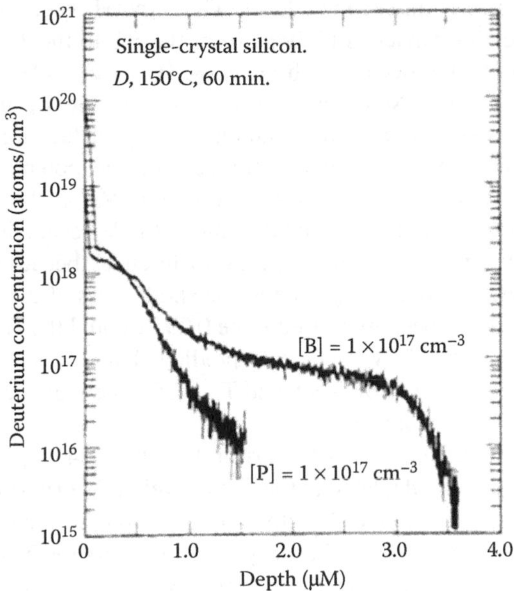

FIGURE 7.2 SIMS depth profile of p-type Si treated in a D plasma (150°C) for 60 min. (Reprinted with permission from N.M. Johnson and M.D. Moyer, *Appl. Phys. Lett.* **46**, 787, 1985. Copyright 1985, American Institute of Physics.)

resonance (EPR),[36–38] IR spectroscopy,[39] and deep level transient spectroscopy (DLTS).[40,41] These data provide information about the structure, electronic properties, and migration of H. Isolated interstitial hydrogen is found in two configurations[42–44] (tetrahedral interstitial (T) and relaxed BC sites) and three charge states[45,46] (+, 0, and –). Since the T site is metastable, there are four states of isolated interstitial H in Si: H^+_{BC}, H^0_{BC}, H^0_T, and H^-_T.

Direct evidence for these four states comes from muon spin rotation (μSR) studies.[47,48] Experiments on muonium (Mu), a light pseudoisotope of H, provide a rich source of information about H in semiconductors. μ^+ is a particle with spin 1/2 and mass about one-ninth that of a proton. When an energetic beam of polarized muons is stopped in a semiconductor, muonium centers are formed whose properties mimic those of H. The lifetime of the muon is only 2.2 μs; therefore, in most cases, it decays before it has time to interact with other defects and impurities, and nonequilibrium configurations can be studied. Elegant methods have been developed to study the muon spin polarization and to determine hyperfine structure for muonium centers. Configurations with the muon at a bond-center site (Mu_{BC}) and at a tetrahedral interstitial site (Mu_T) have been identified as the Mu^+_{BC}, Mu^0_{BC}, Mu^0_T, and Mu^-_T charge states.

In p-type and intrinsic Si, the stable charge state is H^+ and hydrogen acts as a donor. In n-type Si, the stable charge state is H^- and hydrogen acts as an acceptor. Theory predicts that the spin 1/2 states (H^0_{BC} and H^0_T) are not stable for any position of the Fermi level.[45,56] H is a negative-U defect with its donor level above its acceptor level, that is, with inverted order in the gap.[31,45,46,49] However, H_{BC} has only a donor level while H_T has only an acceptor level. Thus, the $H^+ \leftrightarrow H^-$ transition involves overcoming a potential barrier as H^0 hops from the BC to the T site (or vice versa). Thus, the + to – transition occurs in three steps: $H^+_{BC} + e^- \rightarrow H^0_{BC}$, then $H^0_{BC} \rightarrow H^0_T$, and finally $H^0_T + e^- \rightarrow H^-_T$. Note that the second step cannot occur unless the T and BC sites are both local minima of the potential energy surface for H^0.

μSR experiments[47] have provided estimates for the potential energy barriers separating the BC and T sites for neutral muonium (Mu^0). The $Mu^0_{BC} \rightarrow Mu^0_T$ and $Mu^0_T \rightarrow Mu^0_{BC}$ barriers were found to be 0.38 and 0.61 eV, respectively. These experimental values are not true potential energy differences because they include the zero-point energies, which are large due to the small mass of the muon. Indeed, the H^0(BC → T) barrier has been measured to be 0.295 eV and the H^0(T → BC) barrier has been estimated to be 0.2 eV.[41] Theory[50] predicts that H^0_{BC} is lower in energy than H^0_T by 0.14 eV. The calculated BC → T and T → BC potential energy barriers for H^0 are 0.38 and 0.24 eV, respectively.

Note that the calculated T → BC barrier for H^0 (0.24 eV) is higher than the T → T migration barrier calculated at the same level of theory (0.16 eV). Thus, H^0 is unlikely to hop from the T to the BC site unless it is in the immediate vicinity of an impurity or defect that distorts the host crystal and lowers the T → BC barrier, as is the case when H_T becomes H_{BC} near interstitial oxygen (O_i) or substitutional carbon (C_s).[40,41] Thus, even though H^0 is never the stable charge state, it can survive for some length of time while the sample is away from equilibrium. Conventional μSR experiments at low temperatures show two paramagnetic (Mu^0_{BC} and Mu^0_T) centers and one diamagnetic (Mu^+_{BC} or Mu^-_T) center.[7,8,51]

As a negative-U defect, H has an equilibrium occupancy level that lies midway between its donor and acceptor levels and whose position with respect to the Fermi level determines the H charge state. Theory also finds that the H(+/–) occupancy level has a universal alignment in different semiconductor hosts with respect to the vacuum level so that the position of the H(+/–) level for any semiconductor can be determined if the band offsets are known.[52]

Recent μSR studies have focused on the transitions between the different configurations and charge states.[53,54] Where comparable results exist, muonium and H have been found to have similar properties. The properties of muonium have been studied in a variety of semiconductors and are consistent with the existence of a universal Mu(+/–) occupancy level.

Deep-level transient spectroscopy (DLTS) data show that the donor level of H in Si is in the range $E_c - 0.16$ to $E_c - 0.175$ eV.[40,55] Since no change of configuration is involved, this value is very close to the ionization energy of H^0_{BC}. The position of the acceptor level is more difficult to obtain because the equilibrium configurations involved are H^0_{BC} and H^-_T. Since H^0 needs to overcome a barrier to go from the BC to the T site, the $H^0 \rightarrow H^-$ reaction is not a direct ionization. The most reliable value[41] for the acceptor level is $E_c - 0.65 \pm 0.08$ eV. In these experiments, H^-_T was produced by flooding with electrons a sample that contained known concentrations of H^+_{BC} resulting from proton implantation. These experiments confirmed that H is a negative-U center characterized by a (+/–) occupancy level[52] at about $E_c - 0.4$ eV. This value is close to the one deduced from μSR experiments, $Mu^{+/-} = E_c - 0.34 \pm 0.04$ eV.[56]

The most recent calculations[49] of the gap levels of H in Si have been done using the marker method. They involve the lowest-energy configurations in each charge state, namely, H^+_{BC}, H^0_{BC}, and H^-_T. The donor level is found to be at $E_v + 0.96 = E_c - 0.21$ eV and the acceptor level at $E_c - 0.79 = E_v + 0.38$ eV, leading to a (+/–) level at $E_c - 0.50 = E_v + 0.67$ eV.

7.2.2.2 Germanium

Even though the Si−H bond is stronger than the Ge−H bond, one would expect that the configurations of H should be the same in the two hosts except for a difference in the relative stability of the T and BC sites.[57] However, it is the position of the donor and acceptor levels of isolated H in the band gap and, therefore, the H(+/–) occupancy level, that determine which charge state will dominate for different positions of the Fermi level.

In the case of muonium, the acceptor and donor levels for muonium in Si and Ge and the Mu(+/–) occupancy level are shown in Figure 7.3.[56,58] The positions of the levels for isolated H determined by experiment give a very similar picture. For Si, the experimental H(+/–) occupancy level lies in the upper half of the band gap. Therefore, in equilibrium, the H^+ charge state will dominate in Si that is intrinsic (low doping or elevated temperature) with its Fermi level at midgap.[31]

DLTS experiments have also been performed in proton-implanted Ge.[59] In this case, a donor level of H was found at $E_c - 0.11$ eV but no acceptor level was found, consistent with it being resonant with the valence band, as reported in μSR experiments.[56] Thus, in Ge, the experimental H(+/–) occupancy level lies below midgap[56] and the H^- charge state will dominate in intrinsic Ge. In proton-implanted Si and Ge, implantation damage should pin the Fermi level at midgap. Only H^+ has been seen by IR spectroscopy for Si, whereas for Ge, both H^+ and H^- were seen.[60]

The role played in experiments by the occupancy level for H in Ge remains uncertain. Experiments involving defects produced by implantation at low temperatures may reflect nonequilibrium conditions for the implanted proton (or muon). So the charge states that are observed in such experiments may not reflect the position of the true occupancy level. Experiments for muonium in Ge find an occupancy level in the Ge band gap below midgap. Other experimental results and theory suggest

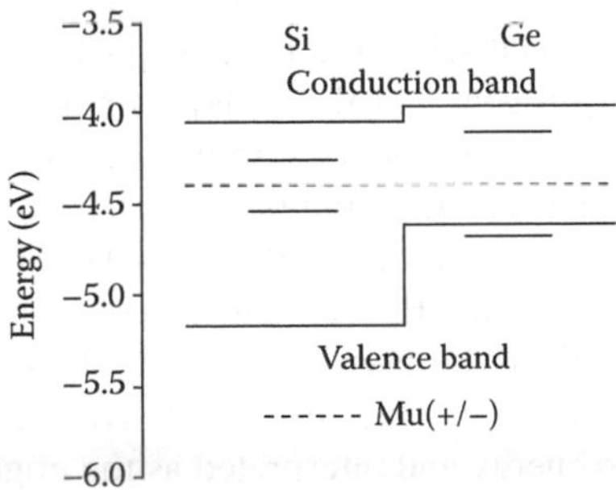

FIGURE 7.3 Donor and acceptor levels of Mu in Si and Ge obtained from μSR. (R.L. Lichti, private communication.)

an occupancy level resonant with the Ge valence band. In either case, the H^- charge state should dominate in *n*-type or intrinsic material. It remains unclear whether or not H^+ can be found in equilibrium in *p*-type Ge.

7.2.2.3 Open Questions: Diffusion Properties of H^0 and H^-

The migration barrier of H_{BC}^+ has been calculated by several authors[61,62] and agrees with the measured value, 0.48 eV (Section 7.2.1). Recent first-principles nudged-elastic-band calculations[50] predict that the BC → BC migration barrier of H_{BC}^+ is 0.47 eV. When corrected for the zero-point energy associated with the vibrational mode along the direction of initial motion (wag mode), the calculated barrier drops to 0.45 eV, about 0.03 eV lower than the measured one.

Thus, the activation energy for diffusion of H_{BC}^+ is well known experimentally over a wide range of temperatures, and the value has been confirmed at various levels of theory. However, the activation energies for diffusion in the 0 and – charge states are difficult to measure and the few theoretical predictions lack experimental confirmation.

The calculated[50] BC → BC migration barrier of H_{BC}^0 is 0.38 eV (lower than that of H_{BC}^+ obtained at the same level of theory). The diffusivity of H_T^0 has not been measured but is suspected to be extremely high. An activation energy smaller than 0.1 eV has been inferred from DLTS experiments with an estimated diffusivity "27 orders of magnitude higher"[41] than that of H_{BC}^+ at 65 K. This remarkable estimate is consistent with the possibility that one state of H (presumably H_T^0) is tunneling at low temperature.[63] The calculated T → T migration barrier of H_T^0 is 0.16 eV. When corrected for the zero-point energy associated with the triply degenerate vibrational mode of H^0 at the T site, this barrier drops to 0.13–0.11 eV. This would be the lowest migration barrier of any impurity in Si.

Finally, the calculated[50] T → T migration barrier of H_T^- is 0.39 eV, about three times higher than that of H_T^0, but still lower than that of H_{BC}^+. This prediction contradicts the commonly held view that H_T^- in Si is a large ion with a high migration barrier, and suggests that the low concentration of H in *n*-Si following in-diffusion at ~100°C is due to the low solubility of H rather than to a high migration barrier of H_T^-.

A measured[64] 0.7 eV activation energy has been assigned to the migration barrier of H_T^-. These experiments were based on the interpretation of the passivation/reactivation kinetics of hydrogenated substitutional phosphorus. The phosphorus shallow donor is passivated by hydrogen, and the {P,H} pair is stable up to about 100°C. However, the electrical activity reappears at room temperature under band gap light or minority carrier injection. The trapping of h^+ by $\{P,H\}^0$ has been interpreted[64] as the dissociation reaction $\{P \cdots Si{-}H_{AB}\}^0 + h^+ \rightarrow P^+ + H^0$. Since minority carriers are present, H^0 would quickly become H^+ and, under bias, diffuse away from P^+ toward the front of the sample. When the bias is removed, H^+ would become H^-, diffuse toward P^+, and passivate it again. This second passivation reaction was associated with the 0.7 eV activation energy and interpreted as the migration barrier of H_T^-.

However, the extreme sensitivity of the $\{P,H\}^0$ pair to the presence of minority carriers suggests that it can trap a hole and, therefore, has a (+/0) level in the gap. Electrically, P^+ cannot be distinguished from $\{P,H\}^+$. Of course, $\{P,H\}^+$ may or may not be stable at room temperature. Theory[65] predicts that $\{P\cdots Si{-}H_{AB}\}^0 + h^+ \rightarrow \{P\cdots H_{BC}{-}Si\}^+$, as H moves from the (less stable) antibonding (AB) to the (more stable) BC configuration. This change of configuration occurs because P^0 has a lone pair pointing along the trigonal axis, thus forcing H into the AB configuration, while P^+ has a singly occupied orbital along this axis making it possible for H to bind. Thus, in the absence of free e^-, $\{P,H\}^+$ should be more stable than $\{P,H\}^0$. The 0.7 eV activation energy would then be associated with the slow BC → AB reverse reaction $\{P\cdots H_{BC}{-}Si\}^+ + e^- \rightarrow \{P\cdots Si{-}H_{AB}\}^0$ rather than with the migration of H_T^-. Overnight room-temperature anneals in the dark confirm that H does not diffuse to the surface of the sample but remains in the bulk following minority-carrier injection.[65–67]

Note that the passivated boron–hydrogen pair exhibits a (slower but) similar behavior under exposure to light.[68] In this case, the result is due to the dissociation of the pair. The $\{B,H\}^0$ pair has H bound to Si in a bond-centered configuration. The structure of the $\{B,H\}^-$ pair has not been calculated.

Several pieces of the puzzle are clearly missing. First, there is no direct experimental evidence for $\{P,H\}^+$ or $\{B,H\}^-$. Second, the gap levels of the {P,H} or {B,H} pairs have not been measured or calculated. Third, the 0.7 eV activation energy that has been associated with the migration of H^- differs substantially from the theoretical 0.39 eV value, in contrast to the migration barrier of H^+ calculated at the same level of theory, which matches the measured value. More experimental work with H^- in Si is needed to clarify the situation.

7.2.3 Vibrational Properties

Figure 7.4 shows the vibrational spectrum measured for H^+ in Si (Figure 7.4a)[69] and Ge (Figure 7.4b),[60] produced by the implantation of protons at cryogenic temperature. The spectrum for H^+ is consistent with the H atom at a BC site, whereas the spectrum for H^- is consistent with the H atom close to the tetrahedral site.

The most recent calculations[50] of the vibrational spectra of H in Si produce the following frequencies. The antisymmetric stretch of H_{BC}^+ is at 1994 cm^{-1}, very close to the measured one at 1998 cm^{-1}.[69] The wag modes are at 309 cm^{-1} and the symmetric stretch of the two Si NNs is at 397 cm^{-1}.

The antisymmetric stretch frequency of H_{BC}^0 is predicted to be 1780 cm^{-1}, reflecting the fact that the antibonding orbital of the three-center two-electron Si–H–Si

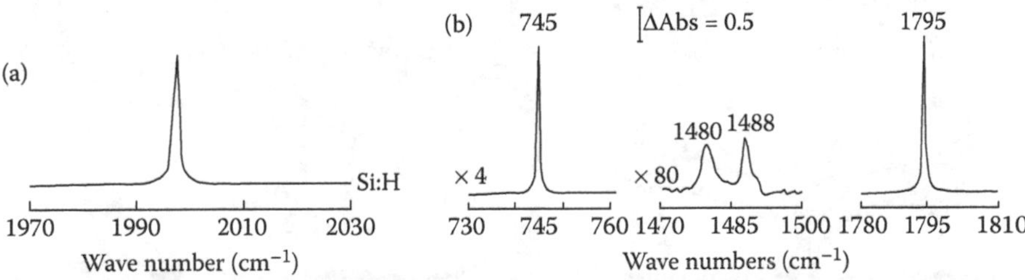

FIGURE 7.4 Low-temperature IR vibrational spectrum of H in proton-implanted Si (a) and Ge (b). ((a) M. Budde, Ph.D. Dissertation, University of Aarhus, October, 1998. (b) Reproduced with permission from M. Budde et al. *Phys. Rev. Lett.* **85**, 2965, 2000.)

bond is now partially populated, thus weakening the bond. The wag modes are at 534 cm^{-1}, almost resonant with the Γ phonon, and the symmetric stretch of the two Si NNs is almost unchanged at 390 cm^{-1}.

The potential energy surface of H_T^0 is very flat all around the T site, with very small potential energy differences between the exact T site and off-T-site locations. This makes it impossible to predict the vibrational frequencies of H_T^0. The calculated frequencies at several sites around T range from 500 to 800 cm^{-1}, leading to zero-point energies in the range 0.03–0.05 eV, which is greater than the energy difference between these sites. Thus, H_T^0 is only at the T site on the average.

H_T^- has a singlet at 528 and a doublet at 514 cm^{-1}, indicating that its lowest-energy site in the Si_{64} supercell is displaced slightly off the T site along a trigonal axis. This could well be a supercell-size effect rather than a true off-T site equilibrium configuration.

7.3 HYDROGEN PAIRS

Two electrically inactive hydrogen pairs (Figure 7.5) are commonly found in Si: the interstitial H_2 molecule and H_2^*, where a single Si−Si bond is replaced by the trigonal antibonding/bond-centered configuration H_{AB}−Si···H_{BC}−Si. The former is the main reservoir of hydrogen in low carbon Si materials. The latter dominates in the H_2^*(C) (also called CH_2^*) form in C-rich Si, with substitutional C replacing one of the two Si atoms involved.

7.3.1 Hydrogen Molecules

In the early 1980s, theory proposed that an H_2 molecule in Si would be stable at the interstitial T site.[70,71] Interstitial H_2 has also been suggested to play an important role during the in-diffusion of H.[33] The stretching vibration of an interstitial H_2 molecule

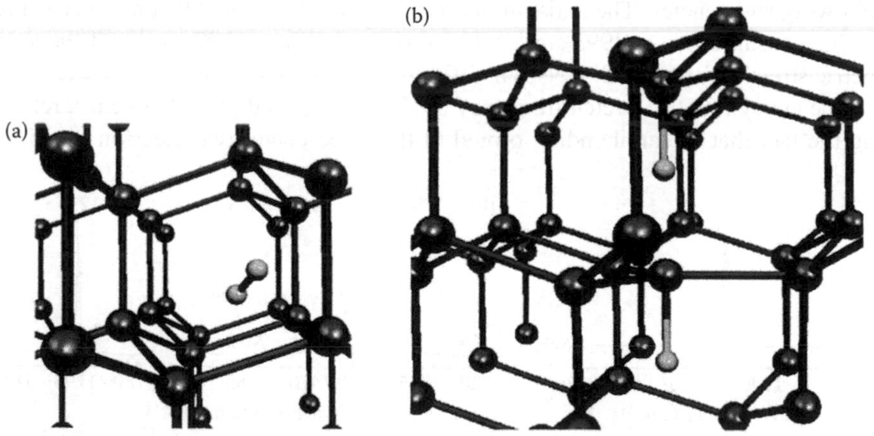

FIGURE 7.5 The interstitial H_2 molecule near the T site (a) and the $H_2^* = Si{-}H_{BC}\ldots Si{-}H_{AB}$ pair (b) in Si or Ge. The two H_2^*(C) defects in Si have one of the Si neighbors in H_2^* replaced by C.

in a semiconductor was observed first in GaAs by Raman spectroscopy[72] at 3934 cm^{-1}. The H_2 line is split by 8 cm^{-1}, leading to the conclusion that this doublet is due to the ortho- and para-H_2 species and, therefore, that the H_2 molecule is freely rotating.

The isolated interstitial H_2 molecule in Si was first observed by Raman spectroscopy[73] (3601 cm^{-1} at room temperature) in plasma-exposed samples at 250°C and by Fourier-transform infrared absorption (FTIR)[74,75] (3618 cm^{-1} at 10 K, with a very narrow 0.1 cm^{-1} line width) in thick samples annealed around 1250°C in a hydrogen ambient and then rapidly quenched. This frequency is about 550 cm^{-1} lower than that of the free H_2 molecule, indicating that the H−H bond length is longer and weaker in Si than it is in free space. The IR line anneals out around 350°C.

The details of the story of H_2 in Si are fascinating and somewhat complicated. At first, it appeared that some of the Raman and IR lines were at the wrong place, had the wrong amplitudes, and the expected ortho/para splitting was nowhere to be found. Further, in the case of H_2 trapped near interstitial oxygen (O_i) in Si, it is the O_i line—not the H_2 line—that exhibits the ortho/para splitting.[76] Indeed, the IR line associated with the antisymmetric stretch of O_i splits into two components with intensity ratios 1:3 when H_2 is used and 2:1 when D_2 is used. It took a lot of very careful work involving the excited ro-vibrational states of the HD molecule to sort things out. The key points are summarized in References 77–79.

Early experiments suggested that H_2 might be trapped as a static defect. Theory, however, found no evidence for a substantial barrier to the nearly free rotation of interstitial H_2.[80–84] For example, *ab initio* molecular dynamics (MD) simulations[83] show that H_2 in Si bounces around rapidly in the tetrahedral cage even at temperatures as low as 30 K, and is only at the T site on the average. In T_d symmetry, the $A_1 \rightarrow A_1$ transition is dipole-forbidden, but not the $T_2 \rightarrow T_2$ transition. Thus, the IR spectra only show ortho-H_2 (or para-D_2). A careful study of the temperature dependences of the HD lines (Figure 7.6)[76,85,86] has revealed

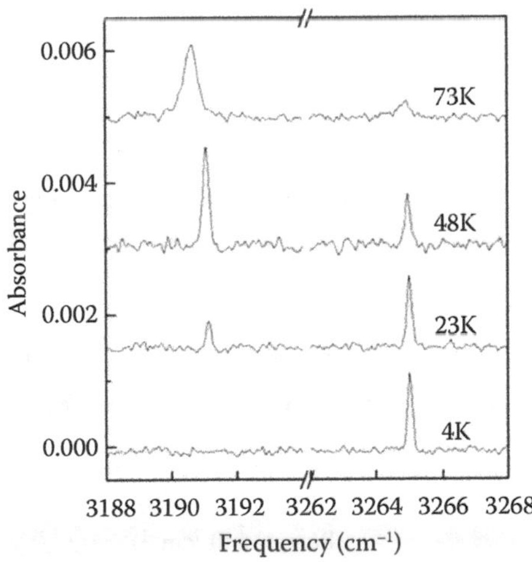

FIGURE 7.6 Annealing behavior of the HD line in Si. (Reproduced with permission from E.E. Chen et al. *Phys. Rev. Lett.* **88**, 105507, 2002.)

the missing lines as well as the excited ro-vibrational states of the molecule. A subsequent Raman study found vibrational lines for both the ortho- and para-H_2 species in Si at 3618 and 3627 cm^{-1}.[87]

When H_2 or D_2 traps near O_i,[88–90] the antisymmetric stretch of O_i splits into two lines with 3:1 (for {O_i,H_2}) or 1:2 ({for {O_i,D_2}) intensities ratios, reflecting the subtle differences in the O_i interactions with the ortho and para species of the H_2 molecule. Annealing studies show that H_2 can be driven away from O_i in a reversible manner, a nice example of configurational entropy at work. These experiments provided the activation energy for diffusion of the H_2 molecule (0.78 ± 0.05 eV) and the H_2−O_i binding energy (0.26 ± 0.02 eV). Since O_i is in a slightly puckered bond-centered configuration, it stretches a Si−Si bond, thus opening up the interstitial cage immediately next to it. H_2 benefits from being in a slightly bigger cage.

A recent study has extended results for H_2 in semiconductors also to Ge. Vibrational lines for the ortho- and para-H_2 species in Ge were found at 3826 and 3834 cm^{-1} by Raman spectroscopy (Figure 7.7).[91] These frequencies are consistent with theoretical predictions. A trend in the vibrational frequencies for H_2 in several semiconductor hosts was predicted by theory.[81] The vibrational frequency for H_2 in a semiconductor was found to be lower than for H_2 in gas phase (4161 cm^{-1}) because the electron density at the T site of the crystal lattice weakens the H−H bond and reduces the vibrational force constant. As the semiconductor lattice constant decreases, the electron density at the T site increases. Thus, the H_2 vibrational frequency of H_2 is reduced as the lattice constant for the semiconductor host is reduced. This trend is qualitatively consistent with the experimental vibrational frequencies for H_2, which are a few hundred cm^{-1} greater in Ge (a_L = 5.646 Å) and GaAs (a_L = 5.653 Å) than in Si (a_L = 5.431 Å).

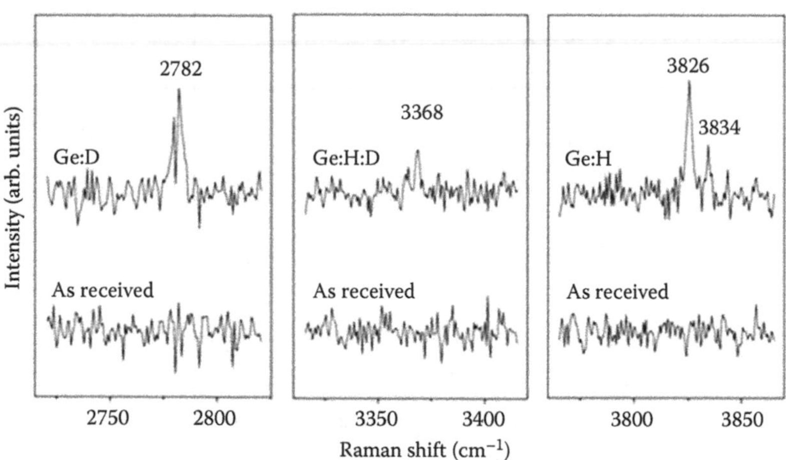

FIGURE 7.7 Raman spectra of H_2, HD, and D_2 in Ge. (Reproduced with permission from M. Hiller et al. *Phys. Rev. B* **72,** 153201, 2005.)

7.3.2 Open Question: Formation of H_2

One issue that is still not understood is how H_2 forms in the first place. Indeed, when hydrogen is introduced from a gas at high temperatures, the solubility of H is proportional to the square root of the gas partial pressure.[92] Thus, H_2 dissociates at the surface of the sample, diffuses in atomic form, and then forms H_2 in the bulk. But how is the molecule formed?

In *p*-type or intrinsic Si, the dominant atomic species is H_{BC}^+. The Coulomb repulsion between the two H_{BC}^+ species should keep them away from each other. Furthermore, there are precious few free e^- available for H_{BC}^+ to become H_{BC}^0, and then a substantial barrier must be overcome for H_{BC}^0 to become H_T^0, before a molecule can form. In *n*-type Si, the dominant species is H_T^-, generating a Coulomb repulsion, and there are few h^+ available for H_T^0 formation. However, this situation depends on both the temperature and the Fermi level. An *n*-type Si sample doped ~10^{16} cm^{-3} has a Fermi level about 150 meV above midgap at 200°C. Since the (+/–) occupancy level is about +100 meV above midgap, the equilibrium ratio[31] of H^+/H^- is in the range 1/10–1/100, and the H_2 formation could result from the interaction of H^+ and H^-. However, in most samples, the direct reaction $H_{BC}^+ + H_T^- \rightarrow H_2$ is unlikely since there is no Si material with high concentrations of *both* charged species at the same time.

We are left with the metastable H_T^0 and H_{BC}^0 states of H, which could, in principle, interact and produce H_2. But the neutral charge states of H are believed to be always a minority species. *Ab initio* MD simulations starting with any combination of two H interstitials in a supercell have so far failed to result in the formation of H_2.

The only MD simulations that provided evidence for spontaneous interstitial H_2 formation in Si involved multiple Hs trapped at a single substitutional impurity, such as a 3*d* transition metal (TM).[93] There is DLTS evidence that a number of substitutional 3*d* TM impurities can trap more than one H.[94–98] Theory also suggests[99] that the substitutional B or P impurities can also trap more than one H, although experimental evidence for this is only indirect.[34,100] Thus, it is possible that a number of substitutional impurities can trap more than one H interstitial and thus act as catalysts for H_2 formation. Experimental evidence for these processes is lacking.

7.3.3 H_2^* in Si and Ge

The second hydrogen pair in Si involves two H interstitials interacting with a single Si–Si bond and forming the antibonding/bond-centered H_2^* configuration (Figure 7.5b). The star is a remnant of the "Mu*" notation originally used for the so-called anomalous muonium before it was identified as bond-centered muonium[43,44] and subsequently renamed Mu_{BC}^0. The H_2^* defect does not form spontaneously upon hydrogenation but always appears in irradiated samples.[74]

The existence of H_2^* was predicted by theory[101,102] and then observed by FTIR in proton-implanted samples.[103,104] The Si–H_{BC} and Si–H_{AB} stretch modes are at 2062 and 1838 cm^{-1}, respectively, and the two lines anneal out together around 200°C. Theorists agree that the H_2 and H_2^* configurations are very close to each other in energy. Thus, isolated H_2 in Si has no incentive to form H_2^*, as this would involve overcoming a potential barrier (replacing a Si−Si bond with two Si−H bonds) with no meaningful gain in energy. The spontaneous formation of H_2^* starting with two isolated H interstitials is unlikely for reasons similar to those mentioned in the case of H_2 in Section 7.3.2.

The dynamics of H_2 interactions with the products of irradiation damage (vacancies *V*s and self-interstitials *I*s) have been studied using *ab initio* MD simulations. It was shown that H_2 molecules cannot survive in the vicinity of either an isolated *V* or *I*. The molecule spontaneously dissociates and traps at the native defect, resulting in the formation of *V*HH or *I*HH defects.[105] MD simulations also show that the interactions of *V* with *I*HH or of *I* with *V*HH result in the formation of H_2^*.[106] These calculations do not prove that such interactions are the only mechanism for H_2^* formation, but they do show that H_2 will not survive in irradiated material and that irradiation provides at least one formation pathway for H_2^* with no potential energy barrier to overcome.

7.4 INTERACTION OF H WITH IMPURITIES AND DEFECTS

Interstitial H diffuses easily in Si. It is chemically active and interacts covalently at or near any strained region in the material. Upon hydrogenation, vacancies or self-interstitials become $\{V,H_n\}$ or $\{I,H_n\}$ complexes; TM impurities often trap multiple H interstitials and form $\{TM,H_n\}$ defects; substitutional dopants become the neutral {B,H} or {P,H} pairs leading to a sharp reduction in the number of charge carriers; interstitial or substitutional carbon traps one or two hydrogen atoms; and the list goes on. The interest in such interactions is that the formation of new covalent bonds at or near a defect shifts the energy levels and thus profoundly affects its electrical activity. The thermal stability of {X,H} complexes varies from below room temperature to over 1000°C. This makes it possible in many cases to reversibly change the electrical properties of a sample by cycles of hydrogenation and anneals.

7.4.1 Passivation of Shallow Impurities

For Si, as for many other semiconductors, the shallow impurities used to control conductivity can be passivated by hydrogen.[1–9] For boron-doped Si, it was found that a high resistivity surface layer was created by the in-diffusion of H (Figure 7.8).[107,108]

In order to explain the passivation of shallow acceptors in Si, it was proposed that atomic H introduces a donor level in the band gap.[109] The electron from the H donor recombines with the hole associated with the acceptor impurity, thereby making *compensation* of the acceptor the first step toward passivation. The positively charged H^+ is mobile and is attracted to the negatively charged acceptor A^-. The two charged species become covalently bound to form a neutral acceptor–H complex, completing the *passivation* of the acceptor. These steps are given in the following expression: $(H^+ + e^-) + (A^- + h^+) \rightarrow H^+ + A^- \rightarrow \{H,A\}^0$.

Because H is amphoteric in Si, shallow donor impurities such as P can also be passivated by H.[110] In *n*-type material, atomic H introduces an acceptor level in the semiconductor band gap, leading to the compensation of the shallow donor.[111,112]

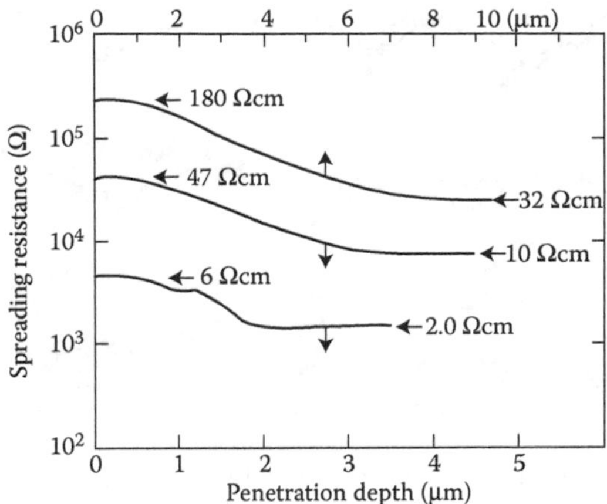

FIGURE 7.8 Spreading resistance profiles of *p*-type Si with the resistivities shown after treatment in an H plasma. (Reproduced with permission from J. Pankove et al. *Phys. Rev. Lett.* **51**, 2224, 1983.)

The mobile H^- is attracted to the positively ionized donor and these charged species become covalently bound to form a neutral complex.

The structures of the acceptor–H and donor–H complexes in Si have been studied by a variety of experimental methods and theory. For the acceptor–H complexes, the H atom sits near a BC site between the acceptor impurity and a neighboring Si atom (Figure 7.9a). For the donor–H complexes, the H atom is located at an antibonding site and is bonded to one of the Si neighbors to the donor impurity (Figure 7.9b).

Unlike for Si, the passivation of shallow impurities in Ge has not been studied exhaustively. It has been suggested that complexes of H with acceptors and donors in Ge are not very stable thermally.[113,114] However, hints of passivation were found; the passivation of shallow impurities has been suggested as a possible cause of the spatial variation of the conductivity found for as-grown crystals of ultrapure Ge.[32,33] The thermal stability of the hydrogen passivation of B in Ge was studied by Hall measurements on B implanted *n*-type samples. The electrical activity of the B acceptor was recovered by a brief anneal at 100°C, indicating a low thermal stability of the acceptor–H complex.[115,116] Theory predicts that the configuration of the {B,H} pair is similar in Ge and in Si.[117,118] No calculation exists about the shallow donor–hydrogen complexes in Ge.

Recently, the hydrogen passivation of P and Sb donors in Ge was reported.[119,120] Hydrogenation of the samples was performed using a remote dc plasma in the range 40°C to 150°C for 2 h. The passivation reduces the donor concentrations close to the surface from 10^{15} to 10^{14} cm^{-3}. A significant fraction of the shallow donors (about 20%) was neutralized up to about 4–5 μm below the junction in the sample treated at 150°C. The carrier profiles change under reverse bias anneals (RBA) at temperatures above room temperature. The donors are reactivated in the high-field region of the depletion layer due to the dissociation of the donor–hydrogen complexes. The high electric field sweeps the H species toward the bulk where the electric field is significantly reduced. Figure 7.10 shows an example for an RBA treatment of a P-doped Ge sample after

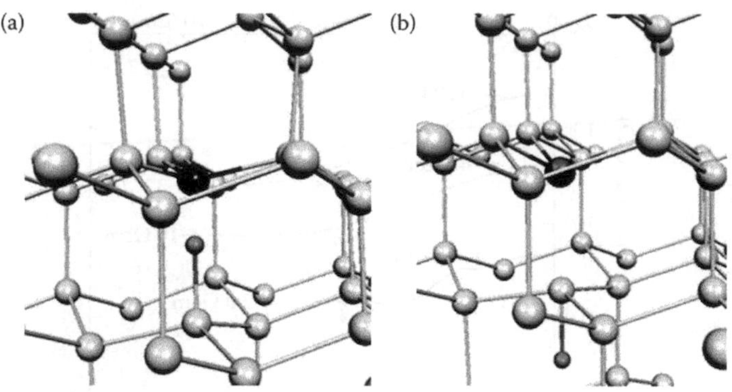

FIGURE 7.9 The {B,H} complex (a) has H in a BC configuration ($Si-H_{BC}\cdots B$) while the {P,H} pair (b) has H in an AB configuration ($H_{AB}-Si\cdots P$).

different annealing durations at 380 K. The change of the profile with RBA treatment is consistent with a negative charge state of H in *n*-type Ge. An analysis of the dopant profiles after RBA at different temperatures gives the enthalpy E_D of the dissociation process $E_D(PH) = 1.45 \pm 0.03$ eV and $E_D(SbH) = 1.36 \pm 0.06$ eV. Taking into account that the migration enthalpy of the H atom was reported to be around 0.4 eV, one finds a binding enthalpy of about 1 eV for the donor–hydrogen complexes.

No acceptor–hydrogen complexes were detected in similar experiments for B-, Al-, or Ga-doped Ge samples.[121] The acceptor profiles support the negative charge and the high mobility of hydrogen at room temperature. The ionized shallow acceptors will drive H^- toward the surface, which adds to the negative surface charge density and leads to a significant reduction of the space charge layer width. There is strong evidence that the presence of hydrogen hinders Schottky barrier formation in *p*-type Ge. Therefore, the absence of a Schottky barrier in as-grown *p*-type

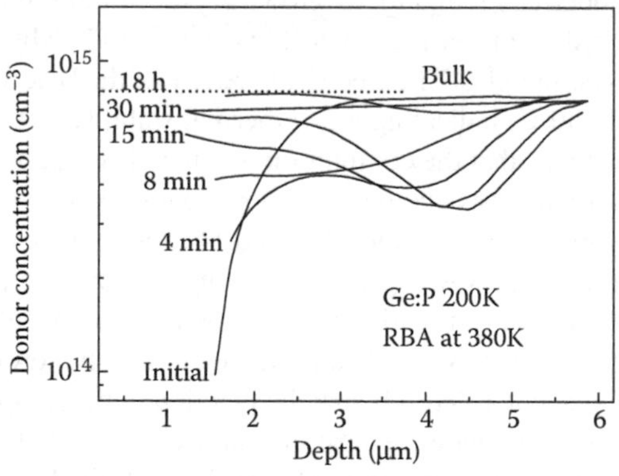

FIGURE 7.10 Depth profiles dopant concentration in P-doped Ge. The hydrogenation was achieved by dc plasma treatment at 40°C during 2 h. The reverse bias anneal ($V_R = -6$ V) was performed at 380 K. The *CV* curves were recorded at 200 K.

samples, in *p*-type Ge treated in the dc H-plasma, or in wet chemical solutions can be explained by the presence of negatively charged H.

7.4.2 Hydrogen Interactions with Native Defects

The native vacancy and interstitial defects whose dangling bonds are terminated by H atoms are an important class of H-containing defects. In a pioneering study, Si was implanted with protons at room temperature, producing many new H vibrational-absorption lines.[122] In these experiments, the energetic protons create vacancy and interstitial defects with dangling bonds to which the implanted H can bind. After many years of study, a number of defects have been identified definitively.[103,104,121,123] There is a family of vacancy–H complexes (VH_n) for which H terminates different numbers of dangling bonds of the vacancy.[104] There is an IH_2 defect for which the dangling bonds of a Si interstitial are terminated by H atoms.[104] One of these defects is shown in Figure 7.5b and two others are shown in Figure 7.11.

The same defects involving native defects and hydrogen that are formed by the implantation of protons into Si also form in Ge. These defects, VH_n, IH_2, and H_2^*, have structures and vibrational properties that are remarkably similar for the two hosts.[68,104,124,125]

An acceptor level at $E_v + 80$ meV that affects the conductivity of dislocation-free, ultrapure Ge has been associated with a V_2H defect.[126,127] Because of its low concentration, structure-sensitive data are not available, and no structure for this defect has yet been proposed.

7.4.3 Hydrogen Platelets and the Smart-Cut® Process

The implantation of high concentrations of protons creates extended defects in most semiconductors. This behavior was utilized to form thin crystalline Si-films on insulators (SOI).[15] The so called Smart-Cut technique involves bonding the implanted wafer surface onto an oxidized Si wafer or another substrate. A thermal treatment leads to the cracking of the donor wafer at the implantation depth of the ions and to an SOI structure on top of the substrate.[128]

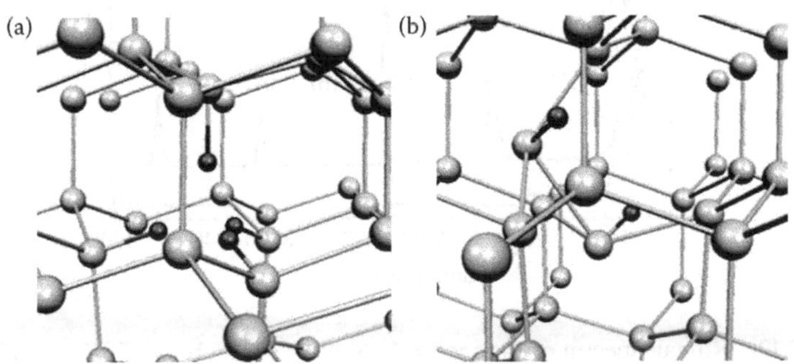

FIGURE 7.11 The fully saturated vacancy VH_4 (a), and the self-interstitial with two hydrogens IH_2 (b).

Hydrogen plasma treatment of Si (Figure 7.12) and Ge samples under remote conditions generates well-defined platelets on (111) lattice planes ({111} platelets).[100,129] The platelets are stabilized by atomic hydrogen. Two broad-structured Raman bands (Si: ~2120 and ~4158 cm^{-1}; Ge: ~1980 and ~4155 cm^{-1}) appear in the samples.[130,131] The two lower-frequency modes are close to the expected frequencies of the Si−H and Ge−H stretching modes, respectively,[100,124] while the two high-frequency modes are close to the stretching mode of free H_2 molecules.[132]

From the polarization dependence of the Raman signals in Si and Ge, information about a very similar platelet formation process was derived.[131–133] At first, the agglomeration of H_2^* defects leads to an extended platelet structure $\{H_2^*\}_n$. An increase in sample temperature during the plasma treatment generates a restructuring that results in an open structure with Si−H bonds and H_2 molecules $\{2Si{-}H + H_2\}_n$. This mechanism for the H-induced platelet formation needs no nucleation sites. A high concentration of H_2^* was found to be sufficient to induce platelet formation.[100] The creation of H_2^* in Si is favored if the Fermi level position at the plasma treatment temperature is close to the hydrogen (+/–) occupancy level.[134] A comparison of the rotational and the stretch modes of H_2 in the {111} platelets reveals the ortho- and para-states of H_2 molecules.[135]

Proton implantation used in the original Smart-Cut process leads to a somewhat more complicated defect formation process. The implantation produces hydrogen defects different from those found after plasma treatment and identified by FTIR spectroscopy.[136] With increasing temperature, some of these structures anneal and H forms more stable configurations, such as V_2H_6 or V_nH_{n+4} clusters. These defects act as the nucleation points for internal cavities. Upon further heating, molecular hydrogen traps in these cavities, the pressure on the internal surfaces causes cracks,

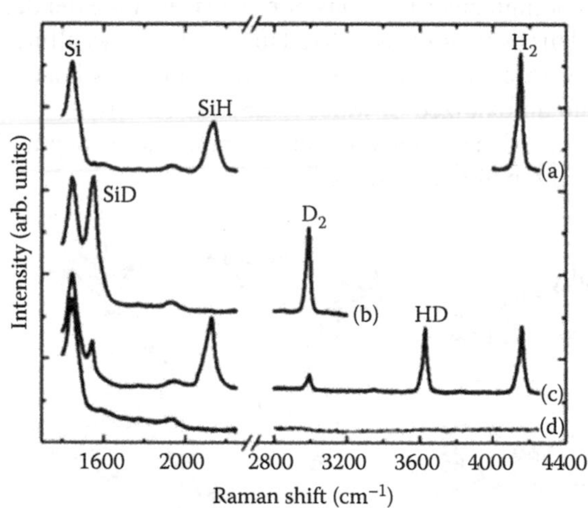

FIGURE 7.12 Raman spectra of P-doped FZ−Si after exposure at 250°C for 8 h to (a) H_2 plasma; (b) D_2 plasma; (c) H_2:D_2 plasma (50:50); and (d) no plasma. The spectra were recorded at room temperature. (Reproduced with permission from J. Pankove et al. *Phys. Rev. Lett.* **51**, 2224, 1983.)

which extend and coalesce.[137,138] Larger vacancy clusters appear to be the key building blocks of the voids.[73]

7.4.4 Hydrogenation of Deep Levels

One of the early motivations for the study of H in semiconductors was its ability to passivate deep-level defects. For Si and Ge, it was shown that the exposure of a sample to a hydrogen plasma can eliminate electrical levels associated with TM impurities.[139,140] However, in these early studies, little was known about the microscopic properties of the hydrogenated defects or about how H affects the electronic states.

The word "passivation" is often used when the known gap level of a defect or impurity X disappears from the gap following hydrogenation. While this strongly suggests that some {X,H} complex forms, this does not necessarily mean that the {X,H} complex is electrically inactive.

In many cases, H passivates (or partially passivates) deep-level defects, in particular dangling-bond-type defects, which are common at surfaces, interfaces, metallic contacts, or grain boundaries. An important example of this behavior involves the interface between Si and the native oxide that commonly forms on its surface. Hydrogen efficiently passivates the Si dangling bond (P_b center) at the Si/SiO_x interface.[141]

Wet chemical etching has been used to introduce H into thin surface layers of Si samples that also contained TM impurities.[142] In these studies, many TM−H complexes were studied by DLTS[94–98] and predicted theoretically.[93,143–145] It was found that there are families of complexes that include different numbers of H atoms, for example, PtH, PtH_2, PtH_3, and PtH_4 (Figure 7.13).[146] Hydrogenated defects such as PtH and PtH_2 were discovered and studied by structure-sensitive EPR and IR spectroscopies. In these studies, it was also found that the defects are electrically active.[147–149] Of the PtH_n centers, only PtH_4 is believed to be passivated; the other centers remain electrically active. Pt and Au impurities can also be hydrogenated by introducing H throughout bulk Si samples by annealing in H_2 gas at elevated temperature (1250°C).

The underlying reason for the partial passivation of TM impurities is related to the fact that at the substitutional site, the H atoms bind directly to the impurity. The substitutional TM is already fourfold coordinated (four TM−Si bonds). Following hydrogenation, the TM becomes five-, six-, or even sevenfold coordinated. Large coordination numbers for TMs are not uncommon in chemical compounds.[150] Such configurations involve numerous *s–p–d* hybrid orbitals and only a few of them are passivated by H. As a result, many {TM,H_n} complexes are electrically active. Typical substitutional TM−H binding energies are of the order of 1.2–1.5 eV.

Hydrogen introduction into Si solar cells is known to improve the minority carrier lifetime. The general explanation is that some defects are passivated by atomic hydrogen. One example is the highly detrimental Fe contamination of solar cell materials. The reported electrically active {Fe,H} defects cannot account for the improvement.[151,152] A more sophisticated—still not fully understood—process could be at work. A first hint could be the dissociation of {Fe,B} pairs and the formation of {B,H} pairs with hydrogen injection at lower temperatures.[153,154] At the higher temperatures used for solar cell passivation, the H^+ concentration could stabilize higher interstitial Fe concentrations, which, during cooling, would enhance precipitation.[155]

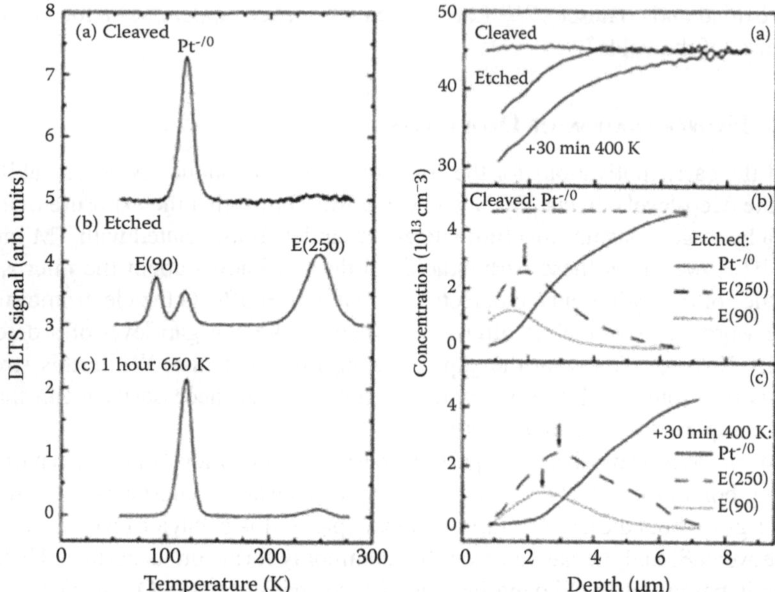

FIGURE 7.13 DLTS spectra and depth profiles of the PtH and PtH_2 complexes in Si. (Reprinted with permission from J.-U. Sachse et al. *Appl. Phys. Lett.* **70**, 1584, 1997. Copyright 1997, American Institute of Physics.)

Under these conditions, hydrogen would indirectly contribute to the improvement of solar cell materials.

For Ge, a hydrogenated TM impurity with particularly interesting properties is Cu.[156] Substitutional Cu is a triple acceptor in Ge that can be passivated in steps by the addition of H atoms. For example, the addition of two H atoms gives rise to a CuH_2 complex that is a single acceptor. Its electronic ground state is split into a manifold of levels. When H is replaced by deuterium or tritium, a single ground-state level is seen, leading to the conclusion that the complicated ground state for the CuH_2 center is due to the tunneling of the H atoms about the Cu atom. There are other centers in both Si and Ge for which the tunneling of the light H atom is important.[157,158] The CuH_3 center is believed to be fully passivated.[159]

The proposed structure[156] for the CuH_2 complex in Ge is similar to the one calculated[144] for the CuH_2 complex in Si (Figure 7.14). Similar structures involving TM−H covalent bonds with high coordination numbers occur for other substitutional TM impurities.

7.4.5 Activation of Impurities by Hydrogen

Unintentional shallow centers were discovered in ultrapure Ge grown in a hydrogen ambient and rapidly quenched from 400°C to room temperature.[33,126] Immediately following quenching, shallow acceptor centers are formed. A donor center is then formed if the sample is held at room temperature, and remains stable up to near

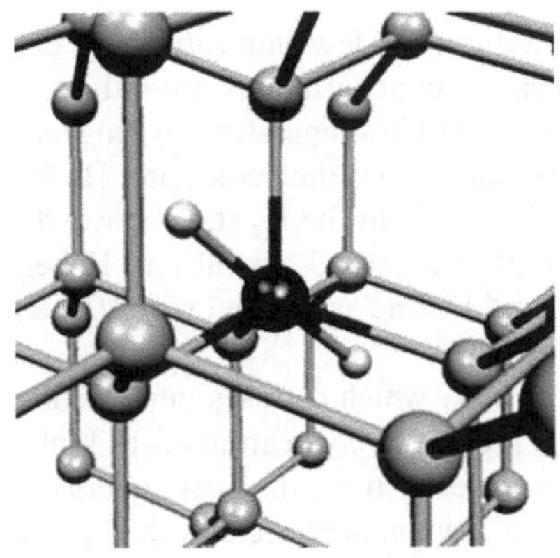

FIGURE 7.14 The calculated structure of the FeH_2 or CuH_2 complexes in Si have the H (white) bound directly to the substitutional transition metal (brown), which is sixfold coordinated. This configuration is very similar to that proposed for CuH_2 in Ge.[156]

100°C. Because the concentration of these defects is small ($<10^{11}$ cm^{-3}), they were studied by the highly sensitive, photo–thermal–ionization spectroscopy in a Fourier spectrometer to reveal their detailed electronic structure.[14,126,160,161] An isotope shift of the electronic lines upon D substitution proved the involvement of hydrogen.[14]

The shallow acceptor centers in the Ge host were labeled A(H,Si) and A(H,C) and assigned to complexes in which H^- is weakly bound to substitutional Si or C. These isoelectronic impurities are introduced into ultrapure Ge during growth from the quartz or graphite crucibles that contain the melt.[160,161] The A(H,Si) and A(H,C) centers have trigonal symmetry,[161] and theory has found that the H atom lies near the T site of the lattice.[117] The donor center in Ge was labeled D(H,O) and assigned to a complex with H bound to interstitial oxygen.[157]

In the Ge host, the Si and C impurities are not electrically active until they are activated by weakly binding a hydrogen atom. In ultrapure (intrinsic) Ge, where the Fermi level lie above the H(–/+) occupancy level, the negative charge state of isolated H dominates and H^- weakly binds to substitutional Si or C impurities to form the shallow acceptors that are observed.

There are examples of impurity activation by H in Si as well. H binds to substitutional C to form a donor with a level at $E_c - 0.16$ eV.[162,163] This defect is reminiscent of the A(H,C) center in Ge except that, in the Si host, it is H^+ that binds to C and the result is a donor rather than an acceptor. Another example of activation in Si is the E_3'' center in which a H atom becomes trapped near interstitial oxygen.[40,41]

7.5 CONTRASTING THE PROPERTIES OF H IN HIGH-PURITY, O-RICH, AND C-RICH Si MATERIALS

Hydrogen behaves differently in floating-zone (FZ), Czochralski (CZ), and cast Si materials, all of which have different grown-in impurities and impurity concentrations.

As-grown FZ−Si contains very few impurities and defects. Substitutional carbon dominates, but typical C_s concentrations are only of the order of a few times 10^{16} cm^{-3}, with even less O_i. High-temperature hydrogenation followed by a rapid quench results in the formation of H_2 molecules and {H,B} pairs in B-doped material.[73,74,85] Atomic H is released from the H_2 state above 200°C. The fully saturated vacancy VH_4 dominates after a 500°C anneal and H begins to leave the sample above 600°C. The 2223 cm^{-1} IR line associated with VH_4 anneals out at 650°C, as H dissociates from the complex.[164]

Typical CZ−Si is rich in O_i, which diffuses into the Si melt during growth from the quartz crucible and is present in concentrations as high as a few times 10^{18} cm^{-3}. The concentration of C_s is lower by more than two orders of magnitude. In CZ−Si, H or H_2 prefer sites near O_i at low temperatures. Both H^+_{BC} and H^-_T are seen by DLTS, isolated as well as in the immediate vicinity of O_i.[41] Related information comes from µSR[7,51] and positron (from muon decay) channeling[165] data obtained at low temperatures under identical conditions in FZ− and CZ−Si. In the high-purity FZ−Si, the paramagnetic Mu^0_T and Mu^0_{BC} lines are seen, in addition to the diamagnetic line, which could be Mu^+_{BC} or M^-_T. However, in CZ−Si, only Mu^0_T is seen.[51] The channeling experiments indicate that the muon is trapped at a (near) tetrahedral site very near O_i. In samples exposed to hydrogen gas near the melting point of Si and then rapidly quenched, H_2 molecules form and make up the reservoir from which interstitial hydrogen can be released following anneals at a few hundred degrees Celsius.[73,74,85]

The most unexpected of all hydrogen–oxygen interactions is the enhanced diffusivity of O_i. Under normal conditions, the annealing of CZ−Si around 400–450°C results in the formation of a series of O-related double donors.[166–170] However, CZ−Si samples exposed to a H plasma during the anneal show a large enhancement in thermal donor formation compared to samples not exposed to atomic H.[17,18] The enhancement factors are 300, 30, and 5 at 350°C, 400°C, and 450°C, respectively, corresponding to an activation energy in the range 1.8–2.0 eV, substantially less than the 2.6 eV[169] activation energy for diffusion of isolated O_i.

It is believed that trace amounts of H suffice to greatly increase the diffusivity of O_i. The role of H is purely catalytic. Theory predicts[171–175] that H stabilizes the saddle-point for diffusion of O_i by partially saturating a Si dangling bond. Thus, a single H interstitial allows O_i to make several successive BC-to-BC jumps. No O−H bond forms because the stronger Si−O bond is always favored over the weaker O−H bond. However, theory has so far only described a fraction of the process and our understanding of H-enhanced thermal donor formation is still fragmentary. More advanced MD simulations and detailed potential energy surfaces, obtained using the nudged-elastic-band method, are also lacking.

Cast Si is C rich. In the top cuts of mc-Si ingots,[176] the substitutional carbon (C_s) content can be as large as 9×10^{17} cm^{-3}. This is nearly a factor of three higher than the solubility limit at the melting point, 3.5×10^{17} cm^{-3}.[177] The as-grown material also contains C_s−C_s pairs. The O content is much lower, a few times 10^{16} cm^{-3} at best. If hydrogenation is done by annealing in an H_2 ambient at 1250°C followed by a rapid quench to room temperature, then most of the H forms complexes with C instead of isolated interstitial H_2 molecules as is the case in FZ− and CZ−Si (Figure 7.15).

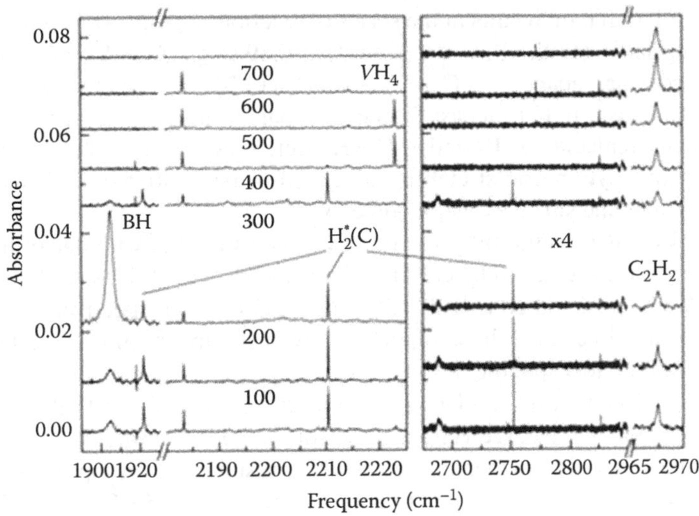

FIGURE 7.15 Temperature evolution of the IR-active H-related centers in C-rich cast Si material. (Reprinted with permission from C. Peng et al. *J. Appl. Phys.* **109**, 053517 (2011). Copyright 2011, American Institute of Physics.)

Three defects dominate at low temperatures in C-rich Si: $H_2^*(C)$, C_2H_2, and VH_3HC. The two $H_2^*(C)$ defects (Figure 7.16), also known as CH_2^*, occur when C_s traps two H atoms along one of the four C_s−Si bonds, leading to the trigonal configurations $C_s-H_{BC}\cdots Si-H_{AB}$ or $Si-H_{BC}\cdots C_s-H_{AB}$. The latter configuration is calculated to be just 0.1 eV higher in energy than the former. These configurations can be viewed as a H_2^*[105] or a H_2 molecule trapped by C_s.

$H_2^*(C)$ is electrically inactive. The measured (calculated) local vibrational modes (LVMs)[164] of $C_s-H_{BC}\cdots Si-H_{AB}$ are at 2752 (2753) cm^{-1} for the C−H stretch and 1922 (1850) cm^{-1} for the Si−H stretch. Those of $Si-H_{BC}\cdots C_s-H_{AB}$ are at 2675 (2574) cm^{-1} for the C−H stretch and 2210 (2184) cm^{-1} for the Si−H stretch.

C_2H_2 involves H_2^* or an H_2 molecule trapped by two adjacent substitutional C atoms. The C_s-C_s bond becomes $C_s-H\cdots H-C_s$. This defect has also been seen in irradiated

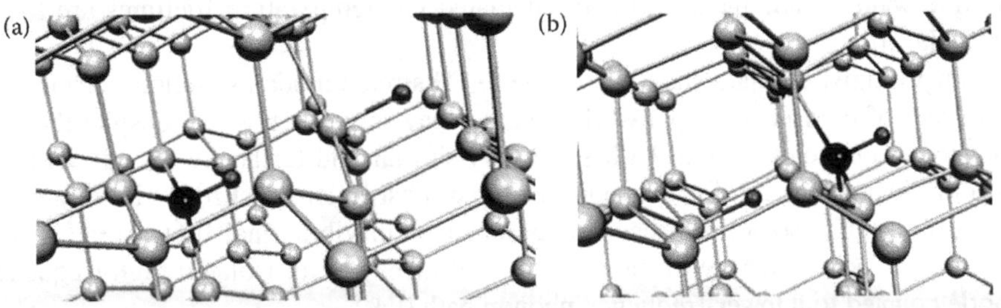

FIGURE 7.16 The two $H_2^*(C)$ defects: $C_s-H_{BC}\cdots Si-H_{AB}$ (a) and $Si-H_{BC}\cdots C_s-H_{AB}$ (b). The C atom is black, the H atoms are small dark grey.

FZ−Si.[178,179] Irradiation results in the series of reactions $C_s + I_{Si} \rightarrow C_i$ (+0.80 eV with a preexisting self-interstitial); $C_i + C_s \rightarrow C_iC_s$ (+0.89 eV); $C_iC_s + V \rightarrow C_sC_s$ (+3.28 eV with a preexisting vacancy); $C_sC_s + H_2 \rightarrow C_2H_2$ (+3.25 eV) or $C_sC_s + 2H_{BC} \rightarrow C_2H_2$ (+5.4 eV). However, in C-rich cast Si, the C_sC_s pair forms in the as-grown material. The measured (calculated) IR-active C_s−H stretch mode[164] is at 2967 (2986) cm^{-1}. The IR-inactive symmetric stretch is calculated to be at 3065 cm^{-1}. This defect is thermally stable and survives 700°C anneals.

Finally, VH_3HC[164] is the fully saturated vacancy VH_4, but with one of the four Si NNs to the vacancy replaced by C.[49] It is characterized by *I*R lines at 2184.3, 2214.4, and 2826.9 cm^{-1} (calculated at 2187, 2215, and 2850 cm^{-1}, respectively). The 2183.4 and 2826.9 cm^{-1} lines have been reported previously under similar conditions, but were not assigned to a specific defect.[180,181]

These C−H complexes evolve as follows during anneals.[164] $H_2^*(C)$ dissociates below 200°C, and the released H traps at substitutional B forming {HB} pairs (if B is present in the material). Further anneals at 400°C result in the disappearance of {HB} and the formation of VH_4. The latter begins to anneal out at 550°C (not at 650°C as in FZ−Si) and the VH_3HC lines grow. Since no interstitial C is present in the sample at these temperatures, VH_4 must trap at C_s. But C_s is not mobile up to much higher temperatures.[176] Thus, VH_4 must become mobile at 550°C. The formation of this defect has been modeled theoretically.[50] The reaction $VH_4 + C_s \rightarrow VH_3HC$ occurs with a 1.03 eV energy gain. All four $\{V,H_n\}$ complexes ($n = 1$, 2, 3, and 4) are mobile. The activation energies for diffusion calculated using the nudged-elastic band method are 1.0, 1.6, 2.6, and 3.4 eV, respectively.

7.6 VIBRATIONAL LIFETIMES

Hydrogen-containing defects always give rise to LVMs associated with the stretch and wag modes of the H atom(s). Many of these modes have been observed by FTIR absorption and/or Raman spectroscopy.[2] Most theoretical papers of the past decade include calculations of vibrational frequencies.

The optical excitation of high-frequency H-related LVMs results in a phonon being trapped[182] at the defect for a length of time called the vibrational lifetime τ. The lifetimes of a number of LVMs have been measured as a function of temperature by transient-bleaching spectroscopy[19,22,39,183] or estimated from their low-temperature IR line width.[21] The measured and calculated low-temperature lifetimes are listed in Table 7.1.

Surprisingly, the data show that some Si−H stretch mode excitations survive for hundreds of picoseconds (ps) while others decay in just a few ps[19] despite the fact that the frequencies of both LVMs are very similar and far above the optical phonon of the host crystal. Furthermore, isotopic substitutions that result in very small frequency shifts sometimes cause huge changes in vibrational lifetime.[21,22] These results are inconsistent with the conventional theory of an isolated high-frequency mode coupled to a lower-frequency phonon bath.[184,185]

Nonequilibrium *ab initio* MD simulations[20,21,186] have shown that the decay of the high-frequency LVMs can be classified as "normal" or "anomalous" depending on the decay mode.[182]

TABLE 7.1
Measured and Calculated Frequencies ω (cm^{-1}) and Vibrational Lifetimes τ (ps) of H-Related Defects in Si

Defect	Mode	ω (expt)	ω (theo)	τ (expt)	τ (theo)
H_2^*					
$H_{ab}-Si\cdots H_{bc}-Si$	$H_{bc}-Si$	2062	2126	~3.8 (75 K) a,b	3.7 (75 K) d,e
$D_{ab}-Si\cdots D_{bc}-Si$	$D_{bc}-Si$	1500	1538	4.8 (<50 K) a,b	18 (125 K) d,e
IH_2^*					
IH_2	H−Si	1990	2065	11^{ir} f	9.2 (150 K) e
ID_2	D−Si	1449	1481	18^{ir} f	29 (225 K) e
V_2H_2					
V_2H_2	H−Si	2072	2092	~175 (225 K) a,b	157(225 K) d,e
V_2D_2	D−Si	1510	1509	93^{ir} f	106 (225 K) e
H_{bc}^+					
$^{28}Si-H_{bc}-^{28}Si$	$H_{bc}-Si$	1998	2014	~6.5 (<50 K) a	8.7 (100 K) d,e
CH_2^*					
$H_{ab}-C\cdots H_{bc}-Si$	$H_{ab}-C$	2688.3	2567	2.4^{ir} c	2.7–4.8 (120 K) c
$H_{ab}-C\cdots D_{bc}-Si$	$H_{ab}-C$	2688.6	2564	39^{ir} c	23 (120 K) c

Note: The temperature at which the lifetimes were measured or calculated is in parenthesis. The ir superscript means that the lifetime was not measured by transient bleaching spectroscopy but estimated from *I*R line width at 4.2 K. The ~ symbol means that the value listed was taken from the published figure giving the measured τ(T). In References a and b, the temperature ignored the $3\hbar\omega/2$ energy added to the excited mode. For the Si_{64} cell, the correct T of the MD simulation is about 25 K higher than the one reported. The references are a = [183]; b = [19]; c = [21] d = [20], e = [186], f = [187].

A slow normal decay involves the coupling of the high-frequency LVM only to bulk phonons, as no other LVM participates in the decay. Normal decay obeys conventional theory[184,185] and no unexpected isotope effect is observed. One example of such a slow decay is the IR-active Si−H stretch mode of the divacancy with two Hs, V_2H_2.[19,20]

However, a fast anomalous decay involves one or more receiving modes that are LVMs themselves. These modes were not included in the conventional theory. Surprisingly, large isotope effects are sometimes observed because the frequency of one of the *receiving* modes experiences an isotope shift and this can cause a (fast) two-phonon decay to become a (much slower) three-phonon decay[187] (Figure 7.17). Examples of such fast decays are the $Si-H_{bc}$ stretch mode of H_2^* [tau1,tau-th1] and the $C-H_{ab}$ stretch mode of the $H_2^*(C)$ defect.[21]

These measurements and calculations show that vibrational energy can be trapped by H-related defects for lengths of time that vary considerably with the defect and sometimes its isotopic composition. Phonon trapping occurs at all defects and involves not just high-frequency LVMs but also low-frequency defect-related modes.[182,188,189]

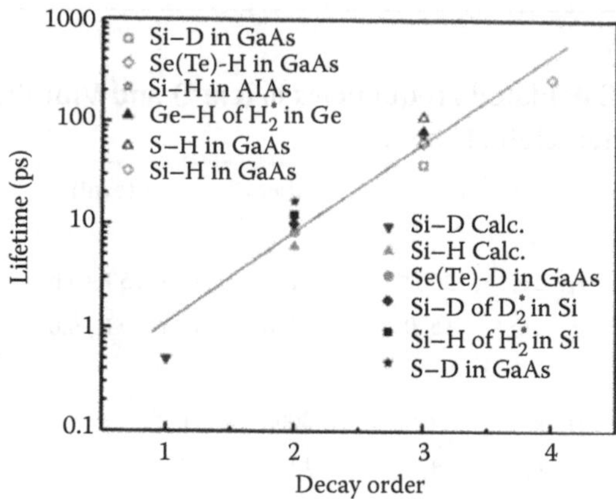

FIGURE 7.17 Vibrational lifetimes versus the number of phonons involved in the decay of the excitation. (Reproduced with permission from B. Sun et al. *Phys. Rev. Lett.* **96**, 035501, 2006.)

7.7 HYDROGEN IN THE PHOTOVOLTAIC INDUSTRY

Crystalline Si, solar cell materials are fabricated by a variety of methods (Czochralski, cast, ribbon, etc.), giving rise to materials with different defect content and properties.[190,191] The Si substrates that are often used for the fabrication of solar cells to reduce cost give rise to defect issues that must be addressed. These Si materials typically have higher concentrations of structural defects, carbon, and TM impurities than can be tolerated for microelectronics applications. To improve the performance of solar cells fabricated from low-cost Si materials, hydrogen is introduced to passivate defects in the Si bulk. A process that is used by industry to introduce hydrogen is by the postdeposition annealing of a hydrogen-rich SiN_x layer that is used as an antireflection (AR) coating.[192–195] A number of questions about this hydrogen introduction process and hydrogen's subsequent interactions with defects have proved difficult to address because of the low concentration of hydrogen that is typically introduced by this method.

A schematic of a simple design for a Si solar cell[196] is shown in Figure 7.18. A shallow, phosphorus-diffused junction is fabricated on a *p*-type Si substrate. Multicrystalline Si with a grain size of order 1 cm^2 is often used for the substrate material as a compromise

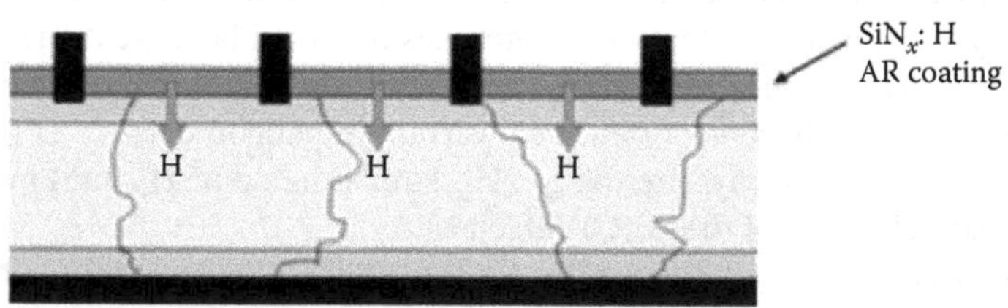

FIGURE 7.18 Schematic of a mc-Si solar cell hydrogenated by the postdeposition annealing of a SiN_x antireflection coating.

between quality and cost. In addition to the front and back contacts, an AR coating is deposited onto the solar cell front surface to increase the absorption of incident sunlight.

Silicon nitride has proved to be an excellent dielectric for the AR-coating layer.[192–195] A SiN_x:H layer that contains ~20 at.% H can be deposited at low temperature (250–450°C) by plasma-enhanced, chemical-vapor deposition (PECVD) from SiH_4 and NH_3 source gases. Layers with an index of refraction near 2 and a thickness of ~75 nm are typical.

What has made the use of a SiN_x:H AR coating attractive is its ability to serve several complementary purposes that can be integrated together with simple processing steps.[192–195] In addition to acting as an AR coating, the SiN_x layer passivates the Si surface, reducing the surface recombination velocity. For the cell design shown in Figure 7.18, the metal contacts are screen printed onto the solar cell after the SiN_x deposition. A firing step, typically several seconds near 800°C,[195–197] burns the front contacts through the SiN_x layer and, on the back surface, forms a p^+/p back-surface field and an Al back contact. Furthermore, and of primary interest here, hydrogen from the SiN_x layer diffuses into the Si substrate during this contact firing step to passivate bulk defects.

Several complementary research achievements came together during a time period of just a few years near 1980 to make the hydrogen passivation of defects in Si solar cells an attractive strategy for improving cell performance. Hydrogen-plasma treatments were found to reduce the electrical activity of grain boundaries in polycrystalline Si and to improve the efficiencies of solar cells fabricated from these materials.[198–200] It was found that many point defects and structural defects like dislocations could also be passivated or partially passivated by hydrogen-plasma exposure.[191,201,202] The use of SiN_x coatings was transferred from the microelectronics industry to the Si solar cell industry, and it was suggested that the SiN_x:H coating provides a source of hydrogen that can passivate defects in the Si bulk.[203]

The effect that the postdeposition annealing of a SiN_x layer has on solar cell performance has been widely studied in order to improve this hydrogenation method.[192–195,204] A difficulty that has been encountered in such studies is that the concentration of H introduced into the Si bulk under ordinary circumstances is too small to detect and quantify by methods such as SIMS, even when D is used to improve the detection limit (~10^{15} cm^{-3}).[205,206] Therefore, the concentration and penetration depth of H were unknown, and it had remained controversial whether H was even introduced in sufficient concentration to passivate bulk defects.[207]

To help address questions for which quantitative information about the H concentration would be insightful, a Si model system was developed in which hydrogenated impurities in the Si bulk could be detected by IR spectroscopy with a sensitivity sufficient to determine the concentration and penetration depth of H introduced by the postdeposition annealing of a SiN_x coating.[176,208–210] It was found that H could be introduced into the Si bulk with a concentration up to ~10^{15} cm^{-3} and that the concentration of H that is introduced is sensitive to processing methods, for example, the SiN_x deposition conditions and annealing procedure. An H penetration depth was found that is consistent with the rapid in-diffusion of isolated H estimated from the diffusivity determined by Van Wieringen and Warmoltz[10] rather than a slower, trap-limited diffusion process.

The hydrogen concentration [H] ≤10^{15} cm^{-3} is an interesting range. It is sufficiently low to be below the detection limit of SIMS but sufficiently high to passivate the typical concentrations of traps found in solar cells fabricated from mc-Si. Indeed, recent studies identify structural defects in mc-Si, which show a reduced recombination after introduction of H by the postdeposition annealing of a SiN_x antireflection coating.[204]

7.8 CONCLUSIONS

At first glance, the effect of H in Si and Ge seems to be quite different. In Si, the passivation of dopants and defects has been the focus of many studies. In Ge, the activation of impurities in ultrapure Ge and the physics of tunneling centers have been emphasized. However, a closer look reveals that the physics of hydrogen in Si and Ge is more similar than the emphasis placed on results in the literature might suggest.

Similar to Si where dopant passivation has been studied extensively, donor passivation has also been confirmed for Ge. However, because of the low thermal stability of dopant–H complexes in Ge, these defects have been studied in detail only recently. The activation of impurities by hydrogen in Ge has been the focus of much attention because of the effect such defects have on the conductivity of ultrapure Ge. Activated impurities have also been reported for Si, but in this case, the activation of impurities has not played an important role in Si technology because of their low thermal stability. The hydrogenation of deep levels appears to be similar in the two hosts, with the proposed structures of TM–hydrogen complexes and native-defect–H complexes being the same. The formation and physics of hydrogen molecules in Si and Ge are also similar.

An interesting difference between Si and Ge is the position of the H(+/–) occupancy level of isolated hydrogen. In intrinsic Si, H^+ should dominate whereas in intrinsic Ge, recent results suggest that H^- will be dominant. In terms of the diffusion and solubility of isolated H at elevated temperature, the lower occupancy level for Ge appears to have little effect. In terms of the physics of the hydrogenated defects that occur, the lower occupancy level in Ge causes donor impurities to be preferentially passivated and for H^- to appear in proton-implanted Ge and in the defect complexes in which H^- is weakly bound to isoelectronic impurities.

Have the very different technological applications of Si and ultrapure Ge rather than the physics of the defects and defect processes that are involved led to the different focuses on hydrogen passivation and activation in these materials? Our review of recent results suggests that the properties of H in the elemental semiconductors can be understood within the same framework.

As Ge becomes the focus of new applications in microelectronics and optoelectronics, it will be fun to see what interesting and unusual properties of H emerge, as they often have for other host materials.

ACKNOWLEDGMENTS

The work of SKE is supported, in part, by the grant D-1126 from the R. A. Welch Foundation. The work of MS is supported by NSF Grant DMR 1160756. MS is grateful for a Humboldt Foundation Award for Senior US Scientists that supports visits to the Dresden University of Technology.

REFERENCES

1. J.I. Pankove and N.M. Johnson (eds.) *Hydrogen in Semiconductors*, Semiconductors and Semimetals, vol. 34 (Academic, Boston, 1991).
2. S.J. Pearton, J.W. Corbett, and M. Stavola, *Hydrogen in Crystalline Semiconductors* (Springer-Verlag, Berlin, 1992).
3. M. Stutzmann and J. Chevallier (eds.) *Hydrogen in Semiconductors* (Elsevier, New York, 1991).
4. S.K. Estreicher, Hydrogen-related defects in crystalline semiconductors: A theorist's perspective, *Mat. Sci. Eng. Reports* **14**, 319, 1995.
5. N.H. Nickel (ed.) *Hydrogen in Semiconductors II*, Semiconductors and Semimetals, vol. 61 (Academic, San Diego, 1999).
6. A.R. Peaker, V.P. Markevich, and L. Dobaczewski, Hydrogen related defects in silicon, germanium and silicon-germanium alloys, in *Defects in Microelectronic Materials and Devices*, D.M. Fleetwood, S.T. Pantelides, and R.D. Schrimpf (eds.) (CRC Press, Boca Raton, 2009), p. 27–56.
7. S.F.J. Cox, Muonium as a model for interstitial hydrogen in the semiconducting and semimetallic elements. *Rep. Prog. Phys.* **72**, 116501, 2009.
8. B. Pajot and B. Clerjaud, *Optical Absorption of Impurities and Defects in Semiconducting Crystal* (Springer-Verlag, Berlin, 2012), Chapt. 8.
9. J. Chevallier and B. Pajot, Defect interaction and clustering in semiconductors. *Solid State Phenomena* **85–86**, 203, 2002.
10. A. Van Wieringen and N. Warmoltz, On the permeation of hydrogen and helium in single crystal Si and Ge at elevated temperatures. *Physica* **22**, 849, 1956.
11. C.T. Sah, J.Y.C. Sun, and J.J.T. Tzou, Deactivation of the boron acceptor in silicon by hydrogen. *Appl. Phys. Lett.* **43**, 204; Deactivation of group III acceptors in silicon during keV electron irradiation. *Appl. Phys. Lett.* **43**, 962, 1983; Effects of keV electron irradiation on the avalanche-electron generation rates of 3 donors on oxidized Si. *J. Appl. Phys.* **54**, 4378, 1983.
12. J.I. Pankove, R.O. Wance, and J.E. Berkeyheiser, Neutralization of acceptors in silicon by atomic hydrogen. *Appl. Phys. Lett.* **45**, 1100, 1984
13. R.C. Frank and J.E. Thomas, The diffusion of hydrogen in single-crystal germanium. *J. Chem. Phys. Solids* **16**, 144, 1960.
14. E.E. Haller, Isotope shifts in the ground state of shallow, hydrogenic centers in pure germanium. *Phys. Rev. Lett.* **40**, 584, 1978.
15. M. Bruel, Silicon on insulator material technology. *Electron. Lett.* **31**, 1201, 1995.
16. S. Kleekajai, L. Wen, C. Peng, M. Stavola, V. Yelundur, K. Nakayashiki, A. Rohatgi, and J. Kalejs, IR study of the concentration of H introduced into Si by the postdeposition annealing of a SiN_x coating. *J. Appl. Phys.* **106**, 123510, 2009.
17. H.J. Stein and SooKap Hahn, Hydrogen introduction and hydrogen-enhanced thermal donor formation in silicon. *J. Appl. Phys.* **75**, 3477, 1994.
18. A.R. Brown, M. Clayburn, R. Murray, P.S. Nandhra, R.C. Newman, and J.H. Tucker, Enhanced thermal donor formation in silicon exposed to a hydrogen plasma. *Semic. Sci. Technol.* **3**, 591, 1988.
19. G. Lüpke, X. Zhang, B. Sun, A. Fraser, N.H. Tolk, and L.C. Feldman, Structure-dependent vibrational lifetimes of hydrogen in silicon. *Phys. Rev. Lett.* **88**, 135501, 2002.
20. D. West and S.K. Estreicher, First-principles calculations of vibrational lifetimes and decay channels: Hydrogen-related modes in Si. *Phys. Rev. Lett.* **96**, 115504, 2006.
21. T.M. Gibbons, S.K. Estreicher, K. Potter, F. Bekisli, and M. Stavola, Huge isotope effect on the vibrational lifetimes of an $H_2^*(C)$ defect in Si. *Phys. Rev. B* **87**, 115207, 2013.

22. K.K. Kohli, G. Davies, N.Q. Vinh, D. West, S.K. Estreicher, T. Gregorkiewicz, and K.M. Itoh, Isotope dependence of the lifetime of the 1136-cm^{-1} vibration of oxygen in silicon. *Phys. Rev. Lett.* **96**, 225503, 2006.
23. M. Stavola, Hydrogen diffusion and solubility in c-Si, in *Properties of Crystalline Si*, R. Hull (ed.) (INSPEC, Exeter, 1999), p. 511.
24. T. Ichimiya and A. Furuichi, On the solubility and diffusion coefficient of tritium in single crystals of silicon. *Int. J. Appl. Rad. Isotopes* **19**, 573, 1968.
25. S.A. McQuaid, M.J. Binns, R.C. Newman, E.C. Lightowlers, and J.B. Clegg, Solubility of hydrogen in silicon at 1300°C. *Appl. Phys. Lett.* **62**, 1612, 1993.
26. M.J. Binns, S.A. McQuaid, R.C. Newman, and E.C. Lightowlers, Hydrogen solubility in silicon and hydrogen defects present after quenching. *Semic. Sci. Technol.* **8**, 1908, 1993.
27. S.A. McQuaid, R.C. Newman, J.H. Tucker, E.C. Lightowalers, R.A.A. Kubiak, and M. Goulding, Concentration of atomic hydrogen diffused into silicon in the temperature range 900–1300°C. *Appl. Phys. Lett.* **58**, 2933, 1991.
28. I.A. Veloarisoa, M. Stavola, D.M. Kozuch, R.E. Peale, and G.D. Watkins, Passivation of shallow impurities in Si by annealing in H_2 at high temperature. *Appl. Phys. Lett.* **59**, 2121, 1991.
29. Yu.V. Gorelkinskii and N.N. Nevinnyi, Electron paramagnetic resonance of hydrogen in silicon. *Physica B* **170**, 155, 1991.
30. Yu.V. Gorelkinskii and N.N. Nevinnyi, EPR of interstitial hydrogen in silicon: Uniaxial stress experiments. *Mater. Sci. Engr. B* **36**, 133, 1996.
31. C. Herring, N.M. Johnson, and C.G. Van de Walle, Energy levels of isolated interstitial hydrogen in silicon. *Phys. Rev. B* **64**, 125209, 2001.
32. W.L. Hansen, E.E. Haller, and P.N. Luke, Hydrogen concentration and distribution in high-purity Ge crystals. *IEEE Trans. Nucl. Sci.* **NS-29**, 738, 1982.
33. R.N. Hall, HP Ge: Purification, crystal growth, and annealing properties. *IEEE Trans. Nucl. Sci.* **NS-31**, 320, 1984.
34. N.M. Johnson and M.D. Moyer, Absence of O diffusion during H passivation of shallow-acceptor impurities in single-crystal silicon. *Appl. Phys. Lett.* **46**, 787, 1985.
35. M. Stutzmann, J.B. Chevalier, C.P. Herrero, and A. Breitschwerdt, A comparison of H incorporation and effusion in doped crystalline Si, Ge, and GaAs. *Appl. Phys. A* **53**, 47, 1991.
36. Y.V. Gorelkinskii and N.N. Nevinnyi, EPR of hydrogen in silicon. *Sov. Tech. Phys. Lett. (USSR)* **13**, 45, 1987.
37. Y.V. Gorelkinskii, Electron paramagnetic resonance of hydrogen and hydrogen-related defects in crystalline silicon, in *Hydrogen in Semiconductors II*, ref. (5), p. 25.
38. B. Bech Nielsen, K. Bonde Nielsen, and J.P. Byberg, EPR experiments on H-implanted Si crystals. *Mat. Sci. Forum* **143–147**, 909, 1994.
39. M. Budde, G. Lüpke, C. Parks Cheney, N.H. Tolk, and L.C. Feldman, Vibrational lifetime of bond-center hydrogen in crystalline silicon. *Phys. Rev. Lett.* **85**, 1452, 2000.
40. K. Bonde Nielsen, B. Bech Nielsen, J. Hansen, E. Andersen, and J.U. Andersen, Bond-centered hydrogen in silicon studied by *in situ* deep-level transient spectroscopy. *Phys. Rev. B* **60**, 1716, 1999.
41. K. Bonde Nielsen, L. Dobaczewski, S. Søgård, and B. Bech Nielsen, Acceptor state of monoatomic hydrogen in silicon and the role of oxygen. *Phys. Rev. B* **65**, 075205, 2002.
42. S.F.J. Cox and M.C.R. Symmons, Molecular radical models for the muonium centres in solids. *Chem. Phys. Lett.* **126**, 516, 1986.
43. T.L. Estle, S.K. Estreicher, and D.S. Marynick, Preliminary calculations confirming that anomalous muonium in C and Si is bond-centered muonium. *Hyperf. Inter.* **32**, 637, 1986 and Bond-centered H or Mu in C: The explanation for anomalous muonium and an example of metastability. *Phys. Rev. Lett.* **58**, 1547, 1987.

44. S.K. Estreicher, Equilibrium sites and electronic structure of interstitial hydrogen in Si. *Phys. Rev. B* **36**, 9122, 1987.
45. C.G. Van de Walle, Y. Bar-Yam, and S.T. Pantelides, Theory of hydrogen diffusion and reactions in crystalline silicon. *Phys. Rev. Lett.* **60**, 2761, 1988.
46. C.G. Van de Walle, P.J.H. Denteneer, Y. Bar-Yam, and S.T. Pantelides, Theory of hydrogen diffusion and reactions in crystalline silicon. *Phys. Rev. B* **39**, 10791, 1989.
47. B. Hitti, S.R. Kreitzman, T.L. Estle, E.S. Bates, M.R. Dawdy, T.L. Head, and R.L. Lichti, Dynamics of negative muonium in n-type silicon. *Phys. Rev. B* **59**, 4918, 1999.
48. I. Fan, K.H. Chow, B. Hitti, R. Scheuermann, W.A. MacFarlane, A.I. Mansour, B.E. Schultz, M. Eglimez, J. Jung, and R.L. Lichti, Optically induced dynamics of muonium centers in Si studied via their precession signatures. *Phys. Rev. B* **77**, 035203, 2008.
49. K.J. Chang and D.J. Chadi, Hydrogen bonding and diffusion in crystalline silicon. *Phys. Rev. B* **40**, 11644, 1989.
50. S.K. Estreicher, A. Docaj, M.B. Bebek, D.J. Backlund, and M. Stavola, Hydrogen in C-rich Si and the diffusion of vacancy–H complexes. *Phys. Stat. Sol. A* **209**, 1872, 2012.
51. B.D. Patterson, Muonium states in semiconductors. *Rev. Mod. Phys.* **60**, 69, 1988.
52. C.G. Van de Walle and J. Neugebauer, Universal alignment of hydrogen levels in semiconductors, insulators and solutions. *Nature* **423**, 626, 2003.
53. K.H. Chow, B. Hitti, and R.F. Kiefl, μSR on muonium in semiconductors and its relation to hydrogen, in *Identification of Defects in Semiconductors*, M. Stavola (ed.) Semiconductors and Semimetals, vol. 51A (Academic Press, Boston, 1998), p.138.
54. R.L. Lichti, Dynamics of muonium diffusion, site changes and charge-state transitions, in *Hydrogen in Semiconductors II*, ref. (5), p. 311.
55. B. Holm, K. Bonde Nielsen, and B. Bech Nielsen, Deep state of hydrogen in crystalline silicon: Evidence for metastability. *Phys. Rev. Lett.* **66**, 2360, 1991.
56. R.L. Lichti, K.H. Chow, and S.F.J. Cox, Hydrogen defect-level pinning in semiconductors: The muonium equivalent. *Phys. Rev. Lett.* **101**, 136403, 2008.
57. S.K. Estreicher and Dj.M. Maric, What is so strange about hydrogen interactions in germanium? *Phys. Rev. Lett.* **70**, 3963, 1993.
58. R.L. Lichti, K.H. Chow, J.M. Gil, D.L. Stripe, R.C. Vilão, and S.F.J. Cox, Location of the H[+/–][+/–] level: Experimental limits for muonium. *Physica B* **376–377**, 587, 2006.
59. L. Dobaczewski, K. Bonde Nielsen, N. Zangenberg, B. Bech Nielsen, A.R. Peaker, and V.P. Markevich, Donor level of bond-center hydrogen in germanium. *Phys. Rev. B* **69**, 245207, 2004.
60. M. Budde, B. Bech Nielsen, C. Parks Cheney, N.H. Tolk, and L.C. Feldman, Local vibrational modes of isolated hydrogen in germanium. *Phys. Rev. Lett.* **85**, 2965, 2000.
61. F. Buda, G.L. Chiarotti, R. Car, and M. Parinello, Proton diffusion in crystalline silicon. *Phys. Rev. Lett.* **63**, 294, 1989.
62. P.E. Blochl, C.G. Van de Walle, and S.T. Pantelides, First-principles calculations of diffusion coefficients: Hydrogen in silicon. *Phys. Rev. Lett.* **64**, 1401, 1990.
63. Ch. Langpape, S. Fabian, Ch. Klatt, and S. Kalbitzer, Observation of 1H tunnelling diffusion in crystalline Si. *Appl. Phys. A* **64**, 207, 1997.
64. N.M. Johnson and C. Herring, Diffusion of negatively charged hydrogen in silicon. *Phys. Rev. B* **46**, 15554, 1992.
65. S.K. Estreicher, C.H. Seager, and R.A. Anderson, Bistability of donor-hydrogen complexes in silicon: A mechanism for debonding. *Appl. Phys. Lett.* **59**, 1773, 1991.
66. C.H. Seager and R.A. Anderson, Reversible changes of the charge state of donor/hydrogen complexes initiated by hole capture in silicon. *Appl. Phys. Lett.* **63**, 1531, 1993.
67. C.H. Seager, R.A. Anderson, and S.K. Estreicher, Comment on "Inverted Order of Acceptor and Donor Levels of Monatomic Hydrogen in Silicon". *Phys. Rev. Lett.* **74**, 4565, 1995.

68. T. Zundel and J. Weber, Boron reactivation kinetics in hydrogenated silicon after annealing in the dark or under illumination. *Phys. Rev. B* **43**, 4361, 1991.
69. M. Budde, Hydrogen-related defects in proton-implanted Si and Ge. Ph.D. Dissertation, University of Aarhus, October, 1998.
70. A. Mainwood and A.M. Stoneham, Interstitial muons and hydrogen in crystalline silicon. *Physica* **116B**, 101, 1983.
71. J.W. Corbett, S.N. Sahu, T.S. Shi, and L.C. Snyder, Hydrogen-related defect-impurity complexes in Si. *Phys. Lett.* **93A**, 303, 1983.
72. J. Vetterhöffer, J. Wagner, and J. Weber, Isolated hydrogen molecules in GaAs. *Phys. Rev. Lett.* **77**, 5409, 1996.
73. A.W.R. Leitch, V. Alex, and J. Weber, Raman spectroscopy of hydrogen molecules in crystalline silicon. *Phys. Rev. Lett.* **81**, 421, 1998.
74. R.E. Pritchard, M.J. Ashwin, J.H. Tucker, and R.C. Newman, Isolated interstitial hydrogen molecules in hydrogenated crystalline silicon. *Phys. Rev. B* **57**, R15048, 1998; R.E. Pritchard, J.H. Tucker, R.C. Newman, and E.C. Lightowlers, Hydrogen molecules in boron-doped crystalline silicon. *Semic. Sci. Technol.* **14**, 77, 1999.
75. M. Suezawa, H_2-related defects in Si quenched in H_2 gas studied by optical absorption measurements. *Jpn. J. Appl. Phys.* **38**, L484, 1999.
76. E.E. Chen, M. Stavola, and W. Beal Fowler, Ortho and para O–H_2 complexes in silicon. *Phys. Rev. B* **65**, 245208, 2002.
77. S.K. Estreicher, The H_2 molecule in semiconductors: An angel in GaAs, a devil in Si. *Acta Phys. Polon.* **102**, 513, 2002.
78. M. Stavola, E.E. Chen, W.B. Fowler, and G.A. Shi, Interstitial H_2 in Si: Are all problems solved? *Physica B* 340–342, 58, 2003.
79. M. Hiller, E.V. Lavrov, and J. Weber, Raman scattering study of H_2 in Si. *Phys. Rev. B* **74**, 235214, 2006.
80. Y. Okamoto, M. Sato, and A. Oshiyama, Comparative study of vibrational frequencies of H_2 molecules in Si and GaAs. *Phys. Rev. B* **56**, R10016, 1997.
81. C.G. Van de Walle, Energetics and vibrational frequencies of interstitial H_2 molecules in semiconductors. *Phys. Rev. Lett.* **80**, 2177, 1998.
82. B. Hourahine, R. Jones, S. Öberg, R.C. Newman, P.R. Briddon, and E. Roduner, Hydrogen molecules in silicon located at interstitial sites and trapped in voids. *Phys. Rev. B* **57**, R12666, 1998.
83. S.K. Estreicher, K. Wells, P.A. Fedders, and P. Ordejón, Dynamics of hydrogen molecules in Si. *J. Phys. Condensed Matter* **13**, 62, 2001.
84. W.B. Fowler, P. Walters, and M. Stavola, Dynamics of interstitial H_2 in crystalline silicon. *Phys. Rev. B* **66**, 075216, 2002.
85. E.E. Chen, M. Stavola, and W.B. Fowler, Key to understanding interstitial H_2 in Si. *Phys. Rev. Lett.* **88**, 105507, 2002.
86. E.E. Chen, M. Stavola, W.B. Fowler, and J.A. Zhou, Rotation of molecular hydrogen in Si: Unambiguous identification of ortho-H_2 and para-D_2. *Phys. Rev. Lett.* **88**, 245503, 2002.
87. E.V. Lavrov and J. Weber, Ortho and para interstitial H_2 in silicon. *Phys. Rev. Lett.* **89**, 215501, 2002.
88. R.C. Newman, R.E. Pritchard, J.H. Tucker, and E.C. Lightowlers, Dipole moments of H_2, D_2, and HD molecules in Czochralski silicon. *Phys. Rev. B* **60**, 12775, 1999.
89. R.E. Pritchard, M.J. Ashwin, J.H. Tucker, R.C. Newman, E.C. Lightowlers, M.J. Binns, S.A. McQuaid, and R. Falster, Interactions of hydrogen molecules with bond-centered interstitial oxygen and another defect center in Si. *Phys. Rev. B* **56**, 13118, 1997.
90. V.P. Markevich and M. Suezawa, Hydrogen–oxygen interaction in silicon at around 50°C. *J. Appl. Phys.* **83**, 2988, 1998.

91. M. Hiller, E.V. Lavrov, J. Weber, B. Hourahine, R. Jones, and P.R. Briddon, Interstitial H_2 in germanium by Raman scattering and ab initio calculations. *Phys. Rev. B* **72,** 153201, 2005.
92. R.C. Newman, R.E. Pritchard, J.H. Tucker, and E.C. Lightowlers, The dipole moments of H_2, HD and D_2 molecules and their concentrations in silicon. *Physica B* **273–274,** 164, 1999.
93. D.J. Backlund and S.K. Estreicher, Structural, electrical, and vibrational properties of Ti-H and Ni-H complexes in Si. *Phys. Rev. B* **82**, 155208, 2010.
94. J. Weber, in *Hydrogen in Semiconductors and Metals*, N. H. Nickel, W. B. Jackson, R. C. Bowman, and R. Leisure (eds.), *MRS Symposia Proceedings* **513** (Materials Research Society, Warrendale, PA, 1998), p. 345.
95. J.-U. Sachsc, E.Ö. Svcinbjörnsson, W. Jost, J. Weber, and H. Lemke, Electrical properties of platinum-hydrogen complexes in silicon. *Phys. Rev*. B **55**, 16176, 1997.
96. J.-U. Sachse, E.Ö. Sveinbjörnsson, N. Yarykin, and J. Weber, Similarities in the electrical properties of transition metal–hydrogen complexes in silicon. *Mater. Sci. Eng. B* **58,** 134, 1999.
97. M. Shiraishi, J.-U. Sachse, H. Lemke, and J. Weber, DLTS analysis of nickel–hydrogen complex defects in silicon. *Mater. Sci. Eng. B* **58**, 130, 1999.
98. J.-U. Sachse, J. Weber, and E. Ö. Sveinbjörnsson, Hydrogen-atom number in platinum-hydrogen complexes in silicon. *Phys. Rev. B* **60**, 1474, 1999.
99. L. Korpas, J.W. Corbett, S.K. Estreicher, Multiple trapping of hydrogen at boron and phosphorus in silicon. *Phys. Rev. B* **46**, 12365, 1992.
100. N.M. Johnson, F.A. Ponce, R.A. Street, and R.J. Nemanich, Defects in single-crystal silicon induced by hydrogenation. *Phys. Rev. B* **35**, 4166, 1987.
101. P. Deák, L.C. Snyder, and J.W. Corbett, State and motion of hydrogen in crystalline silicon. *Phys. Rev. B* **37**, 6887, 1988.
102. K.J. Chang and D.J. Chadi, Diatomic-hydrogen-complex diffusion and self-trapping in crystalline silicon. *Phys. Rev. Lett.* **62**, 937, 1989.
103. J.D. Holbech, B. Bech Nielsen, R. Jones, P. Sitch, and S. Öberg, H_2^* defect in crystalline silicon. *Phys. Rev. Lett.* **71**, 875, 1993.
104. M. Budde, B. Bech Nielsen, P. Leary, J. Goss, R. Jones, P.R. Briddon, S. Öberg, and S.J. Breuer, Identification of the hydrogen-saturated self-interstitials in silicon and germanium. *Phys. Rev. B* **57**, 4397, 1998.
105. S.K. Estreicher, J.L. Hastings, and P.A. Fedders, Defect-induced dissociation of H_2 in silicon. *Phys. Rev. B* **57**, R12663, 1998.
106. S.K. Estreicher, J.L. Hastings, and P.A. Fedders, Radiation-induced formation of H_2^* in silicon. *Phys. Rev. Lett.* **82**, 815, 1999.
107. C.T. Sah, J.Y.C. Sun, and J.J.T. Tzou, Deactivation of the boron acceptor in silicon by hydrogen. *Appl. Phys. Lett.* **43**, 204, 1983; C.T. Sah, J.Y.C. Sun, and J.J.T. Tzou, Deactivation of group III acceptors in silicon during keV electron irradiation. *Appl. Phys. Lett.* **43**, 962, 1983.
108. J. Pankove, D.E. Carlson, J.E. Berkeyheiser, and R.O. Wance, Neutralization of shallow acceptor levels in silicon by atomic hydrogen. *Phys. Rev. Lett.* **51**, 2224, 1983.
109. S.T. Pantelides, Effect of hydrogen on shallow dopants in crystalline silicon. *Appl. Phys. Lett.* **50**, 995, 1987.
110. N.M. Johnson, C. Herring, and D.J. Chadi, Interstitial hydrogen and neutralization of shallow-donor impurities in single-crystal silicon. *Phys. Rev. Lett.* **56**, 769, 1986.
111. A.J. Tavendale, S.J. Pearton, and A.A. Williams, Evidence for the existence of a negatively charged hydrogen species in plasma-treated *n*-type Si. *Appl. Phys. Lett.* **56**, 949, 1990.
112. J. Zhu, N.M. Johnson, and C. Herring, Negative-charge state of hydrogen in silicon. *Phys. Rev. B* **41**, 12354, 1990.

113. E.E. Haller, Hydrogen in crystalline semiconductors. *Semicond. Sci. Technol.* **6**, 73, 1991.
114. E.E. Haller, Stability of H-dopant pairs in Ge, in *Handbook on Semiconductors*, S. Mahajan (ed.) vol. 3 (Elsevier, Amsterdam, 1994), Chapt. 15.
115. Vl. Kolkovsky, S. Klemm, and J. Weber, Dissociation energies of P– and Sb–hydrogen-related complexes in n-type Ge. *Semicond. Sci. Technol.* **27**, 125005, 2012.
116. S.J. Pearton, J.W. Corbett, and T.S. Shi, Hydrogen in crystalline semiconductors. *Appl. Phys. A* **43**, 153, 1987.
117. P.J.H. Denteneer, C.G. Van de Walle, and S.T. Pantelides, Structure and properties of hydrogen-impurity pairs in elemental semiconductors. *Phys. Rev. Lett.* **62**, 1884, 1989.
118. Dj.M. Maric, P.F. Meier, and S.K. Estreicher, {H,B}, {H,C}, and {H,Si} pairs in silicon and germanium. *Phys Rev. B* **47**, 3620, 1993.
119. J. Bollmann, R. Endler, V.T. Dung, and J. Weber, Hydrogen ion drift in Sb-doped Ge Schottky diodes. *Physica B* **404**, 5099, 2009.
120. Vl. Kolkovsky, S. Klemm, M. Allardt, and J. Weber, Detrimental effects of atomic hydrogen on the formation of Schottky barriers in *p*-type Ge. *Semicond. Sci. Technol.* **28**, 025007, 2013.
121. B. Bech Nielsen, L. Hoffman, and M. Budde, Si-H stretch modes of hydrogen-vacancy defects in silicon. *Mat. Sci. Eng. B* **36**, 259, 1996.
122. H.J. Stein, Bonding and thermal stability of implanted hydrogen in silicon. *J. Electr. Mat.* **4**, 159, 1975.
123. R. Jones, B.J. Coomer, J.P. Goss, B. Hourahine, and A. Resende, The interaction of hydrogen with deep level defects in silicon. *Sol. St. Phenom.* **71**, 173, 2000.
124. M. Budde, B. Bech Nielsen, R. Jones, J. Goss, and S. Öberg, Local modes of the H_2^* dimer in germanium. *Phys. Rev. B* **54**, 5485, 1996.
125. M. Budde, B. Bech Nielsen, J.C. Keay, and L.C. Feldman, Vacancy–hydrogen complexes in group-IV semiconductors. *Physica B* **273–274**, 208, 1999.
126. E.E. Haller, Hydrogen-related phenomena in crystalline Ge. *Festkorperpröbleme XXVI/ Adv. Solid State Phys.* **26**, 203, 1986.
127. E.E. Haller, G.S. Hubbard, W.L. Hansen, and A. Seeger, Hydrogen in Ge. *Inst. Phys. Conf. Ser.* **31**, 309, 1977.
128. M. Bruel, The history, physics and applications of the smart cut process. *MRS Bullet.* **23**, 35, 1998.
129. J. Lauwaert, M.L. David, M.F. Beaufort, E. Simoen, D. Depla, and P. Clauws, P. Hydrogen-plasma-induced plate-like cavity clusters in single-crystalline germanium. *Mat. Sci. Semicond Proc.* **9**, 571, 2006.
130. A.W.R. Leitch, V. Alex, and J. Weber, H_2 molecules in c-Si after hydrogen plasma treatment. *Sol. State Commun.* **105**, 215, 1998.
131. M. Hiller, E.V. Lavrov, and J. Weber, Hydrogen-induced platelets in Ge determined by Raman scattering. *Phys. Rev. B* **71**, 045208, 2005.
132. B.P. Stoicheff, High resolution Raman spectroscopy of gases: IX. spectra of H_2, HD, and D_2. *Can. J. Phys.* **35**, 730, 1957.
133. E.V. Lavrov and J. Weber, Evolution of hydrogen platelets in silicon determined by polarized Raman spectroscopy. *Phys. Rev. Lett.* 87, 185502, 2001.
134. N.H. Nickel, G.B. Anderson, N.M. Johnson, and J. Walker, Nucleation of hydrogen-induced platelets in silicon. *Phys. Rev. B* **62**, 8012, 2000.
135. M. Hiller, E.V. Lavrov, and J. Weber, Raman scattering study of H_2 trapped within {111}-oriented platelets in Si. *Phys. Rev. B* **80**, 045306, 2009.
136. N. Rochat, A. Tauzin, F. Mazen, and L. Clavelier, Ge and Si: A comparative study of hydrogenated interstitial and vacancy defects by IR spectroscopy. *Electrochem. Sol. St. Lett.* **13**, G40, 2010.

137. M.L. David, F. Pailloux, D. Babonneau, M. Drouet, and J.F. Barbot, The effect of the substrate temperature on extended defects created by hydrogen implantation in Ge. *J. Appl. Phys.* **102**, 096101, 2007.
138. J.M. Zahler, I. Fontcuberta, A. Morral, M.J. Griggs, H.A. Atwater, and Y. Chabal, Role of hydrogen in hydrogen-induced layer exfoliation of germanium. *Phys. Rev. B* **75**, 035309, 2007.
139. S.J. Pearton, in *Hydrogen in Crystalline Semiconductors*, See Ref. 2, page 65.
140. S.J. Pearton and A.J. Tavendale, Hydrogen passivation of gold-related deep levels in silicon. *Phys. Rev. B* **26**, 7105, 1982.
141. A. Stesmans, Dissociation kinetics of hydrogen-passivated Pb defects at the (111)Si/SiO_2 interface. *Phys. Rev. B* **61**, 8393, 2000 and Interaction of Pb defects at the (111) Si/SiO_2 interface with molecular hydrogen: Simultaneous action of passivation and dissociation. *J. Appl. Phys.* **88,** 489, 2000.
142. J. Weber, Hydrogen in silicon: Evidence for transition-metal hydrogen complexes and hydrogen molecules. *Proc. 24th Int. Conf. on the Physics of Semiconductors*, D. Gershoni (ed.) (World Scientific, Singapore, 1999), p. 209.
143. S.K. Estreicher, D. West, and P. Ordejón, Copper-defect and copper-impurity interactions in Si. *Solid State Phenom.* **82–84**, 341, 2002.
144. D. West, S.K. Estreicher, S. Knack, and J. Weber, Copper interactions with H, O, and the self-interstitial in silicon. *Phys. Rev. B* **68**, 035210, 2003.
145. N. Gonzalez Szwacki, M. Sanati, and S.K. Estreicher, Two FeH pairs in n-type Si and their implications: A theoretical study. *Phys. Rev. B* **78**, 113202, 2008.
146. J.-U. Sachse, E.O. Sveinbjornsson, W. Jost, J. Weber, H. Lemke, New interpretation of the dominant recombination center in platinum doped silicon. *Appl. Phys. Lett.* **70**, 1584, 1997.
147. S.J. Uftring, M. Stavola, P.M. Williams, and G.D. Watkins, Microscopic structure and multiple charge states of a PtH2 complex in Si. *Phys. Rev. B* **51**, 9612, 1995.
148. M. Höhne, U. Juda, Y.V. Martynov, T. Gregorkiewicz, C.A.J. Ammerlaan, and L.S. Vlasenko, EPR spectroscopy of platinum-hydrogen complexes in silcon. *Phys. Rev. B* **49**, 13423, 1994.
149. M.J. Evans, M. Stavola, M.G. Weinstein, and S.J. Uftring, Vibrational spectroscopy of defect complexes containing Au and H in Si. *Mater. Sci. Eng. B* **58**, 118, 1999.
150. A. Streitwieser, Jr. and C.H. Heathcock, *Introduction to Inorganic Chemistry*, 3rd ed. (McMillan, New York, 1985), p. 1089 ff.
151. T. Sadoh, A. Tsukamoto, A. Baba, D. Bai, A. Kenjo, T. Tsurushima, H. Mori, and H. Nakashima, Deep level of iron-hydrogen complex in silicon. *J. Appl. Phys.* **82**, 3828, 1992.
152. M. Sanati, N. Gonzalez Szwacki, and S.K. Estreicher, Interstitial Fe in silicon, its interactions with H and shallow dopants. *Phys. Rev. B* **76,** 125204, 2007.
153. O.V. Feklisova, A.L. Parakhonsky, E.B. Yakimov, and J.Weber, Dissociation of iron-related centers in Si stimulated by hydrogen. *J. Mat. Sci. Eng. B* **71**, 268, 2000.
154. C.K. Tang, L. Vines, B.G. Svensson, and E.V. Monakhov, Hydrogen-induced dissociation of the Fe-B pair in boron-doped P-type silicon. *Sol. Stat. Phenom.* **178–179**, 183, 2011.
155. P. Karzel, A. Frey, S. Fritz, and G. Hahn, Influence of hydrogen on interstitial iron concentration in multicrystalline silicon during annealing steps. *J. Appl. Phys.* **113**, 114903, 2013.
156. J.M. Kahn, L.M. Falicov, and E.E. Haller, Isotope-induced symmetry change in dynamic semiconductor defects. *Phys. Rev. Lett.* **57**, 2077, 1986.
157. B. Joós, E.E. Haller, and L.M. Falicov, Donor complex with tunneling hydrogen in pure germanium. *Phys. Rev. B* **22**, 832, 1980.
158. K. Muro and A.J. Sievers, Proton tunneling with millielectrovolt energies at the Be-H acceptor complex in silicon. *Phys. Rev. Lett.* **57**, 897, 1986.

159. S.J. Pearton, Hydrogen passivation of copper-related defects in germanium. *Appl. Phys. Lett.* **40**, 253, 1982.
160. E.E. Haller, B. Joós, and L.M. Falicov, Acceptor complexes in germanium: Systems with tunneling hydrogen. *Phys. Rev. B* **21**, 4729, 1980.
161. J.M. Kahn, R.E. McMurray, E.E. Haller, and L.M. Falicov, Trigonal hydrogen-related acceptor complexes in germanium. *Phys. Rev. B* **36**, 8001, 1987.
162. A.L. Endrös, Charge-state-dependent hydrogen-carbon-related deep donor in crystalline silicon. *Phys. Rev. Lett.* **63**, 70, 1989.
163. A.L. Endrös, W. Krühler, and F. Koch, Electronic properties of the hydrogen–carbon complex in crystalline silicon. *J. Appl. Phys.* **72**, 2264, 1992.
164. C. Peng, H. Zhang, M. Stavola, W.B. Fowler, B. Esham, S.K. Estreicher, A. Docaj, L. Carnel, and M. Seacrist, Microscopic structure of a VH_4 center trapped by C in Si. *Phys. Rev. B* **84**, 195205, 2011.
165. B.D. Patterson, A. Bosshard, U. Straumann, P. Truöl, A. Wüest, and Th. Wichert, Positron blocking from muon decay in silicon. *Phys. Rev. Lett.* **52**, 938, 1984.
166. W. Kaiser, H.L. Frisch, and H. Reiss, Mechanism of the formation of donor states in heat-treated silicon. *Phys. Rev.* **112**, 1546, 1958.
167. U. Gösele and T.Y. Tan, Oxygen diffusion and thermal donor formation in silicon. *Appl. Phys. A* **28**, 79, 1982.
168. R.C. Newman, Thermal donors in silicon: Oxygen clusters or self-interstitial aggregates. *J. Phys. C* **18**, L967, 1985.
169. F. Shimura (ed.), *Oxygen in Silicon*, Semiconductors and Semimetals, vol. 42 (Academic, Boston, 1994).
170. R. Jones (ed.) *Early Stages of Oxygen Precipitation in Silicon* (Kluwer, Dordrecht, 1996).
171. S.K. Estreicher, Interstitial O in Si and its interactions with H. *Phys. Rev. B* **41**, 9886, 1990.
172. R. Jones, S. Oberg, and A. Umerski, Interaction of hydrogen with impurities in semiconductors. *Mater. Sci. Forum* **83–87**, 551, 1992.
173. S.K. Estreicher, Y.K. Park, and P.A. Fedders, Hydrogen–oxygen interactions in silicon, in *Early Stages of Oxygen Precipitation in Si*, R. Jones (ed.) (Kluwer, Dordrecht, 1996), p. 179.
174. M. Ramamoorthy and S.T. Pantelides, Enhanced modes of oxygen diffusion in silicon. *Sol. St. Commun.* **106**, 243, 1998.
175. R.B. Capaz, L.V.C. Assali, L.C. Kimerling, K. Cho, and J.D. Joannopoulos, Mechanism for hydrogen-enhanced oxygen diffusion in silicon. *Phys. Rev. B* **59**, 4898, 1999.
176. C. Peng, H. Zhang, M. Stavola, V. Yelundur, A. Rohatgi, L. Carnel, M. Seacrist, and J. Kalejs, Interaction of hydrogen with carbon in multicrystalline Si solar-cell materials. *J. Appl. Phys.* **109**, 053517, 2011.
177. The solubility of C in Si is given in, G. Davies and R.C. Newman, in *Handbook on Semiconductors*, T.S. Moss (ed.) vol. 3b (Elsevier, Amsterdam, 1994), p. 1557.
178. J.R. Byberg, B. Bech Nielsen, M. Fanciulli, S.K. Estreicher, and P.A. Fedders, Dimer of substitutional carbon in silicon studied by EPR and ab initio methods. *Phys. Rev. B* **61**, 12939, 2000.
179. E.V. Lavrov, L. Hoffmann, B. Bech Nielsen, B. Hourahine, R. Jones, S. Öberg, and P.R. Briddon, Combined infrared absorption and modeling study of a dicarbon-dihydrogen defect in silicon. *Phys. Rev. B* **62,** 12859, 2000.
180. B. Pajot, B. Clerjaud, and Z.-J. Xu, High-frequency hydrogen-related infrared modes in silicon grown in a hydrogen atmosphere. *Phys. Rev. B* **59**, 7500, 1999
181. N. Fukata and M. Suezawa, Annealing behavior of hydrogen-defect complexes in carbon-doped Si quenched in hydrogen atmosphere. *J. Appl. Phys.* **87**, 8361, 2000.
182. S.K. Estreicher, T.M. Gibbons, By. Kang, and M.B. Bebek, Phonons and defects in semiconductors and nanostructures: phonon trapping, phonon scattering, and heat flow at heterojunctions. *J. Appl. Phys.* **115**, 012012, 2014.

183. G. Lüpke, N.H. Tolk, and L.C. Feldman, Vibrational lifetimes of hydrogen in silicon. *J. Appl. Phys.* **93**, 2317, 2003.
184. A. Nitzan and J. Jortner, Vibrational relaxation of a molecule in a dense medium. *Molec. Phys.* **25**, 713, 1973; A. Nitzan, S. Mukamel, and J. Jortner, Some features of vibrational relaxation of a diatomic molecule in a dense medium. *J. Chem. Phys.* **60**, 3929, 1974.
185. S.A. Egorov and J. L. Skinner, On the theory of multiphonon relaxation rates in solids. *J. Chem. Phys.* **103**, 1533, 1995.
186. D. West and S.K. Estreicher, Isotope dependence of the vibrational lifetimes of light impurities in Si calculated from first-principles. *Phys. Rev. B* **75**, 075206, 2007.
187. B. Sun, G.A. Shi, S.V.S. Nageswara, M. Stavola, N.H. Tolk, S.K. Dixit, L.C. Feldman, and G. Lüpke, Vibrational lifetimes and frequency-gap law of hydrogen bending modes in semiconductors. *Phys. Rev. Lett.* **96**, 035501, 2006; M. Budde, G. Lüpke, E. Chen, X. Zhang, N.H. Tolk, L.C. Feldman, E. Tarhan, A.K. Ramdas, and M. Stavola, Lifetimes of hydrogen and deuterium related vibrational modes in silicon. *Phys. Rev. Lett.* **87**, 145501, 2001.
188. By. Kang and S.K. Estreicher, Thermal conductivity of Si nanowires: A first-principles analysis of the role of defects. *Phys. Rev. B* **89**, 155409, 2014.
189. S.K. Estreicher, T.M. Gibbons, and M.B. Bebek, Heat flow and defects in semiconductors: The physical reason why defects reduce heat flow, and how to control it. *J. Appl. Phys.,* in print.
190. G. Hahn and A. Schönecker, New crystalline silicon ribbon materials for photovoltaics. *J. Phys. Condens. Matter* **16**, R1615, 2004.
191. T. Buonassisi, A.A. Istratov, M.D. Pickett, M. Heuer, J.P. Kalejs, G. Hahn, M.A. Marcus et al. Chemical natures and distributions of metal impurities in multicrystalline silicon materials. *Prog. Photovolt: Res. Appl.* **14**, 513, 2006.
192. The following paper reviews the SiN_x passivation of c-Si solar cells and includes a historical overview: A.G. Aberle, Overview on SiN surface passivation of crystalline silicon solar cells. *Sol. Energy Mater. Sol. Cells* **65**, 239, 2001.
193. F. Duerinckx and J. Szlufcik, Defect passivation of industrial multicrystalline solar cells based on PECVD silicon nitride. *Sol. Energy Mater. Sol. Cells* **72**, 231, 2002.
194. A. Cuevas, M.J. Kerr, and J. Schmidt, Passivation of crystalline silicon using silicon nitride. *Proc. 3rd World Conf. on Photovoltaic Energy Conversion* (IEEE Cat. No. 03CH37497), p. 913, 2003.
195. H.F.W. Dekkers, Study and optimization of dry process technologies for thin crystalline silicon solar cell manufacturing. Dissertation, Catholic University of Leuven, 2008.
196. S.M. Sze, *Physics of Semiconductor Devices*, 2nd ed. (Wiley, New York, 1981).
197. F. Huster, Investigation of the alloying process of screen printed aluminum pastes for the BSF formation on silicon solar cells. *Proc. 20th European Photovoltaic Solar Energy Conference*, Barcelona, 2005, p. 1466.
198. C.H. Seager and D.S. Ginley, Passivation of grain boundaries in polycrystalline silicon. *Appl. Phys. Lett.* **34**, 337, 1979.
199. C.H. Seager, D.J. Sharp, J.K.G. Panitz, R.V. D'Aiello, Passivation of grain boundaries in silicon. *J. Vac. Sci. Technol.* **20**, 430, 1982.
200. J. Hanoka, C.H. Seager, D.J. Sharp, and J.K.G. Panitz, Hydrogen passivation of defects in silicon ribbon grown by the edge-defined film-fed growth process. *Appl. Phys. Lett.* **42**, 618, 1983.
201. J.L. Benton, C.J. Doherty, S.D. Ferris, D.L. Flamm, L.C. Kimerling, and H.J. Leamy, Hydrogen passivation of point defects in silicon. *Appl. Phys. Lett.* **36**, 670, 1980.
202. C. Dubé and J.I. Hanoka, Hydrogen passivation of dislocations in silicon. *Appl. Phys. Lett.* **45**, 1135, 1984.
203. R. Hezel and R. Schörner, Plasma Si nitride—A promising dielectric to achieve high-quality silicon MIS/IL solar cells. *J. Appl. Phys.* **52**, 3076, 1981.

204. M.I. Bertoni, S. Hudelson, B.K. Newman, D.P. Fenning, H.F.W. Dekkers, E. Cornagliotti, A. Zuschlag et al. Influence of defect type on hydrogen passivation efficacy in multicrystalline silicon solar cells. *Prog. Photovolt: Res. Appl.* **19**, 187, 2010.
205. H.F.W. Dekkers, S. DeWolf, G. Agostinelli, J. Szlufcik, T. Pernau, W.M. Arnoldbik, H.D. Goldbach, and R.E.I. Schropp. Investigation on mc-Si bulk passivation using deuterated silicon nitride. *Proc. 3rd World Conf. on Photovoltaic Energy Conversion* (IEEE Cat. No. 03CH37497), p. 983, 2003.
206. Hydrogen could be detected by SIMS in Si samples containing a high concentration of O precipitates. See, G. Hahn, D. Karg, A. Schönecker, A.R. Burgers, R. Ginige, and K. Cherkaoui, Kinetics of hydrogenation and interaction with oxygen in crystalline silicon. *Conf. Rec. 31st IEEE Photovoltaic Specialist Conference* (IEEE Cat. No. 05CH37608), p. 1035, 2005; G. Hahn, A. Schönecker, A.R. Burgers, R. Ginige, K. Cherkaoui, and D. Karg, Hydrogen kinetics in crystalline silicon—PECVD SiN studies in MC and Cz silicon. *Proc. 20th European Photovoltaic Solar Energy Conference*, Barcelona, p. 717, 2005.
207. C. Boehme and G. Lucovsky, H loss mechanism during anneal of silicon nitride: Chemical dissociation. *J. Appl. Phys.* **88**, 6055, 2000; Dissociation reactions of hydrogen in remote plasma-enhanced chemical-vapor-deposition silicon nitride. *J. Vac. Sci. Technol. A* **19**, 2622, 2001.
208. F. Jiang, M. Stavola, A. Rohatgi, D. Kim, J. Holt, H. Atwater, and J. Kalejs, Hydrogenation of Si from SiN_x(H) films: Characterization of H introduced into the Si. *Appl. Phys. Lett.* **83**, 931, 2003.
209. S. Kleekajai, F. Jiang, M. Stavola, V. Yelundur, K. Nakayashiki, A. Rohatgi, G. Hahn, S. Seren, and J. Kalejs, Concentration and penetration depth of H introduced into crystalline Si by hydrogenation methods used to fabricate solar cells. *J. Appl. Phys.* **100**, 093517, 2006.
210. An alternative probe of the hydrogenation of Si from a SiN_x coating has recently been developed. In this method, SIMS is used to detect deuterium that has diffused from a deuterated SiN_x coating through a Si substrate to an amorphous-Si trapping layer deposited on the substrate's back surface. See M. Sheoran, D.S. Kim, A. Rohatgi, H.F.W. Dekkers, G. Beaucarne, M. Young, and S. Asher, Hydrogen diffusion in silicon from plasma-enhanced chemical vapor deposited silicon nitride film at high temperature. *Appl. Phys. Lett.* **92**, 172107, 2008.

8 Point Defect Complexes in Silicon

Edouard V. Monakhov and Bengt G. Svensson

CONTENTS

8.1 Introduction ... 255
8.2 Divacancy (V_2) Center ... 257
 8.2.1 Electronic Properties of V_2 ... 257
 8.2.2 Interaction with Oxygen ... 259
 8.2.3 Unexpected Electronic Properties of V_2O ... 261
 8.2.4 Interaction with Hydrogen ... 261
 8.2.5 Diffusion of V_2 ... 264
 8.2.6 Interaction with Fe ... 265
8.3 Trivacancy (V_3) Center ... 268
 8.3.1 Electronic Properties of V_3 ... 268
 8.3.2 Thermal Stability of V_3 ... 271
8.4 Vacancy–Oxygen Center (VO) ... 271
 8.4.1 Electrical, Optical, and Annealing Properties of VO ... 271
 8.4.2 Interaction with Oxygen and Formation of a Metastable Complex ... 274
 8.4.3 Interaction with Hydrogen ... 274
8.5 Interstitial Defects ... 276
 8.5.1 Carbon-Related Interstitial Defects ... 276
 8.5.2 Boron-Related Complexes ... 279
 8.5.3 Di-Interstitial-Related Complexes ... 281
8.6 Final Remarks ... 281
Acknowledgments ... 282
References ... 283

8.1 INTRODUCTION

Understanding and control of point defect complexes are of decisive importance for the present and future use of silicon in electronics and photovoltaics (PV). Intrinsic defects, formed during crystal growth and/or device processing, interact strongly with common residual impurities such as oxygen (O), carbon (C), hydrogen (H), nitrogen (N), transition metals (e.g., Fe), as well as with dopants (e.g., B and P). Most of these complexes are electrically active with deep states in the bandgap, and in order to minimize their adverse effect on device performance, they need to be controlled at concentrations on the order of ~10^{10} cm^{-3} (or sometimes even lower).

In this chapter, an overview of recent experimental results on the interaction of (i) vacancy complexes (VO, V_2, and V_3) with O, H, and Fe, and (ii) self-interstitials with B, C, O, and Fe, is given. Both *n*-type and *p*-type materials, grown by various techniques, are addressed and the main characterization techniques employed are deep-level transient spectroscopy (DLTS), including minority carrier transient spectroscopy (MCTS), and Fourier transform infrared (FTIR) spectroscopy. The experimental data are corroborated by results from *ab initio* modeling and, for example, the prominent V_3 complex is shown to be a bistable defect appearing in a fourfold coordinated (FFC) configuration or (110) planar structure, where the former is less harmful as recombination center.

Ever since the advent of the semiconductor era in the 1940s, discrete defects in crystalline semiconductor materials have challenged and fascinated generations of scientists. In particular, this holds for silicon, being the workhorse for today's large-volume production of electronics and PV, where the technological impact of such defects has proven to be indisputable. In fact, the research activity on point defects and their complexes has increased considerably during the past years and defect physics in silicon is partly going through a period of renaissance. This is predominantly due to two reasons: (i) the obsession of shrinking device dimensions and reduced cost per unit device area for ultra-large-scale-integrated circuits and the associated increasing demands of high-purity wafers with large-sized diameters, and (ii) high-efficiency and low-cost solar cells in order to enable PV as a truly viable technology for large-volume production of renewable (fossil-free) energy.

Normally, point defect complexes in silicon manifest themselves by affecting the electrical, optical, structural, and mechanical properties of the material. In contrast to the impression possibly given in the preceding paragraph, this is not always necessarily detrimental but can also be taken advantage of in various applications. A prominent example of the so-called defect engineering is the control and optimization of the charge carrier lifetime in switching power devices by irradiation with energetic particles such as MeV electrons and light ions[1] or in-diffusion of Au. Another example is the improvement of mechanical strength of silicon wafers by oxygen impurities, which increases the wafer resistance to warpage and generation of dislocations.[2] A third one is gettering of metallic impurities by irradiation-induced defects.[3]

The concentration of point defect complexes in as-grown monocrystalline silicon wafers is today typically below ~10^{13} cm^{-3} with a few exceptions such as interstitial oxygen (O_i), substitutional carbon (C_s), substitutional nitrogen (N_s), and possibly also hydrogen-related centers. O_i and C_s are the main ones with concentrations reaching ~10^{18} cm^{-3} and ~10^{17} cm^{-3}, respectively, depending on the growth method used. For instance, high values of [O_i] (brackets denote concentration) occur in wafers grown by the Czochralski (Cz) technique where the oxygen originates from partial dissolution of the silica crucible containing the silicon melt at temperatures in excess of 1420°C. In wafers grown by the more costly float-zone (Fz) technique, [O_i] is reduced by two to three orders of magnitude and also [C_s] is usually in the 10^{16} cm^{-3} range or below. Cz wafers are, however, the most commonly used ones for both electronics and PV, primarily because of the cost issue but partly also because of higher mechanical strength. Despite the large variation in residual impurity content between Cz and Fz materials, in both cases, the concentration of intrinsic point defect complexes is usually below the detection limit of most experimental techniques, that is, ~10^{12} cm^{-3} or lower. Here, it should be underlined that defect concentrations even in the 10^{11} cm^{-3} range can have an adverse effect on device performance; a prime example is the so-called light-induced degradation (LID) of Si solar cells commonly attributed to a boron–oxygen complex, although no spectroscopic signal has been identified, with an estimated concentration of ~10^{11} cm^{-3} (see, e.g., Reference 4).

Hence, studies of point defects in silicon require the use of experimental techniques with a high sensitivity and in this regard, the most frequent ones are based on electrical junction spectroscopy such as DLTS, MCTS, and admittance measurements (AS). However, these techniques do not suffice for defect identification and correlation must be made with results obtained by other, less sensitive, methods such as FTIR, photoluminescence (PL), positron annihilation spectroscopy (PAS), and electron paramagnetic resonance (EPR), including optically detected magnetic resonance (ODMR) and electron nuclear double resonance (ENDOR). In order to enable such correlations, the defect concentration needs to be enhanced, which is commonly accomplished via controlled irradiation with swift light particles (MeV electrons or protons) leading to a prevailing generation of point defects rather than extended ones. As will be illustrated later, when discussing some of the prominent defects, the spectroscopic measurements are often performed under the application of uniaxial stress and especially, EPR (and the options thereof) has proven to be powerful with respect to defect identification and determination of detailed atomic structure.

In the following sections, we will critically review recent progress in the understanding of the electronic properties and, to some extent, the optical ones of (i) multivacancy complexes with emphasis on the divacancy (V_2) and the trivacancy (V_3), (ii) multivacancy–oxygen complexes (V_2O, V_2O_2, V_3O), (iii) monovacancy–oxygen complexes (VO and VO_2), (iv) vacancy–hydrogen- and vacancy–iron-related complexes, and (v) interstitial centers related to carbon–oxygen, carbon–hydrogen, boron–oxygen, and di-self-interstitials. Also, the thermal stability and diffusion mechanisms of some of the complexes will be addressed.

8.2 DIVACANCY (V_2) CENTER

8.2.1 Electronic Properties of V_2

V_2 is one of the most common vacancy-related defects in silicon stable at room temperature (RT). V_2 consists of two monovacancies at adjacent atom sites, labeled d and d′ (Figure 8.1). The atomic structure of V_2 is normally described first by introducing an undistorted, neutral complex with six broken bonds and D_{3d} symmetry. Second, owing to Jahn–Teller distortion, the symmetry of the complex is lowered to C_{2h} as atoms b and c pull together to form a "bent" pair bond as do atoms b′ and c′. In a simplified visualization, the electrical activity of V_2 can be related to the remaining unpaired electrons at atoms a and a′. According to this simplistic view, in the positive charge state of V_2, one of the unpaired electrons at a or a′ is taken away. For the single negative state, one of the dangling bonds captures an electron. For the doubly

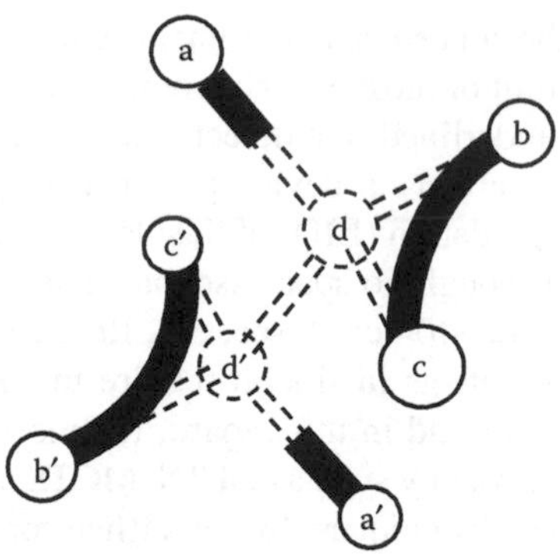

FIGURE 8.1 Simple atomic structure of V_2. Missing atoms d and d′ are indicated with dashed line.

negative state, the second dangling bond captures a second electron. This model was put forward for the first time by Watkins and Corbett in the 1960s.[5]

Only the positive and singly negative charge states of V_2 have unpaired electrons and can be directly observed by EPR. Albeit, already from the first reports,[5] it was concluded through indirect evidence that V_2 can exist in four different charge states (+, 0, –, and 2–), it has been confirmed later by a variety of experimental techniques.[6] However, some controversy has existed about the exact positions in the bandgap of the charge state transitions; in a level scheme deduced from the early EPR studies, the transition from $V_2(-)$ to $V_2(2-)$ was estimated to occur at ~0.4 eV below the conduction band edge (E_C) while DLTS studies yielded positions of ~E_C–0.23 eV, ~E_C–0.42 eV and ~E_V+0.19 eV for $V_2(2-)$ to $V_2(-)$, $V_2(-)$ to $V_2(0)$, and $V_2(0)$ to $V_2(+)$, respectively.[6–8] As mentioned above, V_2 is subject to Jahn–Teller distortion and at low temperatures (≤20 K) where the EPR measurements were performed, the center is frozen in one of the three equivalent configurations corresponding to the C_{2h} state (unpaired electrons accommodated in the "long bond" a–a′, b–b′, or c–c′ in Figure 8.1). At high enough temperatures (≥30 K), a thermally activated reorientation (electronic bond switching) between the three equivalent distortion directions occurs and eventually the jump rate becomes so high that the center does not relax in the low-symmetry configurations; a motional averaged state with the effective symmetry of D_{3d} appears and V_2 becomes a six-Si-atom center.[5,9] As pointed out in Reference 10, the DLTS peaks associated with the different charge states of V_2 occur at high temperatures where the reorientation time for bond switching is several orders of magnitude smaller than the time for carrier emission from the V_2 levels. Thus, the energy level positions obtained by DLTS originate from the motional averaged D_{3d} state and not the distorted C_{2h} one (detected by EPR). In fact, both low-temperature EPR[11] and FTIR[12] measurements from the late 1970s and 1980s, respectively, do not provide even indirect evidence of a $V_2(2-)$ state positioned between ~0.40 and ~0.17 eV below E_C.

Finally, a matter of concern for the assignment of the $E_C - 0.23$ eV level with $V_2(2-)$ has been its large apparent electron capture cross section (~10^{-15}–10^{-16} cm^2).[7,13] This is about 4–5 orders of magnitude higher than anticipated for a repulsive and highly localized single minus charge site (~10^{-19}–10^{-21} cm^2). However, detailed studies by

Hallén et al.[14] of the electron (and hole) capture coefficients and their temperature dependence for deep level defects unveiled a coefficient about 3–4 orders of magnitude smaller for the E_C–0.23 eV level than that for the E_C–0.42 eV level but with a relatively large temperature dependence and high entropy factor. This indicates that the electron capture by $V_2(-)$ to $V_2(2-)$ is most likely governed by a multiphonon process and invokes a large increase in vibrational entropy and/or electronic degeneracy between the two charge states.

8.2.2 Interaction with Oxygen

Despite extensive investigations, the annealing kinetics and the interaction of V_2 with impurities have not been fully understood. The source of controversy has been a difference in annealing behavior observed by different techniques. The early EPR investigations by Watkins and Corbett[5] and FTIR studies by Cheng et al.[15] have led to the conclusion that V_2 anneals via two competing routes: diffusion and dissociation. It has been established that V_2 anneals with a lower activation energy in oxygen-rich Cz Si as compared to that in oxygen-lean Fz Si. The difference was attributed to the different annealing routes: (i) diffusion and interaction with oxygen in Cz Si, and (ii) dissociation in Fz Si. Tentative annealing studies indicated the activation energy for V_2 diffusion to be around 1.3 eV and that for V_2 dissociation to be around 1.9 eV. Later, it has been shown that annealing of V_2 in Cz Si gives rise to another center identified as V_2O.[16]

In contrast, DLTS studies reveal a more complex picture. In some studies, it was found that V_2 annealed with the same activation energy in Cz and Fz Si.[17] Moreover, annealing of V_2 in Cz Si was not accompanied by the formation of new levels related to V_2–oxygen complexes, which contradicted the expectations from EPR studies. Indeed, according to theoretical findings,[18,19] both V_2O and V_2O_2 are expected to be electrically active, with the energy levels in the bandgap close to those of V_2. This is due to the similar atomic structure of these complexes. Figure 8.2 shows the perceived atomic structures of V_2, V_2O, and V_2O_2 in their most stable configurations as seen along the <111> orientation. In these complexes, the oxygen atoms replace the reconstructed Si bonds, while the two dangling bonds remain.

Certain DLTS studies, however, revealed some peculiarities in the V_2 annealing. For instance, Trauwert et al.[20] reported a shift in the position of the $V_2(+)$ peak in *p*-type Cz Si during heat treatment. The position of the corresponding energy level in the bandgap shifted from $E_V + 0.19$ eV to $E_V + 0.24$ eV. One of the speculations

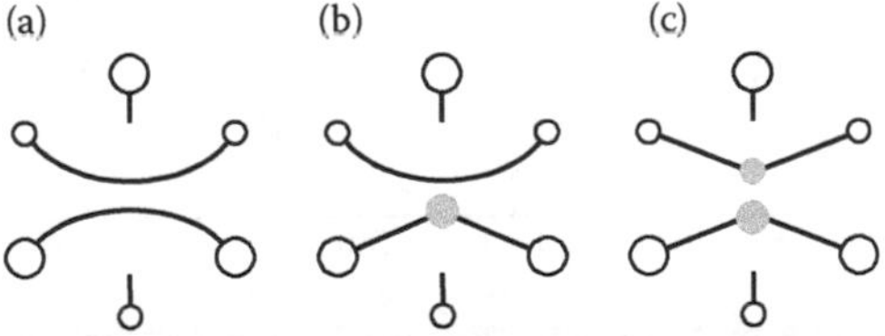

FIGURE 8.2 Atomic structure of V_2 (a), V_2O (b), and V_2O_2 (c) as seen along the <111> axis. The bigger atoms are closer to the viewer. The gray-filled circles indicate oxygen atoms and the open ones silicon atoms.

was the formation of V_2–oxygen complexes. In another DLTS study of defect formation during self-ion implantation at elevated temperatures using *n*-type Cz Si samples, Lalita et al.[21] observed irregularities in the energy position of the DLTS signatures related to $V_2(-)$ and $V_2(2-)$. It was noticed that implantation with low dose ($<10^9$ cm^{-2}) of 5.6 MeV Si ions at 400°C gave rise to DLTS peaks with signatures similar but not identical to those for the two acceptor states of V_2.

Only later, using high-purity diffusion-oxygenated Fz (DOFZ) material, was the formation of a new center during V_2 annealing conclusively observed.[22–24] It was demonstrated that the annealing of the single and the double acceptor states of V_2 was accompanied by an almost one-to-one formation of corresponding acceptor states of the new complex. Moreover, the energy positions of the two acceptor states were close to those for V_2, which indicated a similar electronic structure of the new center. Based on these observations and predictions from theory, the new center, labeled X, was tentatively identified as V_2O. A typical transformation of the two acceptor states of V_2 during annealing is illustrated in Figure 8.3.[23] The transformation from V_2 to V_2O has also been confirmed by Markevich et al. in Cz Si.[25] Later, Mikelsen et al.[26] demonstrated a direct proportionality of the formation rate of X to the concentration of oxygen in the material, which strengthened the identification of X as V_2O. Very recently, a comprehensive study on the transition of *both* donor and acceptor states of V_2 to those of V_2O has been reported.[27] The measurements were performed for *p*-type Cz Si using both conventional DLTS and so-called optical DLTS (ODLTS) where trapping and emission of optically excited minority carriers are studied. It was shown that annealing of the DLTS peak attributed to $V_2(+)$ and the rise of the peak attributed to $V_2O(+)$ are concurrent with the evolution of the peaks attributed to $V_2(-)$ and $V_2O(-)$.

Studies of thermal stability have revealed that V_2O anneals with a first-order kinetics and an activation energy of 2.0 eV and a preexponential factor of 2×10^{13} s^{-1}.[28] Moreover, an increase in the concentration of VO has been detected during the

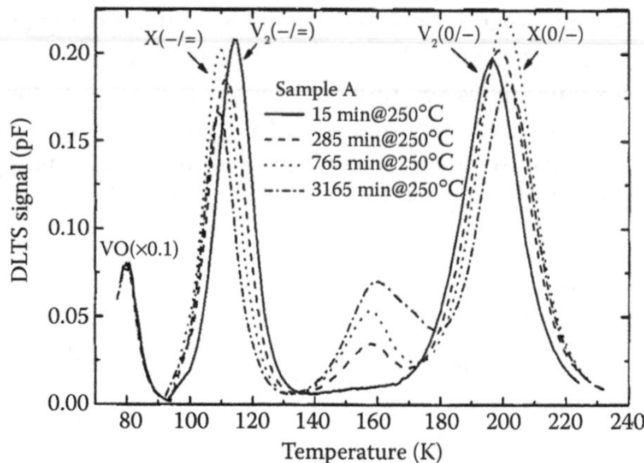

FIGURE 8.3 DLTS spectra for 15 MeV electron-irradiated *n*-type DOFZ-Si after heat treatment at 250°C for different durations. (From G. Alfieri et al. *Phys. Rev. B* 68, 233202, 2003. With permission.)

annealing of V_2O, while no new electrically active centers emerge. These findings have ruled out the diffusion path for the annealing, where V_2O migrates and interacts with O_i forming V_2O_2. This conclusion is also corroborated by the value of the preexponential factor, which is typical for a dissociative process and reflects the maximum lattice vibration (or attempt) frequency. For dissociation of V_2O, two possible routes can be proposed: (i) $V_2O \rightarrow V_2 + O_i$, and (ii) $V_2O \rightarrow V + VO$. Coutinho et al.[29] have shown theoretically that route (ii) has a lower energy barrier and, hence, is more probable.

The results on the annealing mechanism of V_2O have direct implications on the formation mechanisms of V_2O_2. It can be concluded that the most probable route is the interaction of two VO complexes: $VO + VO \rightarrow V_2O_2$. This interaction depends quadratically on the concentration of VO and is unlikely to occur in the case of DLTS studies where the defect concentration is low compared to that of oxygen, strongly promoting VO_2 rather than V_2O_2 formation for the loss of VO centers.

8.2.3 Unexpected Electronic Properties of V_2O

Recent studies by Markevich et al.[30,31] indicate that our understanding of the electronic properties of V_2O may be incomplete. During annealing of V_2 in oxygen-rich p-type samples upon isochronal and isothermal treatments, a correlation between the disappearance of V_2 and the appearance of two hole traps with activation energies of 0.23 and 0.09 eV has been observed. Since the 0.23 eV trap is established to be a donor state of V_2O, that is, $V_2O(+)$, the most reasonable identification of the 0.09 eV trap is another charge state of V_2O. This charge state is most likely a double donor state, that is, $V_2O(2+)$. Figure 8.4 illustrates the transition from $V_2(+)$ to $V_2O(+)$ and the concurrent appearance of $V_2O(2+)$ during the heat treatments.[31] One can also notice peaks labeled as V_3O, which will be discussed in a later section. These findings challenge theoretical predictions, in which the double donor state of V_2O is believed to lie below the valence band edge.[18,19]

Here, it should also be mentioned that Ganagona et al.[32,33] have observed the E_V+0.09 eV level after MeV proton irradiation at elevated temperature (350°C) and after irradiation at RT followed by annealing at 250–300°C using *p*-type samples with a higher boron doping concentration than that of those used by Markevich et al.[30,31] ($\sim 2.2 \times 10^{15}$ cm^{-3} versus $\sim(6–7) \times 10^{14}$ cm^{-3}). In these more highly doped samples, the growth of the E_V + 0.09 eV level does, however, not display a one-to-one correlation with that of $V_2O(+)$ (or the loss of $V_2(+)$) but rather a one-to-five relationship. The reason for this discrepancy between materials with different doping concentrations is not fully understood at the moment but it may be speculated that the hole filling of the E_V + 0.09 eV level during the DLTS measurement sequence is affected/suppressed by a high electric field.

8.2.4 Interaction with Hydrogen

The discovery of the electronic levels of V_2O and the observation of V_2 annealing via interaction with oxygen by DLTS has significantly advanced the understanding of V_2 annealing. Nevertheless, it was still unclear why previous DLTS studies

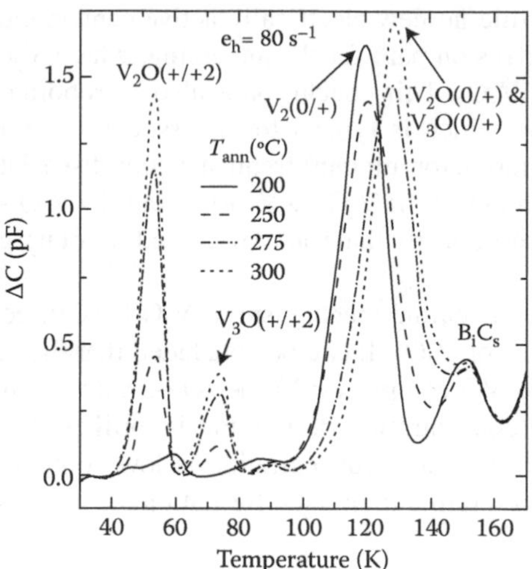

FIGURE 8.4 DLTS spectra for electron-irradiated epitaxial Si after 30-min isochronal annealing at 200°C, 250°C, 275°C, and 300°C. (Reprinted with permission from V.P. Markevich et al. Donor levels of the divacancy-oxygen defect in silicon. *J. Appl. Phys.* 115, 012004. Copyright 2014, American Institute of Physics.)

had not revealed the transformation from V_2 to V_2O. In order to resolve this issue, one can compare the requirements of DLTS and EPR measurements. In EPR, one analyzes bulk material with a relatively high concentration of defects in the range of 10^{17}–10^{19} cm^{-3}. In contrast, DLTS probes the depletion region of Schottky or *p–n* junctions that is relatively close to the surface. In addition, DLTS requires considerably lower defect concentrations, which should not exceed about 10–20% of the doping concentration. In practice, DLTS operates with defect concentrations in the range 10^{10}–10^{15} cm^{-3}. It has been suspected that another impurity, with limited concentration not to affect the EPR analyses, can be responsible for the V_2 annealing observed by previous DLTS studies and here, hydrogen emerges as a main candidate.

This hypothesis motivated comparative studies of hydrogenated and nonhydrogenated DOFZ samples and indeed, a clear effect of hydrogen on the V_2 annealing has been demonstrated.[34] Figure 8.5 shows DLTS spectra for nonhydrogenated (a) and hydrogenated (b) samples. One can see a clear difference in the annealing evolution of the V_2 peaks. While in the nonhydrogenated sample, V_2 exhibits transformation to V_2O, in the hydrogenated sample, V_2 disappears without the formation of V_2O peaks. One can also observe annealing of VO with a corresponding formation of VOH confirming the presence of H in the sample (Figure 8.5b).

Hydrogen is widely known for the ability to passivate electrically active defects, and especially those with dangling bonds. However, it requires six hydrogen atoms to fully passivate V_2, and V_2H_n complexes with $n < 6$ are expected to be electrically active. According to theoretical predictions and considerations based on the atomic structure visualized in Figure 8.6, V_2H should have only one, single-acceptor state with the electronic level close to that of $V_2(-)$. Similarly, the donor state of V_2H is expected to be close to that of $V_2(+)$. The overlap of $V_2H(-)$ and $V_2H(+)$ with $V_2(-)$ and

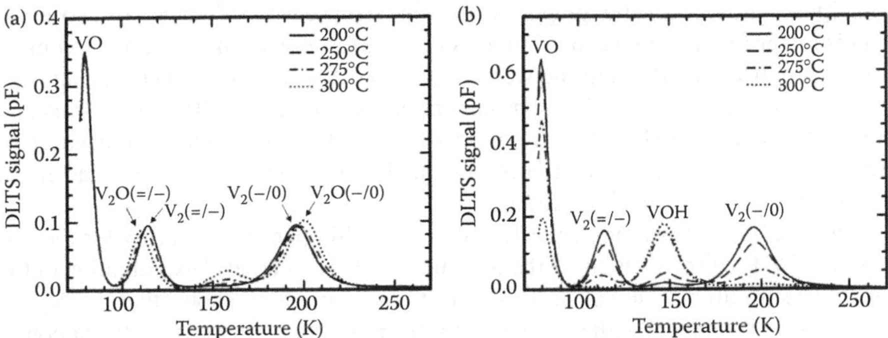

FIGURE 8.5 DLTS spectra of nonhydrogenated (a) and hydrogenated (b) DOFZ-Si after annealing at different temperatures. (From E.V. Monakhov et al. *Phys. Rev. B* 69, 153202, 2004. With permission.)

$V_2(+)$, respectively, complicates greatly the analysis of hydrogen interaction with V_2. Moreover, higher-order V_2H_n complexes with $n = 3$ and 5 are also anticipated to have the same or very similar electronic properties as V_2. However, V_2H_n complexes with $n = 2$ and 4 are expected to have different electronic properties as compared to those with $n = 1$, 3, or 5. This is due to the different nature of the electrical activity that originates from the reconstructed Si bonds in the former case, while the electronic activity for V_2H_n with $n = 1$, 3, or 5 originates from dangling Si bonds (as for V_2).

The experimental observation of the V_2H_n complexes is challenging by several reasons: (i) fast diffusion of hydrogen, (ii) simultaneous presence of both atomic and molecular fractions of hydrogen with similar diffusivities, (iii) considerable depth nonuniformity in hydrogen concentration, and (iv) the overlap of electronic levels of V_2H_n with different n. Presently, we still cannot conclusively identify V_2H_n complexes with DLTS. An indication of the V_2H complex can be found in DLTS spectra for H-implanted samples. It has been observed that the amplitudes of the $V_2(2-)$ and $V_2(-)$ peaks deviate in case of H-implantation from a ratio of one-to-one to more than one-to-two in the implantation peak region with a high content of H.[10] This suggests that the $V_2(-)$ peak has an overlapping contribution from the $V_2H(-)$

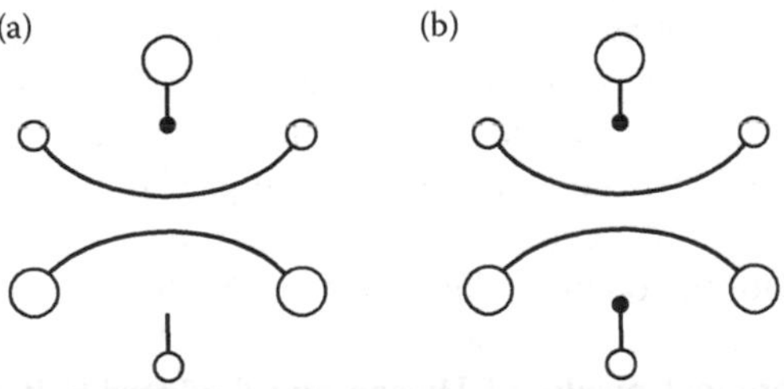

FIGURE 8.6 Anticipated atomic structures of V_2H (a) and V_2H_2 (b) as seen along the <111> axis. The black-filled circles indicate hydrogen atoms and the open ones indicate silicon atoms. The bigger atoms are closer to the viewer.

transition, while no overlapping contribution is present for $V_2(2-)$. This observation is rather unambiguous and should not be confused with the so-called ion mass effect also affecting the ratio between $V_2(2-)$ and $V_2(-)$, where other mechanisms have been discussed.[10,35–37] The ion mass effect is only about 40% for He ions and substantially less for H ions (it is not possible to extract an exact value because of the interaction between V_2 and H but extrapolation of data for heavier (and inert) mass ions gives a value of 10–20%).

The existence of the molecular fraction, H_2, in Si opens a possibility for the formation of V_2H_2 directly, without the intermediate V_2H. A careful examination of the DLTS spectra after different heat treatments in Figure 8.5b shows that the $V_2(2-)$ and $V_2(-)$ peaks anneal with a close one-to-one ratio. This rules out the formation of V_2H, in the configuration predicted by theory, with significant concentration. Indeed, if V_2H was formed, the one-to-one ratio would be violated due to the contribution from $V_2H(-)$ to the $V_2(-)$ peak, as suggested by the results for H-implanted samples. It can then be proposed that V_2 also interacts with hydrogen without the formation of V_2H, and H_2 is the most probable candidate for such an interaction: $V_2 + H_2 \rightarrow V_2H_2$. This hypothesis, however, can be criticized since no new levels to be attributed to V_2H_2 are observed in Figure 8.5b.[34]

Finally, recent efforts by Bleka et al.[38,39] using hydrogenated high-purity samples characterized by both DLTS and MCTS have established correlations between the annealing of the V_2 peaks and the formation of other new ones in the lower part of the bandgap, such as a hole trap at $E_V + 0.23$ eV. The $E_V + 0.23$ eV level has been tentatively assigned to the donor state of V_2H, $V_2H(+)$. However, the final identification of these new levels remains to be confirmed by other techniques.

8.2.5 Diffusion of V_2

The understanding of V_2 annealing in different Si materials has opened up a possibility for detailed and accurate studies on the diffusion of V_2 and hydrogen by DLTS. For instance, by measuring the formation rate of V_2O in materials with known oxygen content, one can readily deduce the diffusivity of V_2. Mikelsen et al.[40] have established that the neutral V_2 diffuses in *n*-type material with an activation energy of 1.30 ± 0.02 eV and a preexponential factor (D^0_{V2}) of $(3 \pm 1.5) \times 10^{-3}$ cm^2/s. Later studies by Ganagona et al.[41] using *p*-type Cz samples have resulted in an almost identical value for the activation energy: 1.31 ± 0.03 eV, and a preexponential factor of $(1.5 \pm 0.7) \times 10^{-3}$ cm^2/s, reflecting the diffusivity of neutral V_2's also in this material (see Figure 8.7).

The ability to measure V_2 diffusion accurately by DLTS can contribute to the discussion on the atomistic mechanism of V_2 diffusion. In the original work, Watkins and Corbett[5] proposed diffusion via the so-called partial dissociation of V_2. In this mechanism, the jump of V_2 occurs via a two-stage process: (i) a jump of one of the vacancies with the formation of a relatively stable V−Si−V configuration and (ii) a subsequent jump of the second vacancy to restore V_2. This model, however, has been challenged by theoretical results of Hwang and Goddard.[42] It was stated that the "split configuration," where the two monovacancies are separated by a Si atom at the lattice site, is not stable. On the contrary, the V−Si−V configuration was found to be the saddle point for V_2 diffusion, which implies that V_2 migrates via a single-step

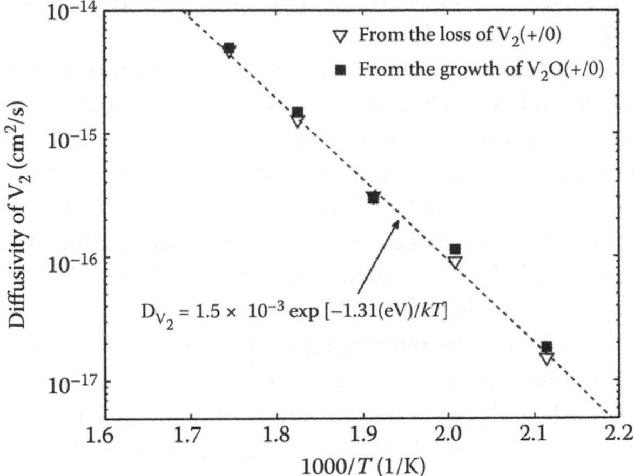

FIGURE 8.7 Absolute diffusivity values for V_2 in the neutral charge state versus the reciprocal absolute temperature. (Reprinted with permission from N. Ganagona et al. Transformation of divacancies to divacancy-oxygen pairs in *p*-type Czochralski-silicon; mechanism of divacancy diffusion. *J. Appl. Phys.* 115, 034514. Copyright 2014, American Institute of Physics.)

process: one-stage hopping of a Si atom along the divacancy with a saddle point in the V–Si–V configuration. According to Swalin,[43] these two mechanisms have a direct impact on the geometrical factor (g) of the prefactor for the diffusivity: $g = 1/8$ for the one-stage process (direct jump) and $g = 1/32$ for the two-stage process with partial dissociation. At the moment, it is difficult to distinguish whether the experimental data prove one or the other of the two mechanisms. Assuming the attempt rate to be equal to the Debye frequency in Si, $\nu = 1.3 \times 10^{13}$ Hz,[24] and $\exp(-\Delta S/k) \approx 1$ (ΔS is the entropy change during the migration step and k is Boltzmann's constant); the theoretical estimate for $D^0_{V_2}$ is 4.8×10^{-3} cm²/s and 1.2×10^{-3} cm²/s for the one-stage and two-stage processes, respectively. Comparing these two estimates for $D^0_{V_2}$ with the value extracted from the experimental data, it is tempting to conclude that the experiment supports the mechanism with partial dissociation as the dominant one for migration of V_2 in the neutral charge state. However, the theoretical values assume the attempt frequency to be equal to the Debye frequency and the entropy factor equal to unity, which can significantly overestimate $D^0_{V_2}$.

8.2.6 Interaction with Fe

Iron is one of the most important metallic impurities in silicon from a technological point of view. Iron is well known for its detrimental effects on the performance of silicon solar cells and integrated circuits. For instance, solar cells based on *p*-type silicon are known to suffer from the so-called LID[4]: a loss in efficiency during the operation under sun illumination. One of the dominant mechanisms of LID is the interaction of Fe with boron dopants. According to this mechanism, the iron–boron pair (FeB) splits due to carrier excitation under illumination to substitutional boron (B_s) and interstitial

Fe (Fe_i), where the latter has an electronic level close to mid-gap and is, therefore, an effective charge carrier lifetime killer.

Theoretical results have recently been reported on the interaction between irradiation-induced defects and Fe.[44] The calculations predict two stable complexes between V_2 and Fe_i. The one with the lowest total energy is labeled as VFeV, where the Fe is situated half-way between two vacancies. This configuration has a single acceptor level at $E_V + 0.38$ eV ($E_C - 0.73$ eV) and a double acceptor level at $E_C - 0.55$ eV. The other configuration is labeled as FeV_2 with the Fe adjacent to V_2, which has one donor level at $E_V + 0.25$ eV and one acceptor level at $E_V + 0.36$ eV ($E_C - 0.75$ eV). The two configurations are depicted in Figure 8.8.

Experimental results on the interaction of V_2 with Fe are scarce and relatively inconclusive. This is mainly due to (i) low concentration of Fe_i, limited by the solid solubility, and (ii) relatively low concentrations of V_2 with DLTS as the prevailing analysis technique used. Thus, V_2–Fe complexes are expected to be observed only as minor peaks in DLTS spectra, and the effect of other impurities, such as hydrogen in particular, should be carefully controlled. For instance, Komarov[45] investigated reactions between irradiation-induced defects with residual impurities in *n*-type silicon. He observed the appearance of a hole trap at 0.184 eV above E_V after annealing at 150°C, which remained stable at 400°C. Tentatively, this defect was assigned

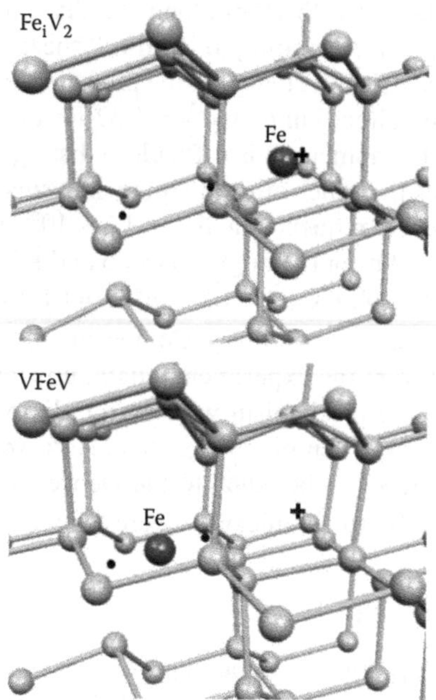

FIGURE 8.8 Atomic structures of Fe_iV_2 and VFeV. The darker atom indicates Fe. The black dots indicate the vacancies, while the black cross indicates a perfect tetrahedral interstitial site. (Reprinted with permission from S.K. Estreicher, M. Sanati, and N. G. Szwacki, Iron in silicon: Interactions with radiation defects, carbon, and oxygen. *Phys. Rev. B* 77, 125214. Copyright 2008 by the American Physical Society.)

to FeV_2, with the basic argument that Fe was present in the samples. The effect of hydrogen and other impurities, however, was not addressed.

Recently, comparative studies of as-grown and Fe-enriched silicon samples have revealed DLTS signatures that are attributed to V_2–Fe-related complexes.[46] In order to carefully control the impurity content, ion implantation was chosen for the introduction of Fe. Both the as-grown and Fe-enriched samples did undergo identical thermal treatments, and the only difference was the Fe-implantation performed at RT for the Fe-enriched ones. Tang et al.[46] revealed the formation of two minor levels at $E_V + 0.25$ eV and $E_V + 0.34$ eV in the Fe-enriched samples, which were absent in the pure, as-grown ones (see Figure 8.9). The former level ($E_V + 0.25$ eV) was

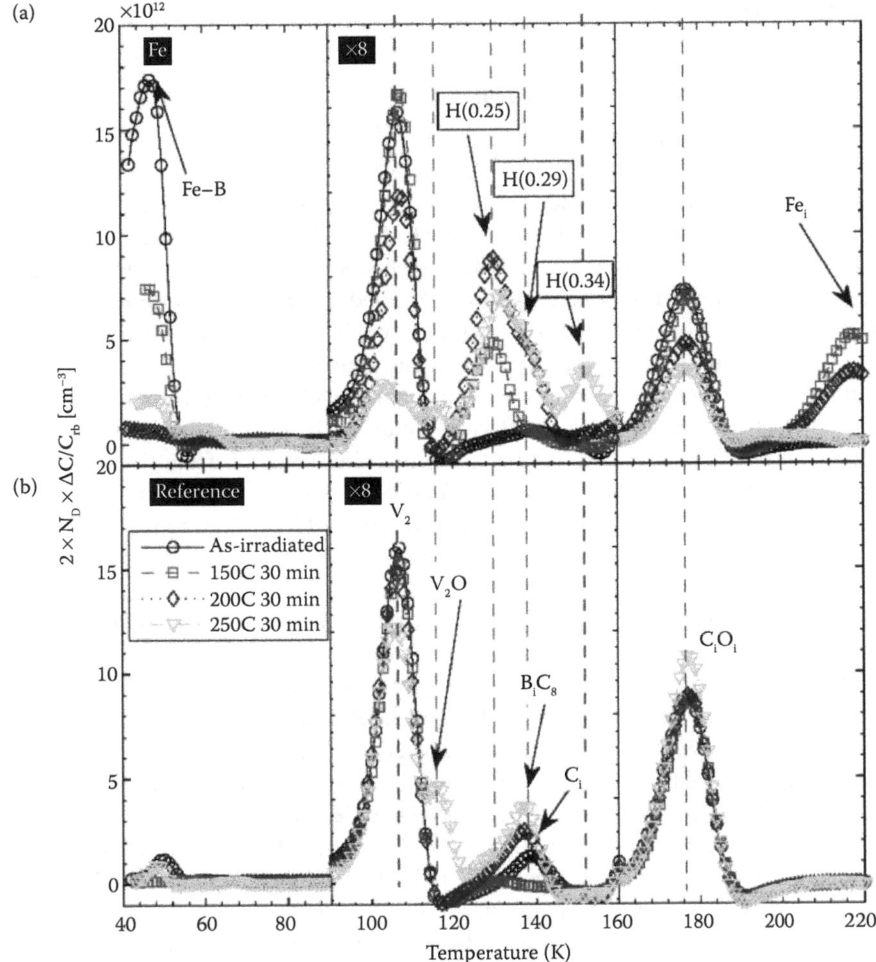

FIGURE 8.9 DLTS spectra for Fz boron-doped Si with (a) and without (b) intentional Fe contamination after 4 MeV electron irradiation and different heat treatments. (Reprinted with permission from C.K. Tang et al. Divacancy-iron complexes in silicon. *J. Appl. Phys.* 113, 044503. Copyright 2013, American Institute of Physics.)

tentatively assigned to the donor state of FeV_2, in accordance with the predications by Estreicher et al.[44] The latter level ($E_V + 0.34$ eV) was attributed to the more stable VFeV complex with the theoretically estimated level position at $E_V + 0.38$ eV for its single acceptor state.[44]

The close association between V_2 and the $E_V + 0.25$ eV and $E_V + 0.34$ eV levels was further substantiated by the fact that the sum of the three peaks ($V_2(+) + E_V + 0.25$ eV + $E_V + 0.34$ eV) remained identical to $V_2(+)$ in the reference sample for annealing temperatures up to 250°C before other major reactions started to occur, such as the formation of V_2O. However, for an unambiguous assignment of the $E_V + 0.25$ eV and $E_V + 0.34$ eV levels to the different V_2–Fe configurations, more work is needed, preferably in combination with uniaxial stress measurements to determine defect symmetries.

8.3 TRIVACANCY (V_3) CENTER

8.3.1 Electronic Properties of V_3

Studies of the electronic properties of higher-order V_n complexes with $n > 2$ are a considerable challenge. First, the concentration of these complexes is expected to be low relative to that of lower-order defects, so it is hard to expect dominant features and signals from such complexes. Second, the higher-order complexes can have multiple stable configurations, which complicate their identification. In the case of DLTS studies, one can expect that the signatures of V_n with $n > 2$ overlap with the signals from $V_2(-)$ and $V_2(+)$. This is intuitively expected from the similarity of the nature of the electrical activity that arises from the dangling bonds of Si atoms. It has been suggested, for instance, that the deviation from the one-to-one ratio between the DLTS peaks for $V_2(2-)$ and $V_2(-)$ in ion-implanted samples, the so-called ion mass effect, is due to the contribution of V_n ($n \geq 3$) to the $V_2(-)$ peak. Although it has been shown later that the ion mass effect can to a large extent be explained by other mechanisms,[35–37] the contribution from V_n could not be ruled out fully. Only relatively recently, signatures that can be conclusively related to V_3 have been established.

Using electron-irradiated high-purity *n*-type magnetic Cz (MCz) and Fz samples, Bleka et al.[47] observed a prominent defect level, labeled as E4 in Figure 8.10, with a position at $E_C - 0.37$ eV and a concentration of ~25% of that of V_2. This peak was found to decrease during storage at RT, and about 50% of the original intensity was lost during 5 weeks. Simultaneously, a correlated decrease in the peak attributed to $V_2(-/0)$ was detected. By subtracting the DLTS spectra after annealing for different durations at RT, another peak, labeled E5 in Figure 8.10, overlapping with $V_2(-)$ was revealed. Moreover, a one-to-one proportionality between the amplitudes of E4 and E5 during the annealing was obtained. Based on such a strong correlation, E4 and E5 were attributed to two charge states of the same complex. This complex was tentatively suggested to be the {110} planar tetravacancy.

Later, it was realized that the E4/E5 peaks originate from V_3. It has been theoretically predicted and experimentally verified that V_3 is a bistable center with two possible configurations (Figure 8.11): {110} planar and fourfold coordinated

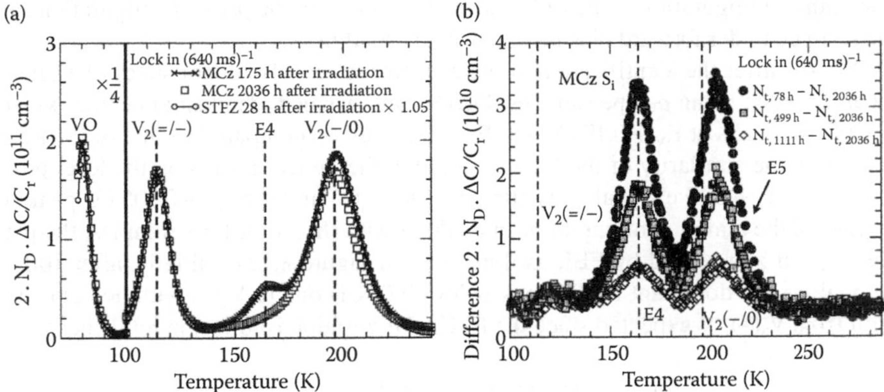

FIGURE 8.10 DLTS spectra for *n*-type MCz and standard Fz Si after irradiation and room temperature annealing (a) and subtracted DLTS spectra after different annealing times (b). Dashed lines indicate positions of the V_2 peaks. (From J. H. Bleka et al. *Phys. Rev. B* 76, 233204, 2007. With permission.)

(FFC).[48] According to the theoretical prediction, V_3 in the planar configuration gives rise to two deep acceptor levels in the upper half of the band gap, while the FFC configuration has trigonal symmetry and one shallow acceptor level close to the conduction band edge. Moreover, the FFC configuration is theoretically found to be the ground state of the neutral charge state of V_3 and is lower in energy by 0.5 eV than the planar one.

Indeed, the results of the DLTS measurements conform closely with the theoretical ones. Upon annealing of the E4/E5 peaks at 125°C for 30 min, a new electron trap emerges positioned at $E_C - 0.075$ eV with a close one-to-one correlation in the amplitude to that of E4/E5.[48] This observation is in excellent accordance with the theory, where the planar configuration is expected to transform into the more stable FFC one. It has been concluded, hence, that V_3 in the planar configuration gives rise to two acceptor levels: $V_3(2-)$ at $E_C - 0.36$ eV (i.e., E4) and $V_3(-)$ at $E_C - 0.46$ eV (E5), while the FFC configuration gives rise to a single acceptor level at $E_C - 0.075$ eV. Moreover, Fleming et al.[37] and Markevich et al.[48] have established that the transformation from

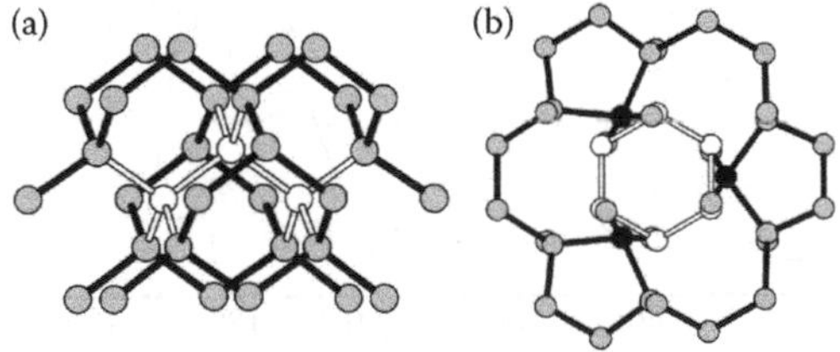

FIGURE 8.11 Atomic structures of {110} planar (a) and fourfold coordinated (b) V_3. The white balls represent vacancies (missing Si atoms). The black atoms represent Si atoms shifted from substitutional sites. (From V.P. Markevich et al. *Phys. Rev. B* 80, 235207, 2009. With permission.)

the planar configuration to the FFC one is reversible, and the planar configuration can be restored under forward bias injection (FBI) conditions.

Shortly after the identification of the acceptor states, also donor states of V_3 have been unveiled using *p*-type samples. DLTS data revealed identical annealing kinetics of two peaks at $E_V + 0.19$ eV and $E_V + 0.105$ eV as compared to that of E4/E5.[49] Based on the similarity of the kinetics and the theoretical results on the level positions, the $E_V + 0.19$ eV peak was attributed to $V_3(+)$ and the $E_V + 0.105$ eV peak to $V_3(2+)$ of the planar configuration. In analogy with that for *n*-type samples, the planar V_3 can be restored by FBI, as illustrated in Figure 8.12a. Initially, after 100°C annealing, the dominant DLTS peak below 160 K is due to $V_2(+)$ with no contribution from $V_3(+)$, as expected since the FFC configuration is lowest in total energy (in

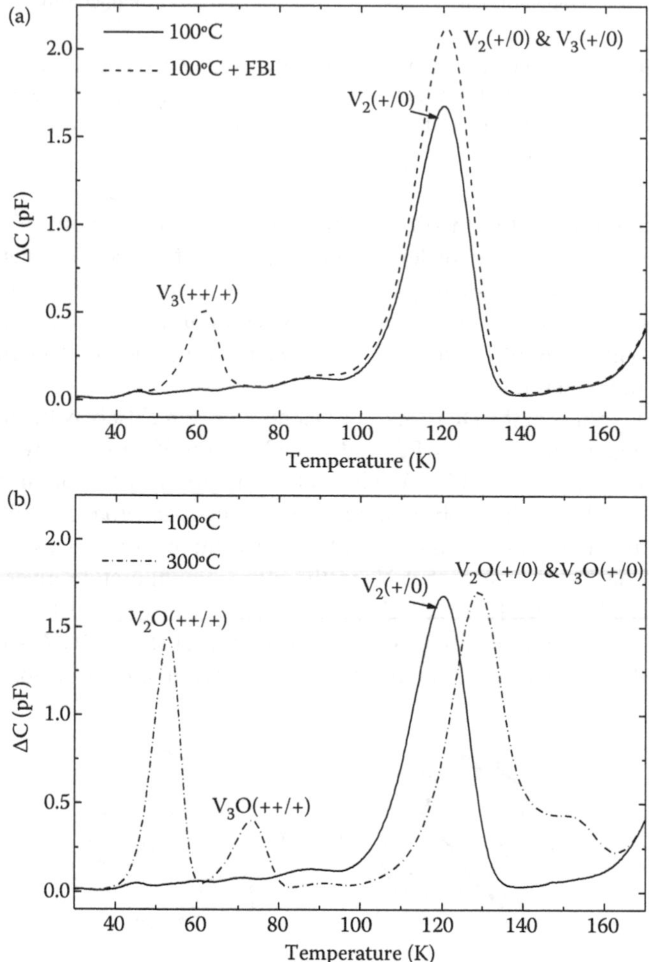

FIGURE 8.12 DLTS spectra for 6 MeV electron-irradiated *p*-type epitaxial samples after heat treatment at 100°C for 30 min and FBI for 10 min at 300 K (a), and after annealing at 100°C for 30 min and annealing at 300°C for 30 min (b).

accordance with the results from n-type samples). However, after FBI at 300 K, a new peak appears at around 60 K, and the one at around 120 K increases in amplitude by a similar amount. This is interpreted as the reverse transformation of FFC to planar V_3 and the appearance of $V_3(+)$, overlapping with $V_2(+)$, and $V_3(2+)$.

In a joint theoretical and experimental effort, the transformation kinetics between the planar and the FFC configurations of V_3 was studied in detail by Coutinho et al.[50] The kinetics obeys first order, and the activation energies for the transformation of the different charge states of V_3 are similar. For instance, the transformation for the neutral charge state of V_3, measured using p^+nn^+ diodes under reverse bias, is governed by an activation energy of 1.16 ± 0.02 eV and a preexponential factor of 2.8×10^{13} s^{-1}; while those for the negative charge state, measured using p^+nn^+ diodes at zero bias, are 1.22 ± 0.02 eV and 1.6×10^{14} s^{-1}, respectively. These results are in close agreement with those obtained from *ab initio* simulations, and in addition, the simulations suggest that the diffusion of V_3 occurs via consecutive FFC-to-planar transformation steps.

8.3.2 Thermal Stability of V_3

Similar to V_2, V_3 can anneal via different routes, such as (i) dissociation, (ii) migration and interaction with O, and (iii) interaction with H. So far, only mechanism (ii) has been investigated in detail. It is established that the annealing of V_3 with concurrent formation of V_3O is governed by a mechanism obeying first-order kinetics having an activation energy of 1.47 ± 0.04 eV, attributed to the migration of V_3 and rather close to that for V_2 diffusion (1.41 ± 0.05 eV).[50] It should be noted that the activation energy found by Coutinho et al.[50] for V_2 diffusion is somewhat higher compared to that found by Mikelsen et al.[40] and Ganagona et al.[41] but almost within the experimental uncertainties. Theoretical estimates give an activation energy of 1.40 eV for diffusion of V_3 in the neutral charge state.[50]

Figure 8.12b illustrates the annealing of V_3 via interaction with O_i and formation of V_3O. After heat treatment at 300°C, V_3 has migrated and formed V_3O with a single donor level overlapping with $V_2O(+)$ and a double donor level at $E_V + 0.11$ eV (around 72 K in the spectrum).

8.4 VACANCY–OXYGEN CENTER (VO)

8.4.1 Electrical, Optical, and Annealing Properties of VO

O_i is the most abundant impurity in essentially all kinds of silicon materials, for example, Cz, MCz, and Fz, and it is a dominant trap for migrating monovacancies, forming the VO center. VO, also commonly known as the A-center, was first reported more than 50 years ago[51–53] and is perhaps the most extensively studied point defect in silicon, especially in materials irradiated at RT where it is almost always the prevailing defect. A variety of techniques have been employed such as EPR, IR, FTIR, DLTS, PAS, and ENDOR, and here, we will provide only a brief overview of the electrical, optical, and annealing properties of VO, and discuss some rather recent results regarding its interaction with O_i.

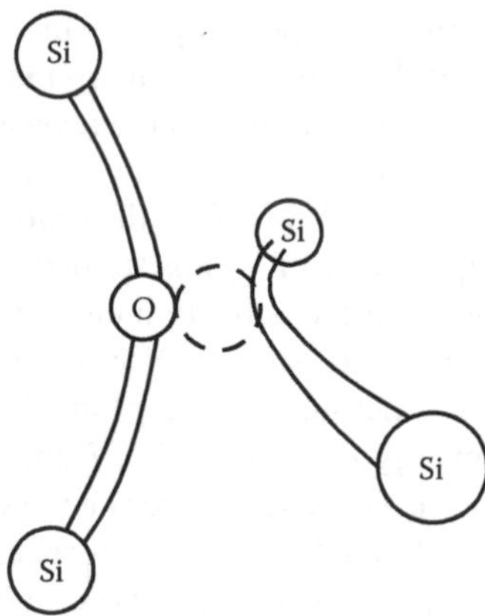

FIGURE 8.13 Atomic structure of the VO center. The dashed circle indicates the substitutional site of the missing Si atom.

The generally agreed atomic configuration of VO is schematically depicted in Figure 8.13; the oxygen atom is slightly displaced from the central tetrahedral substitutional site in a <100> direction and forms strong molecular bonds with only two of the surrounding silicon atoms. The other two silicon neighbor atoms pull together to form a covalent bond where the remaining and additional (accepted) electrons are accommodated. In contrast to O_i, this implies that VO is electrically active and acts as an acceptor. The acceptor level occurs at ~0.17 eV below E_C and when the level is filled, that is, VO is negatively charged, the Si–A center resonance appears in EPR spectra.[53] The defect has C_{2v} point group symmetry. Applying uniaxial stress, the center can be preferentially aligned and by measuring the recovery of this alignment, thermally activated switching of the oxygen atom from one pair of silicon atoms to another in the vacancy can be observed indirectly. Measurements were carried out for VO(0) and an activation energy of ~0.4 eV was obtained for this motion.

At RT, VO gives rise to two local vibrational IR bands at 830 and 877 cm^{-1}, corresponding to the neutral and negative charge state, respectively.[53,54] The vibrational frequencies of VO are substantially lower than for O_i (~1107 cm^{-1}) and reflect the more roomy substitutional environment of the vibrating oxygen atom. The VO center itself is normally stable up to temperatures in excess of ~300°C but can also anneal out at lower temperatures acting as a trap for migrating interstitials released from other less stable complexes.[55] The fast disappearance of VO is particularly pronounced in Cz samples with high concentrations of O_i and C_s. Using FTIR spectroscopy, the annealing kinetics of VO(0) has been studied in some detail for a comprehensive set of MeV electron-irradiated Cz samples.[56]

The annealing of VO can be divided into two components, one fast contribution occurring during the initial stage of the heat treatment and one slow component dominating at the intermediate and final stages of the annealing sequence. As exemplified in Figure 8.14, the relative importance of the first component depends on the electron dose used, that is, the initial concentration of VO, and a second-order

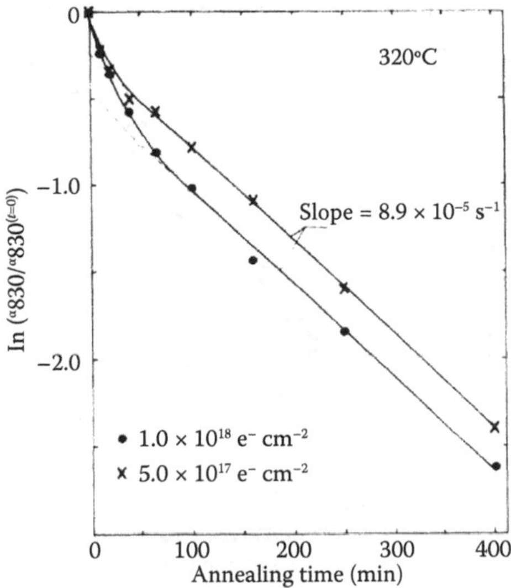

FIGURE 8.14 Logarithm of the absorption coefficient for the 830 cm^{-1} VO band as a function of annealing time at 320°C in two Cz samples irradiated with different doses of 2 MeV electrons. (From B.G. Svensson and J.L. Lindström, *Phys. Rev. B* 34, 8709 1986; B.G. Svensson, J.L. Lindström and J.W. Corbett, *Appl. Phys. Lett.* 47, 841, 1985. With permission.)

reaction is anticipated. Simultaneously with the fast loss of VO centers, a significant increase of [O_i] occurs. These data indicate that an interstitial-related defect of limited concentration relative to that of VO releases I's, which then are captured by VO. This causes the annihilation of VO and a corresponding increase of [O_i]. The second component fulfills first-order kinetics and concurrently the concentration of vacancy–dioxygen pairs (VO_2) grows.[57,58] This is illustrated in Figure 8.15 for annealing at 320°C and where the 889 cm^{-1} absorption band is due to VO_2. However, the second component involves probably more than one process since the proportionality constant between the loss of [VO] and growth of [VO_2] was found to decrease with temperature. The extracted value of the activation energy for the second component was ~2.25 eV, attributed to a mixed process where motion of VO centers and subsequent generation of VO_2 ($VO + O_i \rightarrow VO_2$) constitute a main but not the only contribution. A prime candidate for another contribution to the second component is dissociation ($VO \rightarrow V + O_i$), which fulfills the two most essential criteria for such a contribution: (a) O_i's are released and as a result, [O_i] stays constant despite the generation of VO_2, and (b) the kinetics is first-order. Assuming dissociation as the second contribution to the slow component yielded activation energy values of ~2.0 and ~2.5 eV for the migration and dissociation of VO, respectively. The migration process should then be considered to involve partial dissociation of VO while the dissociation refers to a complete breakup of the complex. Some 20 years later, Mikelsen et al.[28] showed that these values deduced for the diffusion of VO in high-dose electron-irradiated Cz samples also apply for low-dose irradiated and high-purity DOFZ samples studied by DLTS (except for an increase by 25%).

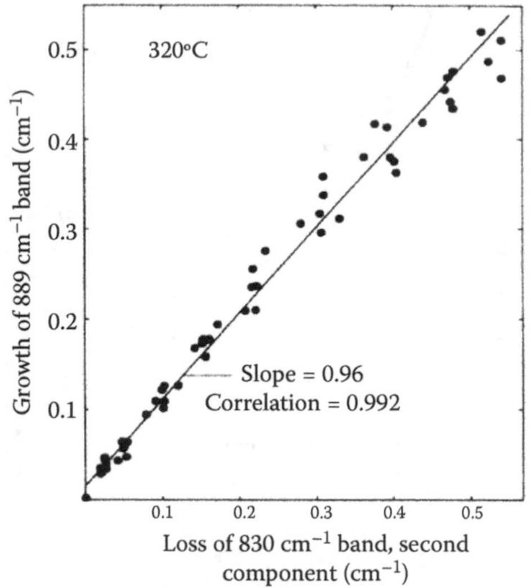

FIGURE 8.15 Growth of VO_2 (889 cm^{-1} band) versus the second component loss of VO (830 cm^{-1} band) during annealing at 320°C. (From B.G. Svensson and J.L. Lindström, *Phys. Rev. B* 34, 8709 1986; B.G. Svensson, J.L. Lindström and J.W. Corbett, *Appl. Phys. Lett.* 47, 841, 1985. With permission.)

8.4.2 Interaction with Oxygen and Formation of a Metastable Complex

With respect to the generation of VO_2 pairs through the interaction between VO and O_i, it is important to point out that O_i can be regarded as practically immobile at temperatures below 350°C. A jumping rate of only 10^{-6} s^{-1} is obtained at 350°C using the well-established diffusion coefficient for O_i.[59] Thus, any contribution from migration and trapping of O_i to the formation of VO_2 pairs is negligible compared to that of VO. VO_2 is electrically inactive in its low-energy configuration (no "unsatisfied" bonds) and not detectable by electrical techniques, that is, VO_2 cannot be monitored at concentrations below the detection limit for local vibrational modes in FTIR spectroscopy (≥10^{14} cm^{-3}). However, rather recently, a metastable configuration of VO_2, labeled VO_2^* and where one of the two O atoms is situated just outside the vacancy as an O_i in a neighboring bond, has been reported by Lindström, Murin, and coworkers.[60] VO_2^* exhibits a shallow acceptor state at E_C–0.06 eV and has a total energy, which is ~0.25 eV higher than that of VO_2. The equilibrium fraction of VO_2^* increases with temperature and after treatment at 350°C followed by rapid quenching to RT, VO_2^* does not revert to VO_2 but remains frozen in. However, the absolute fraction of VO_2^* did not exceed 5–10% of the total and VO_2 was the clearly dominant configuration.

8.4.3 Interaction with Hydrogen

In contrast to the interaction between O_i and VO, considerable DLTS information can be found on the interaction of VO with hydrogen. In fact, owing to a relatively low concentration of hydrogen in Si, limited by the solid solubility, DLTS can perhaps be considered as a particularly suitable technique because of its high sensitivity. Currently, the annealing of VO via the interaction with hydrogen is usually believed to take place in two steps:

$$VO + H \rightarrow VOH,$$

$$VOH + H \rightarrow VOH_2,$$

where VOH is expected to have both acceptor, VOH(–), and donor, VOH(+), states at around $E_C - 0.32$ eV and $E_V + 0.27$ eV, respectively. VOH_2 is considered to be electrically inactive.

Using DLTS, VOH was first identified by Irmscher et al.[61] in an experiment exploring hydrogen-implanted samples, where a level at $E_C - 0.32$ eV was attributed to VOH(–). Several studies have subsequently confirmed this identification. Moreover, the growth and annealing of VOH(0/–) has later also been correlated with the growth and annealing of a hole trap at $E_V + 0.27$ eV. This hole trap has been assigned to VOH(+) and the close one-to-one proportionality between the two levels is illustrated in Figure 8.16 for a number of different Si materials implanted with H and analyzed by both DLTS and MCTS.[62]

Recent studies of hydrogenated and deuterated samples have shown that our understanding of the interaction between VO and H is incomplete. It has been observed, for example, that annealing of VO and V_2 in high-purity *n*-type Fz samples gives rise to several new peaks, where the most prominent one, labeled E1, corresponds to a level at $E_C - 0.37$ eV.[63] Owing to similar amplitude as compared to the V_2 peaks, E1 was first tentatively assigned to V_2D_2. However, it was found later that E1 can, in fact, considerably exceed the amplitude of the V_2 peaks, which rules out an assignment as V_2D_2 or V_2H_2.[38] Moreover, an excellent anticorrelation with the amplitude of VO has been

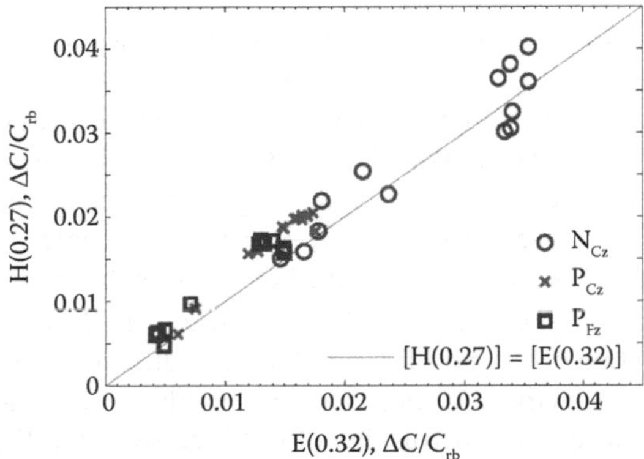

FIGURE 8.16 The amplitude of the $E_V + 0.27$ eV level versus that of the $E_C - 0.32$ eV level in *n*-type Cz, *p*-type Cz, and *p*-type Fz samples implanted with H at RT, as determined by employing both DLTS and MCTS measurements on the same sample. Data for samples in as-implanted stage as well as after annealing up to 300°C are included.

observed for E1: a decrease in [VO] corresponded almost one-to-one with an increase in [E1]. This correlation was also valid along the depth profile of [VO] and [E1]. Detailed kinetics studies at 225°C for prolonged durations revealed that the correlated annealing/formation of VO/E1 was followed by a reverse process: annealing of E1 and regrowth of VO. As the heat treatment continues, VO starts to anneal again and VOH(−) appears. These findings lead to a suggestion that E1 can represent another relatively unstable configuration of VOH, tentatively labeled as VOH*. Such an interpretation, however, needs to be thoroughly examined by further experimental and theoretical studies.

8.5 INTERSTITIAL DEFECTS

Self-interstitials (I) in Si are known to be highly mobile at RT and interact readily with other defects and impurities. According to most recent estimates from a DLTS study by Hallén et al.,[64] the migration energy for I can be as low as 0.065 ± 0.015 eV. Besides the obvious interaction and annihilation with monovacancies, the most prominent interactions are with substitutional carbon (C_s), substitutional boron (B_s), and between the self-interstitials themselves forming multi-interstitial complexes (I_n, $n \geq 2$). In addition, there have been multiple FTIR reports on the formation of self-interstitial–interstitial oxygen (IO_i) pairs (see References 65 and 66 and references therein), although these pairs have a limited stability even at RT and the electrical properties of IO_i are still to be established. Over the years, the interstitial-related defects have been extensively investigated; Davies and Newman have published a comprehensive review on carbon-related complexes[67] while Jones, Estreicher, and coworkers[68–70] have discussed boron-related clusters with some focus on the phenomena of LID of solar cells and transient-enhanced boron diffusion (TED) in conjunction with doping by ion implantation. In this section, we intend to report on some recent developments in the understanding of interstitial-related defects.

8.5.1 CARBON-RELATED INTERSTITIAL DEFECTS

Carbon is one of the dominant impurities in Si grown by different methods. In Cz, Fz, and epitaxial Si, the concentration of carbon has a typical range of ~10^{15}–10^{16} cm^{-3}. Ion implantation or particle irradiation generate I's that interact with C_s forming interstitial carbon (C_i): $I + C_s \rightarrow C_i$. C_i is electrically active and can readily be observed by DLTS. It is an amphoteric defect and can act as both donor and acceptor. The acceptor state, $C_i(-)$, has an electronic level at $E_C-0.10$ eV, and the donor state, $C_i(+)$, occurs at $E_V+0.29$ eV.[71–73] C_i exhibits a diffusion constant of $D_{C_i} \approx 0.44 \exp(-0.87\ (eV)/kT)\ cm^2/s$, as deduced by Newman and Tipping[74] through a combination of IR and EPR data for the neutral and positive charge states of C_i (the negatively charged C_i may have a slightly lower migration barrier by ~0.1 eV). This translates into a diffusivity of ~10^{-15} cm^2/s at 300 K and C_i disappears normally within a few days at RT (except for extremely pure material). The annealing of C_i gives rise to new complexes, and depending on the impurity content, C_i can form complexes with O_i, C_s, or B_s.

One of the most-known and well-documented defects related to C_i is the complex with O_i: C_iO_i. Since oxygen is an abundant impurity in Si, C_iO_i is often a dominant

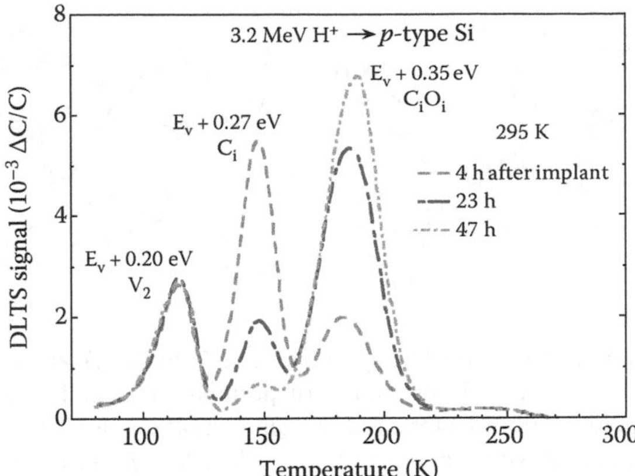

FIGURE 8.17 DLTS spectra of a *p*-type Fz Si sample taken at different times after irradiation with 3.2 MeV protons. The sample was kept at 295 K between the DLTS measurements. (Reprinted from *Nucl. Instrum. Met. B* 120, J. Lalita et al. Defect evolution in MeV ion-implanted silicon, 27. Copyright 1996, with permission from Elsevier.)

interstitial defect center. The complex was first observed in the 1960s using IR[75] and EPR[76] measurements and, in 1977, the DLTS signature was identified by Mooney et al.[77] It has a donor level, $C_iO_i(+)$, at around $E_V + 0.36$ eV and the complex is stable up to temperatures of ~300°C where it starts to dissociate.[78] Figure 8.17 illustrates how C_i, which is initially the dominant feature, gradually transforms into C_iO_i in a low-doped *p*-type Fz sample kept at 295 K subsequent to irradiation with 3.2 MeV protons. The transformation occurs with a one-to-one proportionality and O_i is the prevailing trap even in this high-purity material.[79]

Despite the dominant character of C_iO_i and the amount of studies performed, the exact structure of the complex remained somewhat controversial for a long time (see Reference 80 and references therein). Moreover, even the formation mechanism has appeared to be more complex than was initially thought. Only recently, it has been conclusively shown that the formation of C_iO_i can occur via an intermediate center, or precursor, labeled $C_iO_i^*$ (see References 81 and 82 and references therein). In an extensive study combining experimental and theoretical work, the most probable configuration for $C_iO_i^*$ has been established.[82] Figure 8.18 shows the established and well-recognized structure of C_iO_i, proposed previously,[80] as well as the anticipated structure for the metastable $C_iO_i^*$. The binding energy of $C_iO_i^*$ was calculated to be 0.7 eV, which is considerably less than the 1.7 eV obtained for the ground state of C_iO_i. Comparative DLTS and FTIR measurements are in favor of the proposed configuration of $C_iO_i^*$, and the results are also in good agreement with theoretically predicted values, particularly for the local vibrational modes observed by FTIR.[82]

In addition to a dissociative process for the loss of C_iO_i, different authors have also revealed a reverse process during the initial annealing stage where the C_iO_i concentration increases before the dissociation prevails.[67,78] The reverse process is especially pronounced in samples irradiated with not-too-high doses preventing depletion of C_s and it has been speculated that the growth in C_iO_i may be related to

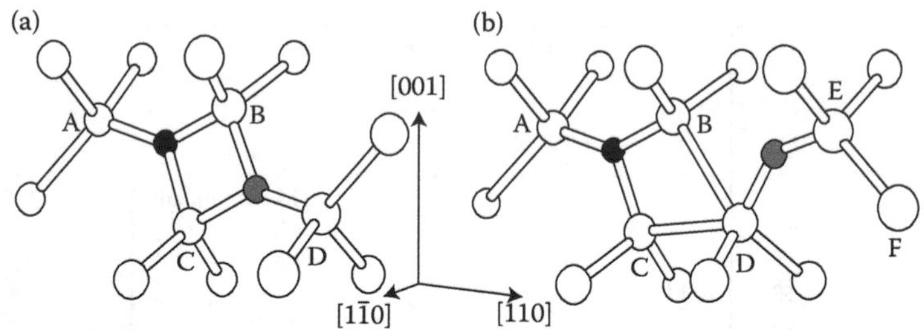

FIGURE 8.18 Atomic structure of C_iO_i (a) and $C_iO_i^*$ (b). Black atoms represent carbon, gray atoms represent oxygen. (Reprinted with permission from L.I. Khirunenko et al., Formation of interstitial carbon–interstitial oxygen complexes in silicon: Local vibrational mode spectroscopy and density functional theory. *Phys. Rev. B* 78, 155203. Copyright 2008 by the American Physical Society.)

dissociation of electrically inactive interstitial clusters. In this scenario, the released I's interact with C_s ($I + C_s \rightarrow C_i$) resulting in the generation of C_i with subsequent formation of C_iO_i. However, very recent results by Ganagona et al.[83] rule out this speculation and instead a mechanism invoking C_iO_iI complexes appears to be more probable; the increase in C_iO_i is ascribed to the dissociation of $C_iO_iI \rightarrow C_iO_i + I$, corroborated by the thermal stability reported for C_iO_iI[84] and the kinetics of the processes involved. In fact, DLTS data for the initial increase of $[C_iO_i]$ in *p*-type Cz samples irradiated with different doses of MeV protons show a close (quantitative) agreement with the simulated concentration of generated C_iO_iI centers, as illustrated by Figure 8.19. Hence, rather strong evidence is obtained for the involvement of C_iO_iI and further studies of this complex are encouraged.

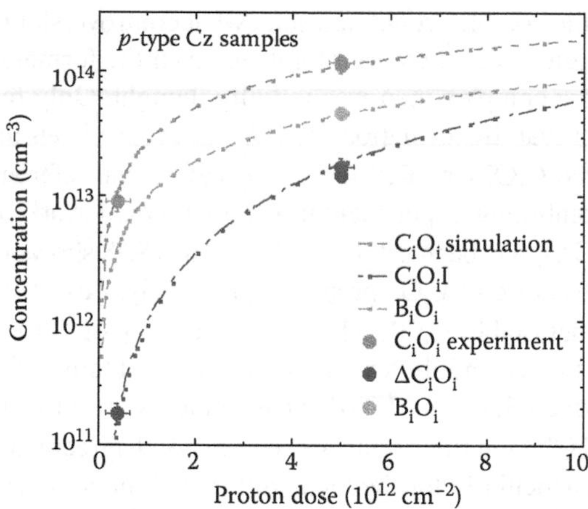

FIGURE 8.19 Comparison between simulated values of C_iO_i, B_iO_i, and C_iO_iI and corresponding experimental ones as a function of dose for boron-doped *p*-type Cz samples irradiated with 1.8 MeV protons. ΔC_iO_i represents the measured initial increase of C_iO_i and it correlates closely with the computed C_iO_iI concentration.

So far, C_iO_iI has escaped detection by DLTS and its electronic properties are not conclusively established. The complex has, however, been identified with FTIR by detecting corresponding local vibrational modes (LVMs) (see References 67, 84, and 85 and references therein). It appears that C_iO_iI can exist in different configurations evolving with annealing temperature,[84] and although the complex may be electrically inactive, it can readily transform into C_iO_i, being a deep and detrimental recombination center. In addition, since the formation of C_iO_iI requires two interstitials, this defect becomes increasingly important at high irradiation or implantation doses. Thus, C_iO_iI may have a substantial technological impact.

The presence of hydrogen can significantly alter the annealing behavior of C_iO_i. In this case, the stability of C_iO_i is determined by fast-diffusing hydrogen species, rather than by dissociation, and annealing can occur at temperatures as low as 125–150°C, depending on the availability of diffusing hydrogen.[86–88] This is a large reduction in thermal stability by up to ~200°C as compared to that normally observed. The interaction between C_iO_i and hydrogen has been studied in detail using *ab initio* calculations[89] and it is established that C_iO_i can capture up to two hydrogen atoms. The complex with one hydrogen atom, C_iO_iH, is expected to be electrically active but not the one with two hydrogen atoms, $C_iO_iH_2$. Two possible stable configurations have been identified for C_iO_iH: (i) a ring configuration, labeled as O, and (ii) an open configuration, labeled as R (see Reference 89 for details on the structure). In the O configuration, C_iO_iH is predicted to be a deep acceptor with the (–) level at around $E_C - 0.27$ eV. In the R configuration, the complex is anticipated to be a shallow donor with the (+) level at around $E_C - 0.05$ eV. It has also been noticed that the properties of C_iO_iH are strikingly similar to the first member of a family of shallow thermal donors that contain hydrogen.[89] Despite the theoretical and experimental indications, neither the O nor the R configurations of C_iO_iH have been conclusively identified.

The C_iC_s pair is formed by a C_s atom capturing a migrating C_i, and it is a prominent irradiation-induced defect in carbon-rich material. This holds especially in Fz (and epitaxial) samples where the competition by O_i as a trap for C_i is much less than in Cz samples. The attraction of C_i to C_s is promoted by the elastic energy stored in the strain fields of the two defects. The C—Si bond is shorter than the Si—Si bond yielding an atomic displacement toward the C_s site while C_i is anticipated to cause a displacement away from the interstitial site. Hence, the strain field can be minimized when C_s and C_i come close enough and the reaction probability is enhanced by this short-range attraction. The complex exhibits bistability with one stable configuration for $C_iC_s(+)$ and $C_iC_s(-)$ and one for $C_sC_i(0)$, as proposed by Song et al.[73] The donor state is located ~0.09 eV above E_V, and the acceptor state occurs at $E_C - 0.17$ eV overlapping with that of the VO center. An extensive discussion of C_iC_s and its properties can be found in Reference 67.

The diffusing C_i can also interact with B_s. The properties of this complex are described in the next section.

8.5.2 Boron-Related Complexes

Interstitial boron (B_i), generated through the interaction $I + B_s \rightarrow B_i$, has been extensively investigated. It is established that B_i is a negative-U defect with a single donor level at $E_C - 0.13$ eV and a single acceptor level at $E_C - 0.37$ eV (see, e.g., References 90 and 91 and references therein). According to *ab initio* calculations, the negative-U behavior is due to considerable structural transitions for the different charge states.[92,93] For instance, Jeong and Ohiyama[93] argued that the most stable configuration of $B_i(+)$ has the boron atom located closely to the substitutional site with a self-interstitial trapped at an adjacent tetrahedral interstitial site. In contrast, for $B_i(-)$, a configuration with the boron atom close to the interstitial site is favored. The neutral charge state is metastable at any position of the Fermi level.

B_i is mobile at RT, and using heavily boron-doped samples irradiated with high doses of MeV electrons, Tipping and Newman[94] extracted a B_i diffusivity of $D_{B_i} \approx 0.04\exp(-0.6\ (\text{eV})/kT)\ \text{cm}^2/\text{s}$ from IR measurements. This translates into D_{B_i} being in the 10^{-12} cm^2/s range at RT, that is, about three orders of magnitude higher than D_{C_i}. B_i interacts readily with other impurities and one of the most abundant boron-related complexes is the pair between B_i and O_i (B_iO_i). B_iO_i is stable up to about 150–175°C and disappears via dissociation. It has a donor level at around $E_C - 0.25$ eV and is observed experimentally by MCTS in boron-doped p-type samples or by conventional DLTS in n-type samples codoped with boron.[77,95,96] The assignment of the $E_C - 0.25$ eV level to B_iO_i is now generally accepted and has been made based on correlation of the peak amplitude with the boron and oxygen concentrations in different types of materials. *Ab initio* calculations of the donor level position appear to be in agreement with the experimental observations[68] and Figure 8.20b shows the obtained atomic structure of B_iO_i. Recently, however, this assignment has been questioned to some extent based on the results obtained by Vines et al.[97] It was found that the production rate of the $E_C - 0.25$ eV level in epitaxial and Cz samples under particle irradiation may not be consistent with the identification as B_iO_i, or at least that our commonly accepted understanding of defect reactions in p-type Si is incomplete. More detailed investigations are, thus, encouraged to resolve this issue.

Upon annealing of B_iO_i in moderately and highly boron-doped samples, the released B_i's can form a complex with B_s: B_iB_s. In the case of sufficiently high boron doping levels where the concentration of other impurities is comparable to or below that of boron, the B_iB_s pair also forms directly after the generation of B_i during particle irradiation ($I + B_s \rightarrow B_i$) followed by migration and trapping. Further, in such materials, B_s and B_i appear in negative and positive charge states, respectively, and the pair formation is promoted by Coulomb attraction. B_iB_s itself is electrically inactive and not detectable by junction spectroscopy techniques, such as DLTS.[69,98] The atomic configuration of B_iB_s can be described as a pair of boron atoms around the substitutional site and oriented along a <001> direction. The structure is very similar to that of B_iC_s shown in Figure 8.20c. B_iB_s is considered as an immobile complex that anneals via dissociation at around 400°C.

As already mentioned in the previous section, B_i can readily interact with C_s forming the B_iC_s complex,[69,98,99] equally well described as C_iB_s.[68] The configuration is depicted in Figure 8.20c with the B–C pair located around the substitutional site and oriented along a <001> axis. The defect possesses a single donor level at $\sim E_V + 0.29$ eV and is stable up to about 350–400°C.[77,95,100] The observed range of annealing temperatures indicates an activation energy of ~2 eV, which is fully

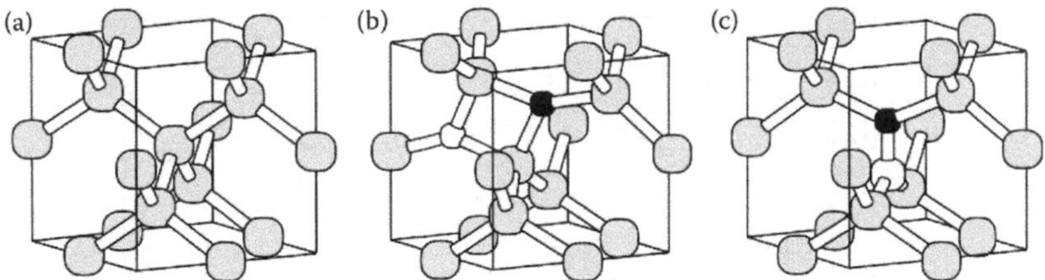

FIGURE 8.20 Atomic structures of bulk Si (a) to aid the reader, B_iO_i (b), and B_iC_s (c). Small black atoms represent boron, small white are oxygen, and the larger white ones are carbon. (Reprinted from *Phys. B: Condens. Matter* 340–342, J. Adey et al. Interstitial boron defects in Si, 505, Copyright 2003, with permission from Elsevier.)

consistent with a process where C_i is released (migration energy ~0.8 – 0.9 eV) and the binding energy equals the calculated value of 1.2 eV[101], yielding a total barrier for dissociation of ~2.0–2.1 eV.

8.5.3 Di-Interstitial-Related Complexes

Similar to vacancies, self-interstitials interact and form di-interstitials (I_2). A number of theoretical works, both *ab initio* and molecular dynamics,[102–107] have been dedicated to I_2. It seems to be widely recognized that I_2 is a fast-diffusing species, although the estimates of the migration energy scatter widely, from 0.3 eV up to 0.9 eV. Despite substantial efforts, there is still some controversy with respect to the exact atomic structure of I_2 and its diffusion path (see, e.g., References 104 and 107 and references therein). Experimental observations of I_2 are also inconclusive. Lee et al.[108] assigned the P6 center in EPR spectra to a positive charge state of I_2 where the two I's are located adjacent to a substitutional atom along a 100 axis (i.e., the 100 split di-interstitial). Some 20 years later, Lee[109] put forward a revised model where the two I's lie in the {100} plane at a position considerably off from two tetrahedral interstitial sites nearby, sharing one Si lattice atom. Based on results from *ab initio* DFT calculations, however, Eberlein et al.[104] concluded that both the models proposed by Lee et al.[108,109] are metastable and have a magnetic field tensor at variance with that of the P6 center.

The diffusing I_2s are believed to interact with O_i and form I_2O_i complexes. Hermansson et al.[110] have tentatively proposed that I_2O_i gives rise to an absorption band at 936 cm^{-1} but further work needs to be pursued for a firm assignment. Also, the electronic properties of I_2O_i are not yet established.

8.6 FINAL REMARKS

"Point defect complexes in crystalline Si" continues to be a vivid field of research, driven by both a tremendous technological impact and scientific curiosity. As illustrated in the four preceding sections of this chapter, substantial advances have been made during the past 10–15 years regarding the understanding of multivacancy complexes and their interaction with abundant impurities, especially oxygen. Also, for self-interstitial complexes and their interplay with dopants and

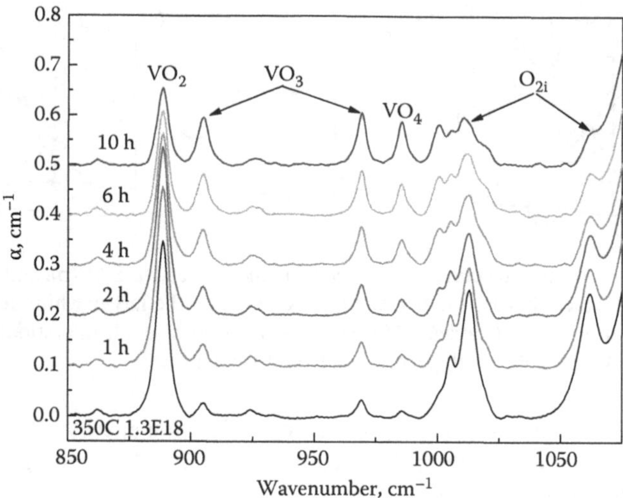

FIGURE 8.21 Fragments of FTIR spectra for *n*-type Cz samples after 2.5 MeV electron irradiation at 350°C and subsequent annealing at 425°C with different duration.

residual impurities, especially carbon and boron, progress has been made albeit at a lower level of sophistication than for the multivacancy complexes. However, many unresolved issues remain and a satisfactory understanding/control of hydrogen-related defects is, indeed, challenging. Nitrogen is another element with relatively unknown properties regarding its importance for the evolution of electrically active point defects. Furthermore, low-temperature processes are desirable for device fabrication in both electronics and PV, implying that we have to control oxygen, and associated thermal donors, more carefully than today; Figure 8.21 displays some new FTIR results on the evolution of VO_2, VO_3, and VO_4 complexes and the O_{2i} dimer during annealing of electron-irradiated Cz samples at 425°C.[111] It can be concluded from these data that both VO_2 and O_{2i} exhibit a much higher diffusivity than O_i, concurred by the rapid growth of the VO_3 and VO_4 complexes, and that they mediate a fast transport of oxygen atoms at temperatures where formation of thermal donors occurs.

Last but not least, *ab initio* defect modeling based on DFT and utilizing high-performance computers (enabling large-sized supercells) has experienced a stunning development and is today an indispensable tool not only for interpretation of measured data but also for predicting defect properties in order to aid experiments.

ACKNOWLEDGMENTS

Extensive collaboration and fruitful discussions with present and past members of our research group are gratefully acknowledged: G. Alfieri, B.S. Avset, J.H. Bleka, A. Galeckas, N. Ganagona, A. Hallén, F. Herklotz, N. Keskitalo, A.Yu. Kuznetsov, J. Lalita, P. Lévêque, H. Malmbekk, M. Mikelsen, P. Pellegrino, V. Quemener,

C.K. Tang, L. Vines, and D. Åberg. The same holds for our international partners: F.D. Auret, J.F. Barbot, the late J.W. Corbett, J. Coutinho, G. Davies, the late L. Dobaczewski, T. Hallberg, C. Jagadish, R. Jones, A. LaMagna, J.L. Lindström, V.P. Markevich, L.I. Murin, A. Nylandsted-Larsen, A.R. Peaker, I. Pintilie, V. Privitera, W. Skorupa, J.S. Williams, J. Wong-Leung, and the CERN-RD50 collaboration.

Partial funding was received from "The Norwegian Research Centre on Solar Cell Technology," a center for environmental-friendly energy research sponsored by the Norwegian Research Council and industry and research partners in Norway.

REFERENCES

1. B.J. Baliga, *Modern Power Devices* (Wiley, New York, 1987), pp. 52–55.
2. K. Sumino and I. Yonenaga, in *Oxygen in Silicon, Semiconductors and Semimetals,* Vol. **42**, eds. R.K. Willardson, A.C. Beer and E.R. Weber (Academic Press, Boston, 1994), pp. 450–511.
3. S.M. Myers, M. Seibt, and W. Schröder, Mechanisms of transition-metal gettering in silicon. *J. Appl. Phys.* **88**, 3795, 2000.
4. T.U. Naerland, H. Angelskår, and E.S. Marstein, Direct monitoring of minority carrier density during light induced degradation in Czochralski silicon by photoluminescence imaging. *J. Appl. Phys.* **113**, 193707, 2013.
5. J.W. Corbett and G.D. Watkins, Silicon divacancy and its direct production by electron irradiation. *Phys. Rev. Lett.* **7**, 314, 1961.
6. B.G. Svensson, in *Properties of Crystalline Silicon*, ed. R. Hull, EMIS Datareview series No. 20 (INSPEC, London, 1999), pp. 763–773, and references therein.
7. A.O. Evwaraye and E. Sun, Electron-irradiation-induced divacancy in lightly doped silicon. *J. Appl. Phys.* **47**, 3776, 1976.
8. B.G. Svensson and M. Willander, Generation of divacancies in silicon irradiated by 2-MeV electrons: Depth and dose dependence. *J. Appl. Phys.* **62**, 2758, 1987.
9. G. de Wit, E.G. Sieverts, and C.A.J. Ammerlaan, Divacancy in silicon: Hyperfine interactions from electron-nucleardouble resonance measurements. *Phys. Rev. B* **14**, 3494, 1976.
10. B.G. Svensson, B. Mohadjeri, A. Hallén, J.H. Svensson, and J.W. Corbett, Divacancy acceptor levels in ion-irradiated silicon. *Phys. Rev. B* **43**, 2292, 1991.
11. E.G. Sieverts, S.H. Muller, and C.A.J. Ammerlaan, On the production of paramagnetic defects in silicon by electron irradiation. *Solid State Commun.* **28**, 221, 1978.
12. J.H. Svensson, B.G. Svensson, and B. Monemar, Infrared absorption studies of the divacancy in silicon: New properties of the singly negative charge state. *Phys. Rev. B* **38**, 4192, 1988.
13. L.C. Kimerling, Defect states in electron-bombarded silicon: Capacitance transient analysis. in *Radiation Effects in Semiconductors 1976, Inst. Phys. Conf. Ser.* No. 31, eds. N.B. Urli and J.W. Corbett (IOP, Bristol, 1977), p. 221.
14. A. Hallén, N. Keskitalo, F. Masszi, and V. Nagl, Lifetime in proton irradiated silicon. *J. Appl. Phys.* **79**, 3906, 1996.
15. L.J. Cheng., J.C. Corelli, J.W. Corbett, and G.D. Watkins, 1.8-, 3.3-, and 3.9-μ bands in irradiated silicon: Correlations with the divacancy. *Phys. Rev.* **152**, 761, 1966.
16. Y.-H. Lee and J. W. Corbett, EPR studies of defects in electron-irradiated silicon: A triplet state of vacancy-oxygen complexes. *Phys. Rev. B* **13**, 2653, 1976.
17. P. Pellegrino, P. Lévêque, J. Lalita, A. Hallén, C. Jagadish, and B. G. Svensson, Annealing kinetics of vacancy-related defects in low-dose MeV self-ion-implanted *n*-type silicon. *Phys. Rev. B* **64**, 195211, 2001.

18. M. Pesola, J. von Boehm, T. Mattila, and R. M. Nieminen, Computational study of interstitial oxygen and vacancy-oxygen complexes in silicon. *Phys. Rev. B* **60**, 11449, 1999.
19. J. Coutinho, R. Jones, S. Öberg, and P. R. Briddon, The formation, dissociation and electrical activity of divacancy-oxygen complexes in Si. *Physica B* 340–342, 523, 2003.
20. M.-A. Trauwaert, J. Vanhellemont, H. E. Maes, A.-M. Van Bavel, G. Langouche, and P. Clauws, Low-temperature anneal of the divacancy in *p*-type silicon: A transformation from V_2 to V_xO_y complexes?. *Appl. Phys. Lett.* **66**, 3056, 1995.
21. J. Lalita, B.G. Svensson, and C. Jagadish, Point defects observed in crystalline silicon implanted by MeV Si ions at elevated temperature. *Nucl. Instrum. Meth. B* **96**, 210, 1995.
22. E.V. Monakhov, B.S. Avset, A. Hallén, and B.G. Svensson, Formation of a double acceptor center during divacancy annealing in low-doped high-purity oxygenated Si. *Phys. Rev. B* **65**, 233207, 2002.
23. G. Alfieri, E.V. Monakhov, B.S. Avset, and B.G. Svensson, Evidence for identification of the divacancy-oxygen center in Si. *Phys. Rev. B* **68**, 233202, 2003.
24. E.V. Monakhov, G. Alfieri, B.S. Avset, A. Hallén, and B.G. Svensson, Laplace transform transient spectroscopy study of adivacancy-related double acceptor centre in Si. *J. Phys. Condens. Matter* **15**, S2771, 2003.
25. V.P. Markevich, A.R. Peaker, S.B. Lastovskii, L.I. Murin, and J.L. Lindström, Defect reactions associated with divacancy elimination in silicon. *J. Phys. Condens. Matter* **15**, S2779–S2789, 2003.
26. M. Mikelsen, E.V. Monakhov, G. Alfieri, B.S. Avset, and B.G. Svensson, Kinetics of divacancy annealing and divacancy-oxygen formation in oxygen-enriched high-purity silicon. *Phys. Rev. B* **72**, 195207, 2005.
27. N. Ganagona, *Point Defects in Solar Silicon*, PhD thesis. University of Oslo, Norway, to be published, 2014.
28. M. Mikelsen, J.H. Bleka, J.S. Christensen, E.V. Monakhov, B.G. Svensson, J. Härkönen, and B.S. Avset, Annealing of the divacancy-oxygen and vacancy-oxygen complexes in silicon. *Phys. Rev. B* **75**, 155202, 2007.
29. J. Coutinho, R. Jones, S. Oberg, and P.R. Briddon, The formation, dissociation and electrical activity of divacancy-oxygen complexes in Si. *Physica B* 340–342, 523, 2003.
30. V.P. Markevich, A.R. Peaker, B. Hamilton, S.B. Lastovskii, L.I. Murin, J. Coutinho, V.J.B. Torres, L. Dobaczewski, and B.G. Svensson, Structure and electronic properties of trivacancy and trivacancy-oxygen complexes in silicon. *Phys. Status Solidi A* **208**, 568, 2011.
31. V.P. Markevich, A.R. Peaker, B. Hamilton, S.B. Lastovskii, and L. I. Murin, Donor levels of the divacancy-oxygen defect in silicon. *J. Appl. Phys.* **115**, 012004, 2014.
32. N. Ganagona, B. Raeissi. L. Vines, E.V. Monakhov, and B.G. Svensson, Formation of donor and acceptor states of the divacancy–oxygen centre in p-type Cz-silicon. *J. Phys. Condens. Matter* **24**, 435801, 2012.
33. N. Ganagona, B. Raeissi. L. Vines, E.V. Monakhov, and B.G. Svensson, Defects in p-type Cz-silicon irradiated at elevated temperatures. *Phys. Status Solidi C* **9**, 2009, 2012.
34. E.V. Monakhov, A. Ulyashin, G. Alfieri, A.Yu. Kuznetsov, B.S. Avset, and B.G. Svensson, Divacancy annealing in Si: Influence of hydrogen. *Phys. Rev. B* **69**, 153202, 2004.
35. B.G. Svensson, C. Jagadish, A. Hallén, and J. Lalita, Generation of vacancy-type point defects in single collision cascades during swift-ion bombardment of silicon. *Phys. Rev. B* **55**, 10498, 1997.
36. E.V. Monakhov, J. Wong-Leung, A.Yu. Kuznetsov, C. Jagadish, and B.G. Svensson, Ion mass effect on vacancy-related deep levels in Si induced by ion implantation. *Phys. Rev. B* **65**, 245201, 2002.

37. R. M. Fleming, C. H. Seager, D. V. Lang, P. J. Cooper, E. Bielejec, and J. M. Campbell, Effects of clustering on the properties of defects in neutron irradiated silicon. *J. Appl. Phys.* **102**, 043711, 2007.
38. J.H. Bleka, I. Pintilie, E.V. Monakhov, B.S. Avset, and B.G. Svensson, Rapid annealing of the vacancy-oxygen center and the divacancy center by diffusing hydrogen in silicon. *Phys. Rev. B* **77**, 073206, 2008.
39. J.H. Bleka, H. Malmbekk, E.V. Monakhov, and B.G. Svensson, Annealing dynamics of irradiation-induced defects in high-purity silicon in the presence of hydrogen. *Phys. Rev. B* 85, 085210, 2012.
40. M. Mikelsen, E.V. Monakhov, G. Alfieri, B.S. Avset, and B.G. Svensson, Kinetics of divacancy annealing and divacancy-oxygen formation in oxygen-enriched high-purity silicon. *Phys. Rev. B* **72**, 195207, 2005.
41. N. Ganagona, L. Vines, E.V. Monakhov, and B.G. Svensson, Transformation of divacancies to divacancy-oxygen pairs in p-type Czochralski-silicon; mechanism of divacancy diffusion. *J. Appl. Phys.* **115**, 034514, 2014.
42. G.S. Hwang and W.A. Goddard, Diffusion and dissociation of neutral divacancies in crystalline silicon. *Phys. Rev. B* **65**, 233205, 2002.
43. R.A. Swalin, *Atomic Diffusion in Semiconductors*, ed. D. Shaw (Plenum, New York, 1973), pp. 65–110.
44. S. K. Estreicher, M. Sanati, and N. G. Szwacki, Iron in silicon: Interactions with radiation defects, carbon, and oxygen. *Phys. Rev. B* **77**, 125214, 2008.
45. B.A. Komarov, Special features of radiation-defect annealing in silicon *p–n* structures: The role of Fe impurity atoms. *Semiconductors* **38**, 1041, 2004.
46. C.K. Tang, L. Vines, V.P. Markevich, B.G. Svensson, and E.V. Monakhov, Divacancy-iron complexes in silicon. *J. Appl. Phys.* **113**, 044503, 2013.
47. J. H. Bleka, E.V. Monakhov, B.G. Svensson, and B.S. Avset, Room-temperature annealing of vacancy-type defect in high-purity n-type Si. *Phys. Rev. B* **76**, 233204, 2007.
48. V.P. Markevich, A.R. Peaker, S.B. Lastovskii, L.I. Murin, J. Coutinho, V.J.B. Torres, P.R. Briddon, L. Dobaczewski, E.V. Monakhov, and B.G. Svensson, Trivacancy and trivacancy-oxygen complexes in silicon: Experiments and *ab initio* modeling. *Phys. Rev. B* **80**, 235207, 2009.
49. V.P. Markevich, A.R. Peaker, B. Hamilton, S.B. Lastovskii, L.I. Murin, J. Coutinho, V.J.B. Torres, L. Dobaczewskiy, and B.G. Svensson, Structure and electronic properties of trivacancy and trivacancy-oxygen complexes in silicon. *Phys. Status Solidi A* **208**, 568, 2011.
50. J. Coutinho, V.P. Markevich, A.R. Peaker, B. Hamilton, S.B. Lastovskii, L.I. Murin, B.G. Svensson, M.J. Rayson, and P.R. Briddon, Electronic and dynamical properties of the silicon trivacancy. *Phys. Rev. B* **86**, 174101, 2012.
51. G. Bemski, Paramagnetic resonance in electron irradiated silicon. *J. Appl. Phys.* **30**, 1195, 1959.
52. G.D. Watkins, J.W. Corbett, and R.M. Walker, Spin resonance in electron irradiated silicon. *J. Appl. Phys.* **30**, 1198, 1959.
53. G.D. Watkins and J.W. Corbett, Defects in irradiated silicon. I. Electron spin resonance of the Si-A center; Defects in irradiated silicon. II. Infrared absorption of the Si-A center. *Phys. Rev.* **121**, 1001, 1961; J.W. Corbett, G.D. Watkins, R.M. Chrenko, and R.S. McDonald, *Phys. Rev.* **121**, 1015, 1961.
54. R.C. Newman, in *Infra-Red Studies of Crystal Defects*, eds. R.R. Coles and Sir N. Mott (Taylor & Francis LTD, London, UK, 1973).
55. G. Davies and R.C. Newman, in *Handbook of Semiconductors*, eds. T.S. Moss and S. Mahajan (Elsevier, Amsterdam, 1994) ch.21, p. 1597.
56. B.G. Svensson and J.L. Lindström, Kinetic study of the 830- and 889-cm^{-1} infrared bands during annealing of irradiated silicon; Growth of the 889 cm^{-1} infrared band in

annealed electron-irradiated silicon. *Phys. Rev. B* **34**, 8709, 1986; B.G. Svensson, J.L. Lindström and J.W. Corbett, *Appl. Phys. Lett.* **47**, 841, 1985.
57. J.W. Corbett, G.D. Watkins, and R.S. McDonald, New oxygen infrared bands in annealed irradiated silicon. *Phys. Rev.* **135**, A1381, 1964.
58. J.L. Lindström, G.S. Oehrlein, and J.W. Corbett, A study of the annealing of the 830 cm^{-1} IR band observed in electron-irradiated silicon. *Phys. Status Solidi A* **95**, 179, 1986.
59. M. Stavola, J.R. Patel, L.C. Kimerling, and P.E. Freeland, Diffusivity of oxygen in silicon at the donor formation temperature. *Appl. Phys. Lett.* **42**, 73, 1983.
60. J.L. Lindström, L.I. Murin, B.G. Svensson, V.P. Markevich, and T. Hallberg, *Physica B* 340, 509, 2003; L.I. Murin, J.L. Lindström, V.P. Markevich, I.F. Medvedeva, V.J.B. Torres, J. Coutinho, R. Jones, and P.R. Briddon, The VO_2*defect in silicon; Metastable VO_2 complexes in silicon: experimental and theoretical modeling studies. *Sol. St. Phenom.* **108**, 223, 2005.
61. K. Irmscher, H. Klose, and K. Maass, Hydrogen-related deep levels in proton-bombarded silicon. *J. Phys. C* **17**, 6317, 1984.
62. H. Malmbekk, L. Vines, B.G. Svensson, and E.V. Monakhov, Comparative study of hydrogen-related defects in p- and n-type silicon. to be published, 2014.
63. J.H. Bleka, E.V. Monakhov, A. Ulyashin, F.D. Auret, A.Yu. Kuznetsov, B.S. Avset, and B.G. Svensson, Defect behaviour in deuterated and non-deuterated n-type Si. *Sol. Stat Phenom.* **108–109**, 553, 2005.
64. A. Hallén, N. Keskitalo, L. Josyula, and B.G. Svensson, Migration energy for the silicon self-interstitial. *J. Appl. Phys.* 86, 214, 1999.
65. H.J. Stein, in: J.W. Corbett, G.D. Watkins eds. Defects in silicon: Concepts and correlations. *Radiation Effects in Semiconductors* (Gordon and Breach, London, 1971), p. 125.
66. A. Brelot, J. Charlemagne, in: J.W. Corbett, G.D. Watkins, eds. Infrared studies of low temperature electron irradiated silicon containing germanium, oxygen and carbon. *Radiation Effects in Semiconductors* (Gordon and Breach, London, 1971), p. 161.
67. G. Davies and R.C. Newman, Carbon in Monocrystalline Silicon, in: T.S. Moss *Handbook on Semiconductors* (Elsevier Science B.V., 1994), p. 1557.
68. J. Adey, J.P. Goss, R. Jones, and P.R. Briddon, Interstitial boron defects in Si. *Phys. B: Condens. Matter* **340–342**, 505, 2003.
69. J. Adey, J.P. Goss, R. Jones, and P.R. Briddon, Identification of boron clusters and boron-interstitial clusters in silicon. *Phys. Rev. B* **67**, 245325, 2003.
70. M. Sanati and S.K. Estreicher, Temperature and sample dependence of the binding free energies of complexes in crystals: The case of acceptor-oxygen complexes in Si. *Phys. Rev. B* 72, 165206, 2005.
71. L.C. Kimerling, H.M. DeAngelis, and J.W. Diebold, On the role of defect charge state in the stability of point defects in silicon. *Sol. State Commun.* **16**, 171, 1975.
72. G.D. Watkins and K.L. Brower, EPR observation of the isolated interstitial carbon atom in silicon. *Phys. Rev. Lett.* **36**, 1329, 1976.
73. L.W. Song, X.D. Zhan, B.W. Benson, and G.D. Watkins, Bistable interstitial-carbon–substitutional-carbon pair in silicon. *Phys. Rev. B* **42**, 5765, 1990.
74. A.K. Tipping and R.C. Newman, The diffusion coefficient of interstitial carbon in silicon. *Semicond. Sci. Technol.* **2**, 315, 1987.
75. A.K. Ramdas and H.Y. Fan, Infrared absorption of neutron irradiated silicon. *J. Phys. Soc. Japan* 18, Suppl. II, **33**, 1963; L.J. Cheng and P. Vadja, 11.6 μ oxygen-associated absorption band in neutron-irradiated silicon. *J. Appl. Phys.* **40**, 4679, 1969; R.C. Newman and A.R. Bean, Irradiation damage in carbon-doped silicon irradiated at low temperatures by 2 MeV electrons. *Radiat. Eff.* **8**, 189, 1971.

76. G.D. Watkins and J.W. Corbett, Defects in irradiated silicon: Electron paramagnetic resonance of the divacancy. *Phys. Rev.* **138**, A543, 1965.
77. P.M. Mooney, L.J. Cheng, M. Süli, J.D. Gerson, and J.W. Corbett, Defect energy levels in boron-doped silicon irradiated with 1-MeV electrons. *Phys. Rev. B* **15**, 3836, 1977.
78. B.G. Svensson and J.L. Lindström, Annealing studies of the 862 cm^{-1} band in silicon. *Phys. Stat. Sol. (a)* **95**, 537, 1986.
79. J. Lalita, N. Keskitalo, A. Hallén, C. Jagadish, and B.G. Svensson, Defect evolution in MeV ion-implanted silicon. *Nucl. Instrum. Met. B* **120**, 27, 1996.
80. R. Jones and S. Öberg, Oxygen frustration and the interstitial carbon-oxygen complex in Si. *Phys. Rev. Lett.* 68, 86, 1992.
81. L. Khirunenko, Yu. Pomozov, N.Tripachko, M. Sosnin, A. Duvanskii, L.I. Murin, J.L. Lindström et al. Interstitial carbon related defects in low-temperature irradiated Si: FTIR and DLTS studies. *Solid State Phenomena* **108–109**, 261, 2005.
82. L.I. Khirunenko, M.G. Sosnin, Yu.V. Pomozov, L.I. Murin, V.P. Markevich, A.R. Peaker, L.M. Almeida, J. Coutinho, and V.J.B. Torres, Formation of interstitial carbon–interstitial oxygen complexes in silicon: Local vibrational mode spectroscopy and density functional theory. *Phys. Rev. B* **78**, 155203, 2008.
83. N. Ganagona, L. Vines, E.V. Monakhov, and B.G. Svensson, Identification of the carbon-dioxygen complex in silicon. To be published.
84. L.I. Murin, J.L. Lindström, G. Davies, and V.P. Markevich, Evolution of radiation-induced carbon–oxygen-related defects in silicon upon annealing: LVM studies. *Nucl. Instr. Meth. B* **253**, 210, 2006.
85. G. Davies, S. Hayama, L. Murin, R. Krause-Rehberg, V. Bondarenko, A. Sengupta, C. Davia, and A. Karpenko, Radiation damage in silicon exposed to high-energy protons. *Phys. Rev. B* **73**, 165202, 2006.
86. O.V. Feklisova and N.A. Yarykin, Transformation of deep-level spectrum of irradiated silicon due to hydrogenation under wet chemical etching. *Semicond. Sci. Technol.* **12**, 742, 1997.
87. H. Malmbekk, L. Vines, E.V. Monakhov, and B.G. Svensson, Hydrogen-related defects in boron doped p-type silicon. *Physica Status Solidi C* **8**, 705, 2011.
88. H. Malmbekk, L. Vines, E.V. Monakhov, and B.G. Svensson, Hydrogen decoration of vacancy related complexes in hydrogen implanted silicon. *Solid State Phenomena* **178–179**, 192, 2011.
89. J. Coutinho, R. Jones, P.R. Briddon, S. Öberg, L.I. Murin, V.P. Markevich, and J.L. Lindström, Interstitial carbon-oxygen center and hydrogen related shallow thermal donors in Si. *Phys. Rev. B* **65**, 014109, 2001.
90. J.R. Troxell and G.D. Watkins, Interstitial boron in silicon: A negative-U system. *Phys. Rev. B* 22, 921, 1980.
91. R.D. Harris, J.L. Newton, and G.D. Watkins, Negative-U defect: Interstitial boron in silicon. *Phys. Rev B* **36**, 1094, 1987.
92. M. Hakala, M.J. Puska, and R.M. Nieminen, First-principles calculations of interstitial boron in silicon. *Phys. Rev. B* **61**, 8155, 2000.
93. J.W. Jeong and A. Oshiyama, Atomic and electronic structures of a boron impurity and its diffusion pathways in crystalline Si. *Phys. Rev. B* 64, 235204, 2001.
94. A.K. Tipping and R.C. Newman, An infrared study of the production, diffusion and complexing of interstitial boron in electron-irradiated silicon. *Semicond. Sci. Technol.* **2**, 389, 1987.
95. P.J. Drevinsky, C.E. Caefer, S.P. Tobin, J.C. Mikkelsen, and L.C. Kimerling, Influence of oxygen and boron on defect production in irradiated silicon. *Mater. Res. Soc. Symp. Proc.* **104**, 167, 1988.

96. L.C. Kimerling, M.T. Asom, J.L. Benton, P.J. Drevinsky, and C.E. Caefer, Interstitial defect reactions in silicon. *Mater. Sci. Forum* **38–41**, 141, 1989.
97. L. Vines, E.V. Monakhov, A.Yu. Kuznetsov, R. Kozłowski, P. Kaminski, and B.G. Svensson, Formation and origin of the dominating electron trap in irradiated p-type silicon. *Phys. Rev. B* **78**, 085205, 2008.
98. J. Zhu, T. Diaz delaRubia, L.H. Yang, C. Mailhiot, and G.H. Gilmer, *Ab initio* pseudopotential calculations of B diffusion and pairing in Si. *Phys. Rev. B* **54**, 4741, 1996.
99. E. Tarnow, Theory of two boron neutral pair defects in silicon. *J. Phys. Condens. Matter* **4**, 5405, 1992.
100. E.V. Monakhov, A. Nylandsted Larsen, and P. Kringhøj, Electronic defect levels in relaxed, epitaxial p-type Si1-xGe_x layers produced by MeV proton irradiation. *J. Appl. Phys.* **81**, 1180, 1997.
101. J. Adey, R. Jones, and P.R. Briddon, Formation of B_iO_i, B_iC_s, and $B_iB_sH_i$ defects in e-irradiated or ion-implanted silicon containing boron. *Appl. Phys. Lett.* **83**, 665, 2003.
102. J. Kim, F. Kirchoff, W.G. Aulbur, F.S. Khan, and G. Kresse, Thermally activated reorientation of di-interstitial defects in silicon. *Phys. Rev. Lett.* **83**, 1990, 1999.
103. S.K. Estreicher, M. Gharaibeh, P.A. Fedders, and P. Ordejón, Unexpected dynamics for self-interstitial clusters in silicon. *Phys. Rev. Lett.* **86**, 1247, 2001.
104. T.A.G. Eberlein, N. Pinho, R. Jones, B.J. Coomer, J.P. Goss, P.R. Briddon, and S. Öberg, Self-interstitial clusters in silicon. *Phys. B: Condens. Matter* **308–310**, 454, 2001.
105. M. Cogoni, B.P. Uberuaga, A.F. Voter, and L. Colombo, Diffusion of small self-interstitial clusters in silicon: Temperature-accelerated tight-binding molecular dynamics simulations. *Phys. Rev. B* **71**, 121203(R), 2005.
106. M. Posselt, F. Gao, and D. Zwicker, Atomistic study of the migration of di- and tri-interstitials in silicon. *Phys. Rev. B* **71**, 245202, 2005.
107. Y.A. Du, R.G. Hennig, and J.W. Wilkins, Diffusion mechanisms for silicon di-interstitials. *Phys. Rev. B* **73**, 245203, 2006.
108. Y.H. Lee, N.N. Gerasimenko, and J.W. Corbett, EPR study of neutron-irradiated silicon: A positive charge stateof the (100) split di-interstitial. *Phys. Rev.* B **14**, 4506, 1976.
109. Y.H. Lee, Silicon di-interstitial in ion-implanted silicon. *Appl. Phys. Lett.* **73**, 1119, 1998.
110. J. Hermansson, L.I. Murin, T. Hallberg, V.P. Markevich, J.L. Lindström. M. Kleverman, and B.G. Svensson, Complexes of the self-interstitial with oxygen in irradiated silicon: a new assignment of the 936 cm^{-1} band. *Physica B* **302–303**, 188, 2001.
111. V. Quemener, B. Raeissi, F. Herklotz, L.I. Murin, E.V. Monakhov, and B.G. Svensson, Kinetics study of vacancy oxygen-related defects in mono-crystalline solar silicon. *Physics Status Solidi C*, 2014.

9 Defect Delineation in Silicon Materials by Chemical Etching Techniques

Bernd O. Kolbesen

CONTENTS

9.1 Introduction ... 289
9.2 General Aspects of Chemistries for Etching of Silicon ... 291
9.3 Classical Chemistries for Preferential Etching of Silicon ... 293
9.4 Novel Chromium-Free Chemistries for Preferential Etching of Silicon Thin Films and Substrates ... 293
9.4.1 FS Cr-Free SOI Etching Solutions ... 293
9.4.2 Etching Mechanism and the Special Role of Bromine or Iodine ... 298
9.4.3 Organic Peracid Etching Solutions ... 300
9.4.4 Etching Solutions with OOE ... 308
9.5 Discussion of Mechanistic Aspects of Preferential Etching at Crystal Defects ... 313
9.6 Determination of Some Characteristic Parameters of Preferential Etching Solutions ... 315
9.7 Summary ... 319
Acknowledgments ... 320
References ... 320

9.1 INTRODUCTION

State-of-the-art Czochralski (CZ) silicon materials and engineered silicon substrates, such as silicon on insulator (SOI) and strained silicon-on-insulator (sSOI), fabricated by the Smart-Cut™ technology[1] with monocrystalline silicon layers some 10 to 100 nm thick still contain a certain density of imperfections and residual crystal defects. These may originate from the crystal growth process of the CZ substrates, such as vacancy agglomerates (D-defects, crystal-originated particles (COPs)) and oxygen precipitate nuclei, or from the fabrication process of the engineered silicon substrates such as oxidation-induced stacking faults (OSF).[2–5] The successful implementation of CZ silicon materials and engineered silicon substrates into the mass production of present or future advanced devices providing high yields and appropriate reliability requires the characterization and monitoring of their crystal quality and defect densities. Chemical etching techniques ("preferential etching") in combination with light optical microscopy are still the workhorse for a quick inspection and simple evaluation of crystal defect types and areal densities. In a recent review,[6] the state of the art of preferential etching "from silicon to germanium" in the view of practical application has been provided. Most of the widely used preferential etching solutions for silicon substrates contain chromium (VI) in the form of potassium dichromate or CrO_3 as oxidizing agent.[7–11] Applied to the very thin films of SOI and sSOI substrates, certain limitations are obvious such as etching times in the seconds regime, very small size of etch features of the defects, and doubtful discrimination of different defects, for example, between threading and misfit dislocations. In order to overcome these limitations, considerable efforts have been undertaken to develop new etching recipes and etching procedures. For SOI and sSOI films, one approach involves two-step etching procedures including one *dilute Secco* etching step.[12] Other approaches for the application on "standard" SOI (films with thickness 50–100 nm) and "thin" SOI (films of <50 nm thickness) are aiming at the reduction of the etch rate and increase of the selectivity between defective and perfect crystal lattices by using dilute versions of the *Secco etch* with adjusted etch rates, which provide superior defect delineation capability and excellent smoothness as well as uniform surface quality.[6] However, chromium (VI) compounds are toxic and carcinogenic and therefore their usage is heavily restricted by law in many countries. Hence, in addition to a few chromium-free etching solutions existing in the literature,[13–17] further chromium-free etch mixtures have been published in recent years.[18–24] These works have primarily aimed at applications on engineered silicon substrates such as SOI and sSOI. Chromium-free defect etching solutions such as the *Frankfurt-SOITEC (FS) Cr-free SOI etches*,[19,21] the *organic peracid etches (OPE)*,[20] and defect etching solutions with organic oxidizing agents, so called *organic oxidizing agent etches (OOE)*,[23] have been developed. The *FS Cr-free SOI etches*[19,21,22] were designed for delineation of crystal defects in standard (50–100 nm) and thin (<50 nm) SOI materials fabricated by the Smart-Cut process. They are based on mixtures of nitric acid (HNO_3), hydrofluoric acid (HF), acetic acid (HAc), and small amounts of bromine (provided by spontaneous comproportionation of NaBr and $NaBrO_3$ in the solution) acting as "additional oxidizing agent" (etch rates nm/s). In the "thin SOI etch,"[21] the bromine is replaced by a certain amount of potassium iodide, which in the solution is instantly oxidized by nitric acid to iodine, the additional oxidizing agent in this mixture. Defect densities determined by the *FS Cr-free etches* are in excellent agreement with those obtained by *dilute Secco etch* on standard SOI. The OPE[20] consist of HF, hydrogen peroxide (H_2O_2), and organic acid such as acetic or propanoic acid. They provide very low removal rates (nm/min at 25°C) and are suitable for the application on SOI films as well as on moderately and highly doped CZ and float-zoned (FZ) silicon substrates where they reveal crystal defects with high sensitivity. A class of novel etching mixtures with OOE[23] based on benzoquinone (BZQ) and its derivatives dissolved in organic solvents miscible with aqueous HF have been shown to be excellent alternatives for replacing *dilute Secco* etching mixtures. Moreover, some basic aspects and characteristic parameters of preferential etching (etch rate, activation energy, selectivity) and potential correlations thereof will be described in more detail in the last section.

9.2 GENERAL ASPECTS OF CHEMISTRIES FOR ETCHING OF SILICON

In an early review, Gatos and Lavine[25] described the chemical etching characteristics of elemental and compound semiconductors, their dissolution processes, and factors affecting chemical etching such as surface orientation, damage, defects, and impurities. Bogenschütz[26] in 1967 published a monograph wherein he summarized the manifold etching recipes for silicon, germanium, and compound semiconductors, which were in use at that time. In addition, he also discussed some general principles of semiconductor etching. Here, we focus on silicon.

Chemistries for etching of silicon consist of

a. An oxidizing agent
b. A fluoride component for dissolving and complexing the oxidized silicon
c. A dilution component such as water or acetic acid (HAc)

The oxidizing agents mainly used are

- Nitric acid (HNO_3)
- Hydrogen peroxide (H_2O_2)
- Chromium (VI) compounds such as CrO_3 or $K_2Cr_2O_7$

Moreover, some chemistries contain oxidizing additives such as bromine or iodine. The etching processes can be

- Diffusion controlled
- Reaction controlled

In a polishing etch, the etching process is diffusion controlled, the transport of the agents to the surface and the transport of the products from the surface is dominating the overall process. In a preferential etch, which is used to delineate crystal defects, the etching process is reaction controlled. The typical ranges of activation energy of the overall etching process of silicon are polishing etches <15–17 kJ/mol < preferential etches.[27] Robins and Schwartz[27–30] conducted basic experimental work on the most important etching system for silicon consisting of mixtures of hydrofluoric acid/nitric acid with water and/or acetic acid. They described the composition regimes where the rate-limiting step is the oxidation process of silicon or the dissolution of the oxidized silicon (Si^{4+}). They point out that a preferential etch can turn into a polishing etch and vice versa by changing the temperature, composition, dilution, or additives.

The HNO_3/HF etching system has been recently revisited and the various etching mechanisms proposed have been discussed by Acker et al.[31–33] In the classical description,[31] at first the silicon is oxidized by nitric acid (standard oxidizing potential $E^0 = +0.96$ V) according to the reaction

$$3Si + 4HNO_3 \leftrightarrows 3SiO_2 + 4NO + 2H_2O \tag{9.1}$$

The dissolution of the oxidized silicon proceeds along the reaction

$$SiO_2 + 6HF \leftrightarrows 2H^+ + SiF_6^{2-} + 2H_2O \quad (9.2)$$

and the overall reaction

$$3Si + 4HNO_3 + 18HF \leftrightarrows 3H_2SiF_6 + 4NO + 8H_2O \quad (9.3)$$

In the dissolution reaction (Equation 9.2), different fluoride species are involved: HF, HF_2^-, and $(HF)_2$. Their concentrations depend on the total HF concentration and acidity. The etch rate depends on the fluoride concentration and acidity. Protons are involved in the formation of silanol groups (Si–OH). If this step is sufficiently fast, the overall etch rate is determined by the oxidation of silicon. After dissolution of the oxidized silicon, the surface is predominantly hydrogen terminated and then resistant to further dissolution by HF.

Another interpretation proposed by Abel and Schmid[34] is based on several equilibria between nitrogen oxides and nitrous acid. The etch mechanism is activated by a reaction of undissociated nitric acid and silicon (Equation 9.4). As a consequence, the nitrogen monoxide created produces nitrous acid (Equation 9.5), which is considered to be the dominant oxidizing species (Equation 9.6). Its formation is the rate-limiting step. This mechanism is considered to be autocatalytic because of the self-regeneration of nitrous acid:

$$2HNO_3 + 3Si \leftrightarrows 3SiO + 2NO + H_2O \quad (9.4)$$

$$H^+ + NO_3^- + 2NO + H_2O \leftrightarrows 3HNO_2 \quad (9.5)$$

$$2HNO_2 + Si \leftrightarrows SiO + 2NO + H_2O \quad (9.6)$$

In a quite different approach of Turner,[35] the etching of silicon is depicted as an electrochemical process. The two partial reactions (cathodic and anodic) occur spatially separated on the silicon surface by electron or hole exchange. According to Turner,[35] the role of nitric acid as an oxidant is to supply holes for the oxidation of the silicon.

Recent experimental work on HF/HNO_3 mixtures in the HF-rich regime (e.g., 70/30) has been carried out at low T (1°C) by Acker and collaborators.[31–33] Using spectroscopic techniques, they observed N_2O_3 and nitrogen (III) oxide species such as NO^+ and $N_4O_6^{2+}$. N_2O_3 serves as reservoir for NO^+, which covers the silicon surface prevailing in a hydrogen-terminated state in acid solution. NO^+ brings about an oxidative attack on Si–H and/or Si–Si back bonds of Si–H groups and is considered as the active species in the rate-limiting step in the oxidation of the H-terminated silicon. As long as the coverage of the surface by NO^+ is complete, the etch rate is constant, resulting in a smooth silicon surface. If the N_2O_3 concentration is insufficient to provide complete coverage of the surface by NO^+, the etch rate decreases and surface roughening occurs. X-ray photoelectron spectroscopy (XPS) measurements[33] provide evidence that there are no silicon oxide species present at the surface, independent of whether the HNO_3/HF mixtures have high or low nitric acid content. Hence, the oxidation of the hydrogen-terminated silicon surface proceeds without silicon oxide formation eventually to SiF_4, which is finally complexed by HF as H_2SiF_6 in the solution.

However, as the authors[33] point out, the chemical etching process is much more complex (a variety of species seems to be involved) and is combined with an electrochemical reaction mechanism in which the injection of holes into the valence band of silicon takes place.[36] The authors did not study effects on silicon containing crystal defects such as dislocations and vacancy clusters. In the NO^+-deficient regime, defects at the surface may act as preferential sites for NO^+ adsorption giving rise to an enhanced etch rate compared to the perfect silicon areas.

9.3 CLASSICAL CHEMISTRIES FOR PREFERENTIAL ETCHING OF SILICON

In the field of defect delineation by chemical etching, overviews have been provided recently[6,18,37] on widely used preferential etching recipes listed in Table 9.1. These overviews include also a discussion of etching principles. Most of the common preferential etching solutions in Table 9.1 contain Cr(VI) as oxidizing agent and provide etch rates in the μm/min regime. The *Sirtl etch*[7] is the "mother" of the mixtures containing Cr(VI) as CrO_3 and provides superior results on (111) substrates but not on (100). *Secco etch*[8] uses $K_2Cr_2O_7$ and works excellently on (100). *Dilute Secco* mixtures also provide best results on thin Si films of standard SOI material.[6] Wright-Jenkins etch[10] is widely applied to reveal bulk microdefects (BMD) on wafer cross sections.

The concentration ratio m of Cr(VI)/HF is the key parameter to obtain preferential etching conditions and an indicator of the oxidizing power of the mixture.[11] The molar ratio m of the oxidizing agent Cr(VI)/HF $\ll 1$ always holds; hence the etching process proceeds in the oxidation reaction-controlled regime. The ratio m is much more critical for (100) than (111) surfaces, preferential etching occurs more easily on (111), the process window is narrower on (100). The etch rate can be reduced by dilution with water[38] or addition of acetic acid. The Cr(VI)-free etching solutions in Table 9.1[13,15,17] (*Dash, Jeita, MEMC*) use HNO_3 as oxidizing agent. The *Dash etch*[13] is characterized by a very low etch rate due to high dilution by acetic acid (HAc). In HAc, the dissociation of nitric acid is strongly reduced, resulting in significant lowering of the concentration of active species. Even though the molar ratio m HNO_3/HF > 1, the oxidation of Si is the rate-determining step. For these chemistries, little progress has been achieved in the understanding of the etching mechanisms, in particular regarding the active chemical species involved in the processes at the interface semiconductor/liquid.

9.4 NOVEL CHROMIUM-FREE CHEMISTRIES FOR PREFERENTIAL ETCHING OF SILICON THIN FILMS AND SUBSTRATES

9.4.1 FS Cr-Free SOI Etching Solutions

The "original" *FS Cr-free SOI etch* has been developed for delineation of crystal defects in standard (50–100 nm) SOI material fabricated by the Smart-Cut process

TABLE 9.1
Most Widely Used Classical Etching Recipes for Defect Delineation

Etching Solution	Composition	Ratio m Oxidizing Agent/HF	Etch Rate (23°C)	Particular Features Comments
Sirtl [7]	(1) 50 g CrO_3 in 100 mL H_2O (c(Cr) = 5 M) (2) HF (48%) (1):(2) = 1:1	0.208	1.3 µm/min	"Mother" of Cr(VI) etch mixtures; high etch rate, ideal for (111)-orientation, makes cloudy surface on (100), simple formula, etching heats up the etchant (bath temperature must be controlled), visualization of dislocations
Secco [8]	(1) 44.1 g $K_2Cr_2O_7$ in 1000 mL H_2O (c(Cr) = 0.3 M) (2) HF (48%) (1):(2) = 1:2	0.006	1.5 µm/min	Simple formula, ideal for (100)-orientation (possible for (111)), not suitable for highly boron-doped material; dilute versions with 0.02–0.08 M Cr(VI) in use for SOI
Schimmel ("standard") [9]	(1) 75 g CrO_3 in 1000 mL H_2O (c(Cr) = 0.75 M) (2) HF (48%) (1):(2) = 1:2	0.016	1 µm/min	Simple formula, ideal for (100)-orientation (possible for (111)); dilute versions in use for SiGe films
Wright [10]	(1) 45 g CrO_3 in 90 mL H_2O (c(Cr) = 5 M) (2) 6 g $Cu(NO_3)_2 \cdot 3H_2O$ in 180 mL H_2O (3) 90 mL HNO_3 (69%) 180 mL CH_3COOH (98%) and 180 mL HF (49%) (1) and (2), then add (3)		1 µm/min	Ideal for (111)- and (100)-orientation, gives sharp definition for crystal defects induced by hot processing; excellent results for defect evaluation on cleaved wafer surfaces

Yang [11]	(1) 150 g CrO_3 in 1000 mL H_2O (c(Cr) = 1.5 M) (2) HF (49%) (1):(2) = 1:1	0.061	1.5 μm/min	Delineating various crystal defects on (100), as well as on (111) and (110) surfaces
Dash [13]	21.4 Vol% HNO_3 (69%) 7.2 Vol% HF (48%) 71.4 Vol% CH_3COOH (98%)	1.356	0.13 μm/h ca.2 nm/min	Cr(VI) free, gives etch pits for all orientations, visualization of dislocations and stacking faults, very slow, not suitable for SOI
MEMC [15]	36 Vol% HF (49%) 25 Vol% HNO_3 (70%) 18 Vol% CH_3COOH (98%) 21 Vol% H_2O To this solution, one must add 1 g of $Cu(NO_3)_2 \cdot 3H_2O$ per 100 mL of mixed acid	0.309	5 μm/min (20°C) 7.6 μm/min (30°C)	Cr(VI) free, ideal for (100)-, (111)-. and (511)-orientation, flat etch rate; not suitable for SOI
JEITA "Japanese Cr-less" [17]	53.57 Vol% HNO_3 (69%) 3,59 Vol% HF (49%) 21.42 Vol% CH_3COOH (100%) 21.42 Vol% H_2O 0.0083 g potassium iodide (KI) added to 100 mL H_2O		0.32 μm/min	Cr(VI) free, iodide oxidized to iodine by HNO_3 Iodine acts as "additional oxdizing agent," necessary for SOI; not needed for bulk silicon

Source: Modified from B.O. Kolbesen, D. Possner and J. Mähliß, *ECS Trans.* **11**, 195–206, 2007. Reproduced by permission of The Electrochemical Society.

to replace a routinely used very well working *dilute Secco etch* with Cr(VI) concentration of 0.08 M.[6] This "original" *FS Cr-free SOI etching solution* consists of 47.7 parts by volume of nitric acid (HNO_3 69%), 5.9 parts by volume of hydrofluoric acid (HF 49%), 46.4 parts by volume of acetic acid (HAc 100%), and a small amount of bromine (0.5 mL bromine added to 100 mL etching solution). Bromine is the key factor in this solution and is absolutely necessary for the etching of SOI as it initiates the oxidation reaction, which is then carried on by nitric acid.

Bromine is a strong oxidizing agent (standard potential $E^0(2\ Br^-/Br_2) = 1.07$ V). Since it is toxic to the environment, hazardous to health when inhaled, and not easy to handle, it has been replaced by adding a combination of the sodium salts of bromide and bromate to the etching solution (to 100 mL etching solution, 1.75 g NaBr and 0.51 g $NaBrO_3$ are added). Bromide and bromate undergo a well-known comproportionation reaction (Equation 9.7) in acid solutions and produce elemental bromine

$$BrO_3^- + 5Br^- + 6H^+ \rightarrow 3Br_2 + 3H_2O \tag{9.7}$$

NaBr and $NaBrO_3$ are by far not as hazardous as bromine and much easier to store and handle. (Elemental bromine requires storage at low temperatures and protection from light. Moreover, measurement of small volumes of liquid bromine is difficult.)

The etch rate of the *FS Cr-free SOI etch* is 3.5 nm/s at room temperature (RT) on SOI. It is in the same range as that of the *dilute Secco etch* (Cr(VI) concentration 0.08 M) used as a reference (2.0 nm/s etch rate at 23°C). This is low enough for application on standard SOI substrates (50–100 nm). The etch rate is reduced gradually with a decrease in temperature. At 13°C, the etch rate is ~2.1 nm/s and at 7°C ~1.6 nm/s. The solution shows a linear dependence of its etch removal on the etching time. For further reduction of the overall etch rate to 1–2 nm/s, part of the hydrofluoric acid can be substituted by acetic acid to slow down the dissolution of the oxidized silicon layer. However, increasing the acetic acid content decreases the capability of the *FS Cr-free SOI etch* to delineate crystal defects in SOI material due to the change of the etch from reaction-controlled to diffusion-controlled.[19,21]

The *FS Cr-free SOI etch* provides a performance similar to the *dilute Secco etch*: it produces a mirror-like surface with well-developed etch pits at defect sites. In combination with a subsequent HF (49%) dip (also used after the *dilute Secco etch*), the etch figures are clearly visible under an optical microscope by their bright halo (size ~4 μm) due to the underetching of the buried oxide (BOX) and a dark dot in the center (Figure 9.1). In the scanning electron micrograph (SEM) in Figure 9.2, the halo is revealed as dark gray contrast surrounding disk-like the etch pit in the center. A halo is only formed when the etch pit protrudes down to the BOX. A residual Si film thickness of ~30 nm and HF dipping times between 60 and 90 s turned out to be optimum for the SOI material studied. The maximum allowed thickness of the remaining Si layer after etching is limited and related to the size of the original defects. No defects were made visible when the remaining Si film was thicker than ~55 nm after etching. The principle of the defect delineation process for SOI is the same for *FS Cr-free SOI* etching and *dilute Secco* etching (Figure 9.3). It can be divided into two steps (Figure 9.3 left). In the first step (Figure 9.3 center), enhanced

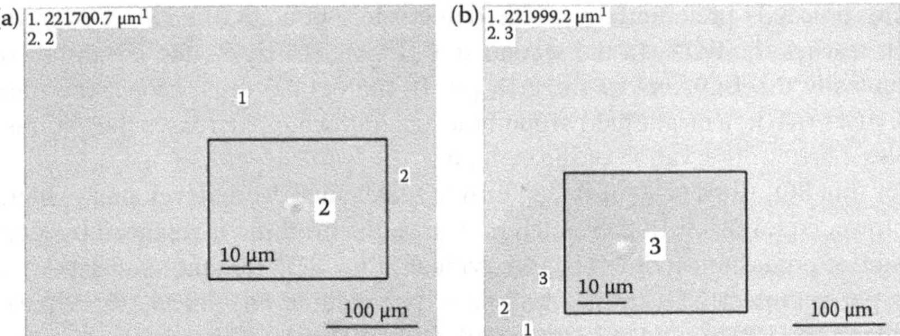

FIGURE 9.1 Optical micrographs displaying etch pits after *FS Cr-free SOI* etching (a) and *dilute Secco* etching (b) on SOI samples etched down to 30 nm. Insets: The etch pit in the center originating from the crystalline defect is surrounded by a bright halo produced by the underetching of the BOX.[18,19] (From J. Mähliß, A. Abbadie and B.O. Kolbesen, *ECS Trans.* **6**(4), 271, 2007. Reproduced by permission of The Electrochemical Society.)

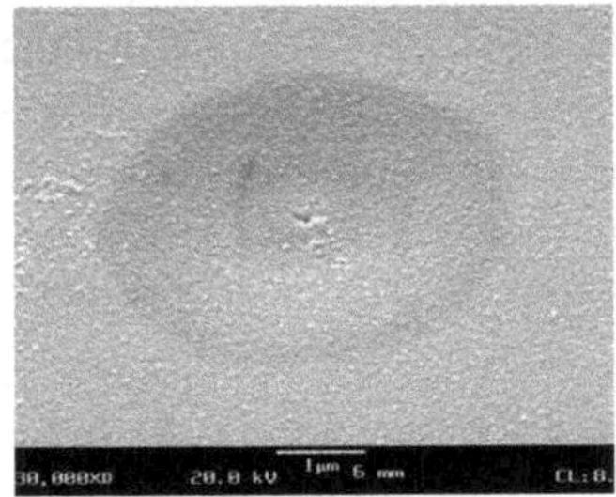

FIGURE 9.2 In the scanning electron micrograph (SEM) after *FS Cr-free SOI* etching, the halo is revealed as dark gray contrast surrounding disk-like the etch pit in the center.

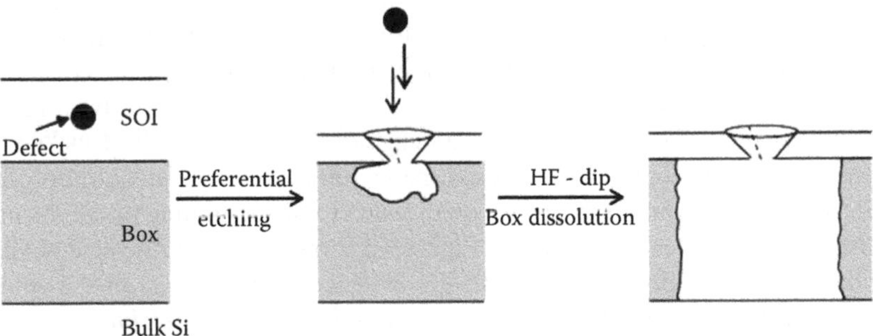

FIGURE 9.3 Sketch of the defect delineation process in SOI[22]: (left) original defect, (center) defect etching solution proceeds preferentially at defect site, (right) subsequent dip in HF dissolves the BOX below defect producing a bright halo in the optical micrograph (Figure 9.1) and a dark gray contrast in the SEM image (Figure 9.2).

etching proceeds preferentially at the defect site and a small etch pit is produced, which reaches the BOX. In the second step (Figure 9.3 right), the HF dip dissolves isotropically the BOX below the etch pit. In the optical microscope, the removed BOX gives rise to a bright halo around the etch pit, which increases the contrast and enables a better observation of the etch pit.

For thin SOI films (<50 nm) the "thin SOI etchants" were developed, which provide lower etch rates.[21] In these etch mixtures, the bromine is replaced by a certain amount of potassium iodide (KI). A typical "thin SOI etchant" consists of 12.24 parts per volume HNO_3 (69%), 1.47 parts per volume HF (49%), 86.29 parts per volume HAc (100%), and 0.232 g KI added to 100 mL of the etching solution. In the solution, the iodide is instantly oxidized by nitric acid ($E^0 = +0.96$ V) to iodine, which is the additional oxidizing agent in this mixture ($E^0 = +0.54$ V). Its etch rate of ~0.8 nm/s at 23°C is low enough even for application on very thin SOI films. The etch rate may be further decreased to ~0.4 nm/s at 7°C. Similar as for the bromine in the case of the *FS Cr-free SOI etch* without iodine, no oxidation/etching of SOI could be observed. Compared to the "original" *FS Cr-free SOI etching* solution, the HAc component in the "thin SOI etchant" is distinctly increased, resulting in a very uniform etch removal and a very smooth and homogeneous surface obtained after etching. As in the case of the *FS Cr-free SOI etch* and the *dilute Secco etch* (0.08 M Cr(VI)), the defects are revealed clearly after a subsequent dip in HF and can be readily counted under an optical microscope. The defect densities determined with the *FS Cr-free SOI etches* correlate very well with those of the *dilute Secco etch* on SOI with different film thicknesses and defect densities.[19,21]

9.4.2 Etching Mechanism and the Special Role of Bromine or Iodine

Each component of the mixtures has a specific purpose[19]: nitric acid is the oxidizing agent for silicon, hydrofluoric acid is necessary to dissolve and complex the oxidized silicon produced, and acetic acid acts as a diluent, which is needed for a uniform etch removal. Bromine and iodine, respectively, are the key factors in these solutions as additional oxidizing agent. They are absolutely necessary for the etching of SOI as they initiate the oxidation reaction, which is then carried on by nitric acid. On the contrary, for etching of bulk silicon, the bromine or iodine additive is not necessary.

As studied by Mähliß et al.,[19] an increase in the bromine content increases the etch rate and vice versa (Figure 9.4). However, while an increase in bromine does not influence the uniformity of etching, a reduction to less than half of its "normal" concentration (equivalent to 0.5 mL bromine in 100 mL etching solution) produced significant deterioration of the surface quality due to the beginning of the formation of hydrogen gas bubbles at the surface. Figure 9.5 shows that the nitric acid obviously contributes to the oxidation and drastically enhances the etch rate of the solution. Furthermore, nitric acid is indispensable to the *FS Cr-free SOI etching solution* as crystal defects are not revealed in its absence. The variations in bromine or iodide concentrations did not influence the number of etch pits revealed and, hence, the defect densities determined.

Why the *FS Cr-free SOI etching solutions* without bromine or iodine are not able to oxidize/etch SOI is not fully understood, though the same solutions proved to etch

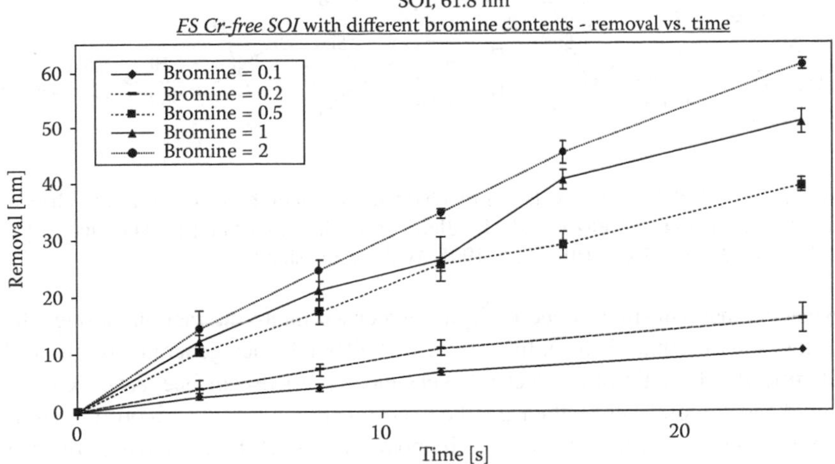

FIGURE 9.4 *FS Cr-free SOI etching solution* with different bromine concentrations (in mL bromine per 100 mL etching solution) illustrating the increase in etch rate with rising bromine concentration. (From J. Mähliß, A. Abbadie and B.O. Kolbesen, *Mater. Sci. Eng. B* **159–160**, 309–313, 2009. With permission.)

bulk silicon. It may be supposed that the Si film/BOX interface, which lies very near to the Si surface/liquid interface, might act as a recombination zone for electrons and holes and therefore prevent the electron/hole transfer.

The *FS Cr-free SOI solution* without nitric acid (Br_2/HF/HAc) proved to etch silicon. As far as the mechanism of oxidation by bromine is concerned, Meltzer and Mandler[39]

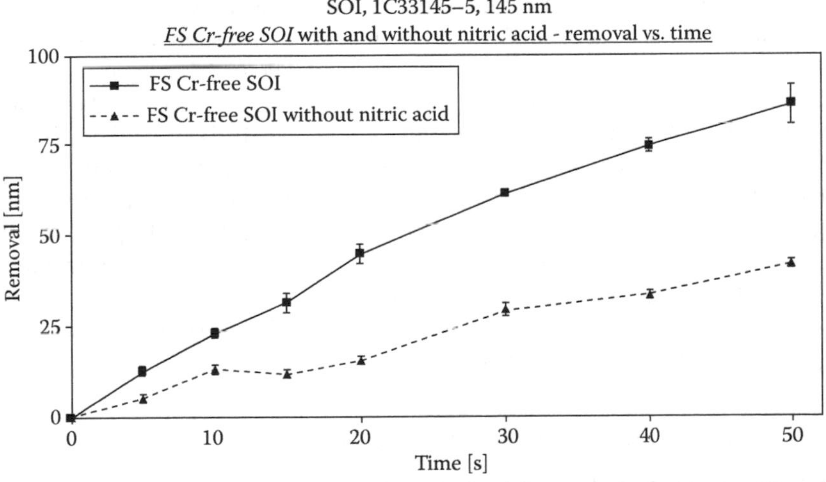

FIGURE 9.5 *FS Cr-free SOI etching solution*: removal with and without nitric acid as a function of time. (From J. Mähliß, A. Abbadie and B.O. Kolbesen, *Mater. Sci. Eng. B* **159–160**, 309–313, 2009. With permission.)

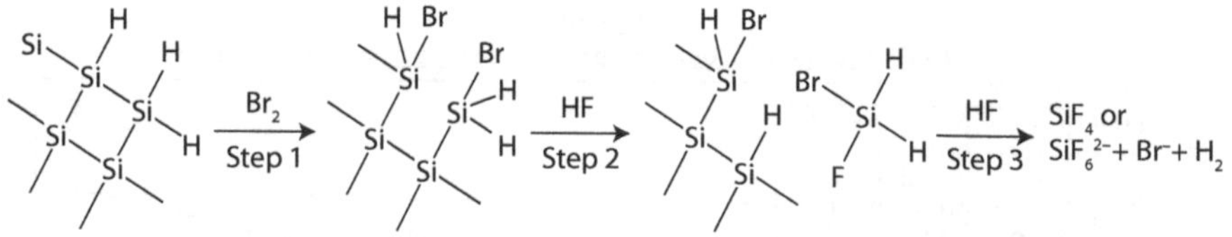

FIGURE 9.6 Model for the oxidation mechanism of silicon by bromine in the absence of nitric acid according to Meltzer and Mandler. (From S. Meltzer and D. Mandler, *J. Chem. Soc. Faraday Trans.* **91**(6), 1019–1024, 1995. With permission.)

suggested a mechanism outlined in Figure 9.6 based on an interaction among silicon, bromine, and hydrofluoric acid. In a mixture of HBr/HF, they generated bromine electrochemically at the tip of an electrode very close to the Si surface. After removal of the oxidized silicon layer by fluoride, the exposed silicon reacts with bromine to form two Si−Br bonds (Figure 9.6, step 1). The polarized Si−Br bond increases the chemical reactivity of the other Si back bonds, which are, therefore, more easily attacked by HF (step 2). Finally, the activated Si−H bonds, which are also unstable in aqueous and fluoride media, will undergo nucleophilic attack evolving hydrogen (step 3).

Ex situ experiments were conducted by Mähliß et al.[21,22] using laser mass spectroscopy (LAMAS) and XPS to determine the chemical state of the bromine and to confirm the existence of Si−Br bonds, which were assumed to form during etching. For the experiment, SOI samples were etched with *FS Cr-free etching solution* for an appropriate length of time and subsequently dried in a flow of nitrogen. In the LAMAS, measurement peaks for the two isotopes of bromine were found at $m/z = 79$ and 81, but no $(Si-Br)^-$ ion pairs could be detected (The energy of the laser could have been so high as to immediately break the Si−Br bonds). Nevertheless, the experiment proved that bromine was present on the surface. The XPS data,[22] similar to LAMAS, provided evidence that bromine was present at the silicon surface in the ultra-high vacuum. (Two peaks appeared corresponding to Br 3d5/2 at 69.4 eV and to Br 3d3/2 at 70.5 eV with the intensity ratio of 3:2 as expected for d5/2/d3/2). Nevertheless, those peaks did not support clear evidence for the existence of Si–Br bonds. After treatment with a mixture of Br_2/HF/HAc (without HNO_3), a peak at 66.8 eV was detected, which may be assigned to Si−Br bonds but may also be caused by some residues of BrO_3^-, which may have been still adsorbed at the surface. After dipping in a mixture of Br_2/HNO_3/HF/HAc no Si−Br peaks could be discovered, the oxidation may have proceeded driven by the nitric acid. In addition, the complete absence of a SiO_2 (Si^{4+}) peak at ~103.5 eV suggests that the Si surface was covered with a Si−Br layer or was terminated by hydrogen both of which may suppress the formation of a silicon oxide layer.

9.4.3 Organic Peracid Etching Solutions

OPE[20] are a novel class of preferential etching solutions with attractive features. These etches are free of toxic Cr(VI) species and, in addition, contain no nitric acid. They consist of a solution of hydrofluoric acid, an organic acid (e.g., acetic acid or propanoic acid), and hydrogen peroxide. The development of these etching solutions is based on the following idea: In the fabrication of sSOI wafers, H_2O_2/HF/HAc etch mixtures are used to remove SiGe transfer layers selectively from the strained Si film.[40] Such solutions have removal rates of 30–50 nm/min on Si/Ge alloys but very low removal rates on bulk silicon (0.3–2 nm/min at 25°C). The removal rate of a widely used solution containing H_2O_2/HF/HAc in the ratio 2/1/3 is about 38 nm/min on $Si_{0.75}Ge_{0.25}$ and only 0.3 nm/min on bulk Si.[40] In particular, it was expected that due to this low etch rate on silicon such mixtures might have the potential for defect delineation in silicon with reasonable etching times in thin and very thin (<50 nm) films typical for state-of-the-art SOI and sSOI materials, respectively. (For comparison: standard preferential etching solutions such as *Secco*[8] or *Wright*[10] with removal rates in the nm/s range (100 times higher) would result in etching times of some 10 s, which is not convenient for practical use.)

The *Dash etchant*,[13] which is one of the first and best known of the chromium-free etchants, was used as starting point for developing the *OPE*. The original *Dash formulation* contains nitric acid as oxidizing agent: 43 mL HNO_3 (65%), 14.5 mL HF (49%), and 143 mL HAc (100%). In the modified *Dash etch* labeled *Organic Peracid Etch A* (*OPE A*), the nitric acid of the *Dash etch* is replaced 1:1 by hydrogen peroxide: 43 mL H_2O_2 (30%), 14.5 mL HF (49%), and 143 mL HAc (100%). In the *modified Dash mixture*, the hydrogen peroxide reacts with the organic acid, forming the corresponding organic peracid (e.g., peracetic acid (PAA), perpropanoic acid (PPA)) in a slow process taking several hours, as outlined for acetic acid in Equation 9.8[20]:

$$H_3C-COOH + H_2O_2 \rightarrow H_3C-COOOH + H_2O \tag{9.8}$$

The organic peracid, for example, the PAA, then acts as the predominant oxidizing agent for silicon in this solution. *OPE A* has a very low removal rate (0.6 nm/min, 25°C), which, compared to the *original Secco solution*,[8] is about a factor 100 lower. As expected, *OPE A* is capable of making defects visible. Based on the *OPE A* recipe, various etching solutions with different H_2O_2, HF, and HAc ratios were tested on manifold SOI and silicon substrate materials and three solutions were qualified eventually for practical use. Table 9.2 shows the compositions and properties

TABLE 9.2
Properties of Selected *Organic Peracid Etches*

Etching Solution	$V(H_2O_2)$ (mL)	$c(H_2O_2)$ (mol/L)	V(HF) (mL)	c(HF) (mol/L)	V(Hac) (mL)	V(Propanoic Acid) (mL)	c(Peracid) (mol/L)	Removal Rate (nm/min)
OPE A	43	2.1	15	2.11	142	–	1.64	0.6
OPE B	43	2.1	43	6.24	114	–	1.6	0.32
OPE C	50	1.55	50	7.26	100	–	2.85	1.34
OPE D	50	1.82	50	7.26	–	100	2.59	1.67
OPE F	75	2.97	5	0.73	120	–	3.67	1.79

Source: Modified from D. Possner, Organic peracid etches: A new class of chromium-free etching solutions for the delineation of defects in thin silicon films. PhD thesis, Goethe-University, Frankfurt/Main, Germany, 2009. With permission.

of these *OPE* solutions. The peracid content can be determined by iodometry,[41] and the removal rates at 25°C by ellipsometry on SOI material. *OPE* solutions provide very low removal rates (0.6–1.8 nm/min at 25°C). By increasing the hydrogen peroxide content and using a hydrogen peroxide of higher concentration (50% instead of 30%), the etch rate is raised. *OPE* solutions produce an intense, sharp odor and irritate the skin and mucous membranes, but they are less toxic than chromium (VI) compounds and can easily be destroyed by reduction.[42]

Figure 9.7 shows the time dependence of the hydrogen peroxide concentration and the PAA content for *OPE A*. The maximum PAA concentration formed is ~1.6 mol/L. In the case of *OPE A*, which has a relatively low HF content, the maximum PAA content is attained after 20 h (Figure 9.7). *OPE* solutions with higher HF content (*OPE B* and *C*) attain the maximum PAA concentration faster. Aqueous solutions containing PAA are relatively stable. PAA decomposes slowly; a notable decay sets in after 2 weeks. The removal rate is almost directly proportional to the PAA concentration as shown in Figure 9.8.

As already outlined, *OPE* solutions are capable of revealing defects with high sensitivity and exhibit a very high selectivity between defective and defect-free regions of the crystal lattice. Crystal defects such as stacking faults, dislocations, swirl-defects, and vacancy agglomerates (D-defects, COPs) are delineated in bulk silicon and SOI films. In the case of the SOI films (thickness range 50–200 nm fabricated by the Smart-Cut process) used in the studies conducted by Possner et al.,[20] bright etch figures of circular shape (halos) with a dark dot in the center were discovered after a subsequent HF dip in optical micrographs (Figure 9.9). The SEM image (Figure 9.10a) also shows the circular shape of the halos. In close view (Figure 9.10b), a hole of 50–100 nm diameter in the center of the defects corresponding to the etch pit produced at the defect is visible. In the early stages of etching, hillocks were observed at defect sites as checked by atomic force microscopy (AFM) (Figure 9.11). These turned into etch pits after prolonged etching and resulted eventually in the undercutting of

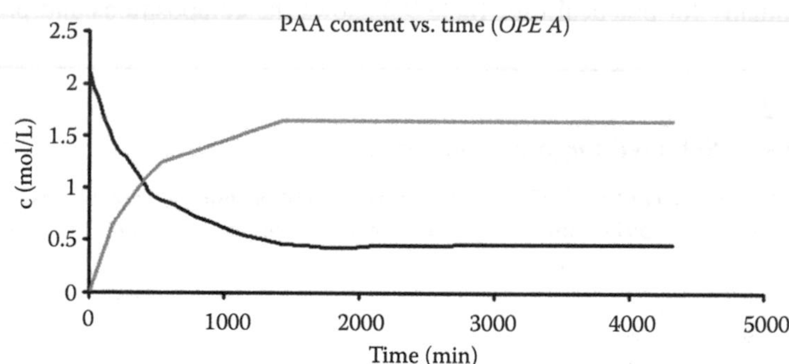

FIGURE 9.7 Increasing peracetic acid (PAA) and decreasing H_2O_2 concentration as a function of time determined for *OPE A*. (From D. Possner, Organic peracid etches: A new class of chromium-free etching solutions for the delineation of defects in thin silicon films. PhD thesis, Goethe-University, Frankfurt/Main, Germany, 2009. With permission.)

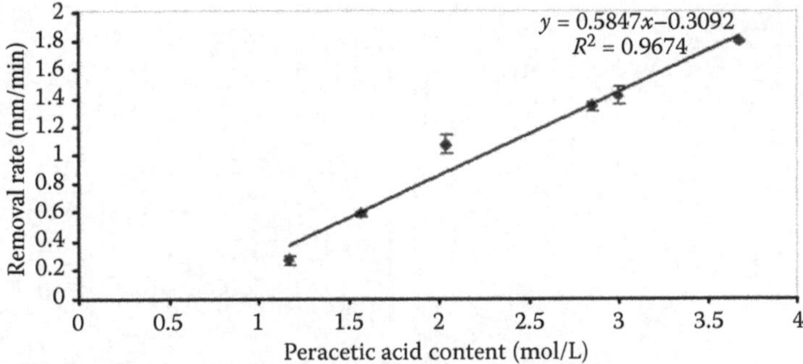

FIGURE 9.8 The removal rate as a function of peracetic acid content. (From D. Possner, Organic peracid etches: A new class of chromium-free etching solutions for the delineation of defects in thin silicon films. PhD thesis, Goethe-University, Frankfurt/Main, Germany, 2009. With permission.)

the BOX producing the halos revealed in the optical microscope. Detailed studies[20] by SEM and transmission electron microscopy (TEM) after focused ion beam (FIB) sample preparation revealed that all hillocks exhibited a hole in the center corresponding to the original defect, which was etched away. Under the hillocks, a cavity was found due to the removal of the BOX by subsequent HF dip resulting in bowing up of the undercut SOI film appearing eventually as hillock in the AFM. Compared to

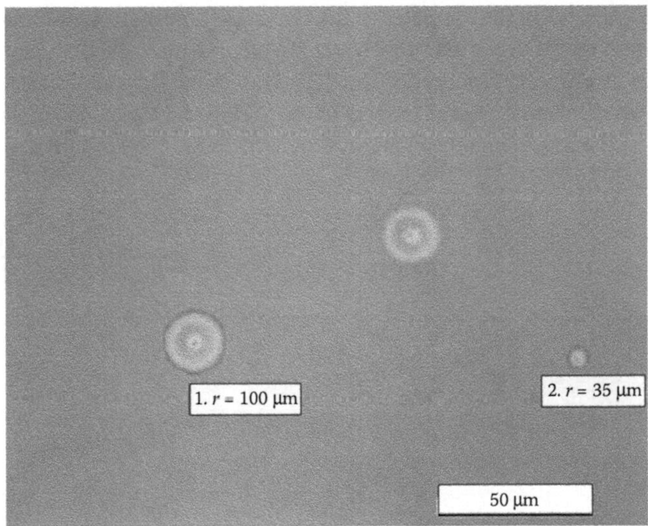

FIGURE 9.9 Optical micrograph of a SOI sample after etching with *OPE D* (90 min): ring-like shaped etch figures (halos) with a dark dot in the center are displayed. (From D. Possner et al. *ECS Trans.* **10**(1), 2007 21. Reproduced by permission of The Electrochemical Society.)

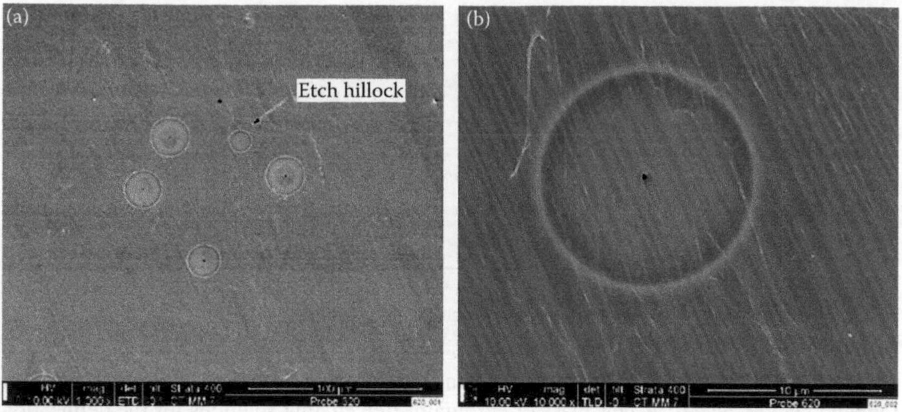

FIGURE 9.10 Defects in SOI (a) SEM image after etching with *OPE D* (90 min): ring-like shaped halos with hole in the center. (b) In the close view, a hole of 50–100 nm diameter in the center corresponding to the etch pit produced at the defect is visible. (From D. Possner et al. *ECS Trans.* **10**(1), 2007 21. Reproduced by permission of The Electrochemical Society.)

the *dilute Secco* reference, the defect densities found after etching are from 10 to 100 times higher, indicating the distinctly increased sensitivity of the *OPE etches*.

When *OPE* mixtures (e.g., *OPE A* and *F*) with a low hydrofluoric acid content are used, a subsequent dip in HF is necessary to produce halos of a size detectable in the optical microscope. For *OPE* mixtures with higher HF content (*OPE B*, *C*, and *D*), there is no need for an HF step. Here, the size of the halos increases with the duration of the etch (Figure 9.12): the diameter of the etch pits (holes) grows, producing a wider channel down to the BOX for the subsequent HF step; as also, the diameter of the halos increases due to the longer action of the HF. This effect provides some information about the depth distribution of the defects in the SOI film. Defects located near the surface are attacked early. When the etch pits reach the BOX, the HF in the mixture starts

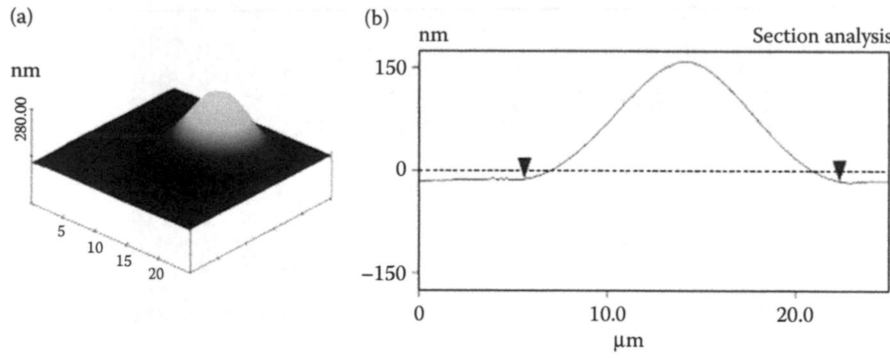

FIGURE 9.11 SOI sample after etching with *OPE D* solution (90 min at 22°C): hillock clearly revealed at defect site by AFM: (a) image, (b) section analysis. (From D. Possner et al. *ECS Trans.* **10**(1), 2007 21. Reproduced by permission of The Electrochemical Society.)

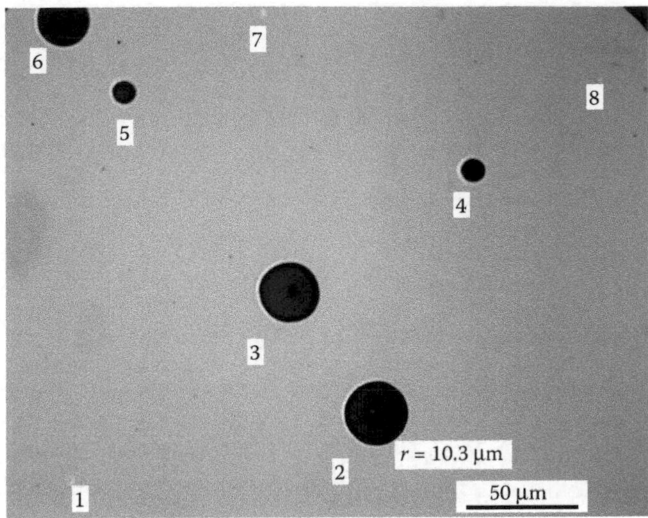

FIGURE 9.12 SOI sample after etching with *OPE C* (90 min) without HF dip: initial thickness 145 nm, thickness after etching 35 nm. Defects located near the surface are attacked earlier by the etch, the size of the halos increases with the duration of the etch due to the longer action of the HF. Defects located near the BOX/SOI interface are attacked relatively late. The etch pits are, therefore, smaller and the action of the HF is shorter. This effect provides some information about the depth distribution of the defects.[43] (From D. Possner, Organic peracid etches: A new class of chromium-free etching solutions for the delineation of defects in thin silicon films. PhD thesis, Goethe-University, Frankfurt/Main, Germany, 2009. With permission.)

to etch off the BOX below, producing a halo of larger size than at a defect located near the BOX/SOI interface, which is attacked relatively late, resulting therefore in a rather short action of the HF on the BOX. The halos produced are then smaller.

The *OPE C*, *D*, and *F* were also tested on some experimental Smart-Cut SOI materials[43] containing OSF that may arise during processing under nonoptimized process conditions. Two different materials, one with a high OSF density and the other with a low OSF density, were used. All the *OPE* solutions were found to reveal OSFs. In Figure 9.13a (etched with *OPE F* (30 min), subsequent dip in HF (45 s)), and in Figure 9.13b (etched with *dilute Secco* (90 s), subsequent dip in HF (45 s)), the OSFs are displayed in a light optical microscope as oval-shaped halo-like etch features with two dark dots in the middle, which correspond to the etch pits formed at the points where the Frank partial dislocation bounding the OSF extends to the surface. The distance between the etch pits and the size of the oval halos is the same for the defects revealed. This is characteristic of OSF, which altogether start to grow at the beginning of the oxidation. On the contrary, grown-in defects such as D-defects or oxygen precipitates appear as etch figures of circular shape (halos) with one dark dot in the center as displayed in Figures 9.9, 9.10, and 9.12.

The OPE also possess the capability to reveal crystal defects in CZ and FZ silicon substrates ("bulk silicon").[20,43] The optical micrograph (Figure 9.14a) and SEM

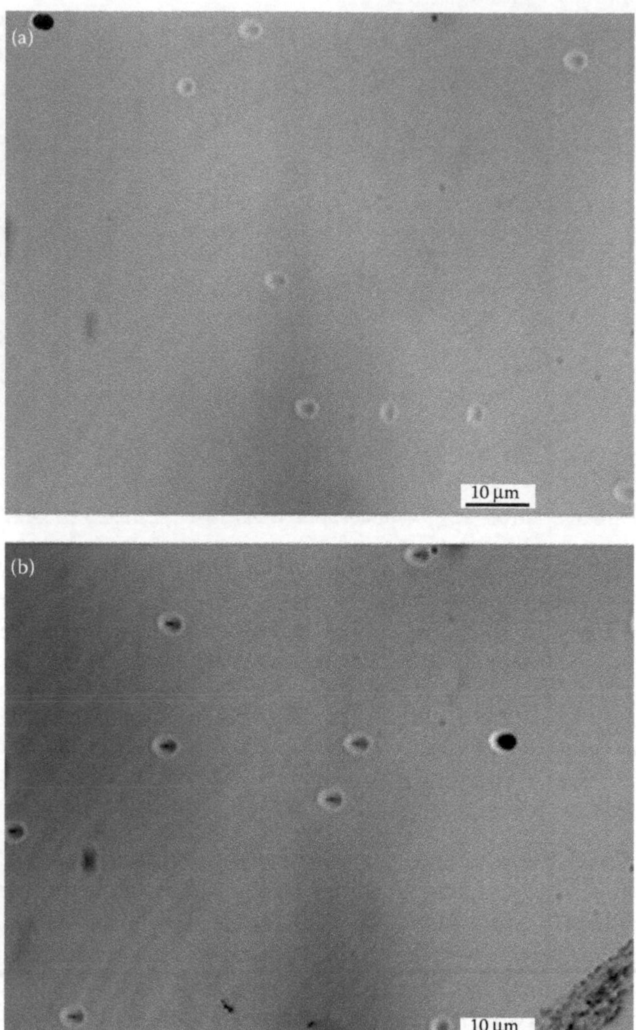

FIGURE 9.13 Oxidation-induced stacking faults in SOI material with a high OSF density. (a) Sample etched with *OPE F* (30 min), subsequent dip in HF (45 s). (b) Sample etched with dilute *Secco* (90 s), subsequent dip in HF (45 s). The OSFs are revealed as oval-shaped halo-like etch figures of roughly the same length and geometry with two dark dots in the middle. (From D. Possner, Organic peracid etches: A new class of chromium-free etching solutions for the delineation of defects in thin silicon films. PhD thesis, Goethe-University, Frankfurt/Main, Germany, 2009. With permission.)

image (Figure 9.14b) display etch pits with the characteristic shape of a square or superposed double squares after etching with *OPE D* (24 h). These correspond to the octahedral cavities formed by the vacancy agglomerates of D-defects (COPs).

In contrast, by *Secco* etching primarily round etch pits are found in the optical micrographs (Figure 9.15) in the same samples often in conjunction with the

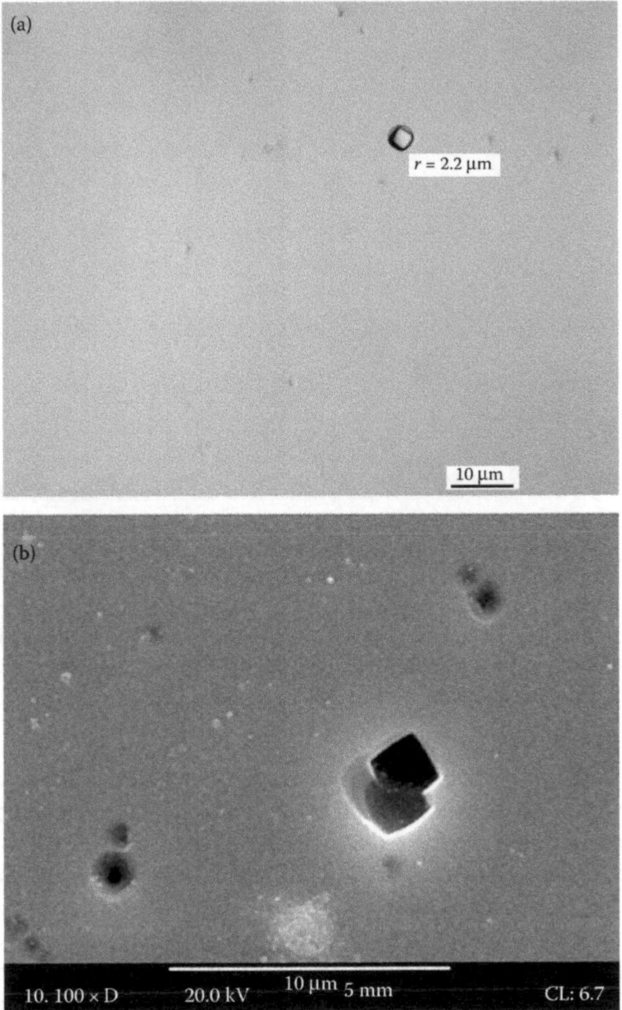

FIGURE 9.14 D-defects (COPs) formed by vacancy agglomerates in CZ bulk material after etching with *OPE D* (24 h, removal 2.5 μm). (a) Light optical micrograph reveals square-shaped etch pit in a sample with a low density of D-defects. (b) SEM: etch pits with the characteristic shape of a square or superposed double squares, which correspond to the octahedral cavities formed by the vacancy agglomerates in a sample with a high density of D-defects. The SEM also reveals D-defects below the surface not exposed to the etchant. (From D. Possner et al. *ECS Trans.* **10**(1), 2007 21. Reproduced by permission of The Electrochemical Society.)

so-called flow pattern.[44] The formation of flow patterns is caused by evolution of hydrogen during the etching process.[44,45] At first, a small hydrogen bubble located at the etch pit is formed. During the etching process, the gas bubble grows and moves upwards across the silicon surface. The hydrogen ambient interferes with the etching process, which leads to the formation of the flow pattern.

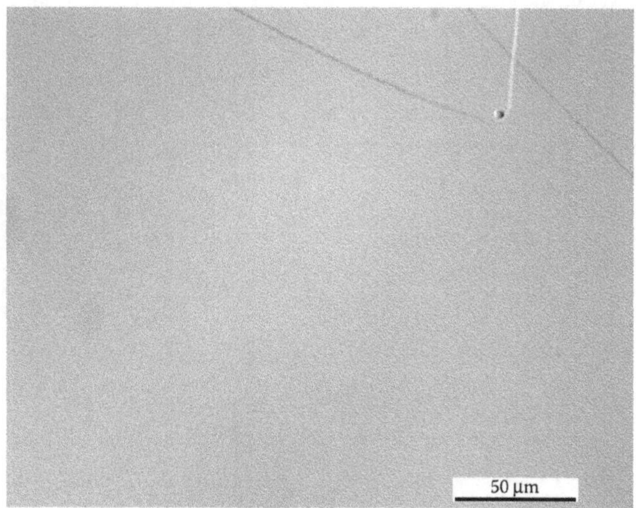

FIGURE 9.15 D-defect in CZ bulk material with a low density of D-defects (COPs) delineated by *Secco* etching (16 min, removal 12 μm) as round etch pit in conjunction with a V-shaped line feature assigned to the so-called flow pattern (light optical micrograph). (From D. Possner, Organic peracid etches: A new class of chromium-free etching solutions for the delineation of defects in thin silicon films. PhD thesis, Goethe-University, Frankfurt/Main, Germany, 2009. With permission.)

The *OPE* solutions work on both types of highly doped silicon material (*p*+ (B); *n*+ (As)) and delineate dislocations and precipitates in heat-treated substrates.[18,20,43] Examples are given in the optical micrograph of Figure 9.16. A CZ silicon sample was scratched and heated in a copper-contaminated furnace and subsequently etched with *OPE C* for 24 h at room temperature. Different kinds of defects can be recognized in Figure 9.16: dislocations as black oval-shaped etch pits, vacancy agglomerates (COPs) as square-shaped pits, and copper-decorated D-defects as small hillocks. In Figure 9.17, swirl-defects, also termed A defects,[46,47] consisting of single interstitial-type dislocation loops or clusters of loops according to TEM studies[48] are delineated in a small zone of the CZ silicon substrate at the edge of a SOI wafer. As in the case of the SOI material, the defect densities found by *OPE* were distinctly higher than for *Secco* etching in the CZ and FZ substrates too. This is explained by the very low etch rates (nm/min) of *OPE* mixtures in contrast to the rather high *Secco* etch rates (nm/s), which provide the capability of the *OPE* mixtures to reveal D-defects of the size of some nanometers as well as tiny oxygen precipitates in CZ substrates.

9.4.4 Etching Solutions with OOE

Several etching solutions containing organic oxidizing agents based on BZQ and its derivatives were explored and have been shown to be excellent alternatives for replacing *dilute Secco* etching mixtures.[23] 1,4-Benzoquinone (*para*-benzoquinone, *p*-BZQ) was the first organic oxidizing agent investigated and etching solutions with

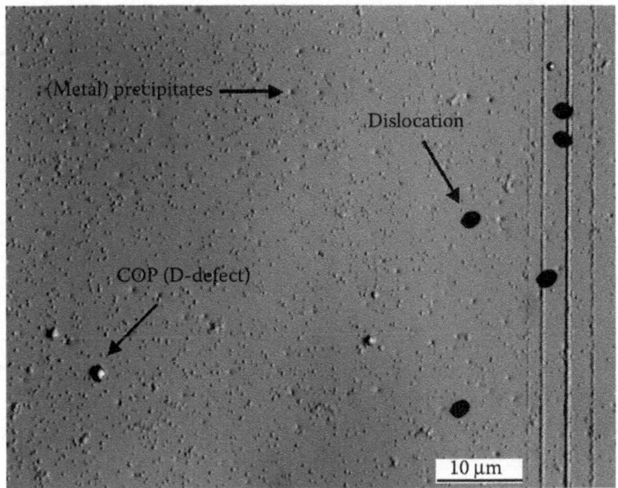

FIGURE 9.16 CZ Si sample after scratching and heating in a copper-contaminated furnace etched with *OPE C* (24 h, removal ~2 μm). Different kinds of defects can be recognized in the light optical micrograph: dislocations as black oval-shaped etch pits, vacancy agglomerates (D-defects, COPs) as square-shaped pits and copper-decorated D-defects as small hillocks. (From D. Possner, Organic peracid etches: A new class of chromium-free etching solutions for the delineation of defects in thin silicon films. PhD thesis, Goethe-University, Frankfurt/Main, Germany, 2009. With permission.)

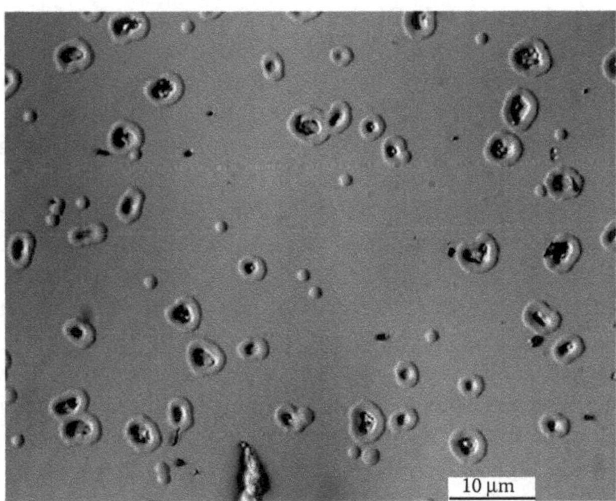

FIGURE 9.17 Swirl-defects (A defects) delineated by etching with *OPE C* (15 min, removal ~20 nm) in a small zone at the edge of a SOI wafer in the bulk of the CZ carrier wafer (light optical micrograph). A defects consist of single interstitial-type dislocation loops or clusters of loops. (From D. Possner, Organic peracid etches: A new class of chromium-free etching solutions for the delineation of defects in thin silicon films. PhD thesis, Goethe-University, Frankfurt/Main, Germany, 2009. With permission.)

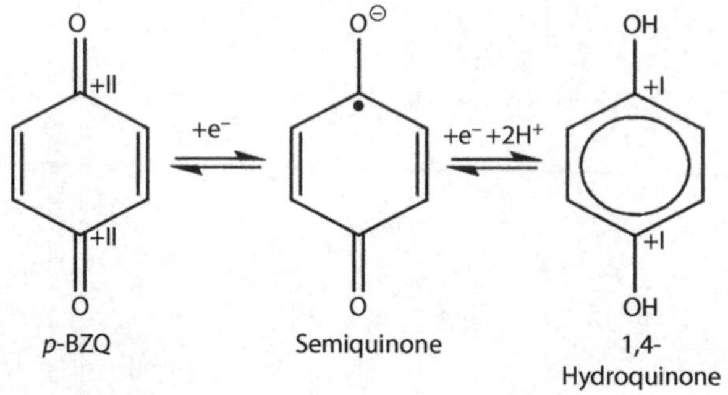

FIGURE 9.18 Redox equilibrium between *p*-BZQ and its reduced form 1,4-hydroquinone. (From J. Mähliß, R. Hakim, A. Abbadie, F. Brunier and B.O. Kolbesen, *J. Electrochem. Soc.* **158**(2), D107–1132011. Reproduced by permission of The Electrochemical Society.)

p-BZQ have been shown to work well. *p*-BZQ is widely used in organic chemistry. Its redox equilibrium with 1,4-hydroquinone is well understood. The reduction takes place gradually by a single-electron transfer in which an intermediate semiquinone radical is formed (Figure 9.18). This process is coupled with several acid–base equilibria.[49] Its standard oxidizing potential of $E^0 = 699.8$ mV[50] is high enough to oxidize semiconductor materials such as silicon. However, *p*-BZQ is toxic and forms a dark-colored charge-transfer complex with its reduced form in solution,[51] which starts to precipitate after a certain period of time.

To prevent such shortcomings, a number of BZQ derivatives with different sterical groups were investigated. Of the derivatives tested, 2,3,5,6-tetrachloro-1,4-benzoquinone (*para*-chloranil, *p-CA*) has been found to work best. *p-CA* is nontoxic, readily available, and inexpensive. Its standard oxidizing potential E^0 of 712.0 mV[52] is slightly higher than that of *p*-BZQ ($E^0 = 699.8$ mV).

As BZQ and its derivatives are only slightly soluble in water organic solvents, miscible with aqueous HF, have to be used. The organic solvents tested exerted a considerable influence on the etch rate of the composed solutions[23] due to their differences with regard to viscosity, dipole moment, and acidity (Figure 9.19). The etch rate increased with the amount of dissolved BZQ or derivative until its saturation limit for the respective solvent was reached. As displayed for acetone in Figure 9.19, the increase in the etch rate stopped when the saturation limit of *p*-BZQ for the solvent was reached. Etch rates in the range of 1–12 nm/min at 20°C were obtained with the various mixtures. This makes them highly suitable for etching thin films.

When the concentration of HF in the etching solution was increased, the etch rate increased significantly; however, the etch removal was not uniform anymore and at high HF concentrations (2 parts HF to 1 part organic solvent), the formation of stain films could be observed.[23] Some of the best results in terms of preferential etching and homogeneity of the surface were obtained with acetone. An etching solution made by dissolving 2 g *p*-BZQ in 100 mL acetone (c(*p*-BZQ) = 0.19 M) and adding 50 mL HF (49%) to the resulting solution (referred to hereafter as *p-BZQ etching solution*) has a linear etch rate of 2.7 nm/min at 20°C and shows a very uniform etch removal. This makes it highly suitable for etching thin films.

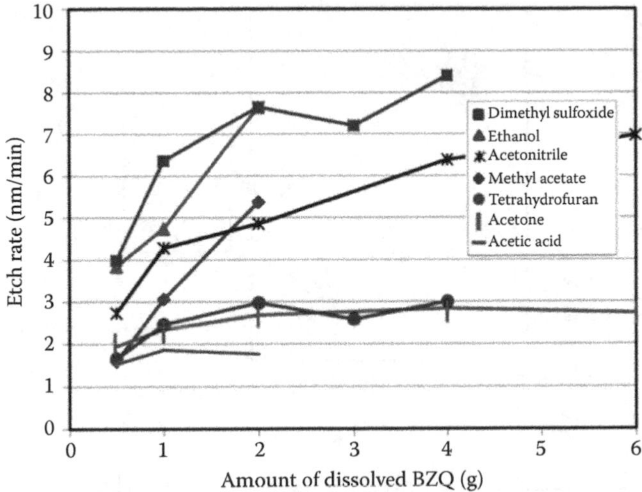

FIGURE 9.19 Plot of the etch rate versus organic solvent used and amount of *p*-BZQ dissolved. (From J. Mähliß et al. *J. Electrochem. Soc.* **158**(2), D107–1132011. Reproduced by permission of The Electrochemical Society.)

The etching solutions using *p-BZQ* or *p-CA* delineate both grown-in and process-induced defects clearly and provide a very smooth and homogeneous surface confirming that these solutions offer similar preferential etching characteristics as *dilute Secco*.[23] Figure 9.20a shows an optical micrograph of a SOI surface after etching and subsequent dip in HF (49%). OSF can be seen as oval-shaped etch figures with two dark dots in the center, while grown-in defects such as D-defects or oxygen precipitates appear as etch figures of circular shape with one dark dot in the middle. The bright halo surrounding the inner edge of each etch figure indicates that the BOX below has been dissolved during a subsequent HF dip, whereas the dark area inside containing the dark dot or dots points to the breakdown of the silicon film. The dark dots correspond to the etch pits produced at the defects in the thin silicon film. Figure 9.20b shows a SEM image of an OSF revealed after etching. It displays clearly two etch pits (corresponding to the dark dots in Figure 9.20a) at the points where the Frank partial dislocation extends to the surface. Moreover, the connecting stacking fault line is revealed. The slight asymmetry in the size of the etch pits at the emerging partial dislocation indicates a different degree of metal decoration. However, the reason is not known.

The good correlation to the defect densities produced with the *dilute Secco* (0.04 M) is outlined in Figure 9.21. To assure a best possible comparison of the preferential etching characteristics between *p-BZQ* and the *dilute Secco* (0.04 M) etching solution, standard SOI substrates with different production background, comprising different levels of OSF densities, were used. The very high OSF density revealed in sample 3 resulted from experiments with nonadapted thermal treatment.

In *p-CA* solutions similar to *Secco* solutions, elemental hydrogen is released during silicon etching.[23] Owing to the low etch rate of *p-CA* solutions (nm/min) in contrast to *Secco* solutions (nm/s), no flow pattern was observed on bulk silicon surfaces

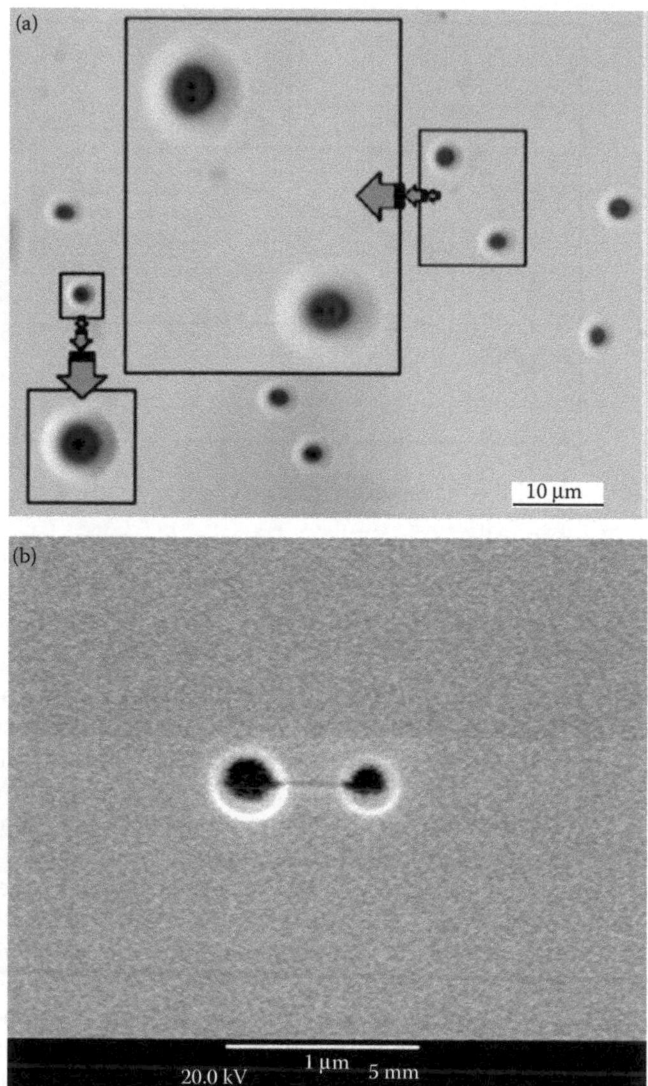

FIGURE 9.20 SOI surface after etching with *p*-BZQ etching solution and subsequent HF dip. (a) The light optical micrograph displays oval-shaped features with two dark dots in the center corresponding to OSF and round features with one dark dot in the middle pointing to D-defects (COPs), oxygen precipitates, or the like. (b) SEM image of an OSF revealed by the *p*-BZQ etching solution without HF dip: at the points where the Frank partial dislocation bounding the OSF extends to the surface two etch pits are clearly displayed and the connecting stacking fault line between these points. (From J. Mähliß et al. *J. Electrochem. Soc.* **158**(2), D107–1132011. Reproduced by permission of The Electrochemical Society.)

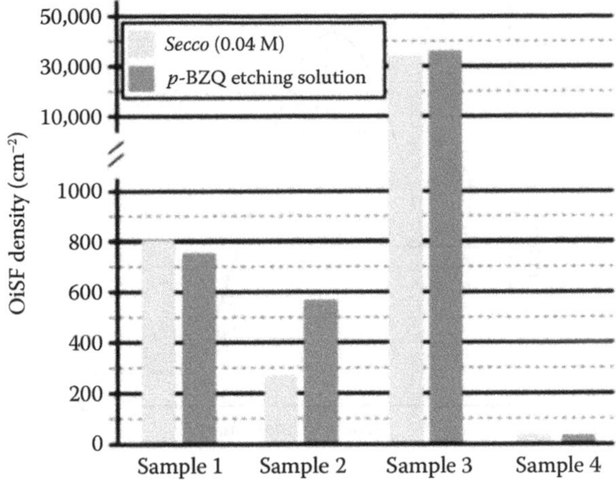

FIGURE 9.21 Comparison of OSF densities obtained after etching SOI substrates with different defect densities with dilute *Secco* or *p*-BZQ etching solution. (From J. Mähliß et al. *J. Electrochem. Soc.* **158**(2), D107–1132011. Reproduced by permission of The Electrochemical Society.)

after etching. For the etching of silicon by *p-chloranil* and HF, a mechanism has been proposed by Mähliß et al.[23] which very likely can also be applied to solutions such as *Secco* using dichromate as oxidizing agent.

9.5 DISCUSSION OF MECHANISTIC ASPECTS OF PREFERENTIAL ETCHING AT CRYSTAL DEFECTS

In principle, two main factors contribute to dislocation etch pit formation:

a. The strain field produced by a dislocation or extended lattice defect
b. Impurities, in particular metals, segregated to the dislocation ("decoration")

In physical chemistry, the course of a general reaction is described by plotting the potential energy versus the reaction coordinate as outlined in Figure 9.22. In the initial state, the educts possess the potential energy A, and in the final state, the potential energy E. Going from A to B, the system in the normal case has to overcome the potential barrier E_a (activation energy). In the case of an etching process in the perfect lattice, the potential energy may be A^P. At a crystal defect, we have more or less distorted bonds between the silicon lattice atoms giving rise to a strain field with higher potential energy A^D in the initial state. (A^D–A^P) is the surface potential φ at the defect site, which is determined by the strain field produced by distorted bonds and, in addition, metals precipitated at the defect (metal decoration), such as a perfect dislocation or partial dislocation of a stacking fault. The increased surface potential at the defect results in reduced activation energy E_a^D and an enhanced etch rate at the defect compared to the perfect lattice, or in other words, increased chemical reactivity of the atoms with distorted bonds. This gives rise to the formation of etch pits at the defect. Moreover,

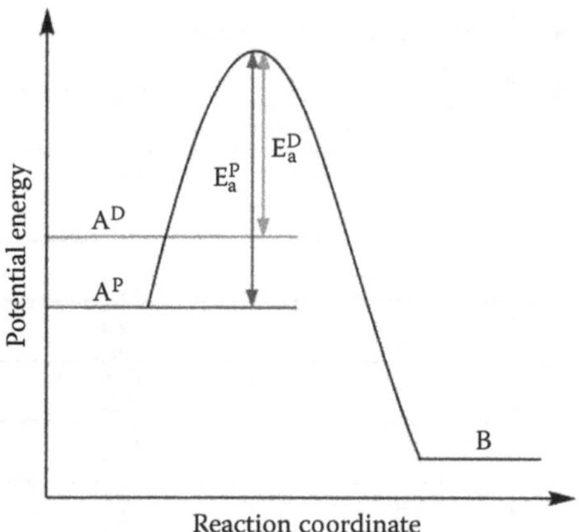

FIGURE 9.22 Potential energy plot for the perfect crystal and crystal defect case[24]: Potential energy A^P = for perfect lattice, A^D = for defect, and E_a^D and E_a^P = activation energies for reaction at defect and at perfect lattice, respectively. (From B.O. Kolbesen, J. Mähliß and D. Possner, *Phys. Stat. Sol. A* **208**(3), 584–587, 2011. With permission.)

the metal silicide precipitates on their own provide a reaction path with reduced energy barrier resulting in drastic increase of the etch rate, a phenomenon that is well known from investigations after preferential etching of partial dislocations of stacking faults with different degrees of metal decoration by TEM.[53]

The influence of the surface orientation of the silicon wafer is brought about by the different surface energies of silicon, which are in the order (100) > (110) > (111).[54] Hence, the surface potential between defect D and perfect lattice P is greater on (111), resulting in more prominent preferential etching behavior of (111) surfaces. Thus, the concentration ratio m of oxidizing agent/HF is less critical on (111) than on (100).

In the widely used preferential etching solutions using hexavalent chromium,[7–11] the active and intermediate species of chromium in the oxidation process are little understood, that is, how Cr(VI) is finally reduced to Cr(III). It is known that in aqueous equilibria, the distribution of the different species H_2CrO_4, $HCrO_4^-$ CrO_4^{2-}, and $Cr_2O_7^{2-}$ is strongly dependent on pH. From studies of the adsorption of dichromate ($Cr_2O_7^{2-}$) on silica or silica gel,[55,56] it is known that dichromate strongly adsorbs at/reacts reversibly with the surface at very low pH (acid several mol/L). Maatman and Kramer[55] suggested a reaction scheme for a silica surface in which the chromate/dichromate ion adds to a surface site produced by the reaction of H^+ with a Si−OH group via elimination of water. The same may be applied to a silicon surface:

$$\text{Si{-}OH} + H^+ \leftrightarrows \text{Si}^+ + H_2O \tag{9.9}$$

$$\text{Si}^+ + Cr_2O_7^{2-} \leftrightarrows \text{Si{-}O{-}}Cr_2O_6^{2-} \tag{9.10}$$

Although the adsorption of the chromate species to the silicon surface seems plausible, it is still not known how the electron transfer from silicon to chromium

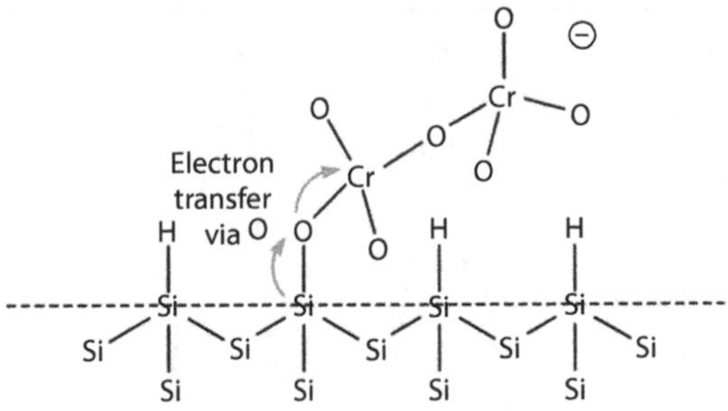

FIGURE 9.23 Model illustrating the electron transfer from silicon to chromium (VI) via the oxygen bridge in the oxidation of silicon by dichromate. (From J. Mähliß, A. Abbadie and B.O. Kolbesen, *Mater. Sci. Eng. B* **159–160**, 309–313, 2009. With permission.)

(VI) occurs. A transfer of three electrons to Cr(VI) in one step to form Cr(III) is very unlikely and requires intermediate steps and species. Figure 9.23 shows a simple model, illustrating the electron transfer from silicon to chromium (VI) via the oxygen bridge in the oxidation of silicon by dichromate.[22] According to the above simple model, the energy barrier for the electron transfer is reduced at the defect, resulting in the enhanced etch rate. In addition, the reduction of Cr(VI) to Cr(III) may be promoted by the formation of the very stable $[CrF_6]^{3-}$ complex between Cr(III) and F^- ions in the solution, resulting in effective removal of free Cr^{3+} ions.

In order to obtain information about possible chromium species at the silicon surface in the *Secco* etching process, near-edge x-ray absorption fine structure spectroscopy (NEXAFS) measurements were performed by Mähliß et al.[22] at the synchrotron BESSY II in Berlin on a CZ silicon sample freshly prepared directly before measurement. The sample was etched for 30 s in *Secco* solution ($K_2Cr_2O_7$ (0.15 M):HF (49%) = 1:2) followed by a water dip of 15 s. The NEXAFS spectrum in Figure 9.24 displays two main peaks with maxima at 578.7 and 586.4 eV. These peaks correspond to the L_2 and to the L_3 edge of Cr(III) according to values published by Garvie et al.[57] (Cr(III): $L_2 = 578.6 \pm 0.2$ eV and $L_3 = 586.6 \pm 0.2$ eV). The multiplet structure of the Cr(III) $L_{2,3}$ edges is consistent with the crystal field strength of Cr(III) compounds where Cr(III) is bonded to oxygen.[58]

The shortcoming of this experiment is that the sample was measured *ex situ*. No information about intermediate chromium species on the surface was obtained as Cr(V) and Cr(IV) are unstable in aqueous solution and disproportionate into Cr(III) and Cr(VI).

9.6 DETERMINATION OF SOME CHARACTERISTIC PARAMETERS OF PREFERENTIAL ETCHING SOLUTIONS

It is still not known why one preferential etching mixture is more selective and sensitive to reveal different types of crystal defects than the other and which parameters are crucial. A deeper understanding of the mechanisms of chemical etching of semiconductors is still very limited, in particular regarding the active chemical

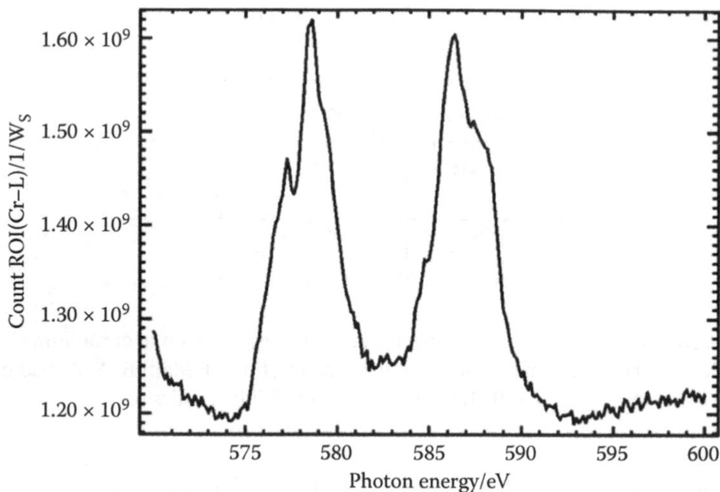

FIGURE 9.24 The NEXAFS spectrum[22] of the chromium $L_{2,3}$ edges displays two main peaks with maxima at 578.7 and 586.4 eV. These peaks correspond to the L_2 and to the L_3 edge of Cr(III) according to values published by Garvie et al. (1994).

species involved in the processes at the interface semiconductor/liquid. As outlined, preferential etching solutions are capable of differentiating between local levels of the potential energy at defect sites and the normal level of the potential energy in the perfect silicon lattice.[24] For the etching reaction at the perfect lattice, the energy barrier E_a^P (activation energy) is substantially higher than E_a^D at the defect (Figure 9.22). Hence, etch rates at defect sites are higher than etch rates of the perfect lattice, giving rise to the formation of etch pits at defect sites.

As a general rule, the activation energy for a reaction-controlled process is higher than that for a diffusion-controlled one. In the preferential etching of silicon, the oxidation of the silicon is the rate-determining step and should control the overall activation energy. Hence, the overall activation energy might be an important parameter in the comparison of the etching characteristics of preferential etching solutions. Moreover, the strain field due to the distorted bonds of the silicon atoms at a dislocation or extended lattice defect and impurities, in particular metals, segregated to the dislocation forming tiny silicide precipitates will affect the local reaction, the oxidation of the silicon, in a reaction-controlled process, and thus the removal rate at the defects. In principle, the overall activation energy of the etching reaction E_a^P can be evaluated by measuring the etch removal rate r_p of the perfect material as a function of temperature. An influence of crystal defects at low or moderate areal density on the average etch rate can be neglected. The activation energy E_a^P can be calculated from the slope of the Arrhenius plot (Equation 9.11):

$$r = A^{*}\exp(-E_a^P/RT)$$

$$\ln r = \ln A - E_a/RT \tag{9.11}$$

TABLE 9.3
Activation Energies E_a and Selectivities S of Defect Etching Solutions

Etching Solution	r_e (nm/min) Perfect Crystal	r_e (nm/min) Crystal Defect	S[a]	E_a (kJ/mol) Perfect Crystal	ΔE_a (kJ/mol) (Perfect/Defect)	ΔE_a (%)
Secco (0.31 M)	770	1320.5	1.72	23.0	1.9	8.2
Dilute Secco (0.04 M)	44.4	81.6	1.84	29.3	1.34	4.6
OPE C	1.32	2.85	2.16	48.7	2.2	4.6
OPE D	1.75	3.79	2.16	46.3	2.1	4.4
OPE F	1.8	3.8	2.11	42.8	1.9	4.4
Jeita [17]	320	470	1.46	36.5	0.9	2.5
MEMC [15]	470	677	1.44	28.6	0.6	2.0
p-CA [23]	12.0	22	1.83	32.5	1.4	4.3

[a] Selectivity $S = \text{removal}_{\text{crystal defect}}/\text{removal}_{\text{perfect crystal}}$.
$\text{Removal}_{\text{crystal defect}} = \text{removal}_{\text{perfect crystal}}$ + depth of etch pit.
r_e = removal rate, M = mol/L Cr(VI).

r = removal rate (nm/min); E_a = activation energy (kJ/mol); A = pre-exponential factor; R = gas law constant = 8.414 J/K mol; T = temperature (K).

The selectivity of a preferential etching solution is defined by the formula given in Table 9.3. It can be determined by measuring the removal rate of the material at the defect and the perfect crystal. The overall activation energies and the selectivities were determined experimentally for various preferential etching solutions (Table 9.3). The etch removal rate was measured with high accuracy by ellipsometry on SOI material (SOI films 80–120 nm thick). The selectivity was determined experimentally on dislocations for various defect etching chemistries. The dislocations were generated by damaging silicon substrates by controlled indentation with a diamond tip and subsequent annealing at 1000°C. After preferential etching the depth of the dislocation etch pits (Figure 9.25) was measured by an atomic force microscope (Figure 9.26). The etching solutions investigated may be grouped in four classes with regard to the oxidizing agent: dichromate (*Secco*), organic peracid (*OPE C, D*, and *F*), nitric acid (*Jeita*,[17] *MEMC*[15]), and p-chloranil (*p-CA*).

A comparison of the removal rates and activation energies of the different etching systems (oxidizing agents) outlined in Table 9.3 does not reveal any straightforward correlation between these parameters. The nitric acid systems (*Jeita*,[17] *MEMC*[15]) provide rather high removal rates with regard to their activation energies, the organic etching system with the oxidizing agent *p-CA* shows for a similar activation energy a removal rate that is more than one order of magnitude lower.

From the data of Table 9.3, the following conclusions can be drawn, however:

1. The mixtures with the same oxidizing agent exhibit common features with regard to their increase of etch rates at the dislocations, their selectivity S, their overall activation energy E_a of the perfect crystal, and the decrease of activation energy ΔE_a at the dislocations.

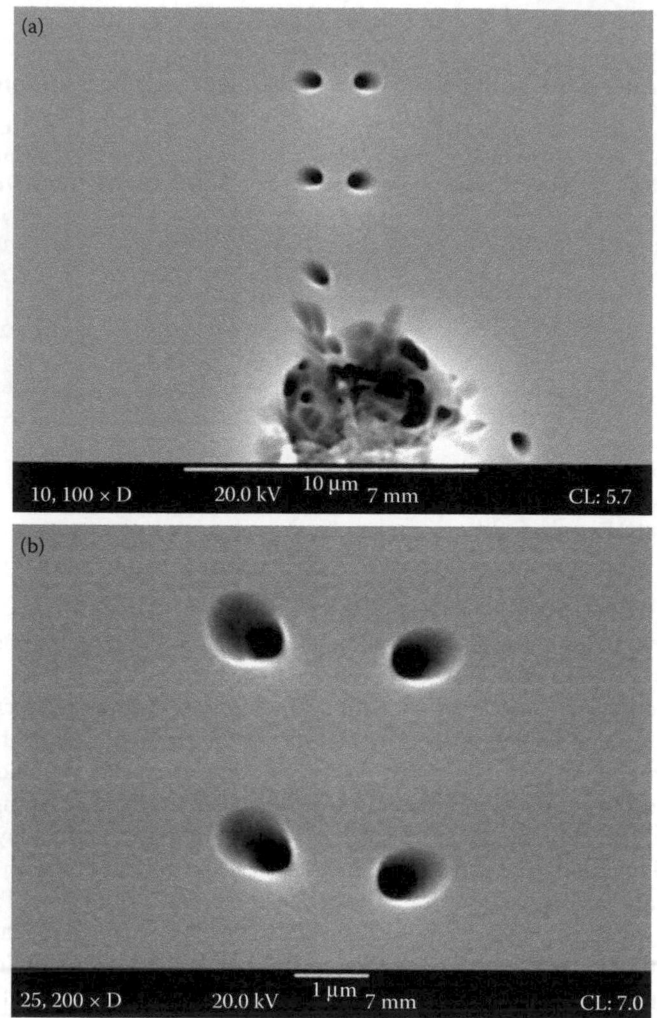

FIGURE 9.25 Indentation with dislocation half loops generated during annealing at 1000°C.[24] (a) SEM images of dislocation etch pits. (b) Magnified view. (From B.O. Kolbesen, J. Mähliß and D. Possner, *Phys. Stat. Sol. A* **208**(3), 584–587, 2011. With permission.)

2. At dislocations, the etch rates increase up to a factor of 2, the effect is highest for *OPE* solutions (111–115%), medium for *Secco* (71% and 83%) and *p-CA* (83%), and lowest for *Jeita* (47%) and *MEMC* (44%).
3. The decrease of the activation energy ΔE_a at the dislocations is nearly 5% for *dilute Secco*, *OPE* solutions, and *p-CA* and 2–2.5% for *MEMC* and *Jeita*. A rather high ΔE_a was determined for the *original Secco* etch (8.2%).

For dislocations, an additional copper decoration step prior to preferential etching raises the selectivity by 50% resulting in values up to 3.0.[59] The copper silicide

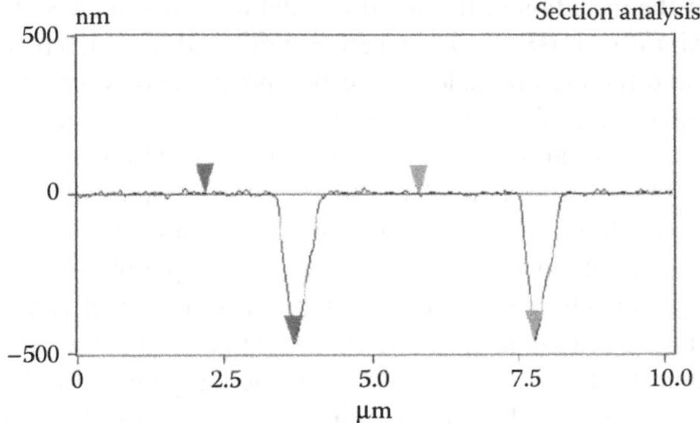

FIGURE 9.26 AFM measurement of the depth of two etch pits by section analysis (line scan). (From B.O. Kolbesen, J. Mähliß and D. Possner, *Phys. Stat. Sol. A* **208**(3), 584–587, 2011. With permission.)

formed at the dislocations provides a reaction path with distinctly enhanced etch rate due to the significantly reduced activation energy as sketched in Figure 9.22.

The missing of a straightforward correlation between the parameters measured (etch rate, activation energy, selectivity) for the different etching systems is attributed to the different interface active species of the etching systems investigated and their microscopic etching mechanisms. In the case of nitric acid, several small molecular N−O species stack to the silicon surface, taking part in the oxidation of the silicon.[32,33] In the *OPE*, *Secco* and *p-CA* etching systems, in a first step, medium-sized molecules and/or their ions (organic peracid, dichromate, and *p*-chloranil) interact with the (hydrogen-terminated) silicon surface, and in a second step, accomplish the electron transfer from the silicon to the oxidizing species. *Secco* and *p-CA* etching solutions show some similarities in their chemistry although their normal potentials and etch rates deviate distinctly. In both mixtures, elemental hydrogen is generated in the etching process of silicon.[23] The model outlined for the etching mechanism of the *p-CA* etching solution[23] may be applied in an analogous manner to the dichromate in the *Secco* etching solution.

9.7 SUMMARY

Chemical etching techniques ("preferential etching") in combination with light optical microscopy are still the workhorse for a quick and simple evaluation of crystal defect types and area densities in state-of-the-art silicon materials and engineered silicon substrates such as SOI. Most of the common solutions for defect etching contain toxic and carcinogenic dichromate or CrO_3 as oxidizing agent. In order to replace Cr(VI), etching chemistries free of chromium have been developed in recent years. Aiming at applications on thin silicon films such as on SOI, we have developed three classes of chromium-free preferential etching solutions: the *FS Cr-free SOI etches*, the *OPE*, and *OOE*. The *FS Cr-free SOI etches* consist of HNO_3, HF, acetic acid (HAc), and small amounts of a bromine or iodine component. The latter is necessary for the initiation of the silicon oxidation reaction on SOI material. *OPE* solutions are mixtures of HF, H_2O_2, and an organic acid in which by a slow reaction between H_2O_2 and the organic acid, the respective peracid is formed, the dominating oxidizing agent in *OPE* mixtures. *OOE* solutions contain an organic oxidizing agent such as *para*-chloranil (*p-CA*) dissolved in an organic solvent and HF. These three classes of chromium-free preferential etching solutions provide etch rates in the nm/min range (*OPE*), some nm/min (*p-CA*) and nm/s (*FS Cr-free SOI etches*). Preferential etching solutions can be characterized by parameters such as removal rate, overall activation energy for the etching process and selectivity. The selectivity describes the ratio of the removal at the crystal defect and at the perfect crystal. In this work, the selectivity was determined experimentally on dislocations for various defect etching chemistries. The dislocations were generated by damaging silicon substrates by controlled indentation with a diamond tip and subsequent annealing at 1000°C. After preferential etching, the depth of etch pits was measured by an atomic force microscope. At the dislocations, the activation energies determined are reduced by about 5%. For the various defect etching chemistries, selectivities between 1.4 and 2.2 were found, in copper decorated samples up to 3.0. No straightforward correlation between the parameters measured (etch rate, activation energy, selectivity) could be confirmed for the different etching systems. This is attributed to the different interface active species (organic peracid, dichromate, *p*-chloranil, and N−O species) of the etching systems investigated and their microscopic etching mechanisms.

ACKNOWLEDGMENTS

Part of this work was supported by SOITEC S.A., Parc technologique des Fontaines, Bernin. The author would like to thank François Brunier, Dr. Alexandra Abbadie, and Dr. Oleg Kononchuk from SOITEC for manifold support and stimulating discussions. The author thanks Doris Ceglarek from Goethe-University/Frankfurt-Main for the SEM and AFM work, Eric Mankel of Technical University of Darmstadt for the XPS measurements, and Dr. Burkhard Beckhoff and Falk Reinhard from the Physikalisch-Technische Bundesanstalt (PTB) for the NEXAFS measurement at BESSY II/Berlin. The excellent collaboration of my former PhD students Dr. Jochen Mähliß, Dr. Daniel Possner, Dr. Hanan Idrisi, and Tito Sanetti is gratefully acknowledged.

REFERENCES

1. M. Bruel, US Patent No. 5,374,564, 1994.
2. S. Sama, M. Porrini, F. Fogale and M. Servidori, *J. Electrochem. Soc.* **148**(9), G517–G523, 2001.
3. P. Papakonstantinou, K. Somasundram, X. Cao and W.A. Nevin, *J. Electrochem. Soc.* **148**(29), G36–G42, 2001.
4. G. Dornberger, D. Temmler, W.v. Ammon and S. Sama, *J. Electrochem. Soc.* **149**(4), G226–G231, 2002.
5. G. Pfeiffer, M. Haag, M. Schmidt, R. Krause, P. Tsai and J.D. Lee, *ECS Trans.* **2**(2), 167, 2006.

6. A. Abbadie, S.W. Bedell, J.M. Hartmann, O. Kononchuk, D.K. Sadana, F. Brunier, C. Figuet and I. Cayrefourq, *J. Electrochem. Soc.* **154**(8), 2007.
7. E. Sirtl and A. Adler, *Z. F. Metallkunde* **52**, 529–532, 1961.
8. F. Secco d'Aragona, *J. Electrochem. Soc.* **119**(7), 948–951, 1972.
9. D.G. Schimmel, *J. Electrochem. Soc.* **123**(5), 734–741, 1976.
10. M. Wright Jenkins, *J. Electrochem. Soc.* **124**(5), 757–762, 1977.
11. K. Yang, *J. Electrochem. Soc.* **131**, 1140–1145, 1984.
12. S.W. Bedell, H. Chen, D.K. Sadana, K. Fogel and A. Domenicucci, *Electrochem. Solid State Lett.* **7**(5), G105–G107, 2004.
13. W.C. Dash, *J. Appl. Phys.* **27**, 1193, 1956.
14. B.L. Sopori, *J. Electrochem. Soc.* **131**, 667, 1984.
15. T.C. Chandler, *J. Electrochem. Soc.* **137**(3), 944, 1990.
16. K. Graff and P. Heim, *J. Electrochem. Soc.* **141**, 2821–2825, 1994.
17. JEITA EM-3603 E 3.0.
18. B.O. Kolbesen, D. Possner and J. Mähliß, *ECS Trans.* **11**, 195–206, 2007.
19. J. Mähliß, A. Abbadie and B.O. Kolbesen, *ECS Trans.* **6**(4), 271, 2007.
20. D. Possner, B.O. Kolbesen, V. Klüppel and H. Cerva, *ECS Trans.* **10**(1), 2007 21.
21. J. Mähliß, A. Abbadie, F. Brunier and B.O. Kolbesen, *ECS Trans.* **16**(6), 309, 2008.
22. J. Mähliß, A. Abbadie and B.O. Kolbesen, *Mater. Sci. Eng. B* **159–160**, 309–313, 2009.
23. J. Mähliß, R. Hakim, A. Abbadie, F. Brunier and B.O. Kolbesen, *J. Electrochem. Soc.* **158**(2), D107–1132011.
24. B.O. Kolbesen, J. Mähliß and D. Possner, *Phys. Stat. Sol. A* **208**(3), 584–587, 2011.
25. H.C. Gatos and M.C. Lavine, Chemical behavior of semiconductors: Etching characteristics. *Prog. Semicon.* **9**, 1–46, 1962.
26. A.F. Bogenschütz, *Ätzpraxis für Halbleiter*, Hanser, Munich, 1967.
27. H. Robbins and B. Schwartz, *J. Electrochem. Soc.* **106**, 505, 1959.
28. H. Robbins and B. Schwartz, *J. Electrochem. Soc.* **107**, 108–111, 1960.
29. B. Schwartz and H. Robbins, *J. Electrochem. Soc.* **108**, 365–372, 1961.
30. B. Schwartz and H. Robbins, *J. Electrochem. Soc.* **123**(12), 1903–1909, 1976.
31. M. Steinert, J. Acker, A. Henßge and K. Wetzig, *J. Electrochem. Soc.* **152**(12), C843–C850, 2005.
32. M. Steinert, J. Acker, M. Krause, S. Oswald and K. Wetzig, *J. Phys. Chem. B* **110,** 11377–11382, 2006.
33. M. Steinert, J. Acker, S. Oswald and K. Wetzig, *J. Phys. Chem. C* **111,** 2133–2140, 2007.
34. E. Abel and H. Schmid, *Z. Phys. Chem. Stoechiom. Verwandtschaftsl.* **132**, 55, 1928.
35. D.R. Turner, On the mechanism of chemically etching germanium and silicon. *J. Electrochem. Soc.* **107**, 810–816, 1960.
36. V. Lehmann, *The Chemical Dissolution of Silicon, The Electrochemistry of Silicon: Instrumentation, Science, Materials and Applications*, Wiley-VCH, Weinheim, Germany, pp. 23–38, 2002.
37. G.A. Rozgonyi, in *Encyclopedia of Materials: Science and Technology*, S. Mahajan (ed.), Elsevier Science Ltd, Amsterdam, pp. 8524–8533, 2001.
38. D.G. Schimmel, *J. Electrochem. Soc.* **126**(3), 479–762, 1979.
39. S. Meltzer and D. Mandler, *J. Chem. Soc. Faraday Trans.* **91**(6), 1019–1024, 1995.
40. T.K. Carns, M.O. Tanner and K.L. Wang, Chemical etching of $Si_{1-X}Ge_X$ in $HF:H_2O_2:CH_3COOH$, *J. Electrochem. Soc.* **142**(4) 1260–1266, 1995.
41. European Patent Application Publ. No. 0 452 120 A1, 1991.
42. D. Swern, *Chem. Rev.* **45**, 11949.
43. D. Possner, Organic peracid etches: A new class of chromium-free etching solutions for the delineation of defects in thin silicon films. PhD thesis, Goethe-University, Frankfurt/Main, Germany, 2009.

44. H. Yamagishi, I. Fusegawa, N. Fujimaki and M. Katayama, *Semicond. Sci. Technol.* **7**, A135–A140, 1992.
45. W. Wijaranakula, *J. Electrochem. Soc.* **141**(12), 3273–3277, 1994.
46. T. Abe and S. Maryama, *Denki Kagaku* **35**, 149, 1967.
47. A.J.R. deKock, *Philips Res. Repts. Suppl.* **1**, 1973.
48. H. Föll and B.O. Kolbesen, *Appl. Phys.* **8**, 319–331, 1975.
49. L. Michaelis and M.P. Schubert, *Chem. Rev.* **22**(3), 437–470, 1938.
50. H.S. Rzepa and G.A. Suñer, *J. Chem. Soc. Chem. Comm.* 1743–1744, 1993.
51. H.G. Cassidy, *J. Polym. Sci. Macromol. Rev*. **6**(1), 1–58, 1981.
52. A.A. Kutyrev and V.V. Moskva, *Russ. Chem. Rev.* **56**, 1028–1044, 1987.
53. B.O. Kolbesen and H.P. Strunk, Analysis, electrical effects and prevention of process-induced defects in silicon integrated circuits, in *VLSI Electronics: Microstructure Science* N.G. Einspruch (ed.), Vol. **12**, *Silicon Materials* H.R. Huff (ed.), Academic Press, 1985.
54. R.J. Jaccodine, *J. Electrochem. Soc.* **110**, 524–527, 1963.
55. R.M. Maatman and A. Kramer, *J. Phys. Chem.* **72**, 104–108, 1968.
56. R.K. Iler, *The Chemistry of Silica*, Wiley, New York, p. 676, 1979.
57. L.A.J. Garvie, A.J. Craven and R. Brydson, *Am. Mineral.* **79**, 411–425, 1994.
58. A.B.P. Lever, *Inorganic Electronic Spectroscopy: Studies in Physical and Theoretical Chemistry*, Vol. **33**, Elsevier, Amsterdam, 1984, 863 p.
59. H. Idrisi, V. Sinke, D. Gerhardt, D. Ceglarek and B.O. Kolbesen, *Phys. Stat. Sol.* **C8**(3), 788–791, 2011.

10 Investigation of Defects and Impurities in Silicon by Infrared and Photoluminescence Spectroscopies

Simona Binetti and Adele Sassella

CONTENTS

10.1 Introduction 323
10.1.1 Basics of *Infrared Spectroscopy* 324
10.1.2 Basics of *Photoluminescence Spectroscopy* 327
10.2 Nondoping Impurities and Related Defects 331
10.2.1 Oxygen 331
10.2.1.1 Luminescence Lines Related to Old Thermal Donors 334
10.2.1.2 Luminescence Lines Related to Oxygen Precipitates 335
10.2.2 Nitrogen 336
10.2.3 Carbon 338
10.2.4 Hydrogen 341
10.3 Dopants 342
10.4 Dislocations 345
10.5 Summary 348
Acknowledgments 348
References 348

10.1 INTRODUCTION

Spectroscopic studies of defects and impurities in silicon have been a key factor in the development of silicon-based electronic devices. Silicon is still the material of choice for microelectronic and photovoltaic applications and the candidate for future applications in optoelectronics. For this reason, it is still the subject of dedicated experimental and theoretical research, in particular for clarifying the role of defects in view of the complete exploitation of its potentialities for the development of advanced technological processes.

When addressing such a wide research subject, having strong connections with technology and a large number of different applications, it is difficult but absolutely necessary to restrict the discussion to the most relevant aspects for the specific goal of the chapter. In this respect, this chapter about the use of spectroscopic techniques for the analysis and study of impurities and defects in silicon requires to be clearly focused, specifying the type of silicon materials as well as the techniques of interest.

Single crystal silicon wafers are used for nearly all applications in microelectronics, thanks to their highly controlled and reproducible properties, while polycrystalline or microcrystalline materials are of common use in solar cell applications, where lowering the purity brings the advantage of lowering costs without significant limitations to performances. These types of silicon materials are therefore considered here, being the most studied and applied ones. However, other silicon materials, where crystallinity is limited or absent, for example, amorphous and porous silicon with their wide variety, as well as radiation-treated silicon, are excluded from the present work, due to their strong process-dependent properties, which prevents any general discussion. Among the rich group of impurities in silicon, oxygen, nitrogen, carbon, and hydrogen are considered (the latter is just cited here, being the subject of Chapter 7), with related defects. In addition, since doping is fundamental in silicon applications, dopants cannot be neglected, as they influence not only the transport properties, but also the physical and chemical properties, such as density and type of defects and defect-related properties. Finally, a specific discussion is devoted to dislocations, the unique extended defects which can be studied in depth by spectroscopy.

As for the spectroscopic tools, this chapter deals with the optical techniques most widely diffused in the field of silicon-related research, namely, infrared (IR) spectroscopy and photoluminescence (PL), of common use both for sample characterization and monitoring and for deeper studies of production processes and of their influence on silicon's physical and chemical properties. Raman spectroscopy, giving results less straightforward to interpret, being less sensitive and not quantitative, is much less applied on crystalline silicon as a characterization tool and is therefore neglected here.

10.1.1 Basics of *Infrared Spectroscopy*

IR spectroscopy is among the most widely used optical techniques for the study of impurities and defects, thanks to its characteristic advantages. Indeed, it couples a very high sensitivity to the possibility to obtain quantitative results about the species detected; in addition, it is rather easy to use, at least in the standard experimental configurations, nondestructive, as most of optical techniques, and cheap. IR spectroscopy[1,2] is sensitive to the vibrational properties of molecules and solids and consists of detecting the intensity loss of an IR light beam after its interaction with the sample, where it excites resonantly some characteristic vibrations. In crystalline solids, such vibrations can be intrinsic lattice phonons or localized vibrations, again intrinsic or related to the presence of impurities, the latter being those of interest for the present chapter. Localized vibrations rise when few atoms, say an impurity and the atoms of the host crystal, move one with respect to the others due to the excitation by the light electromagnetic field. Such localized vibrational modes do not propagate into the crystal, they may possess different symmetry, which makes them IR-active or Raman-active, following precise rules. In particular, when the atomic displacements originate some dipole moments, the related vibration is infrared-active. A good model for the physical description of a vibrational mode consists of a harmonic oscillator, whose frequency ν is related to an effective mass μ for the oscillating atoms and to an effective elastic constant k accounting for the atomic bond strength:

$$\nu = \frac{1}{2\pi}\sqrt{\frac{k}{\mu}} \tag{10.1}$$

When the frequency of the IR light interacting with the oscillators matches ν, resonance occurs and, therefore, part of the light intensity gets absorbed. Figure 10.1a displays a typical experimental configuration for optical transmission measurements: a sample of thickness d is represented in cross section (gray rectangle) and a light beam of intensity I_0 is impinging on it from the left; part of the light is reflected back, with intensity I_R, and part is transmitted, with intensity I_T. Under resonance between the light and a specific vibrational mode, part of the incident light is absorbed and I_0 gets reduced. As a consequence, the spectrum measured collecting I_T as a function of the energy of the light beam, and defining the spectroscopic unit transmittance T, used to draw the spectrum, as

$$T = \frac{I_T}{I_0} \tag{10.2}$$

displays a minimum. In principle, I_R should also be measured, and this needs specific experimental configurations, but its interpretation is less straightforward and is much less used for the study of impurities and defects in silicon, being in addition particularly low. From the transmission measurements, absorbance can be evaluated and used as a spectroscopic unit useful to quantify the amount of light intensity absorbed by the sample, with the absorbance A defined as

$$A = \log_{10}\frac{I_0}{I_T} = -\log_{10} T \tag{10.3}$$

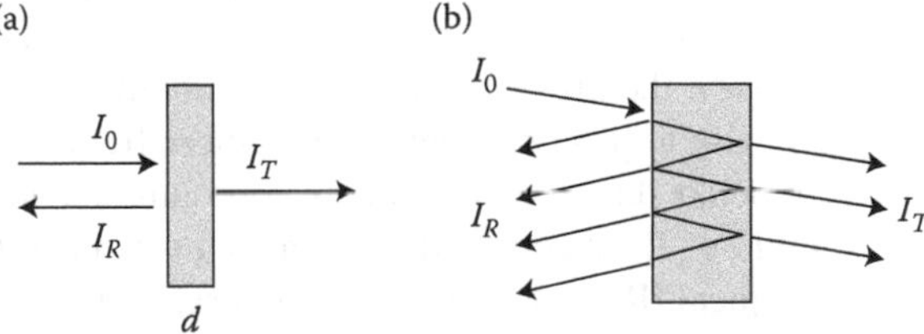

FIGURE 10.1 (a) Experimental configuration for optical measurements in transmission mode, where a light beam of intensity I_0 impinges on a sample (gray) of thickness d and is partly reflected and partly transmitted, then originating two beams of intensity I_R and I_T, respectively. (b) Detail of the light beam inside the sample, where it undergoes multiple reflections which, under proper conditions, may give rise to interference.

Given this definition, A is always positive and shows maxima (much expanded due to the $\log_{10}$ in Equation 10.3) where T shows minima.

It should be noted that absorbance correctly represents the light intensity absorbed by the sample only when reflected intensity can be neglected (or after a proper correction for reflection). Indeed, conservation of energy would require I_0 to equal the sum of reflected, absorbed, and transmitted intensity, so that low transmittance could also originate from a high reflection. In general, I_R also include contributions from multiple reflections inside the sample (see Figure 10.1b), when it possesses plane and parallel surfaces. In this case, for certain values of d such reflections give rise to interference, so that the resulting spectra are affected by a wavy background, which prevents a proper study of its characteristics. To avoid this problem, usually the back surface of the samples for IR characterization is made rough at the μm-scale, so that it scatters light in that range of wavelength, which therefore is not reflected back and forth within the sample. Another way to avoid the problem of interference is collecting low-resolution spectra, where the signal modulations cannot be recorded. Of course, this latter method works when the resolution needed for the specific study is not too high.

Given the atomic masses and the binding energies in silicon and in most solids, typically from 0.5 eV down to few meV, vibrational modes are detected in the IR spectral range. In this range, the wavenumber $\bar{\nu}$, that is, the reciprocal wavelength $\bar{\nu} = 1/\lambda$, is usually used as a spectroscopic unit, expressed in cm^{-1}. In terms of $\bar{\nu}$, the medium IR (MIR) range, where most vibrational modes can be detected, spans from 5000 to 500 cm^{-1}; for vibrations of atoms possessing a high mass, frequencies are lower (see Equation 10.1) and wavenumbers higher, so that the spectroscopic response is detected in the far IR (FIR) range, from 500 down to few cm^{-1}. These are the commonly used spectral limits and are therefore the reference values for this chapter.

Besides the characteristics spectral position $\bar{\nu}$, a vibrational mode is characterized by the intensity and the width of the corresponding absorption peak in the IR spectrum, either represented in transmittance or in absorbance. The intensity is related to the number of bonds (or atoms) of the specific type absorbing the IR light and, as such, can be used to quantify those impurities or defects. The bandwidth, often measured in terms of the peak full-width at half-maximum (FWHM), is related to slight differences in the resonance energy of the specific absorbing species, typically related to the local lattice disorder surrounding them (and expressing a measure of crystallinity). In some cases, the presence and relative amount of isotopes of a particular impurity can give rise to peak broadening, and more frequently, under proper resolution, the different peaks due to each single isotope can be distinguished.

To provide an example of IR spectra, which is also a reference for this chapter, Figure 10.2 shows two typical IR absorption spectra of crystalline silicon in the MIR range: the spectrum in (a) is collected at room temperature with a Fourier-transform spectrometer by averaging 100 scans performed using 32 cm^{-1} resolution on a 500-μm-thick float-zone (FZ) silicon sample; in (b) a similar spectrum is reported, as collected under the same experimental conditions on a Czochralski (CZ) silicon sample of the same thickness, which is known to contain, among other impurities, interstitial oxygen (O_i, here with 10.13×10^{17} cm^{-3} concentration). In both spectra,

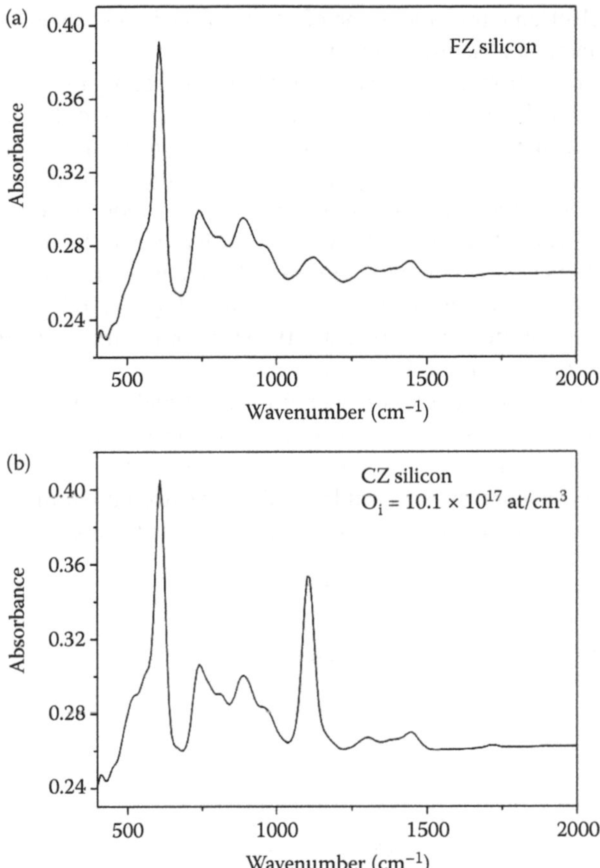

FIGURE 10.2 Absorbance spectra of (a) a float-zone and (b) a Czochralski Si sample with 10.1×10^{17} cm^{-3} interstitial oxygen concentration in the spectral range from 2000 to 400 cm^{-1}.

the typical response of single crystal silicon can be observed, with several peaks due to the crystal phonons from about 500–1000 cm^{-1}; in addition, in the spectrum of CZ silicon in Figure 10.2b, peaks related to localized vibration related to the presence of O_i are clearly observed, first of all the strong peak at 1107 cm^{-1}.

10.1.2 Basics of *Photoluminescence Spectroscopy*

All the phenomena that involve absorption of energy and the subsequent emission of light are classified generically under the term "luminescence." In the case of photon excitation, this luminescence is called *photoluminescence*, the other most common phenomena being cathodoluminescence (after excitation by accelerated electrons) and electroluminescence (after excitation by carrier injection). Basically, photoluminescence rises from an electronic transition from an excited state to a lower state. Therefore, it is a powerful, contactless, nondestructive method of probing both intrinsic electronic transitions and electronic transitions related to impurities or defects in insulators and semiconductors.

If the incident photon has energy greater than the band gap, it can be absorbed by an electron, which gets excited from the valence band up to the conduction band across the forbidden energy gap. In this process of photoexcitation, the electron generally gets excess energy, so that they tend to relax to the band edges, reaching thermal equilibrium with the lattice (the so-called thermalization process). Relaxation of excited carriers generally occurs in times of the order of picoseconds.[3] The produced electron–hole pairs can recombine in several different ways, radiatively or not radiatively. The main radiative decay processes stimulated by above band gap photoexcitation are depicted in the Figure 10.3, shortly described in the following (for a more exhaustive description, refer to Reference 4).

In high-purity semiconductors (or in insulators) and at low temperatures, the excess electrons and holes can form excitons, hydrogen-like complexes where the electron and the hole are bound together by the Coulomb attraction. Excitons can recombine emitting a narrow spectral line called free-exciton luminescence.

The energy of the emitted photon ($h\nu$) in a direct band gap semiconductor is

$$h\nu = E_g - E_x \tag{10.4}$$

and in an indirect band gap semiconductor is

$$h\nu = E_g - E_x - E_p \tag{10.5}$$

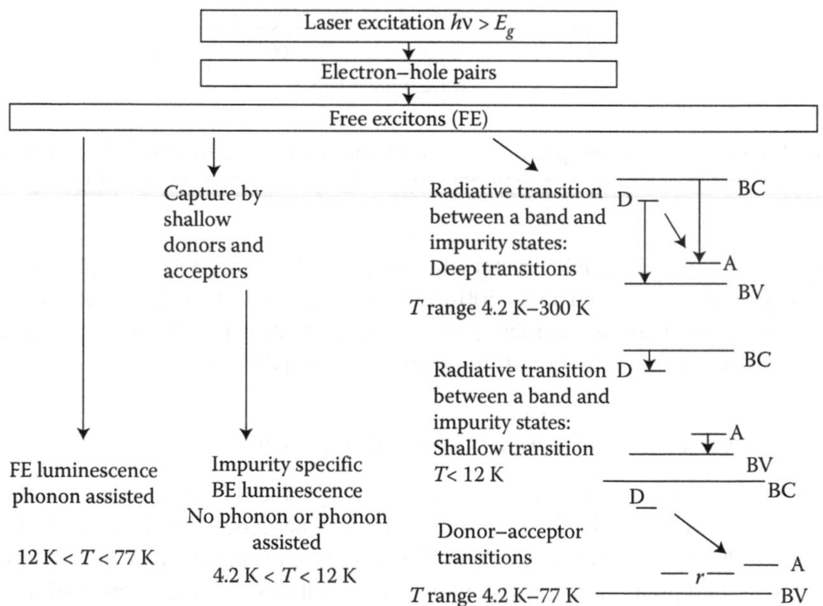

FIGURE 10.3 Main luminescence decay processes stimulated by above band gap excitation. The temperature range in which they are detectable is also reported.

where E_x is the ionization energy of the exciton (necessary to separate the $e^-–h^+$ couple), E_g is the energy gap value, and E_p is the energy of the phonons involved for momentum conservation.

In the presence of impurities (usually dopants), excitons can be bound to them and the energy of the emitted photons is given by

$$h\nu = E_g - E_x - E_{loc} \tag{10.6}$$

where E_{loc} is the binding energy of the exciton to the impurity. The binding energy of excitons at neutral or ionized impurities depends on the electron-to-hole effective mass ratio. The experimentally observed reduction in exciton energy by formation of complexes with neutral donors (acceptors) in many case is proportional to the energy level of the impurity (E_B) and range between 0.05 and 0.3 times E_B (the Haynes rules).[5]

At such a temperature at which $kT > E_x$, where the excitons are broken into free carriers, such free carriers can recombine radiatively in a band-to-band transition. In this case, in a direct band gap semiconductor, the emission spectrum is given by

$$I(\nu) = A(h\nu - E_g)^{1/2} \tag{10.7}$$

where A is the transition probability, related to the reduced mass, and the gap value.

In an indirect band gap semiconductor, band-to-band transitions must be mediated by a phonon emission (more likely) or phonon absorption processes to conserve momentum, so that the emission spectrum is given by

$$I(\nu) = A'(h\nu - E_g + E_p)^{1/2} \tag{10.8}$$

where E_p is the energy of the emitted phonon and A' the transition probability, much smaller than in the direct transitions.

When semiconductors contain impurities or defects, the radiative recombination process can involve intragap energy levels and the recombination path from band to impurity levels is usually named “free-to-bound transition.” Free-to-bound transitions in strongly doped crystals can produce efficient light emission even at room temperature. At low temperature, these transitions are usually with lower intensity than other impurity-assisted radiative recombination processes, such as the bound-exciton recombination and the donor–acceptor pair transitions.

The band to impurity transitions can be distinguished in shallow transitions and deep transitions, depending on their energy. Shallow transitions, detectable in the far infrared, are transitions of an electron from the conduction band to dopant levels or of a hole from its acceptor level to the valence band. Deep transitions are either the transition of an electron from the conduction band to an acceptor state or a transition from a donor to the valence band. Such transitions emit a photon with energy

$$h\nu = E_g - E_i \tag{10.9}$$

where E_i is the impurity binding energy level.

If both types of charge carriers are present, donor–acceptor (DA) transitions can also occur. The recombination energy of this emission is given by

$$h\nu = E_g - (E_D - E_A) + \frac{e^2}{4\pi\varepsilon_r\varepsilon_0 r} - \frac{a}{r^6} \tag{10.10}$$

where E_D and E_A are the donor and acceptor ionization energies. The last term in Equation 10.10 accounts for the van der Waals (VdW) interaction, where a is the effective VdW constant for a given DA pair, r is the DA pair separation, ε_0 is the relative dielectric constant of vacuum, and ε_r is the low-frequency relative dielectric constant of the semiconductor.

Since r is restricted to discrete values, the emission spectra exhibiting many sharp lines correspond to the different discrete values of r between donors and acceptors. Therefore, E_D and E_A can be obtained accurately from a comparison of theoretical calculations and the observed fine structure of the DA pair luminescence.[6,7]

All instruments for PL measurements contain four basic items: a source of light, a sample holder, a monochromator, and a detector. In addition, to be of analytical use, the wavelength of the excitation radiation needs to be selectable. The most convenient excitation source is a laser with $h\nu > E_g$. For silicon, the 514 nm line from an Ar ion laser, the 633 nm line of He–Ne laser, or the 810 nm line of solid-state lasers are the most used excitation lines. If AC amplification and phase-sensitive detection (lock-in amplifier) are employed, the excitation light needs to be modulated with a chopper. A filter in front of the detector is useful to prevent the exciting laser light from entering the monochromator.

Most photoluminescence measurements are carried out at low temperature to prevent thermal ionization of the optically active centers and to minimize the broadening of the spectral lines by lattice vibrations. Therefore, the sample holder is usually inside a cryostat using liquid helium or nitrogen; usually, the temperature range between 12 and 300 K can be scanned, but measurements down to 4 K are also possible.

In summary, the main advantages of photoluminescence spectroscopy are

- It is a nondestructive technique and only small quantities of material are needed.
- It does not require particular sample preparation and handling.
- Impurities and defects can be detected even at low concentration.
- It is sensitive to the chemical species.

The main disadvantage is the difficulty to have quantitative information. For practical purposes, the application of the PL spectroscopy to the quantitative analysis of impurities is of great concern. Generally, luminescence intensity associated with a particular process depends on the relative and absolute concentration of various impurities and defects, on the excitation power density, on surface recombination channels, on sample temperature, and also involves nonradiative decay process. In some cases, such as the case of dopants in silicon, quantitative data from PL can be obtained but very accurate calibrations are required (see Section 10.3).

Another critical point is that defect and impurity identification is not straightforward because photoluminescence provides information only on defects involved in radiative transitions; finally, PL is not a bulk characterization technique. The interested sample regions can be estimated as the largest one between the minority carrier diffusion length and the light penetration depth.

In more than 50 years of investigations, photoluminescence spectroscopy has yielded a large amount of data about the defects responsible for particular luminescence lines. Nonetheless, drawing a detailed structural model of the defect responsible for a particular luminescence feature requires a combination of results from several techniques (such as electron paramagnetic resonance (EPR), infrared absorption, and electrical measurements). A very exhaustive review work on the luminescence centers in silicon is the one by G. Davies,[8] where the physics of the radiative transitions in silicon and the correlation with defect models are described with great care. More recently, supplementary data on optical defects have been published in the volume devoted to silicon of the Landolt–Börnstein Group III Condensed Matter series.[9]

10.2 NONDOPING IMPURITIES AND RELATED DEFECTS

10.2.1 Oxygen

Crystalline silicon for microelectronics or photovoltaic application is always a supersaturated solid solution of oxygen in silicon. In the case of CZ silicon, oxygen concentration at 300 K is equal to the solubility of oxygen in liquid silicon at 1412°C, the temperature of ingot pulling. In addition, it is well known that a long dwelling at temperatures around 450°C during the cooling cycle of oxygen-rich CZ silicon ingots, or any thermal annealing of CZ silicon in the same temperature range, leads to the generation of oxygen-related donors, often referred to as old thermal donors (OTD).

Oxygen is, therefore, the main nondoping impurity in CZ silicon and, as such, its IR response has been deeply investigated and is currently the main tool for measuring the amount of interstitial oxygen (O_i), with precise standards of common use for different temperatures.[10]

The presence of O_i is manifested by the IR absorption peaks due to the symmetric and antisymmetric stretching of the defect-molecule Si−O−Si at 1107 cm^{-1} and at 1205 cm^{-1}, respectively, and one at 515 cm^{-1} due to the bending vibration of the same structure. These peaks are shown in Figure 10.4, as obtained by normalizing the spectrum of a CZ silicon sample to that of an FZ silicon sample; the inset better shows the 1205 cm^{-1} peak, hardly visible in the spectrum over the whole range due to its rather low intensity. The peak at 1107 cm^{-1} has been calibrated and used for the quantitative measurements of oxygen concentration: the main problem in doing such a calibration was the fact that the response of silicon crystal phonons is present in the same spectral region (see Figure 10.4), so that comparison with FZ O-free silicon and with spectra at low temperatures had to be considered.[11,12]

O-related defects, first of all oxide precipitates, but also thermal donors of different origin and displaying different electrical activity, have been the subject of many studies,[10] having a clear influence on the electrical and mechanical properties

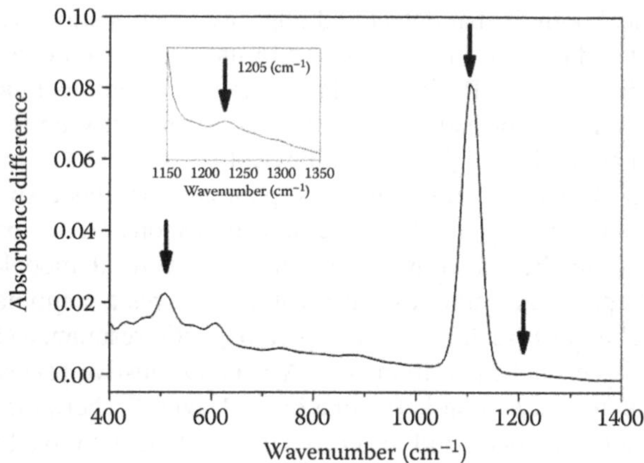

FIGURE 10.4 IR absorbance spectrum of a CZ silicon sample with 10.1×10^{17} cm^{-3} interstitial oxygen concentration, normalized to the spectrum of an FZ silicon sample of the same thickness, collected under the same experimental conditions; inset: zoom in the spectral range from 1150 to 1350 cm^{-1}.

of the host silicon crystal and on the devices built on it. In the typical IR spectra of silicon, some absorption bands are attributed to oxide precipitates, typically from about 1200–1250 cm^{-1}, where they can be easily detected by spectroscopy, even if a quantitative estimate is hard to be reached. Actually, indirect precipitate quantification by measuring the decrease of interstitial oxygen has been sometimes proposed, but it is known to present several limits. Direct detection and study of the precipitate response also creates some problems, as demonstrated by detailed studies, which could go deeper into the interpretation of the origin of the typical IR bands rising with oxygen precipitation; indeed, their spectral position and line shape depend on both the shape and the composition of the oxide precipitates, not easy to be determined, as well as on temperature.[13] Figure 10.5a demonstrates the dependence on precipitate stoichiometry by showing calculated spectra, where different suboxides, from SiO up to $SiO_{1.9}$, are considered as the amorphous material composing the particles of a specific shape, occupying the same volume within the host Si crystal. When different SiO_2 polymorphs are also considered, even larger differences are found.[14] When the precipitate shape is considered, the typical ellipsoidal particles are considered and modeled, with various aspect ratio; in Figure 10.5b the results are shown for SiO_2 precipitates occupying the same volume fraction within the silicon crystal, but displaying a different shape.[15]

Finally, it is important to note how the precise knowledge of the precipitate-related response also permits a more precise determination of O_i content[11,16] even if this procedure requires low temperature measurements.

In silicon for solar cells,[17] oxygen tends to segregate close to structural defects, giving rise to IR absorption bands in the range from about 980 to 1100 cm^{-1}; low-temperature measurements are needed to distinguish those contributions from the absorption by interstitial oxygen, making quantitative evaluation of both difficult.

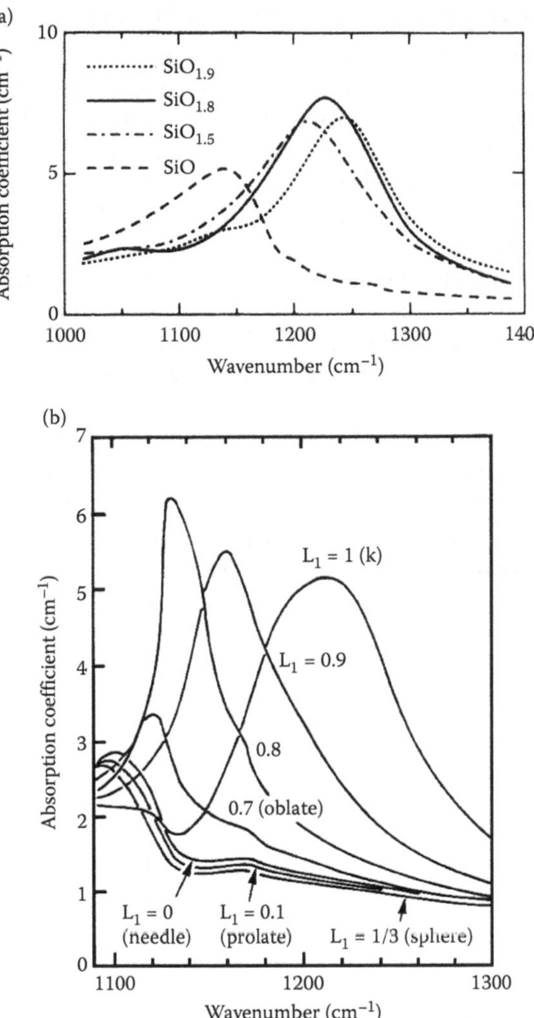

FIGURE 10.5 IR absorbance spectrum calculated for a CZ silicon sample where the response due to platelet precipitates is observed to be around 1200 cm^{-1}, changing with the precipitate (a) composition and (b) shape. ((a) Reprinted with permission from A. Borghesi et al. *Phys. Rev.* B 46, 4123–4127, 1992. Copyright 1992 by the American Physical Society. (b) Reprinted with permission from S.M. Hu, *J. Appl. Phys.* 51, 5945–5948, 1980. Copyright 1980, American Institute of Physics.)

As for other O-related defects, for example, thermal donors, a direct IR response cannot be found. However, IR absorption lines associated with electronic transitions from the ground state into excited states of thermal donors can be observed at low temperature[18–20] in the region between 300 and 900 cm^{-1}. On the contrary, while interstitial oxygen does not introduce any PL lines, oxygen agglomerates complexes are involved in different photoluminescence centers.

10.2.1.1 Luminescence Lines Related to Old Thermal Donors

Since the discovery of thermal donors (TD), many studies have been devoted to their microscopic structure, the growth and decomposition kinetics, as well as to the electrical and optical properties, using all the techniques suitable for the study of electrically and optically active, paramagnetic centers. Despite this, a defect model that consistently explains all their features, including their optical properties, is still lacking. Anyway, it has been generally accepted that TDs are a family of similar shallow double donor defects.

After the first work of Tajima et al.,[21] bound-exciton lines originating from species bound to thermal donors (TD) were studied in detail and compared to EPR analysis on the same CZ silicon samples.[22] The TD-bound-exciton PL spectra are reported in Figure 10.6 where separate lines can be resolved and related to recombination at different TDs.

A series of lines, marked O-lines, near 1175 meV appear after annealing at 450°C for times longer than that generally required to generate the TDs (around 100 h or more).[23,24] Similarly, a series of 13 sharp lines, labeled the S lines, in the range between 1108 and 1014 meV appear in the PL spectrum of silicon with both high C and O content, annealed at 500°C for more than 100 h.[23] In addition to TD-related bound-exciton lines, oxygen-rich silicon annealed at 450°C for several hours exhibits

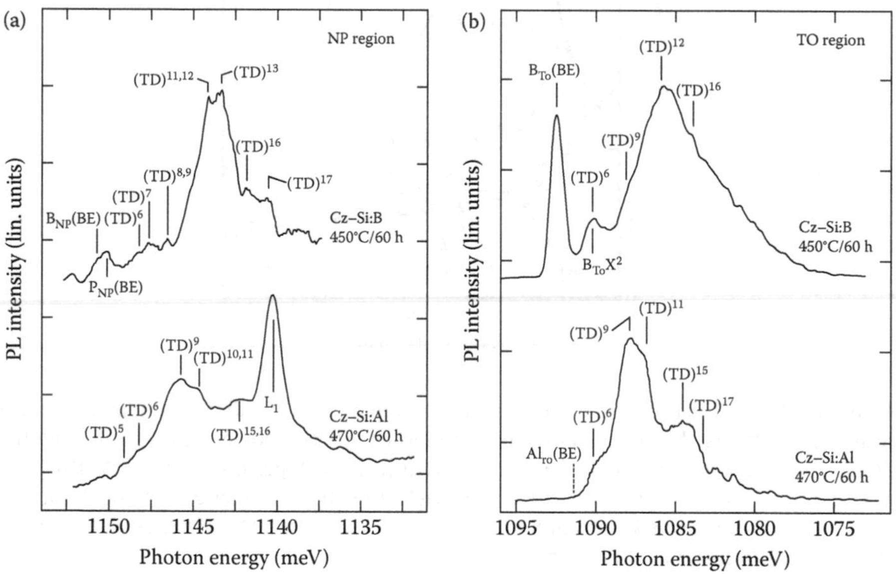

FIGURE 10.6 TD-bound-exciton photoluminescence lines in boron- and aluminum-doped CZ silicon samples submitted to thermal treatment at 470°C for 60 h. Apart from the boron-related luminescence lines, separate lines can be resolved and related to bound-exciton recombination at different TD's market with $(TD)^j$. (a) No phonon (NP) range, (b) TO phonon-assisted range. Spectra were recorded at 4.2 K. (Reprinted with permission from B.J. Heijmink, T. Gregorkiewicz, C.A. Ammerlaan, *Phys. Rev. B* 46, 2034–2040, 1992. Copyright 1992 by the American Physical Society.)

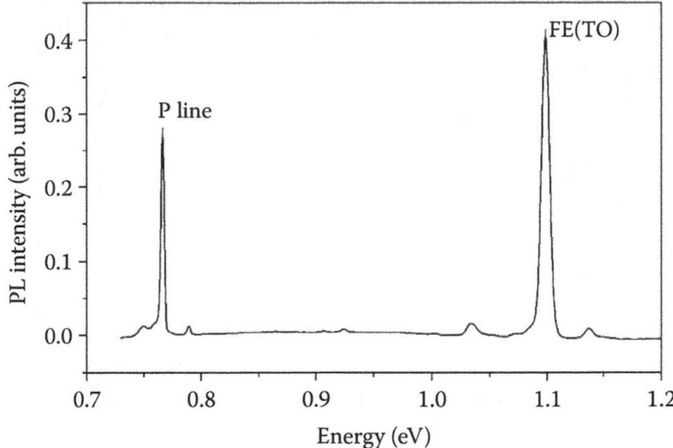

FIGURE 10.7 PL spectrum of a CZ silicon samples annealed at 450°C for 24 h. The spectrum was recorded at 12 K.

a prominent photoluminescence spectrum with a narrow no-phonon line at 0.767 eV, generally labeled P line[25,26] (see Figure 10.7).

A carbon isotopic effect experimentally determined in the P line demonstrated the involvement of carbon–oxygen complexes in the center.[27] The P line is still visible at room temperature[28] and it has also been used as a "diagnostic" line. The P line, for instance, was used to study the thermal history of a CZ-pulled ingot.[29]

Based on the PL intensity of the peak around 0.76 eV observed at room temperature in a crystal pulled at a low growth rate, it was found that the effect of the thermal history of a crystal pulled at 0.8 mm/min was equivalent to an annealing at 500°C for 3 h.

10.2.1.2 Luminescence Lines Related to Oxygen Precipitates

When studying other O-related defects, distinguishing between luminescence lines associated with dislocation or with oxygen precipitates is a problem of nontrivial solution. Indeed, it is well known that the growth of silicon oxide precipitates from silicon oxide embryos at temperatures as high as 800°C, or more, occurs with the simultaneous emission of self-interstitials, which, after clustering, are the precursors of dislocations. In addition, heterogeneous precipitation of oxygen at dislocations is also a favorite process, so that dislocations in CZ silicon are very often decorated with oxygen or oxygen precipitates.

Tajima et al.[30] were the first to assign to oxygen precipitates luminescence lines detected at 0.768 eV at room temperature and around 0.820 eV in the T range from 77 to 150 K; the band labeled D_b was identified as being due to oxygen precipitates on the basis of its correlation with the oxygen precipitation.

In a series of articles,[31–33] Pizzini and coworkers supported the suggestion by Tajima[30,34] with a wide set of experimental results and demonstrated how it is possible to discriminate between emission lines due to oxygen-related centers and those due to dislocations. In addition, they showed that the setup of a band around 0.8 eV

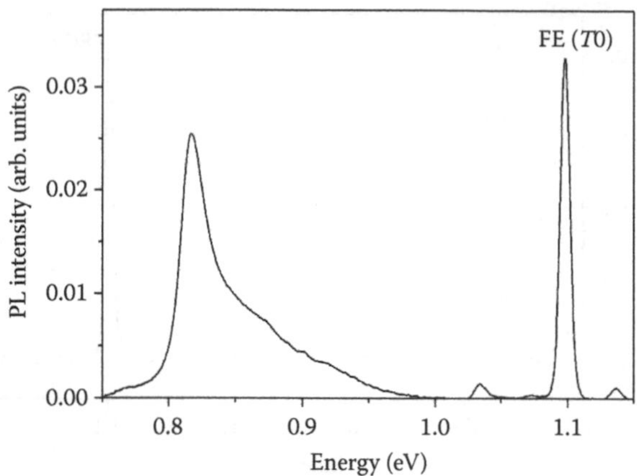

FIGURE 10.8 PL spectrum, collected at 12 K, of a CZ silicon sample submitted to a nucleation annealing at 450°C for 4 h and to a growth annealing step at 800°C for 4 h, followed by 4 h at 1000°C.

is not necessarily connected with dislocations, assigning the band at 0.817 eV to SiO_x precipitates (see Figure 10.8).

It has also been suggested that the strain field associated with oxide precipitates might be responsible for broad bands, systematically present in most of the samples examined. A qualitative conclusion was proposed that the narrow PL bands (like the one lying at 0.817 eV) observed in dislocation-free samples are the fingerprints of oxide mesophases, with a relatively narrow size distribution, while the broad featureless bands are the fingerprints of a broad distribution of oxide nanoparticles.[35] Finally, it has been shown that photoluminescence spectroscopy is much more sensitive to oxygen complexes than IR spectroscopy and even more convenient to use than TEM when the density of oxide nuclei and precipitates is as low as 10^{10}–10^{11} cm^{-3}.

Recently, the presence of a dislocation-related component and a component due to oxygen precipitates has been investigated in PL spectra collected on multicrystalline (mc) Si at room temperature around a small-angle grain boundary.[36]

10.2.2 Nitrogen

Nitrogen is a rapidly diffusing impurity in silicon, studied in N-implanted FZ crystals and in CZ crystals grown by adding Si_3N_4 into the melt during growth.[37] The main N-related defects are N—N pairs, where the two atoms occupy interstitial sites in a locally distorted silicon lattice, and various N—O complexes, some of these being electrically active too. N—N pairs give two main vibrational bands at about 770 and 965 cm^{-1}, with good agreement in their calculated and experimental positions. The peak at 965 cm^{-1} has been calibrated to be used for the quantitative evaluation of nitrogen concentration in both FZ and CZ silicon (in the latter, the conversion factor is different, since the peak is affected by a contribution from N—O complexes,[38,39] down to very low concentrations[40]). Interaction of N−N pair with voids and oxide precipitates is reported in CZ silicon.[39]

Isolated interstitial nitrogen also exists, with three main peaks related to vibrations of the Si−N bond, at 550, 773, and 885 cm^{-1}. Also for nitrogen, isotopic effects are observed, used for supporting the attributions.

Interaction of nitrogen with interstitial oxygen in CZ silicon leads to three bands, at 802, 996, and 1027 cm^{-1}, showing different intensities depending on annealing, interpreted as the interaction with N−N pairs on the basis of the comparison between FZ and CZ crystals and of calculations[41]; an example of a spectrum of an N-doped CZ silicon,[42] normalized to FZ silicon, is reported in Figure 10.9.

Another center gives rise to two further IR bands at 810 and 1018 cm^{-1}, related to groups of two N and two O atoms described with different possible models. Finally, other weak bands are related to centers involving one or two N atoms, also connected to shallow thermal donors.[43,44]

In silicon for solar cells,[17] N has been found to play a role in the formation of SiC precipitates and can also interact with other defects, in particular, dislocations.

Some sharp PL lines in the region between 746 and 773 meV have been associated with nitrogen carbon and nitrogen–carbon–oxygen complexes, labeled with N_1−N_5 centers. These lines can be observed when introduced in various ways, mainly by nitrogen implantation or by a sequential nitrogen and carbon ion implantation, followed by thermal annealing. The N1 (746 meV) and N5 (773 meV) centers have been confirmed as having monoclinic I symmetry. Dornen et al.[45] confirmed that the N1 optical defect also incorporates carbon, while the N3 (762 meV), N4 (767.4 meV), and N5 (772.4 meV) centers also involve an oxygen atom.

A photoluminescence line at 1.1223 eV has been associated with nitrogen, being observed in intentionally nitrogen-doped FZ silicon and its relative intensity

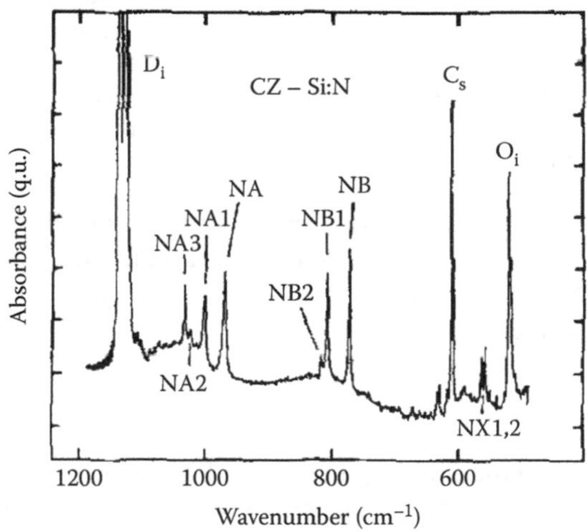

FIGURE 10.9 IR absorbance spectrum of N-doped CZ silicon, normalized to FZ silicon. (With kind permission from Springer Science+Business Media: *Appl. Phys. A*, Nitrogen-oxygen complexes in Czochralski silicon. 46, 1988: 73–76, Wagner, P. et al.)

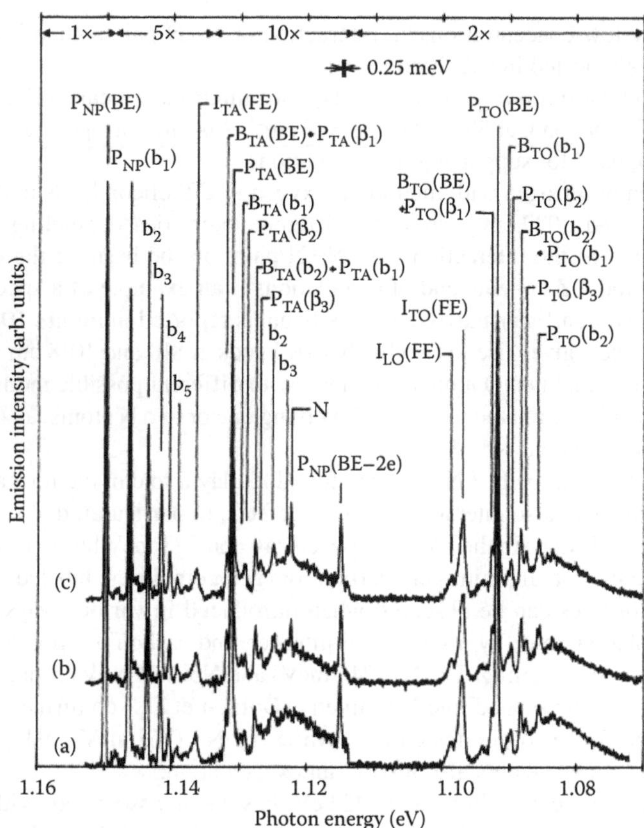

FIGURE 10.10 PL spectra from Si doped with P and N. The N concentrations are respectively (a) 8×10^{14} at/cm^3; (b) 1.4×10^{15} at/cm^3; (c) 3.2×10^{15} at/cm^3. Each PL line is labeled with the chemical symbol of the associated impurity where I denotes the intrinsic luminescence. (Reprinted with permission from Tajima, M. et al. Photoluminescence associated with nitrogen in silicon. *Jpn. J. Appl. Phys.* 20, 1981: L423–L425. Copyright 1992. The Japan Society of Applied Physics.)

increases with nitrogen concentration.[46] In Figure 10.10, the PL spectra of N-doped single crystal are reported,[46] as collected at 4.2 K under 514 nm excitation. It was assumed that the N-line is the no-phonon line of the bound-exciton line associated with the N-related center.

10.2.3 Carbon

Substitutional carbon impurities (C_s) in CZ and FZ silicon give rise to the IR peak detected at about 605 cm^{-1} at RT, which has been calibrated to be used for the quantitative evaluation of C concentration at different temperatures[47]; such an evaluation needs some care, since this peak is at the edge of the strong Si lattice band (see Figure 10.2). The peak related to C_s is observed in Figure 10.11 for silicon crystals with different isotopes of carbon, therefore, showing isotopic shift.

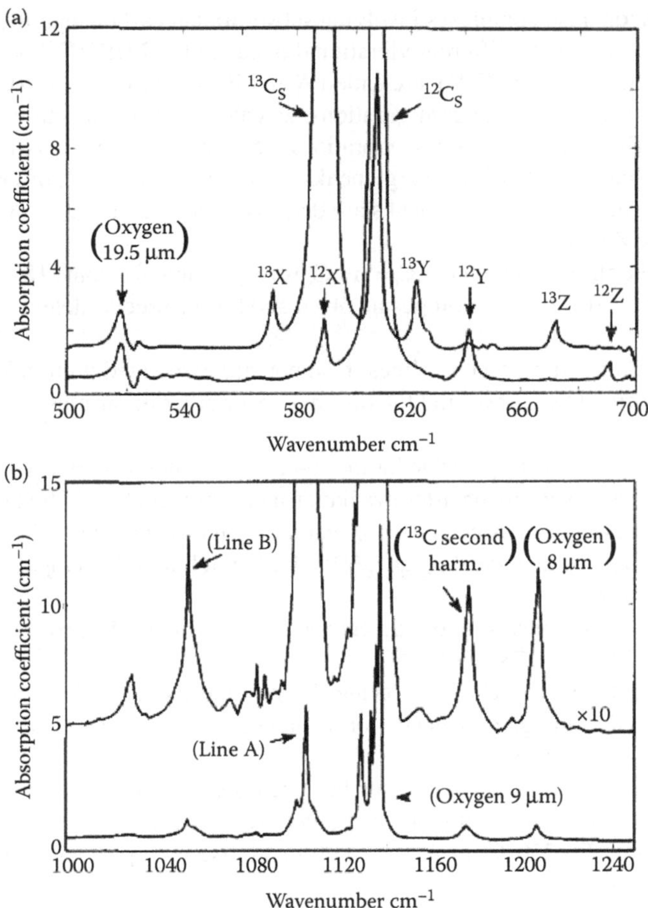

FIGURE 10.11 IR absorption spectra collected at 4 K on different CZ silicon samples, showing several C-related bands; in (a), the main band due to C_s is observed, with a clear isotopic shift, together with several X- and Y-bands; in (b), lines A and B are present, see text. (Reprinted from *Mat. Sci. Eng. B* 36, R.C. Newman, Light impurities and their interactions in silicon, 1–12, Copyright 1996, with permission from Elsevier.)

Carbon impurities in silicon also originate defects and complexes involving the host silicon atoms, mostly studied in crystals where irradiation by electrons forms extrinsic defects of different kinds. Precise atomic structures have been recognized for these defects, involving Si and one or two C atoms and having electric activity as donor or acceptor states. The dominant one is composed of a self-interstitial Si atom and an interstitial C atom[48] and gives rise to two peaks, detectable at 922 and 932 cm^{-1}, for ^{12}C impurities in spectra collected at 4 K, with 2:1 intensity ratio; isotopic effects and *ab initio* calculations fully confirmed attributions and spectral positions.[49] Also, pairs of C impurities, called di-carbon centers,[50,51] give localized vibrational modes observed, in agreement with calculations, at 502 and 706 cm^{-1} for the ^{12}C isotope (a rather strong isotopic shift is observed), depending on the charge.

In CZ silicon, also complexes involving substitutional carbon and interstitial oxygen form, giving rise to different vibrational bands in the MIR[40,52,53]: the main two, at 1104 and 1052 cm^{-1} (at 77 K), are called A and B lines (see Figure 10.11b) and are attributed to precise atomic configurations, as calculated by Kaneta et al.[54] giving rise to perturbed O_i modes or new vibrations from specific centers. All these peaks have been related to other low-energy peaks around 600–700 cm^{-1}, often referred to as X- and Y-bands. All of these peaks are detected at low temperature with very low intensity in CZ silicon.

Finally, the three-atom defect C_i-Si-C_s is responsible for some IR bands, again different for the different C isotopes involved and for the precise defect structure and symmetry.[51]

Carbon-related luminescence lines in silicon are associated with different defect complexes formed by carbon and the host matrix or other impurities, mainly in irradiated silicon.

A $Si-C_s$ complex is responsible for the so-called no-phonon G line at 0.969 eV,[55–57] which usually arises in silicon after irradiation and subsequent annealing up to 200°C. Experimental evidence has been gained that the G line center can be formed without radiation damage by an annealing at 1200°C for 7 h, followed by a rapid quenching at room temperature.[57]

Other defect complexes involving carbon and oxygen are believed to be responsible for the C line at 0.789 eV.[8,58–62]

Isotope shifts due to ^{18}O on the 65 and 72 meV modes and a ^{13}C isotope splitting of the no-phonon line give evidence for a single carbon atom and one oxygen atom incorporated in this optical center.

A series of electronic excited states has been observed in PL spectrum,[60] which can be modeled in terms of donor-like bound exciton. The C center is destroyed by an annealing at $T > 350°C$ for about 20 min and also by a prolonged irradiation at room temperature.

Experimental evidence and strong arguments suggest that the C(3) center at 865 cm^{-1}, and the 789 meV line originate from the same optical center.[63,64] Both G and C lines have been used to determine the carbon content in silicon[61,65]: in the case of Si samples with known oxygen content, the former allowed carbon concentrations of 10^{16}–10^{17} at/cm^3 to be determined, nonetheless within an uncertainty as high as ±30%, while the latter enabled the determination of the lowest levels of carbon (≈10^{14}–10^{15} at/cm^3), almost independently of oxygen concentration. A further defect complex involving carbon and oxygen has been associated with the so-called H line (or K line) at 925.5 meV (see Figure 10.12) observed both in irradiated mono-Si and in mono- and mc-Si submitted to any thermal treatments or dwelling in the temperatures range around 450°C.[25,58]

In a work by Buyanova et al.,[66] the effect of hot electron irradiation on carbon-related defects was investigated. For carbon-rich Si wafers, the increase of irradiation temperature up to 300°C enhances the formation of the C, G, and H lines; on the contrary, for carbon-lean Si wafers high-temperature electron irradiation stimulates the formation of extended defects, such as dislocations and precipitates.

The performances of silicon solar cells are affected by the presence of C impurities,[17] leading to the formation of traps and to gettering of metal impurities; also, SiC

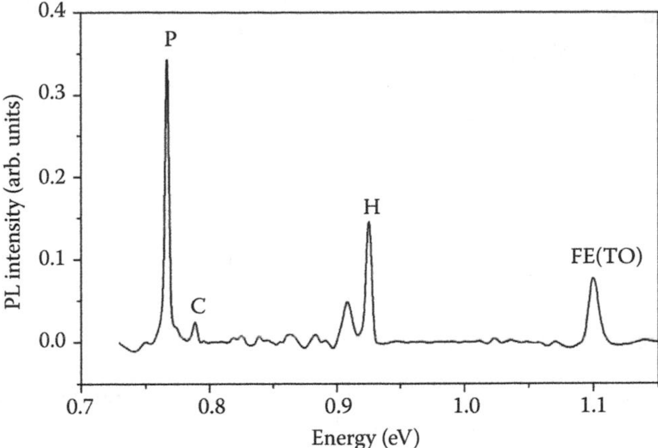

FIGURE 10.12 Photoluminescence spectrum measured at 12 K of a heat-treated CZ silicon sample. The P, C, and H lines are indicated while FE(TO) refers to free exciton recombination with TO phonon emission involved.

precipitates can be found in solar-grade silicon,[67] where they affect the crystal properties. The effects of annealing on such C-induced defects are limited with respect to what happens in CZ silicon.

10.2.4 Hydrogen

Hydrogen in silicon raised a relevant interest since it is an impurity passivating the electrical activity of different defects, this effect being particularly relevant in materials for solar cells, where hydrogenation processes have been studied and optimized.[17,68,69] In addition, hydrogen has a clear interaction with other impurities, mostly oxygen and carbon, inducing the formation of donor centers and complexes.[37,70]

In the IR spectra of Si, the presence of hydrogen impurities can be detected mainly through peaks due to isolated interstitial atoms, giving rise to localized vibrational modes. The main ones, involving a single H atom, are observed at few K at about 1990 and 1450 cm^{-1}, with a very small isotopic shift. Vibrational modes due to H dimers, where each atom is bonded to one or two host Si atoms, were also identified and detected thanks to protonation and deuteration of FZ silicon samples.[70,71] Also, H-related defects, mainly vacancies, can provide some contributions to the IR spectra. In particular, vacancy-H centers of various kinds have been identified,[70] one of them giving a peak detectable at few K at about 2220 cm^{-1}, with spectral shifts related to the different Si isotopes, as expected and clearly observed in Figure 10.13.

Studies of the PL of silicon report that the I-line (0.9650 eV) and the T-line (0.9351 eV), created by annealing at 450°C in silicon deliberately doped with hydrogen, incorporate one hydrogen atom. The presence of hydrogen was inferred from the observed deuterium–hydrogen isotope shift.[25,58,72] Furthermore, it has been shown that hydrogen can be incorporated in these centers by annealing at 450°C in an atmosphere containing water vapor or hydrogen gas, from a hydrogen-passivated surface or a thin oxide layer containing water.[73]

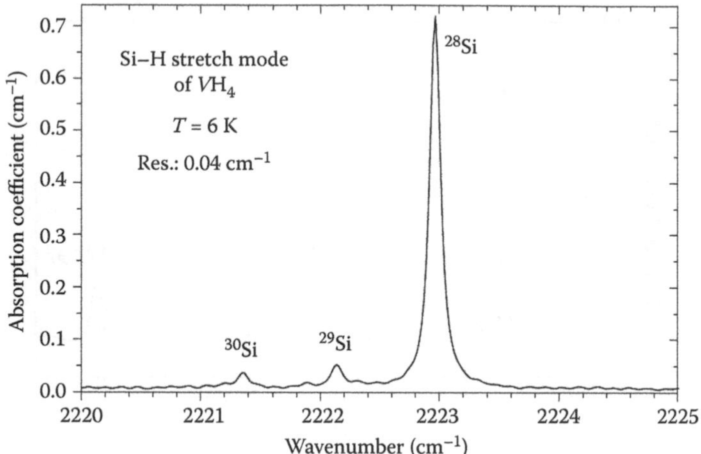

FIGURE 10.13 IR absorbance spectrum collected at 6 K on an FZ silicon sample grown in H_2 atmosphere; the Si–H vibration is detected through the peaks at about 2221, 2222, and 2223 cm^{-1} for the different Si isotopes. (With kind permission from Springer Science+Business Media: *Optical Absorption of Impurities and Defects in Semiconducting Crystals*. 2013, Pajot, B., B. Clerjaud.)

In radiation-damaged silicon deliberately doped with hydrogen, then annealed at 450°C, photoluminescence measurements show that the centers responsible for the X-lines around 1115 meV contain two, three, or four hydrogen atoms. The substitution of hydrogen by deuterium results in a relatively large shift of the X-lines, up to 5% of the exciton-binding energy, so that the structures of many of the lines in materials with a hydrogen–deuterium mixture can be easily resolved.[72]

10.3 DOPANTS

Boron, phosphorus, and arsenic are the most commonly used dopants in silicon for microelectronics, where they act as acceptor (B) or donor impurities (P, As). Low-temperature IR measurements permit to determine their concentration, since below 15 K the concentration of free carriers is negligible. The early approach to such an analysis was proposed in 1967, then developed for high-resistivity silicon.[74,75]

The low-temperature ($T < 15$ K) silicon spectrum consists of a series of intense absorption bands due to electron or hole transitions from the ground state of neutral impurities to a series of hydrogenic-like levels lying close to the respective band edge.[76] The standard procedure for the determination of electrically active boron content and also of other dopants in crystalline silicon is reported in the ASTM standard F1630-95.[77]

Boron is highly soluble in the silicon lattice, where it occupies substitutional positions. Both isolated B atoms and B–B pairs originate vibrational modes, detectable in the MIR spectral range when high B concentrations are present; more precisely, B is detectable at RT through a peak at 644 or 621 cm^{-1} for the two ^{10}B and ^{11}B isotopes, shifting with temperature, while the pair-related response is detected at RT from about 550 to 570 cm^{-1}, again depending on the two isotopes involved and on temperature.[37]

Substitutional boron atoms can interact with donor atoms producing a center with precise symmetry properties and localized vibrations. For example, the B–P and B–As pairs are detectable from 600 to 650 cm^{-1}, depending on the isotopic combinations and, again, on temperature; at slightly higher energy, also localized vibrations of the B–Li pairs can be detected.

Boron also interacts with defects in the silicon lattice, in particular, self-interstitial atoms, forming various electrically active centers with specific IR peaks, which present isotopic shift. These centers have been studied in irradiated samples and display limited stability.

As for other donor impurities, P and As can be observed[37] when present in high concentrations at 441 and 491 cm^{-1} and at 315 and 336 cm^{-1}, respectively.

The first work dealing with radiative recombination processes involving donors and acceptors in silicon dates back to 1967; there the recombination radiation bands from silicon associated with the most important neutral donor and acceptor centers (As, P, Sb, Bi, Ga, In, Al) are reported and discussed in detail.[78]

Tajima[79] was the first to show that quantitative information on donor and acceptor concentrations can be derived from the ratios of bound-exciton to free-exciton luminescence features. This method was standardized by Semiconductor Equipment and Materials International (SEMI).[80]

At very low concentrations, this is more sensitive than any other approach and has the advantage of being impurity-specific and capable of measuring both the acceptor and donor concentrations. Nonetheless, this method is rarely applied to samples with dopant concentrations higher than 10^{15} cm^{-3}, because in such a case, the free-exciton emission line (FE) was almost undetectable.

If the semiconductor contains donors and acceptors with concentration higher than 10^{15} cm^{-3}, at 4.2 K, virtually all of the free excitons are captured giving rise to impurity-specific bound-exciton luminescence. At high excitation densities, the acceptors and the donors bind to more than one exciton and additional lines can be detected in the spectrum. For instance, P is related to the P bound-exciton no-phonon line at 1150 meV and to the multi-bound-exciton lines involving m excitons at 1146.5 meV ($m = 2$); 1143.7 meV ($m = 3$); 1141.7 meV ($m = 4$); 1140.5 ($m = 5$); and 1139.3 meV ($m = 6$).

Recently, as the concentration of residual donor and acceptor in the low-cost solar-grade silicon is in the range between 10^{15} and 10^{17} cm^{-3}, an implementation of the Tajima method[79,80] was further developed[81]; it consists of increasing the measurement temperature to enhance the FE emission, therefore extending the concentration range detectable. As a consequence, new calibration factors were deduced, obtaining a satisfactory agreement among the PL method, secondary ion mass spectrometry, and inductively coupled plasma mass spectroscopy in a rather wide concentration range, between 1×10^{14} and 2×10^{16} cm^{-3}.

These methods work well for compensated CZ and FZ silicon, but their extension to multicrystalline (mc)-Si can be difficult when defect-related PL bands are present, as usually occurs in solar-grade silicon (SoG). In such cases, the calibration factor deduced in Reference 81 cannot straightforwardly be applied.

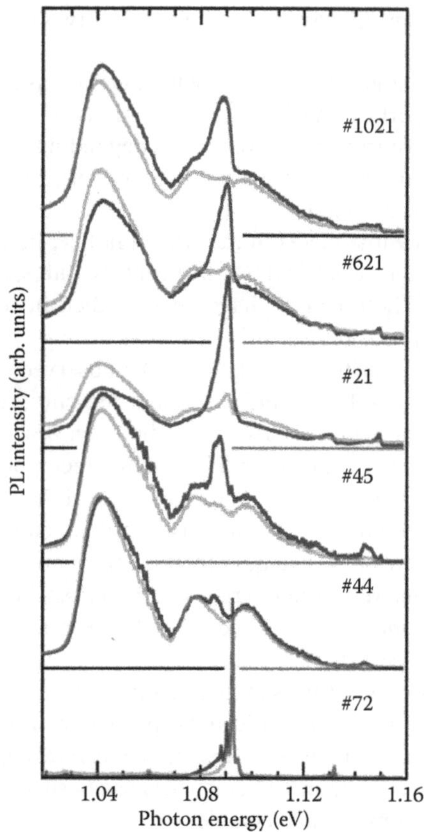

FIGURE 10.14 PL spectra of compensated Si involving P donors and B acceptors (samples no. 1021, no. 621, no. 21, no. 45, and no. 44) and uncompensated B-doped Si (sample no. 72) at 4.2 K under high and low excitation conditions (black and gray curves, respectively). The concentrations of dopants in the examined samples are: sample no. 1021 $N_A = 2.0 \times 10^{16}$ cm^{-3}, $N_D = 1.8 \times 10^{16}$ cm^{-3}; sample no. 621 $N_A = 1.3 \times 10^{16}$ cm^{-3}, $N_D = 0.61 \times 10^{16}$ cm^{-3}; sample no. 21 $N_A = 1.1 \times 10^{16}$ cm^{-3}, $N_D = 0.34 \times 10^{16}$ cm^{-3}; sample no. 45 $N_A = 3.0 \times 10^{16}$ cm^{-3}, $N_D = 2.2 \times 10^{16}$ cm^{-3}; sample no. 44 $N_A = 4.8 \times 10^{16}$ cm^{-3}, $N_D = 1.2 \times 10^{16}$ cm^{-3}; sample no. 72 $N_A = 1.3 \times 10^{16}$ cm^{-3}, N_D = N.D. (Reprinted with permission from M. Tajima et al. Donor-acceptor pair luminescence in compensated Si for solar cells, *J. Appl. Phys.* 110, 2011: 043506–5. Copyright 2011, American Institute of Physics.)

Furthermore, in highly compensated silicon, the possible transition between acceptor and donor levels should be taken into account, which gives the donor–acceptor (DA) pair luminescence.

DA pair luminescence is a well-studied recombination process in compound semiconductor (i.e., GaAs, GaP, CdTe, SiC), but, after the first works appeared in the 1960s,[78,82] very little work has been performed on Si. The main reason for this is that Si crystals for electronic devices have very low residual impurities, resulting in low compensation ratio. Recently, a DA pair emission involving shallow P donors and B acceptors with a fine structure in compensated Si for solar cells has been identified.[83]

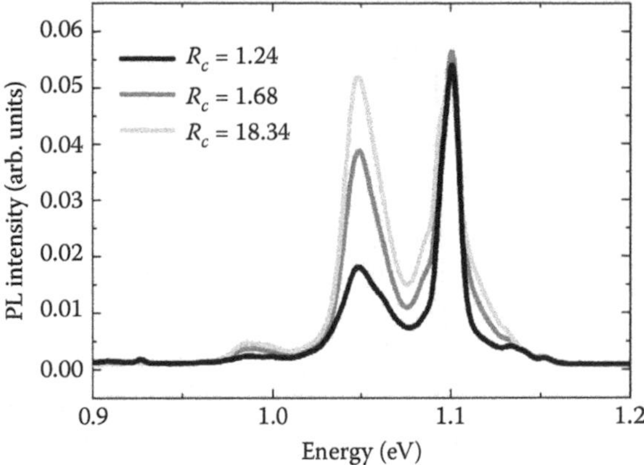

FIGURE 10.15 PL spectra collected 12 K of CZ Si samples with different compensation ratios (R_c = 1.24, 1.68, 18.34, respectively).

PL spectra of compensated CZ samples are discussed in detail in Reference 84. In B–P compensated samples, at 4.2 K, three bands appear at 1.098, 1.079, and 1.041 eV under low excitation conditions, with a similar shape having a high-energy tail (see Figure 10.14). The highest, middle, and lowest bands are a no-phonon band, labeled P:BNP, and their TA and TO phonon sidebands, labeled as P_BNP, P_BTA, and P_BTO, respectively. The peak positions of the three bands shifted equally to the high-energy side under high excitation condition; this is characteristic of DA pair luminescence.[4]

The theoretical spectrum of DA pair luminescence has been calculated starting from the density distribution of pairs as a function of their transition energy, with separation ranging from 1.9 to 3.3 nm. A close agreement was obtained between the observed spectral structure at 4 K and the theoretical curve using the generally accepted P donor and B acceptor ionization energies. If the spectra of compensated silicon are collected at higher temperature, around T = 12 K, a broad band appears around 1.05 eV (see Figure 10.15) whose intensity depends on compensation ratio, $R_c = (N_A + N_D)/(N_A - N_D)$, and can be used as a fingerprint of the compensation itself.

Takima and coworkers[85] have also identified the P-donor and Ga-acceptor pair luminescence in B and P highly compensated Si co-doped with Ga for photovoltaic applications. The bands around 1.07, 1.05, 1.02 eV were assigned to no-phonon band and TA and TO phonon sidebands of P-Ga emissions. By a detailed analysis of the DA band fine structure, the gallium ionization energy (E_A = 74 meV) has been determined.

10.4 DISLOCATIONS

Dislocations are the only extended defects that can be detected by PL spectroscopy as grain boundaries do not introduce radiative recombination centers.

At low temperature, four PL bands, labeled D1, D2, D3, and D4, with photon energies 0.807, 0.870, 0.935, and 1.0 eV, respectively, have been associated with the presence of dislocations in silicon[86] (see Figure 10.16); D1 line being visible already at room temperature (see Figure 10.16b).

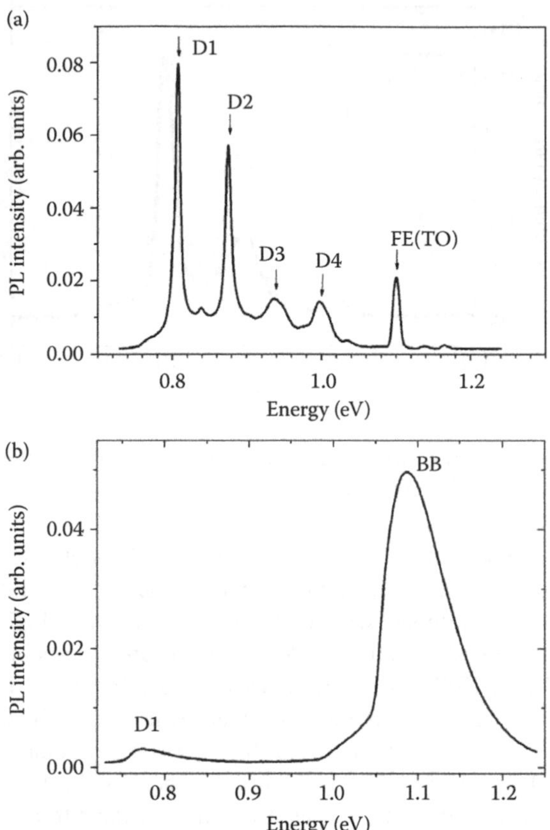

FIGURE 10.16 PL spectra of a dislocated FZ Si sample (a) collected at 12 K (b) collected at 300 K (BB refers to band-to-band emission).

The relative intensity of D1–D4 depends on the dislocation density and can vary along the same samples (see Figure 10.17).

Since 1976, PL studies of dislocations have been extensively carried out and correlated with the results of DLTS and EPR measurements,[88–90] aiming at achieving a better understanding of the electronic properties of dislocations.

For the sake of clarity, it should be remarked that despite several years of investigations, very little has been definitively established about the influence of impurities on the electrical activity of dislocations and their related effects on D bands.

A generally accepted idea is that the D3/D4 lines are associated with the intrinsic nature of dislocations and in particular with the carrier recombination at straight segments of splitted 60° dislocations.[90–94] Furthermore, the D3 line has been considered as a replica of the D4 line,[95] despite the fact that in many cases the intensity of the D4 line exceeds that of the D3 line.

As far as D1 and D2 emission lines are concerned, they are detected even in the absence of D3 and D4, supporting the hypothesis that the D1/D2 and the D3/D4 pairs have different origin.

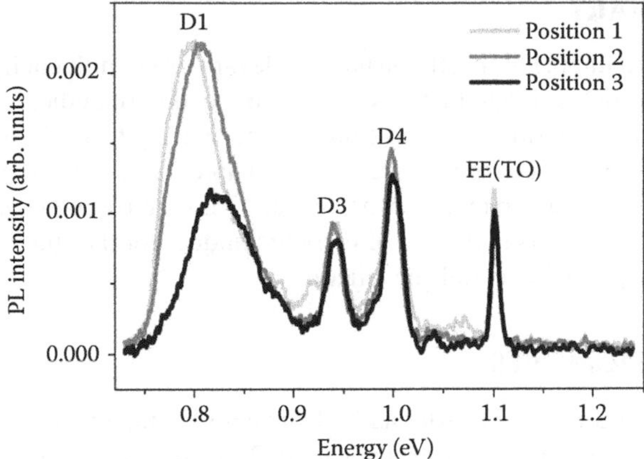

FIGURE 10.17 PL spectra collected on different position on the same EFG multicrystalline silicon sample. Details about sample in Reference 87.

However, the real origin of the D1 and D2 lines is still unclear. It has been shown that their features are affected by the presence of impurities such as metals,[96,97] oxygen,[31,36,98] and nitrogen.[99,100] Furthermore, the origin of D1 and D2 was also associated with staking fault between dislocations and dislocation jogs.

The high-temperature stability of D1 luminescence has been modeled in Reference 101; in this work, D2, D3, and D4 lines are attributed to the excitation of electrons and holes from the D2, D3, and D4 centers to the mobility edge of the dislocation-related energy band. The activation energy of this excitation is rather small, so that the D2, D3, and D4 PL lines disappear rather quickly with increasing temperature, since all carriers are trapped at the D1 centers that are the deepest.

Recently, in samples containing dislocations induced by electron irradiation, a band at 0.78 eV has been identified and studied.[102] It is found that this line only exists in FZ silicon samples, not in CZ ones, with a good thermal stability. It seems that the 0.78 eV line luminescence center has no relationship with oxygen atom clusters; neither thermal donors nor oxygen precipitation. It is assumed that the 0.78 eV line originates from the specific reconstructed dislocation structures, which could be easily affected by point defects and temperature.

As previously mentioned, D1 is visible also at room temperature and it has an appropriate spectral position for the optical communication applications. Therefore, at the end of the last century, luminescence from dislocations in silicon has been actively investigated in order to develop monolithic circuits including optoelectronic components on a silicon substrate.[103,104] Fabrication of light-emitting structures at about 1.5 μm wavelength was at that time and still is of great interest for optoelectronic applications. In spite of the extensive work devoted to this issue, very little has been, however, definitively established about the electrical activity of dislocations and their effects on the D band. Nevertheless, it should be remarked that the solution of this question is of great importance in enabling proper engineering of silicon LEDs based on dislocations and any new findings on this topic could help in limiting the deleterious effects of dislocations on the efficiency of mc-Si solar cells.

10.5 SUMMARY

Since defects and impurities in silicon have a relevant role in influencing its properties and a number of defect-related processes, the application of dedicated experimental techniques to their identification and study has been a key factor in the development of silicon-based microelectronic devices. This chapter has provided a critical review of the main results obtained by IR absorption and PL spectroscopy and illustrates the most recent developments in the use of these techniques for the study of the common impurities and defects in crystalline silicon.

ACKNOWLEDGMENTS

The authors are grateful to Dr. Alessia Le Donne for having recorded many PL spectra on silicon and to their former group leaders Professors S. Pizzini and A. Borghesi for their support.

REFERENCES

1. Wooten, F. *Optical Properties of Solids* (New York: Academic Press, 1972).
2. Brüesch, P. *Phonons: Theory and Experiments II* (Berlin: Springer-Verlag, 1986).
3. Pavesi, L., M. Guzzi. Photoluminescence of $Al_xGa_{1-x}As$ alloys. *J. Appl. Phys.* 75, 1994: 4779–4842.
4. Pankove, J.L. *Optical Processes in Semiconductors* (New York: Dover Publications, 1975).
5. Haynes, J.R. Experimental proof of the existence of a new electronic complex in silicon. *Phys. Rev. Lett.* 4, 1960: 361–363.
6. Thomas, D.G., M. Gershenzon, F.A. Trumbore. Pair spectra and edge-emission in gallium phosphide. *Phys. Rev.* 133, 1964: A269–A279.
7. Dean, P.J. Inter-impurity recombinations in semiconductors. *Prog. Solid State Chem.* 8, 1973: 1–126.
8. Davies, G. The optical properties of luminescence centers in silicon. *Phys. Rep.* 3&4 1988: 83–188.
9. *Impurities and Defects in Group IV Elements, IV-IV and III-V Compounds. Part A: Group IV Elements Landolt–Börnstein-Group III Condensed Matter* (Berlin Heidelberg: Springer, Vol. 41, pp 1–3. 2002).
10. Borghesi, A., B. Pivac, A. Sassella, A. Stella. Oxygen precipitation in silicon. *J. Appl. Phys.* 77, 1995: 4169–4244 and Refs. therein.
11. Borghesi, A., A. Sassella, B. Pivac, M. Porrini. Evaluation of the precipitate contribution to the infrared absorption in interstitial oxygen measurements in silicon. *Appl. Phys. Lett.* 79, 2001: 4106–4108.
12. Sassella, A. Measurement of interstitital oxygen concentration in silicon below 10^{15} atoms/cm^3. *Appl. Phys. Lett.* 79, 2001: 4339–4341.
13. Sassella, A., A. Borghesi, G. Borionetti, P. Geranzani. Optical absorption of precipitated oxygen in silicon at liquid helium temperature. *Mat. Sci. Eng. B* 73, 2000: 224–229.
14. Borghesi, A., A. Piaggi, A. Sassella, A. Stella, B. Pivac. Infrared study of oxygen precipitate composition in silicon. *Phys. Rev. B* 46, 1992: 4123–4127.
15. Hu, S.M. Infrared absorption spectra of SiO_2 precipitates of various shapes in silicon: calculated and experimental. *J. Appl. Phys.* 51, 1980: 5945–5948.
16. Borghesi, A., A. Sassella, P. Geranzani, M. Porrini, B. Pivac. Infrared characterization of oxygen precipitates in silicon wafers with different concentrations of interstitial oxygen. *Mat. Sci. Eng. B* 73, 2000: 145–148.

17. Pivac, B., A. Sassella, A. Borghesi. Non-doping light impurities in silicon for solar cells. *Mat. Sci. Eng. B* 36, 1996: 53–62.
18. Götz, W., G. Pensl, W. Zulehner. Observation of five additional thermal donor species TD12 to TD16 and of regrowth of thermal donors at initial stages of the new oxygen donor formation in Czochralski-grown silicon. *Phys. Rev. B* 46, 1992: 4312–4315.
19. Stein, H.J., S.K. Hahn, S.C. Shatas. Rapid thermal annealing and regrowth of thermal donors. *J. Appl. Phys.* 59, 1986: 3495–3502.
20. Binetti S., M. Acciarri, A. Brianza, C. Savigni, S. Pizzini. Effect of oxygen concentration on diffusion length in Czochralski and magnetic Czochralski silicon. *Mat. Sci. Technol.* 11, 1995: 665–669.
21. Tajima, M., A. Kanamori, T. Iizuka. Photoluminescence analysis of annealed silicon crystals. *Jpn. J. Appl. Phys.* 18, 1979: 1403–1404.
22. Heijmink, B.J., T. Gregorkiewicz, C.A. Ammerlaan. Photoluminescence studies on thermal donors in boron and Aluminum doped silicon. *Phys. Rev. B* 46, 1992: 2034–2040.
23. Nakayama H., T. Nishino, Y. Hamakawa. Luminescence of thermally induced defects in Si crystals. *Appl. Phys. Lett.* 38, 1981: 623–625.
24. Steele, AG., M.M.L.W. Thewalt, S.P. Watkins. A second isoelectronic multiexciton center in annealed Czochralski silicon. *Solid State Commun.* 63, 1987: 81–84.
25. Minaev, N.S., A.V. Mudry. Thermally-induced defects in silicon containing oxygen and carbon. *Phys. Status Solidi (a)* 68, 1981: 561–565.
26. Tajima, M., P. Stallhofer, D. Huber. Deep level luminescence ralated to thermal donors in silicon. *Jpn. J. Appl. Phys.* 22, 1983: L586–L588.
27. Kürner, W., R.Sauer, A. Dornen, K. Thonke. Structure of the 0.767 eV oxygen-carbon luminescence defect in 450°C thermally annealed Czochralski-grown silicon. *Phys. Rev. B* 39, 1989: 13327–13337.
28. Pizzini, S., S. Binetti, E. Leoni, A. LeDonne, M. Acciarri, A. Castaldini, A. Cavallini. Radiative recombination processes of thermal donors in silicon. *MRS Symp. Proc.* 692, 2002: 275–281.
29. Hamada, M., T. Katoda. Characterization of Czochralski silicon wafers grown at a low growth rate by photoluminescence spectroscopy. *Jpn. J. Appl. Phys.* 35, 1996: 182–185.
30. Tajima, M., H. Takeno, T. Abe. Characterization of point defects in Si crystals by highly spatially resolved photoluminescence *Mater. Sci. Forum* 83–87, 1991: 1327–1332.
31. Binetti, S., S. Pizzini, E. Leoni, R. Somaschini, A. Castaldini, A. Cavallini. Optical properties of oxygen precipitates and dislocations in silicon. *J. Appl. Phys.* 92, 2002: 2437–2445.
32. Leoni, E., S. Binetti, B. Pichaud, S. Pizzini. Dislocation luminescence in plastically deformed silicon crystals: Effect of dislocation intersection and oxygen decoration. *Eur. Phys. J. Appl. Phys.* 27, 2004: 123–127.
33. Castaldini, A., D. Cavalcoli, A. Cavallini, S. Pizzini, S. Binetti. Electronic transitions at defect states in CZ p-type silicon. *App. Phys. Lett.* 86, 2005: 162109-1–162109-3.
34. Tajima, M., M. Tokita, M. Warashina. Photoluminescence due to oxygen precipitates distinguished from the D lines in annealed Si. *Mater. Sci. Forum* 196-201, 1995: 1749–1754.
35. Leoni, E., L. Martinelli, S. Binetti, G. Borionetti, S. Pizzini. The origin of the photoluminescence from oxygen precipitates nucleated at low temperature in semiconductor silicon. *J. Electrochem. Soc.* 151, 2004: G866–G869.
36. Tajima, M., Y. Iwata, F. Okayama, H. Toyota, H. Onodera, T. Sekiguchi. Deep-level photoluminescence due to dislocations and oxygen precipitates in multicrystalline Si. *J. Appl. Phys.* 111, 2012: 113523-1–113523-6.
37. Pajot, B., B. Clerjaud. *Optical Absorption of Impurities and Defects in Semiconducting Crystals.* (Berlin: Springer, 2013) and Refs. therein.
38. Itoh, Y., T. Nozaki, T. Mazui, T. Abe. Calibration curve for infrared spectrophotometry of nitrogen in silicon. *Appl. Phys. Lett.* 47, 1995: 488–489.

39. Tanahashi, K., H. Yamada-Kaneta. Technique for determination of nitrogen concentration in Czochralski silicon by infared absorption measurements. *Jpn. J. Appl. Phys.* 42, 2003: L223–L225.
40. Akhmetov, V.D., H. Richter, N. Inoue. Determination of low concentrations of N and C in CZ–Si by precise FT-IR spectroscopy. *Mat. Sci. Eng. B* 134, 2006: 207–212.
41. Berg Rasmussen, F., S. Öberg, R. Jones, C. Ewels, J. Goss, J. Miro, P. Deák. The nitrogen-pair oxygen defect in silicon. *Mat. Sci. Eng. B* 36, 1996: 91–95.
42. Wagner, P., R. Oeder, W. Zulehner. Nitrogen-oxygen complexes in Czochralski silicon. *Appl. Phys. A* 46, 1988: 73–76.
43. Inoue, N., M. Nakatsu, H. Ono, Y. Inoue. Infrared absorption peaks in nitrogen doped CZ silicon. *Mat. Sci. Eng. B* 134, 2006: 202–206.
44. Alt, H.Ch., H.E. Wagner, W. v. Ammon, F. Bittersberger, A. Huber, L. Koester. Chemical composition of nitrogen-oxygen shallow donor complexes in silicon. *Physica B* 401–402, 2007: 130–133.
45. Dornen, A., G. Pensl, R. Sauer. Vibrational mode nitrogen and carbon isotope shifts on the N1(0.746 eV) photoluminescence spectrum in silicon. *Solid State Commun.* 57, 1986: 861–864.
46. Tajima, M., T. Masui, T. Abe, T. Nozaki. Photoluminescence associated with nitrogen in silicon. *Jpn. J. Appl. Phys.* 20, 1981: L423–L425.
47. ASTM Standards F 1391–93.
48. Beans, A.R., R.C. Newman. Low temperature electron irradiation of silicon containing carbon. *Solid State Commun.* 8, 1970: 175–177.
49. Leary, P., R. Jones, S. Öberg, V.J.B. Torres. Dynamic properties of interstitial carbon and carbon-carbon pair defects in silicon. *Phys. Rev. B* 55, 1997: 2188–2194.
50. Lavrov, E.V., B. Bech Nielsen, J.R. Byberg, B. Hourahine, R. Jones, S. Öberg, P.R. Briddon. Local vibrational modes of two neighboring sustitutional carbon atoms in silicon. *Phys. Rev. B* 62, 2000: 158–165.
51. Song, L.W., X.D. Zhan, B.W. Benson, G.D. Watkins. Bistable interstitial-carbon-substitutional-carbon pair in silicon. *Phys. Rev. B* 42, 1990: 5765–5783.
52. Newman, R.C., R.S. Smith. Vibrational absorption of carbon and carbon-oxygen complexes in silicon. *J. Phys. Chem. Sol.* 30, 1969: 1493–1505.
53. Newman, R.C. Light impurities and their interactions in silicon. *Mat. Sci. Eng. B* 36, 1996: 1–12 and Refs. therein.
54. Kaneta, C., T. Sasaki, H. Katayama-Yoshida. Atomic configuration, stabilizing mechanism, and impurity vibrations of carbon-oxygen complexes in crystalline silicon. *Phys. Rev. B* 46, 1992: 13179–13185.
55. O'Donnell, K.P., K.M. Lee, G.D. Watkins. Origin of the 0.97 eV luminescence in irradiated silicon. *Physica B* 116, 1983: 258–263.
56. Davies, G., E.C. Lightowlers, R.C. Newman, A.S. Oates. A model for radiation damage effects in carbon-doped crystalline silicon. *Semicond. Sci. Technol.* 2, 1987: 524–532.
57. Thonke, K., H. Klemisch, J. Weber, R. Sauer. New model of the irradiation-induced 0.97-eV (G) line in silicon: A C_s-Si* complex. *Phys. Rev. B* 24, 1981: 5874–5886.
58. Jones, C. E, E.S. Johnson, W.D. Compton, J. R. Noonan, B.G. Streetman, Temperature, stress, and annealing Effects on the luminescence from electron-irradiated silicon. *J. Appl. Phys.* 44, 1973: 5402–5410.
59. Noonan, J.R., C.G. Kirkpatrick, B.G. Streetman. Photoluminescence from Si irradiated with 1.5 MeV electrons at 100°K. *J. Appl. Phys.* 47, 1976: 3010–3015.
60. Thonke, K., A. Hangleiter, J. Wagner, R. Sauer. 0.79 eV (C line) defect in irradiated oxygen-rich silicon: Excited state structure, internal strain and luminescence decay time. *J. Phys. C: Solid State Phys.* 18, 1985: L795–L801.

61. Davies, G., A.S. Oates, R.C. Newman, R. Woolley, E.C. Lightowlers, M.J. Binns, J.G. Wilkes. Carbon-related radiation damage centres in Czochralski silicon. *J. Phys. C: Solid State Phys.* 19, 1986: 841–855.
62. Trombetta, M., G.D. Watkins. Identification of an interstitial carbon interstitial oxygen complex in silicon. *Appl. Phys. Lett.* 51, 1987: 1103–1105.
63. Newman, R.C., A. R. Bean. Irradiation damage in carbon-doped silicon irradiated at low temperatures by 2 MeV electrons. *Radiat. Eff.* 8, 1971: 189–193.
64. Davies, G., E.C. Lightowlers, R.C. Newman, A.S. Oates. A model for radiation damage effects in carbon-doped crystalline silicon. *Semicond. Sci. Technol.* 2, 1987: 524–532.
65. Nakamura, M., E. Kitamura, Y. Misawa, T. Suzuki, S. Nagai, H. Sunaga. Photoluminescence measurement of carbon in silicon crystals irradiated with high energy. Electrons. *J. Electrochem. Soc.* 141, 1994: 3576–3580.
66. Buyanova I.A., Monemar B., Lindstrom J.L., T. Hallberg, L.I. Murin, V.P. Markevich. Photoluminescence characterization of defects created in electron-irradiated silicon at elevated temperatures. *Mat. Sci. Eng. B* 72, 2000: 146–149.
67. Rajendran, S., M. Larrousse, B.R. Bathey, J.P. Kalejs. Silicon carbide control in the EFG system. *J. Cryst. Growth* 128, 1993: 338–342.
68. Pearton, S.J., A.J. Tavendale. Hydrogen passivation of gold-related deep levels in silicon. *Phys. Rev. B* 26, 1982: 7105–7108.
69. Muller, J.C., Y. Ababou, A. Barhdadi, E. Courcelle, S. Unamuno, D. Salles, P. Siffert, J. Fally. Passivation of polycrystalline silicon solar cells by low-energy hydrogen ion implantation. *Solar Cells* 17, 1986: 201–231.
70. B. Pajot. *Optical Absorption of Impurities and Defects in Semiconducting Crystals. H-Like Centres.* (Berlin: Springer, 2010).
71. Holbech, J.D., B. Bech Nielsen, R. Jones, P. Sitch, S. Öberg. H_2* defect in crystalline silicon. *Phys. Rev. Lett.* 71, 1993: 875–878.
72. Safonov, A.N., E.C. Lightowlers. Luminescence centers containing two, three and four hydrogen atoms in radiation-damaged silicon. *Mat. Sci. Eng. B* 36, 1996: 251–254.
73. Lightowlers, E.C., R.C. Newman, J. H. Tucker. Hydrogen-related luminescence centers in thermally treated Czochralski silicon. *Semicond. Sci. Technol.* 9, 1994: 1370–1374.
74. White, J.J. Absorption-line broadening in boron-doped silicon. *Can. J. Phys.* 45, 1967: 2797–2804.
75. Kolbesen, B.O. Simultaneous determination of the total content of boron and phosphorus in high resistivity silicon by IR spectroscopy at low temperatures. *Appl. Phys. Lett.* 27, 1975: 353–355.
76. Baldereschi, A., N.O. Lipari. Cubic contributions to the spherical model of shallow acceptor states. *Phys. Rev. B* 9, 1974: 1525–1539.
77. ASTM Standard F1630–95, Standard Test Method for Low Temperature FT-IR Analysis of Single Crystal Silicon for III–V Impurities, Annual Book of ASTM Standards, Electronics (II), 1996.
78. Dean, P.J., J.R. Haynes, W.F. Flood. New radiative recombination processes involving neutral donors and acceptors in silicon and germanium. *Phys. Rev.* 161, 1967: 711–729.
79. Tajima, M. Determination of boron and phosphorus concentration in silicon by photoluminescence analysis. *Appl. Phys. Lett.* 32, 1978: 719–721.
80. SEMI MF 1389–0704, 2004.
81. Iwai, T., M. Tajima, A.Ogura. Quantitative analysis of impurities in solar-grade Si by photoluminescence spectroscopy around 20 K. *Phys. Status Solidi C* 8, 2011: 792–795.
82. Enck, R.C., A. Honig. Radiative spectra from shallow donor-acceptor electron transfer in silicon. *Phys. Rev.* 177, 1969:1182–1193.

83. Tajima, M., T. Iwai, H. Toyota, S. Binetti, D. Macdonald. Fine structure due to donor-acceptor pair luminescence in compensated Si. *Appl. Phys. Express* 3, 2010: 071301–3.
84. Tajima, M., T. Iwai, H. Toyota, S. Binetti, D. Macdonald. Donor-acceptor pair luminescence in compensated Si for solar cells, *J. Appl. Phys.* 110, 2011: 043506–5.
85. Tajima, M., K. Tanaka, M. Forster, H. Toyota, A. Ogura. Donor-acceptor pair luminescence in B and P compensated Si co-doped with Ga, *J. Appl. Phys.* 113, 2013: 243701–5.
86. Drozdov, N.A., A.A. Patrin, V.D. Tkachev. Recombination radiation on dislocations in silicon. *JETP Lett.* 23, 1976: 597–599.
87. Slunjiki, R., B. Pivac, A. Le Donne, S. Binetti. Effects of low-temperature annealing on polycrystalline silicon for solar cells. *Solar Energy Mater. Solar Cells* 95, 2011: 559–563.
88. Alexander, A., P. Haasen. Dislocations and plastic flow in the diamond structure. *Solid State Phys.* 22, 1969: 27–158.
89. Schroeter, W., J. Kronewitz, U. Gnauert, F. Riedel, M. Seibt. Bandlike and localized states at extended defects in silicon. *Phys. Rev. B* 52, 1995: 13726–13726.
90. Sauer, R., Ch. Kisielowski-Kemmerich, H. Alexander. Dissociation-width-dependent radiative recombination of electrons and holes at widely split dislocations in silicon. *Phys. Rev. Lett.* 57, 1986: 1472–1475.
91. Suesawa, M., Y. Sasaki, K. Sumino. Dependence of photoluminescence on temperature in dislocated silicon crystals. *Phys. Status Solidi (a)* 79, 1983: 173–181.
92. Sauer, R., J. Weber, J. Stolz. Dislocation-related photoluminescence in silicon. *Appl. Phys.* A 36, 1985:1–13.
93. Izotov, A.N., E.A. Steinman. Reconstruction of optical dislocation centres under the action of shear stresses. *Phys. Status Solidi (a)* 104, 1987: 777–784.
94. Sekiguchi, T., K. Sumino. Cathodoluminescence study on dislocations in silicon. *J. Appl. Phys.* 79, 1996: 3253–3260.
95. Kveder, V.V., E.A. Steinman, H.G.Grimmeiss. Dislocation related electroluminescence at room temperature in plastically deformed silicon. *Solid State Phenom.* 47–48, 1996: 419–424.
96. Higgs, V., M. Goulding, A. Brinklow, P. Kightley. Cathodoluminescence imaging and spectroscopy of dislocations in Si and Si1 – x Ge x alloys. *Appl. Phys. Lett.* 61, 1992: 1082–1087.
97. Seibt, M., R. Kahlil, V. Kveder, W. Schroter. Electronic states at dislocations and metal silicide precipitates in crystalline silicon and their role in solar cell materials. *Appl. Phys. A* 96, 2009: 235–253.
98. Pizzini, S., S. Binetti, A. Le Donne, E. Leoni, M. Acciarri, G. Salviati, L. Lazzaroni. Beam injection studies of dislocations and oxygen agglomeration in semiconductor silicon. *Solid State Phenom.* 57, 2001: 78–79.
99. Binetti S., R. Somaschini, A. Le Donne, E. Leoni, S. Pizzini, D. Li, D. Yang. Dislocation luminescence in nitrogen-doped Czochralski and float zone silicon. *J. Phys.: Condens. Matter* 14, 2002: 13247–13254.
100. Binetti S., S. Pizzini, R. Somaschini, A. Le Donne, D. Li, D. Yang. Effect of heat treatment on the optical and electrical properties of nitrogen-doped silicon samples. *Microelectr. Eng.* 66, 2003: 297–304.
101. Kveder V.V., E.A. Steinman, S.A. Shevchenko, H.G. Grimmeiss. Dislocation-related electroluminescence at room temperature in plastically deformed silicon. *Phys. Rev. B* 51, 1995: 10520–10526.
102. Xiang, L., D. Li, L. Jin, B. Pivac, D. Yang. The origin of 0.78eV line of the dislocation related luminescence in silicon. *J. Appl. Phys.* 112, 2012: 063528-1–063528-4.
103. Ng, W.L., M.A. Lourenco, R.M. Gwilliam, S. Ledain, G. Shao, K.P. Homewood. An efficient room-temperature silicon-based light-emitting diode. *Nature* 410, 2001:192–194.
104. Kveder, V., Badylevich, E. Steinman, A. Izotov, M. Seibt, W. Schroter. Silicon light emitting diodes based on dislocation luminescence. *Appl. Phys. Lett.* 84, 2004: 2106–2108.

11 Device Operation as Crystal Quality Probe

Erich Kasper and Wogong Zhang

CONTENTS

11.1 Introduction 353
11.2 Current–Voltage Characteristics 354
11.3 Alternating Current Response 361
11.4 Photonic Properties 363
11.5 Microwave Measurements: Experimental Setup, Scattering Parameters, Frequency Limits 364
11.6 Summary 374
References 374

11.1 INTRODUCTION

For a complete characterization of heterostructures, a variety of different structural, physical, optical, and electrical methods are available. A complementary treatment looks for the output of device characteristics. This is especially valuable for properties that are not easy to measure with conventional techniques, for example, point defects (vacancies, interstitials), minority carrier densities, and recombination processes.

Device realization needs a starting material of high quality and several critical technological processes such as epitaxy, ion implantation, diffusion, oxidation, insulator deposition, and metallization influence the final result. Devices are tested for their direct current (DC) and alternating current (AC) characteristics, for their capacitance, and photonic and microwave response as a function of their various terminal bias conditions.

A wealth of data are generated, which is primarily used to control the stability of technological conditions.

In this chapter, we focus on the physical background of different outputs and how we can use the data to extract the basic properties of heterostructures from which the devices are built.

Special emphasis is given to the art of microwave characterization, which allows extracting device parameters from wafer scattering parameter measurements.

The reader is referred to textbooks on device physics,[1] ultra large-scale integration,[2] and advanced microelectronics directions[3] for more background information on these topics. Our basic understanding focuses on analytical approximations but for more complicated structures, numerical solutions of the semiconductor equations are needed. Several programs are available in the market; we use Silvaco®[4] because of its clear structure and the ease of obtaining simple description to more sophisticated ones by adding effects and structural details.

The text and figures follow partly the content of one of the author's (E.K) lectures on "Semiconductor Engineering" at the University of Stuttgart.

11.2 CURRENT–VOLTAGE CHARACTERISTICS

The basic units of devices are semiconductor junctions, which create depletion layers where carriers (electrons, holes) are swept out. In a popular approximation (Schottky approximation), complete depletion is assumed, which allows easy calculation of electric fields and potential distribution by applying Poisson's law. Technically, the junction can be fabricated by doping transitions (*p–n* junction), by metal–semiconductor interfaces (Schottky contact), and by metal oxide semiconductor (MOS) sandwiches. We consider for DC measurements mainly the *p–n* junctions as model.

Figure 11.1 shows the typical current–voltage (*I–V*) characteristics from pure diffusion current consideration given as a broken line. The width W_d of the depletion layer varies with the applied voltage V:

$$W_d^2 = \frac{2\varepsilon}{q} \cdot \frac{1}{N}(V_{bi} - V) \tag{11.1}$$

where the prefactor $2\varepsilon/q$ is defined by the material with dielectric constant ε and electron charge q, the effective doping N is given by the donor concentration N_D and the acceptor concentration N_A on either side of the junction

$$\frac{1}{N} = \frac{1}{N_D} + \frac{1}{N_A} = \frac{N_D + N_A}{N_D N_A} \tag{11.2}$$

and the voltage V increases the width with negative bias (reverse voltage) and decreases the width with positive voltage (forward bias). The built-in voltage V_{bi} of the junction results from the charge exchange to obtain an equal Fermi energy level on both sides of the junction under equilibrium. As long as Boltzmann approximation holds for the Fermi–Dirac statistics, the built-in voltage is given by

$$V_{bi} = V_T \ln\left(\frac{N_A N_D}{n_i^2}\right) \tag{11.3}$$

with the temperature voltage $V_T = k_B T/q$ (where k_B is the Boltzmann constant and T the temperature) and the intrinsic carrier concentration n_i

$$n_i^2 = N_C N_V e^{-\frac{E_g}{k_B T}} \tag{11.4}$$

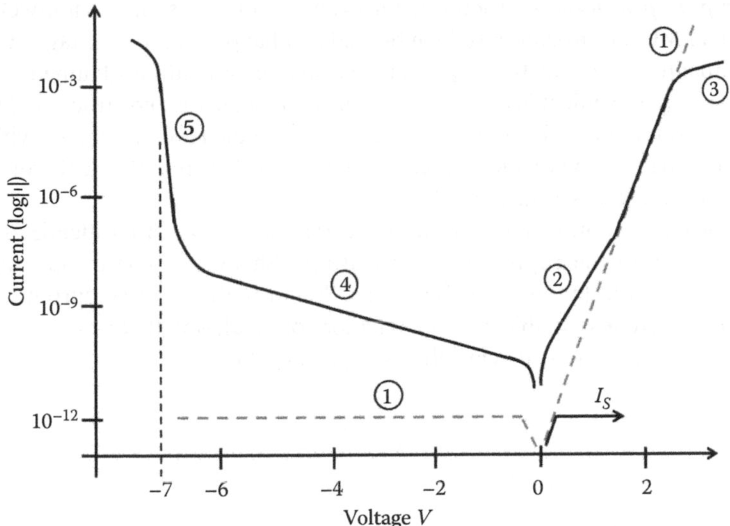

FIGURE 11.1 Current–voltage characteristics of a *p–n* junction in a semilogarithmic plot ln *I* versus V. Notice the different scale on the voltage axis for forward and reverse currents. The different regions (1–5) are explained in the text.

1. $I_S = 10^{-12}\,A, \eta = 1.0$
2. $I_{RS} = 10^{-10}\,A, \eta = 1.8$
3. $R_S = 10\,\Omega$
4. $I_{\text{Rev}} = 10^{-10}\left[\exp\left(-\frac{V}{\eta_r V_r}\right) - 1\right], \eta_r = 50$
5. $V_B = -7\,V$

with bandgap E_g, and effective densities of states N_c, N_v for conduction band and valence band, respectively.

The effective density of conduction band states N_c is given by

$$N_C = 2\left(\frac{2\pi m_n k_B T}{h^2}\right)^{\frac{3}{2}} \tag{11.5}$$

with Planck's constant h and effective mass of electrons m_n.

A similar equation holds for the effective density of valence band states N_v but with the effective mass m_p of holes. The straightforward set of Equations 11.1 through 11.5 determines the width W_d of the depletion layer as a function of the natural constant q, h, k_B, of the specific semiconductor properties m_n, m_p, E_g, of the doping concentrations N_D, N_A, and of the electrical bias V. On the contrary, these semiconductor properties and the doping may be extracted from the depletion layer variation with voltage. This will be discussed later in connection with AC measurements. A measurement of depletion layer width is possible with DC measurements only in special cases. In p–n–p or n–p–n structures, for example, in a p^+ source, an n-well, and a p bulk of an MOS transistor, a sudden breakdown happens at that voltage at which a punch through of the depletion layer between source and bulk is obtained.

Much more convenient is the extraction of semiconductor and structure data from the I–V characteristics. The depletion layer is large at the reverse voltage, which suppresses the current flow to a low level but at a forward current the depletion shrinks and the current increases steeply.

The current is dominated by minority carrier diffusion in an idealized theory, which stems from the beginning of transistors (Shockley). These first transistors were from Ge with its lower bandgap than Si and higher reverse currents, so that other effects were less visible, a good situation for idealizing theories.

This "ideal" diffusion current follows Equation 11.6:

$$I = I_s(e^{-\frac{V}{V_T}} - 1) \tag{11.6}$$

with the prefactor I_s given by

$$\frac{I_s}{Aq} = \frac{D_n n_{p0}}{L_n} + \frac{D_p p_{n0}}{L_p} \tag{11.7}$$

(A is area, D diffusion constant, L diffusion length, n_{p0} and p_{n0} are minority carrier densities at equilibrium.)

Diffusion constant and diffusion length are related to the more well-known properties of mobility μ and carrier recombination time τ_r by

$$D = V_T\mu \tag{11.8}$$

and

$$L^2 = D\tau_r = \mu\tau_r \tag{11.9}$$

Indices n and p are related to electrons and holes, respectively.

The equilibrium minority carrier concentrations depend on the majority doping by

$$n_{p0}N_A = n_i^2 \text{ and } p_{n0}N_D = n_i^2 \tag{11.10}$$

The device dimensions in earlier days were rather large so that the nonequilibrium minority carrier concentration at the junction could decay within a diffusion length to the equilibrium. In modern devices, the dimensions are frequently smaller than the diffusion length, and hence the diffusion length in the original theory (Equation 11.7) has to be replaced by thicknesses d_n, d_p of the n- and p-layers. The ideal current characteristics of the diffusion theory is marked in the semilog plot of Figure 11.1 by a straight line (exponential function in semilog plot) at forward bias and a low dark current $|I| = I_s$ at reverse current for $|V| > V_T$.

The most dramatic changing value in heterostructures is the intrinsic carrier concentration. The value of n_i^2 in Ge is more than six orders of magnitude higher than in Si. So, one sees a strong increase in n_i^2 and in forward current I by switching from Si to SiGe and Ge, because Equations 11.4, 11.8, and 11.10 deliver for a one-sided abrupt n^+p junction

$$\frac{I_s}{Aq} = V_T\mu_n \frac{N_C N_V}{N_A d_p} e^{-\frac{E_g}{k_B T}} \tag{11.11}$$

where we assumed a short layer (L_n replaced by d_p) and a negligible hole minority carrier ($N_D > \gamma > N_A$).

Both the absolute value of I_s and—more reliable—the temperature dependence of I_s give the bandgap value E_g.

$$\ln \frac{I_s}{T^4} = \text{const.} - \frac{E_g}{k_B T} \tag{11.12}$$

Note that the Arrhenius plot ln Y versus $1/T$ should not be taken with $Y = I_s$ but with $Y = I_s/T^4$ because V_T, N_C, and N_V are also temperature dependent.

A typical I–V characteristic of an experimental SiGe heterostructure device is shown in Figure 11.1. One sees many areas where the characteristics deviate (2–5) from the ideal diffusion current (1) behavior.

Let us start with the forward branch (0–1 V; note the different voltage scale compared with the reverse branch). At lower current densities, another current component appears with experimental but less steep increase in voltage. This additional current component is called recombination current

$$I_R = I_{RS}(e^{-\frac{V}{\eta_R V_T - 1}}) \tag{11.13}$$

with the recombination saturation current I_{RS} and the ideality factor $\eta_R > 1$ describing the lower steepness on a semilog plot.

The term recombination current relates to the recombination solely within the depletion region. The recombination rate is strongly enhanced by the presence of impurities and defects with localized energy levels near the middle of the bandgap. A simple model with energy levels exactly in the middle of the bandgap delivers for the ideality factor η_R and the saturation current I_{RS} with $\eta_R = 2$

$$I_{RS} = \frac{A \cdot q \cdot n_i \cdot W_d}{\tau_R} \tag{11.14}$$

Comparison of Equation 11.14 with Equation 11.10 shows a linear increase of the recombination current with n_i versus a parabolic increase with n_i^2 of the diffusion current. Both increase with lower E_g (from Si to SiGe to Ge) but the relative weight of the recombination current decreases, which makes easier the ideal theoretical treatment of the first Ge-based transistors. More recent SiGe and Ge devices are mainly on Si substrates where dislocations to accommodate lattice mismatch and low-temperature budget processing provide recombination centers from dislocations, point defects, and imperfect surface oxide structures. These defects often have energy levels outside the middle of the bandgap resulting in ideality factors η_R below two, for example, 1.5 for dislocations in Ge and 1.6–1.8 for surface defects.

Series resistance R_s is responsible for the forward current deviation at high currents (3). The outer clamp voltage V_c is divided between the device and the series resistance.

$$V_c = V + I \cdot R_s \tag{11.15}$$

At very high currents, the I–V characteristic turns from exponential to linear if the second term in Equation 11.15 dominates. An easy overview about the different regimes is obtained by analyzing the differential slope of the measured semilog plot ln I-V_c.

$$\frac{d\ln I}{dV_c} = \frac{1}{V_T\eta} = \frac{1}{I} \cdot \frac{dI}{dV_c} \tag{11.16}$$

with the differential ideality factor η. Starting with the ideal plot (Equation 11.6) and adding the voltage drop at the series resistance (Equation 11.15), we obtain for the differential ideality factor

$$\eta = 1 + R_s \cdot \frac{1}{V_T} \tag{11.17}$$

Figure 11.2 shows the voltage dependence of the differential ideality factor. At voltage below $V = V_T$, the slope of the semilog plot ln I versus V move toward infinity (because of the mathematical property of the logarithmic function, $\ln x \to \infty$ for $x \to \infty$!), which corresponds to $\eta \to 0$.

$$\eta = \eta_R \cdot \frac{V}{V + V_T} \quad \text{for } V < V_T \tag{11.18}$$

Then, at the recombination current regime, the differential ideality factor η saturates at η_R, and drops toward 1 in the diffusion current regime. At higher currents, the ideality factor η increases again under the curve-flattening influence of R_s. A quantitative extraction of R_s needs the presentation of $\eta(I)$ as shown in Figure 11.3.

Equation 11.18 is represented by a slope

$$\frac{d\eta}{dI} = \frac{R_s}{V_T} \tag{11.19}$$

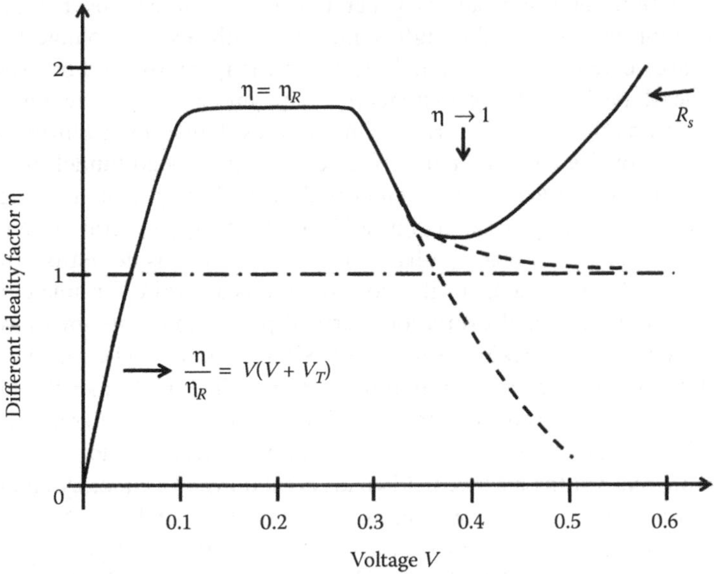

FIGURE 11.2 Ideality factor η versus voltage V_c on the outer clamp.

which is used to extract the value of R_s from the high current portion of the I–V curve. Care has to be taken to avoid confusion from current influenced series resistance (this happens with carrier injection in low-doped regions) and from junction heating at very high current densities.

The very small diffusion current component (1) in reverse direction is often overrun by the generation current (4) from the depletion layer. The midgap energy states, which are responsible for the recombination current (2) under forward

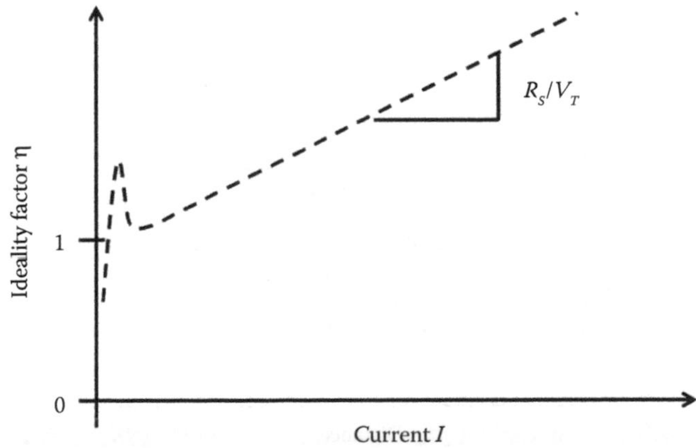

FIGURE 11.3 Ideality factor as a function of current. The slope of the straight line is proportional to the series resistance R_s.

carriers injection condition, act as generation centers under reverse conditions. This generation current should slightly increase with reverse voltage because of depletion layer extension (Equation 11.1) but in many nanoscaled heterostructure devices, the slope looks more exponential with much smaller steepness (ideality factor >10) as the forward current. This was explained by a Philips Research Laboratories group[5] as recombination including trap-assisted tunneling. Consider a trap (a recombination/generation center within the bandgap) in a strong electric field of a reverse-biased junction (Figure 11.4). The energy diagram of a trap in an electric field shows two escape paths for electrons (we consider emission of electrons in the conduction band; similar considerations are valid for hole emission in the valence band). Thermal excitation (vertical path) represents the conventional path described by the Shockley–Read–Hall (SRH) theory.[6–8] The second path (lateral) is tunneling of electrons through the barrier, which is thinner the higher the electrical field is. This trap-assisted tunneling path causes the increase in reverse current, which is favorably analyzed as a function of electric field.

Further increase of the electric field yields breakdown (5), either by band-to-band tunneling or by impact ionization leading to avalanche breakdown. A rough picture of avalanche breakdown is obtained by assuming fixed field strength E_{max} for abrupt onset of impact ionization (The onset of impact ionization is indeed very steep but not perfectly abrupt). In a bulk material, the width of the depletion layer and also the maximum field E_{max} is a function of doping. Breakdown voltages V_B are given as function of doping in the textbooks on semiconductor physics.[1]

The depletion layers in devices are often limited by the layer thickness d. Consider for instance a p–n junction with n^+ top contact and p layer (doping N_A, thickness d) contacted from the backside by a p^+ buried layer. The electric field at the junction is E_{max}, the potential difference at breakdown is $V_D + V_{bi}$ (Figure 11.5).

The electric field at breakdown may deplete to full layer thickness d (punch through). Then the breakdown voltage V_B is a function of doping N_A and layer thickness d.

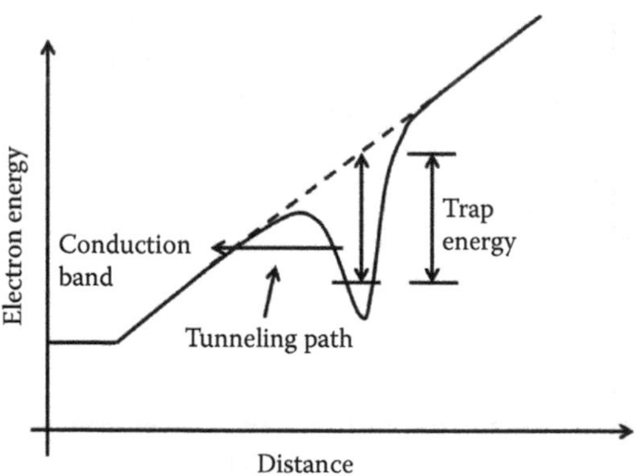

FIGURE 11.4 Energy scheme for electron emission of a trap under high electric field (reverse voltage).

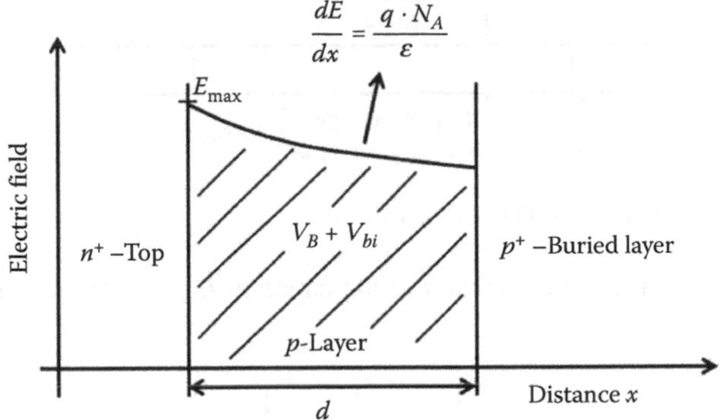

FIGURE 11.5 Electric field in a p–n junction under punch through condition ($d < W_d$).

$$V_B + V_{bi} = d\left(E_{max} - \frac{1}{2}\frac{q \cdot N_A \cdot d}{\varepsilon}\right) \quad \text{with } d < W_d \tag{11.20}$$

At very high current densities, the avalanche breakdown bends to higher voltages (heating effect, positive temperature coefficient). This temperature coefficient allows experimental discrimination from tunneling (negative temperature coefficient).

11.3 ALTERNATING CURRENT RESPONSE

AC can take all the paths of the DC but additionally paths where charges Q may be stored. The equivalent circuit description of a charge storing element is the capacitance with a capacity C

$$C(V) = \frac{dQ}{dV} \tag{11.21}$$

The plate capacitor has a voltage-independent capacity, which leads to $C = (Q/V)$. In such a plate capacitor, the value of the capacity is simply connected with geometrical data by

$$C = A \cdot \frac{\varepsilon}{W} \tag{11.22}$$

where A is the area, s the dielectric constant, and W the plate separation.

The basic capacitance element in a semiconductor device is the depletion layer of a junction. Let us consider for description a MOS varactor (Figure 11.6) with negligible interface charges.

For MOS varactors, the AC current is easy to measure because all DC components are blocked by the insulating oxide. The measured capacity C_{meas} is composed

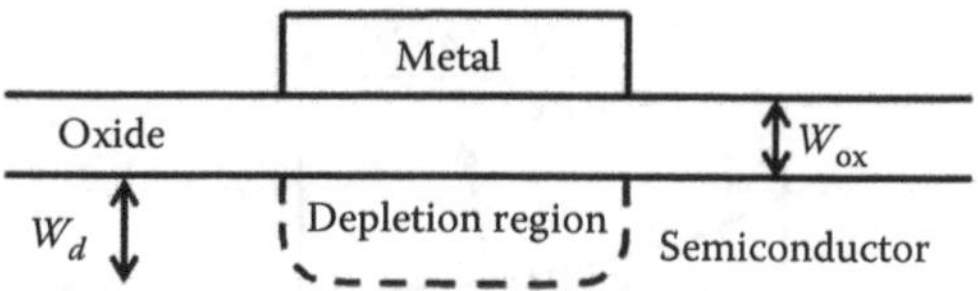

FIGURE 11.6 Scheme of a MOS varactor (n-SiGe).

of two series capacities, the constant oxide capacity C_{ox} and the variable depletion layer capacity C_{dep}

$$\frac{1}{C_{meas}} = \frac{1}{C_{ox}} + \frac{1}{C_{dep}} \tag{11.23}$$

The dopant profile $N(x)$ is extracted from

$$\frac{d(A^2/C_{meas}^2)}{dV} - \frac{2}{\varepsilon q} \cdot \frac{1}{N(x)} \tag{11.24}$$

$$x = A \cdot \varepsilon \cdot \left(\frac{1}{C_{meas}} - \frac{1}{C_{ox}} \right) \tag{11.25}$$

The depletion layer exists as a single source of space charge between the flat band voltage V_{fb} and the threshold voltage V_{th} (Figure 11.7). Beyond the flat band voltage, an accumulation layer of majority carriers (electrons in Figure 11.7) beneath the oxide semiconductor interface confines the capacity value to the oxide capacity C_{ox}.

Below the threshold voltage V_{th}, the curve splits into three parts. An inversion layer (holes) is created below the interface. This leads at low frequencies (LF) to a return of the curve to the oxide capacity value. The inversion layer is created from minority carriers the build-up of which needs more than an order of magnitude of the

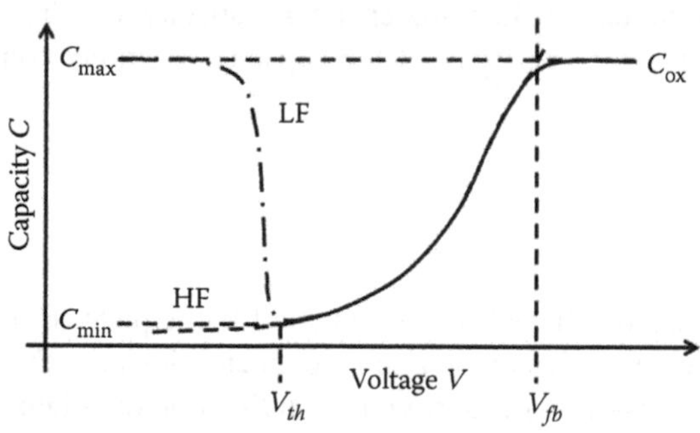

FIGURE 11.7 C–V characteristic of a MOS varactor. LF: low-frequency curve; HF: high-frequency curve; broken line: pulsed bias-deep depletion curve.

generation lifetime τ of the semiconductor. For high frequencies (HF), the inversion layer cannot follow the AC signal and the capacity is fixed at the inversion capacity C_{inv} defined by the constant voltage bias V_{th}. The capacity follows the depletion curve (broken line) if a pulsed voltage bias suppresses the formation of the inversion layer (deep depletion). The C–V characteristic is given for a perfect interface. Fixed interface charge shifts the curve (V_{fb} and V_{th} shift) but does not change shape. Interface states that are charged depending on the Fermi energy position change the shape[9] by increasing the difference between V_{th} and V_{fb}.

From the *C–V* measurements, we obtain information about the carrier density profile, the carrier lifetime and about heterostructure interface properties.[10] The carrier density profile follows the dopant profile with a smearing of about the Debye length L_D.

$$L_D = \sqrt{\frac{\varepsilon_s kT}{q^2 N}} \tag{11.26}$$

where ε_s is the permittivity of semiconductor, kT the thermal energy, q the elementary charge, and N the dopant level. That means, even with an abrupt doping profile, the carrier profile follows with a softer slope.

11.4 PHOTONIC PROPERTIES

All optical properties[11] that can be measured from small areas are considered for device characterization. Frequently applied methods are Raman spectroscopy, reflectometry, and photoluminescence.

Device structures are needed for optical junction space charge techniques (opt JSCT) as there are photocurrent, photo capacity, and electroluminescence (EL).

Photocurrents are closely related to the absorption properties with a dominant contribution from band-to-band transitions. Consider for simplicity a pin structure with a well-defined depletion layer by the geometrical width W_i of the intrinsic layer and a top contact layer of thickness W_t. The internal quantum efficiency (IQE) describes the fraction of photons that are absorbed and converted in electron–hole pairs, for the photocurrent.

$$\text{IQE} = e^{-\alpha_t W_t}\left(1 - e^{-\alpha_i W_i}\right) \tag{11.27}$$

where α_t, α_i are the absorption coefficients and W_t, W_i the thickness of the top layer and the intrinsic layer, respectively. The *IQE* may be calculated from the photocurrent responsivity R_{opt} measured in ampere/watt (A/W).

$$\text{IQE} = \frac{R_{\text{opt}}}{1 - R} \cdot \frac{hc}{q\lambda}$$
$$\text{with } \frac{hc}{q} = 1.24\ \mu\text{m} \tag{11.28}$$

The bandgap of the SiGe heterostructure will be extracted from the absorption coefficient–energy relation.

For indirect transitions

$$\alpha(\mathrm{ind}) = C_{\mathrm{ind}}(E - E_g)^2 \tag{11.29}$$

and for direct transitions

$$\alpha(\mathrm{dir}) = C_{\mathrm{dir}}(E - E_g)^{1/2} \tag{11.30}$$

One has to consider that the optically determined bandgap may differ from the electrical one by the binding energy of excitations (a few meV) and by the existence of phonon replica for indirect transitions.[12]

A more direct evidence of the bandgap energies is obtained from EL, which however needs to drive the junction into heavy forward conditions whereas the photocurrent functions even under zero bias (solar cell operation mode). The bandgap luminescence response of indirect semiconductors is rather weak and limited to observation at very low temperatures. Two effects were responsible; with SiGe and Ge the luminescence was stronger and observable at room temperature. Alloy disorder allows no phonon replica, and the lowest-lying direct transition gets near to the indirect one with increasing Ge content. In Ge, the intensity of the direct transition is already higher,[13] although more electrons populate the lower-lying indirect conduction band but the emission probability for direct transitions is orders of magnitude higher. The line shape of the direct EL peak is influenced by the Fermi–Dirac statistics and the available density of states. In the case of Boltzmann approximation for the Fermi–Dirac statistics, this results in the following relation[14] for the intensity I as a function of the photon energy $E = h\nu$.

$$I(E) = Const.(E - E_g)^{1/2} \cdot e^{-\frac{E-E_g}{k_B T}} \tag{11.31}$$

The first term reflects the increasing number of band states and the second term reflects the decreasing occupation of the states with increasing energy. The maximum of intensity appears at an energy $E_{\max}$

$$E_{\max} = E_g + \frac{1}{2}k_B T \tag{11.32}$$

11.5 MICROWAVE MEASUREMENTS: EXPERIMENTAL SETUP, SCATTERING PARAMETERS, FREQUENCY LIMITS

During the past half century, Moore's law has been well kept successfully.[15] The improvements in many facets, for example, integration complexity, power consumption, yield, fabrication cost, speed, and so on, benefit a lot from this progress. For microwave applications (typically covers the frequency band between 3 and

300 GHz), the most intuitive effect of the speed progress is the shrink of the passive circuit dimensions because electrical wavelength (10 cm–1 mm for the above-mentioned frequency band) scales inversely with frequency. The low area junction capacity then leads to a high cut-off frequency of the active devices, which leaves the design limitation confined mainly by the passive parts.

Owing to the high frequency ($f = 3 \sim 300$ GHz) in the microwave world, with corresponding electrical wavelength in free space $\lambda = c/f = 10$ cm ~ 1 mm (for monolithic design on high-k substrates, the effective guided wavelength λ_g goes even to submillimeter), the classical circuit theory is not sufficient enough to explain microwave phenomena. The fact is that the important parameters of lumped circuits such as voltages and currents cannot be defined for a general case anymore, because the circuit dimensions are on the same level of the electrical wavelength. According to Maxwell's equations,[16] the intuitive expression is that the voltage or the current across the circuit changes much with both time and position (basically due to the phase variation), which is far beyond the valid condition of the classical circuit theory.

The concept of "scattering matrix" was first proposed by Kaneyuke Kurokawa of Bell Labs in 1965[17] to describe the complex microwave networks in the behavior of a simple "black box." This concept skips the tough analysis of the complex microwave networks and offers a convenient solution for the characterization of microwave components in a unique way. Nowadays, this concept is well known as "S-parameters" ("S" refers to "scattering") and it is widely applied for most microwave designs.

Generally speaking, the S-parameter is a mathematical description that quantifies the high frequency (HF) energy propagation through a multiport network. For instance, HF signals incident on one port, and part of the signals are reflected back out of the port immediately, while part of the signals are scattered into the network and finally exit from other ports. Some of the signals may disappear in the form of heat loss or even electromagnetic (EM) radiation. Some of the signals may also be amplified. The "scattering matrix" for an N-port network contains N^2 coefficients (S-parameters), each one representing an input–output path of all possible HF energy flows in the network.

$$\begin{pmatrix} V_{r,1} \\ V_{r,2} \\ \cdot \\ \cdot \\ \cdot \\ V_{r,N} \end{pmatrix} = \begin{pmatrix} S_{11} & S_{12} & \cdots & S_{1N} \\ S_{21} & S_{22} & \cdots & S_{2N} \\ \cdots\cdots & \cdots\cdots & \cdots\cdots & \cdots\cdots \\ S_{N1} & S_{N2} & \cdots & S_{NN} \end{pmatrix} \begin{pmatrix} V_{t,1} \\ V_{t,2} \\ \cdot \\ \cdot \\ \cdot \\ V_{t,N} \end{pmatrix} \tag{11.33}$$

where $V_{r,N}$ is the voltage amplitude of the reflected wave at port N and $V_{t,N}$ the voltage amplitude of the transmitted incident wave at port N.

Both the magnitude and phase of the input signals are changed by the network. Hence, S-parameters are complex numbers. Depending on the design demands, S-parameters could be expressed in other forms, for example, Z-parameters (impedance-matrix), Y-parameters (admittance-matrix), H-parameters (hybrid-matrix),

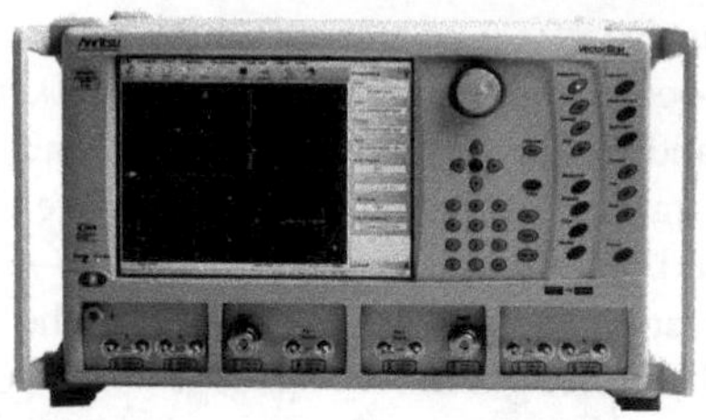

FIGURE 11.8 Photograph of the vector network analyzer (Anritsu Vector Star) with available broadest frequency range from 70 kHz to 70 GHz. The dynamic range is more than 100 dB over full range. This instrument is also available as four-port unit with pulse and true differential options and with frequency extensions for single sweep 70 kHz ~ 145 GHz.

ABCD-parameters (chain- or cascade-matrix), or *T*-parameters (chain-transfer- or chain-scattering-matrix) by using linear algebraic matrix transformation.[18]

To measure the *S*-parameters of the microwave components, the vector network analyzer (VNA)[19] offered by commercial companies, for example, Agilent®, Anritsu®, Rohde & Schwarz®, and so on, is usually employed. This is a small-signal characterization, which means the signals are small enough (typical value: –10 ~ –20 dBm) and have only linear effects on the network. The VNA is designed to extract the magnitude and phase information of the transmitted and reflected waves from the measured active or passive network under small-signal excitations (Figure 11.8).

For one-port network (Figure 11.9), Equation 11.33 is specified as follows:

$$S_{11} = \frac{V_{r,1}}{V_{t,1}} \tag{11.34}$$

In practice, to match the ground–signal–ground (GSG) configuration of the HF probe head, the device under test (DUT) of small footprint has normally contact pads formed by coplanar waveguide structure (Figure 11.10).

Notice that, after calibration, the reference plane of the measurement system can be only shifted to the probe tips. That means, the raw measurement *S*-parameters still include informations, both of DUT and the coplanar contact pad. The procedure that extracts the *S*-parameters of DUT from the raw measurement data is called "de-embedding." Two of the most common methods "open-short" (OS) and

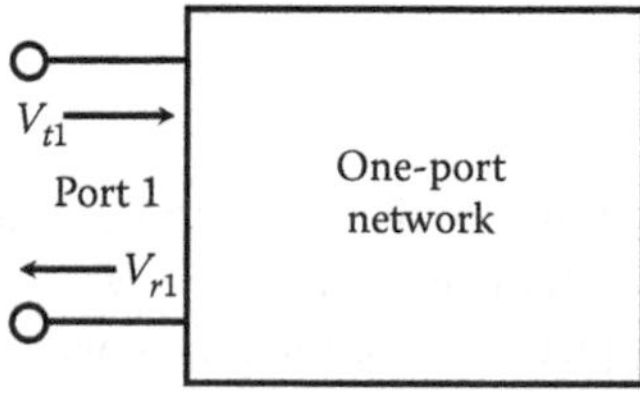

FIGURE 11.9 One-port network with the voltage amplitude of the reflected wave $V_{r,1}$ and the voltage amplitude of the transmitted incident wave $V_{t,1}$.

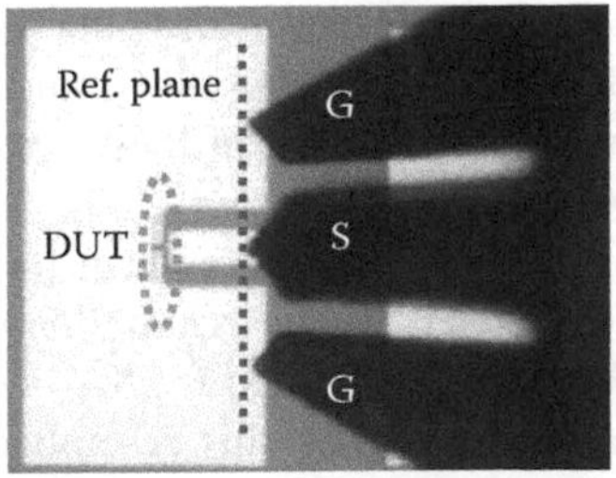

FIGURE 11.10 DUT with GSG contact pad and probe tips.

"short-open" (SO) that base on discrete lumped element model are widely employed for one- or two-port network de-embedding (Figure 11.11) due to its simplicity. For the OS lumped model approach, all parallel parasitic admittances Y_p are assumed to be located in the signal path and all serial parasitic impedances Z_s in the interconnect path. For the SO approach, Z_s is assumed to be located in the signal path and Y_p in the interconnect path. For the OS case, the impedance of DUT $Z_{\text{DUT\_OS}}$ results in

$$Z_{\text{DUT\_OS}} = \frac{1}{\dfrac{1}{Z_{\text{MEAS}}} - \dfrac{1}{Z_{\text{OPEN}}}} - \frac{1}{\dfrac{1}{Z_{\text{SHORT}}} - \dfrac{1}{Z_{\text{OPEN}}}} \tag{11.35}$$

and for the SO case, the impedance of DUT $Z_{\text{DUT\_OS}}$ results in

$$Z_{\text{DUT\_OS}} = \frac{1}{\dfrac{1}{Z_{\text{MEAS}}} - \dfrac{1}{Z_{\text{SHORT}}} - \dfrac{1}{Z_{\text{OPEN}} - Z_{\text{SHORT}}}} \tag{11.36}$$

where Z_{MEAS} is the raw measured impedance of the one-port network, Z_{OPEN} the impedance of the measured "open" test structure, and Z_{SHORT} the impedance of the measured "short" test structure.

The transformation of the Z-parameter to S-parameter for one-port is defined by

$$S_{11} = \frac{Z - Z_0}{Z + Z_0} \tag{11.37}$$

where Z is the input impedance of the one-port network and Z_0 the normalizing or reference impedance, which is normally set to 50 Ω by most HF measurement systems.

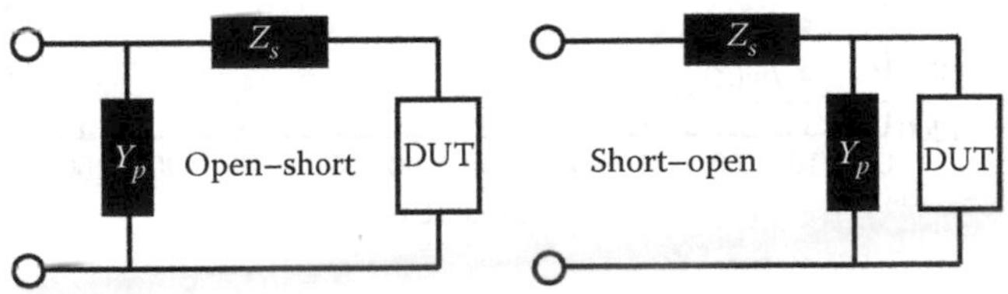

FIGURE 11.11 Equivalent lumped models for OS (left) and SO (right) de-embedding.

As mentioned, both OS and SO de-embedding approaches base on the discrete lumped element models. Because this assumption could not fully describe the real situation for wave propagation, a limitation of these two approaches is noticed for frequency range above 40 GHz.

A more reliable one-port de-embedding approach for higher frequency range is proposed and verified up to 110 GHz under assumption of a symmetry condition.[20] This approach provides a broader valid range for the one-port de-embedding algorithm in microwave applications. Based on this de-embedding approach, let us take impact-ionized avalanche transit time diode (IMPATT-diode) as an example for one-port active device HF characterizations. The IMPATT diode is well known as microwave power source. Different from most diodes, it operates in the avalanche breakdown region. Near that working point, a negative differential resistance can be provided, which means an amplifying effect. This amplifying effect can be well characterized by the small-signal VNA *S*-parameter measurement. In Figure 11.12, the real and imaginary parts of the IMPATT-diode with mesa size 30 μm × 2 μm under bias current of 40 mA are shown. The avalanche frequency has an obvious shift (~30 GHz) after de-embedding. In addition, the resonance amplitudes of the real and imaginary parts of impedance are greatly influenced by the parasitic impedance due to the contact pads. This example shows the necessity of a proper de-embedding for a precise HF characterization of microwave components or even for the further circuits design.

For two-port network (Figure 11.13), Equation 11.33 is specified as follows:

$$\begin{pmatrix} V_{r,1} \\ V_{r,2} \end{pmatrix} = \begin{pmatrix} S_{11} & S_{12} \\ S_{21} & S_{22} \end{pmatrix} \begin{pmatrix} V_{t,1} \\ V_{t,2} \end{pmatrix} \tag{11.38}$$

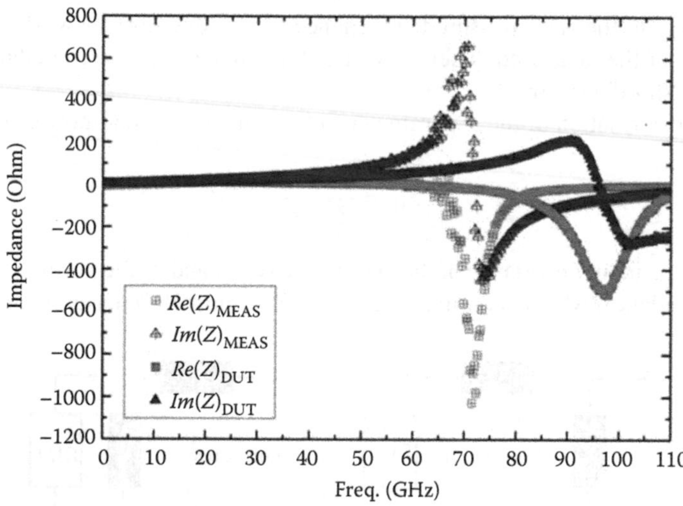

FIGURE 11.12 Real and imaginary parts of the evaluated IMPATT-diode over frequency before and after de-embedding.

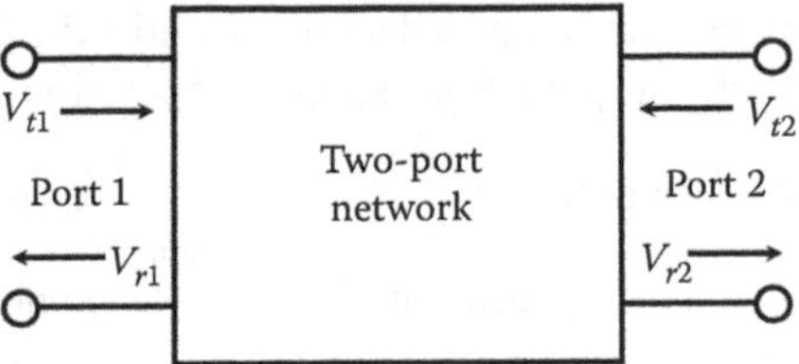

FIGURE 11.13 Two-port network with the voltage amplitudes of reflected wave $V_{r,1}$ at port 1, $V_{r,2}$ at port 2, and the voltage amplitudes of the transmitted incident wave $V_{t,1}$ at port 1, $V_{t,2}$ at port 2.

Same as the de-embedding procedure of one-port network, the OS and SO algorithms are also quite often employed for two-port network de-embedding. Only the expressions of Z_{MEAS}, Z_{OPEN}, and Z_{SHORT} are not single values but Z-matrices transformed from corresponding measured S-matrices. The same limitation in HF range exists, of course, also for OS or SO two-port network de-embedding. In addition, the complete two-port S-parameters of the contact pads cannot be (sometimes impossible for microwave applications above 60 GHz) directly measured due to the small contact area to the DUT side. This makes the two-port de-embedding more challenging compared to one-port case. Various publications about two-port de-embedding based on other complex algorithms could be well found.[21–28] However, there is no universal procedure for a precise two-port de-embedding that is valid for all cases in practice. To achieve an approach for complete two-port matrix information of the contact pads, the EM simulation tools are often used to predict the parasitic impedance.

A theoretical two-port de-embedding procedure based on *ABCD*-matrix (chain-matrix) is introduced to explain the principle (Figure 11.14).

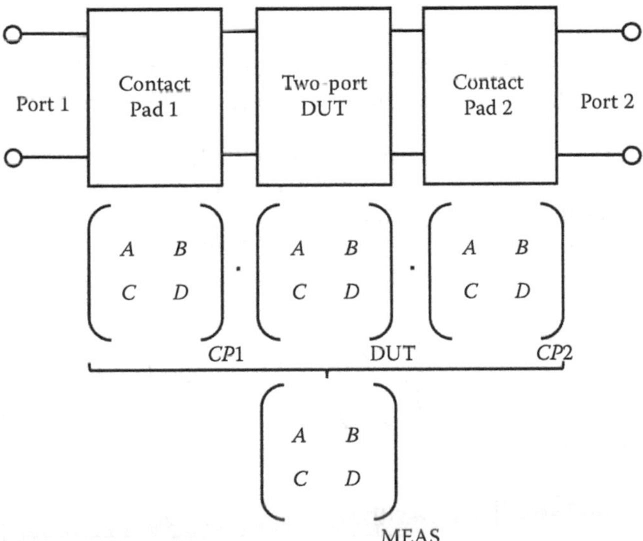

FIGURE 11.14 Two-port network in the form of cascaded *ABCD*-matrices.

ABCD-matrices can be easily cascaded to provide the performance of many cascaded networks, which the pure *S*-parameters operation cannot do. Hence, the raw measured two-port *S*-parameters $\begin{pmatrix} S_{11} & S_{12} \\ S_{21} & S_{22} \end{pmatrix}_{\text{MEAS}}$ is transformed to *ABCD*-parameters $\begin{pmatrix} A & B \\ C & D \end{pmatrix}_{\text{MEAS}}$. The equation in Figure 11.13 could be rewritten as

$$\begin{pmatrix} A & B \\ C & D \end{pmatrix}_{\text{DUT}} = \begin{pmatrix} A & B \\ C & D \end{pmatrix}_{CP1}^{-1} \begin{pmatrix} A & B \\ C & D \end{pmatrix}_{\text{MEAS}} \begin{pmatrix} A & B \\ C & D \end{pmatrix}_{CP2}^{-1} \tag{11.39}$$

where $\begin{pmatrix} A & B \\ C & D \end{pmatrix}_{\text{DUT}}$ is the *ABCD*-matrix of DUT, $\begin{pmatrix} A & B \\ C & D \end{pmatrix}_{CP1}^{-1}$ the inverse *ABCD*-matrix of contact pad 1, $\begin{pmatrix} A & B \\ C & D \end{pmatrix}_{CP2}^{-1}$ the inverse *ABCD*-matrix of contact pad 2, and $\begin{pmatrix} A & B \\ C & D \end{pmatrix}_{\text{MEAS}}$ the *ABCD*-matrix of the measured two-port network.

The reference impedance $Z_0 = 50\ \Omega$ is taken at the reference plane of port 1 and port 2, after proper calibration of course. Thus, the S-parameters to *ABCD*-parameters can be described as follows:

$$\begin{pmatrix} A & B \\ C & D \end{pmatrix} = \begin{pmatrix} \dfrac{(1+S_{11})(1-S_{22})+S_{12}S_{21}}{2S_{21}} & \dfrac{(1+S_{11})(1+S_{22})+S_{12}S_{21}}{2S_{21}} \\ \dfrac{(1-S_{11})(1-S_{22})-S_{12}S_{21}}{2S_{21}} & \dfrac{(1-S_{11})(1+S_{22})+S_{12}S_{21}}{2S_{21}} \end{pmatrix} \tag{11.40}$$

Based on Equation 11.40, all the *ABCD*-matrices on the right side of Equation 11.39 can be calculated from the measured corresponding *S*-matrices, which can deliver the final $\begin{pmatrix} A & B \\ C & D \end{pmatrix}_{\text{DUT}}$ just after two times multiplication.

The transformation from the calculated $\begin{pmatrix} A & B \\ C & D \end{pmatrix}_{\text{DUT}}$ back to $\begin{pmatrix} S_{11} & S_{12} \\ S_{21} & S_{22} \end{pmatrix}_{\text{DUT}}$ can be performed by using

$$\begin{pmatrix} S_{11} & S_{12} \\ S_{21} & S_{22} \end{pmatrix} = \begin{pmatrix} \dfrac{A+\dfrac{B}{Z_0}-CZ_0-D}{A+\dfrac{B}{Z_0}+CZ_0+D} & \dfrac{2(AD-BC)}{A+\dfrac{B}{Z_0}+CZ_0+D} \\ \dfrac{2}{A+\dfrac{B}{Z_0}+CZ_0+D} & \dfrac{-A+\dfrac{B}{Z_0}-CZ_0+D}{A+\dfrac{B}{Z_0}+CZ_0+D} \end{pmatrix} \tag{11.41}$$

For the HF characterization of the typical two-port active device metal oxide semiconductor field effect transistors (MOSFETs), the transit frequency f_t and the maximum oscillation frequency f_{max} are most interesting. The transit frequency f_t defines the cut-off frequency of the unity current gain β for current-amplifying applications. The HF measurement setup for MOSFETs with the simplest equivalent lumped model is shown in Figure 11.15.

The current gain β is defined as

$$\beta = \frac{I_D}{I_G} \tag{11.42}$$

where I_D is the drain current and I_G the gate current.

For general two-port network, the H-parameters (hybrid-matrix) are defined as

$$\begin{pmatrix} V_1 \\ I_2 \end{pmatrix} = \begin{pmatrix} H_{11} & H_{12} \\ H_{21} & H_{22} \end{pmatrix} \begin{pmatrix} I_1 \\ V_2 \end{pmatrix} \tag{11.43}$$

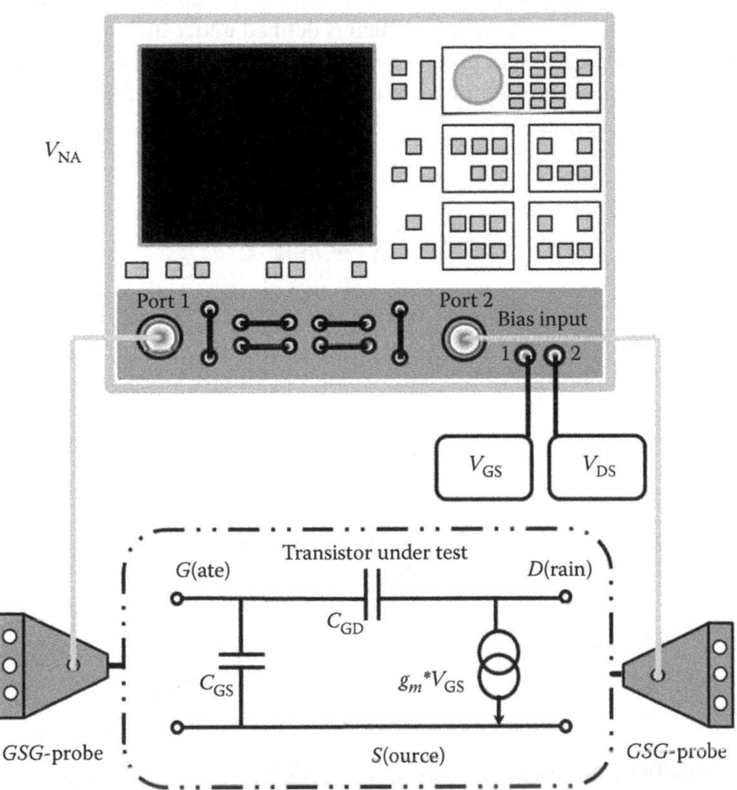

FIGURE 11.15 MOSFET under VNA characterization with its small-signal equivalent circuit.

where V_1, V_2 are the voltage values at port 1 and port 2 and I_1, I_2 the current values at port 1 and port 2. That means the current gain β can be expressed in the form of the *H*-parameter:

$$\beta = \frac{I_D}{I_G} = |H_{21}| \tag{11.44}$$

H_{21} can be calculated from the measured DUT S-parameters conveniently as

$$H_{21} = \frac{-2S_{21}}{(1 - S_{11})(1 + S_{22}) + S_{12}S_{21}} \tag{11.45}$$

and presented in the frequency domain (Figure 11.16).

The extrapolation of the –20 dB/decade line till 0 dB gain then gives the transit frequency f_t of the characterized transistor. For amplifier design, the desired frequency is normally around one-fifth of the transit frequency f_t.

The maximum oscillation frequency f_{max} represents the cut-off frequency of the power gain for power amplifications, which is defined under the condition of a conjugate-matched input and output ports for maximum power transfer. To explain the f_{max} analytically, the small-signal equivalent circuit for MOSFET in Figure 11.17 is taken.

The f_{max} is then defined as[29]

$$f_{max} = \sqrt{\frac{f_t}{8\pi R_G C_{GD}}} \tag{11.46}$$

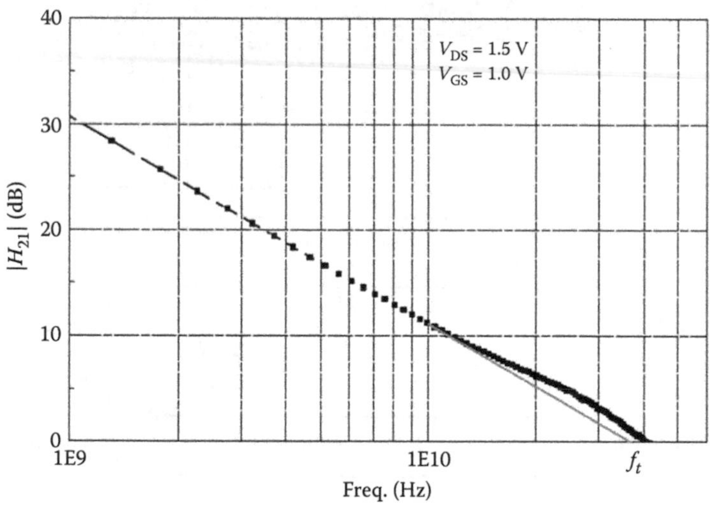

FIGURE 11.16 Transit frequency f_t extraction from $|H_{21}|$ curve.

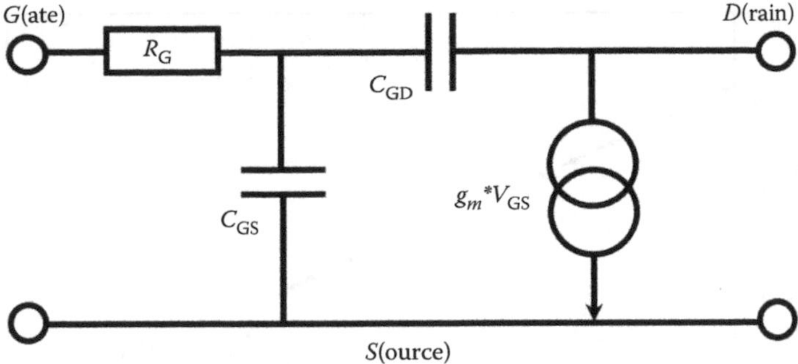

FIGURE 11.17 Small-signal circuit of the MOSFET with gate resistance.

where R_G is the effective gate resistance[30] and C_{GD} the gate drain capacity. From Equation 11.46, it is clear that the effective gate resistance R_G greatly influences f_{max}. In reality, a possible small R_G for MOSFET is decisive for its HF power applications.

The power amplification for input and output matching case is called maximum available gain (MAG), which is calculated from the two-port S-parameters of the MOSFET:

$$\mathrm{MAG} = \left| \frac{S_{21}}{S_{12}} \left(k - \sqrt{k^2 - 1}\right) \right| \tag{11.47}$$

with

$$k = \frac{1 + |\Delta S|^2 - |S_{11}|^2 - |S_{22}|^2}{2|S_{12}S_{21}|} \tag{11.48}$$

and

$$\Delta S = S_{11}S_{22} - S_{12}S_{21} \tag{11.49}$$

as stability factor.

If the stability factor $k < 1$, the two-port network (MOSFET) is not stable for the power matching. The maximum stable gain (MSG) is defined as

$$\mathrm{MSG} = \left| \frac{S_{21}}{S_{12}} \right| \tag{11.50}$$

Thus, it is important that only in the frequency range with $k > 1$ the f_{max} can be extracted. As shown in Figure 11.18, the f_{max} is read from the extrapolation of the curve in the MSG region at the 0 dB gain.

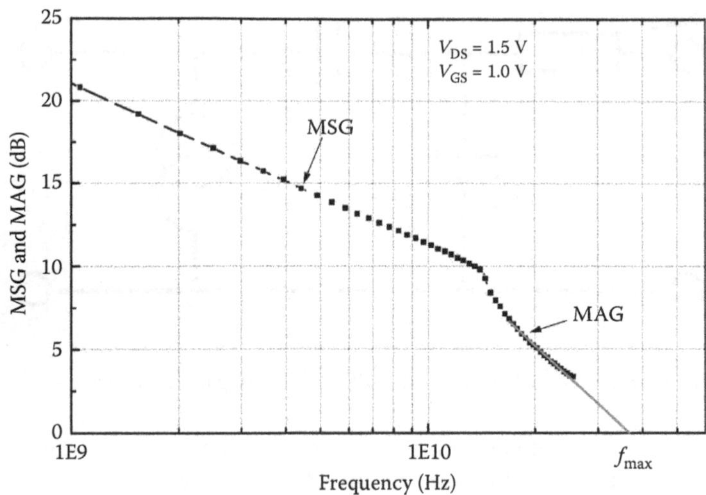

FIGURE 11.18 Maximal oscillation frequency f_{max} extraction from |MSG| and |MAG| curves.

11.6 SUMMARY

Device properties are dependent on many dimensional and structural properties; vice versa, essential material properties of the device layers may be extracted from the test devices with well-defined dimensions. In this chapter, we addressed the wealth of information we can obtain from DC current analysis if we consider the full range from strong forward current to reverse current up to the breakdown, from depletion layer variation under AC voltage modulation as monitored with CV methods, from optoelectronic response inherent to light absorption in junctions, and from microwave response of devices connected to an outer measurement port by coplanar waveguide structures. Especially, the latter configuration was considered in more detail. Microwave and mm-wave (frequencies above 30 GHz) measurements of scattering parameters (*S*-parameters) on input and output ports allow to extract the frequency-dependent impedance of the DUT. This gives, in comparison with models based on the material properties, a choice of preferred data.

REFERENCES

1. S. Sze and K. Ng, *Physics of Semiconductor Devices*, Wiley-Interscience, Hoboken, 2006.
2. C.Y. Chang and S.M. Sze (Eds), *ULSI Devices*, John Wiley & Sons, New York, 2000.
3. E. Kasper, H.-J. Muessig and H.G. Grimmeiss (Eds), *Advances in Electronic Materials*, Trans Tech Publications, Zürich, 2009.
4. Silvaco International, *ATLAS User's Manual*, 2010. http://www.silvaco.com/.
5. G. Hurkx, D. Klaassen and M. Knuvers, A new recombination model for device simulation including tunneling, *IEEE Transactions on Electron Devices* 39, 331–340, 1992.

6. C.T. Sah, R.N. Noyce and W. Shockly, Carrier generation and recombination in p-n junction and p-n junction characteristics, *Proceedings of the IREE* 45, 1228, 1957.
7. R.N. Hall, Electron–hole recombination in germanium, *Physical Review* 87, 387, 1952.
8. W. Shockley and W.T. Read, Statistics of the recombination of holes and electrons, *Physical Review* 87, 835, 1952.
9. S.P. Voinigescu, K. Iniewski, R. Lisak, C. Salama, J.-P. Noel and D. Houghton, New technique for the characterization of Si/SiGe layers using heterostructure MOS capacitors, *Solid State Electronics* 37, 1491–1501, 1994.
10. C. Lamberdi, *Characterization of Semiconductor Heterostructures*, Oxford Press, Oxford, 2008.
11. C. Klingshirn (Ed.), *Semiconductor Quantum Structures: Optical Properties*, Landolt-Boernstein, New Series, Group III, Springer, Berlin/Heidelberg, Vol. 34C, 2007 (ISBN 978-3-540-29647-8).
12. E. Kasper and K. Lyutovich (Eds), *Properties of Germanium and SiGe: Carbon, EMIS Datareview*, no. 24, IEE, England, 2000.
13. E. Kasper, M. Oehme, J. Werner, T. Aguirov and M. Kittler, Direct band gap luminescence from Ge on Si pin diodes, *Frontiers of Optoelectronics* 5, 256–260, 2012.
14. E. Schubert, *Light Emitting Diodes*, Cambridge University Press, New York, 2006.
15. Intel reveals 14 nm PC, declares Moore's law "alive and well". *The Register*. September 10, 2013.
16. J.C. Maxwell, A dynamical theory of the electromagnetic field, *Philosophical Transactions of the Royal Society of London* 155, 459–512, 1865.
17. K. Kurokawa, Power waves and the scattering matrix, *IEEE Transactions on Microwave Theory and Techniques* 12(2), 194–202, 1965.
18. D.A. Frickey, Conversions between S, Z, Y, H, *ABCD*, and T parameters which are valid for complex source and load impedance, *IEEE Transactions on Microwave Theory and Techniques* 42(2), 205–211, 1994.
19. http://www.hpmemory.org/wb_pages/wall_b_page_05.htm.
20. H. Xu and E. Kasper, A de-embedding procedure for one-port active mm-wave devices, *2010 10th Topical Meeting on Silicon Monolithic Integrated Circuits in RF Systems*, New Orleans, Digest of Papers, 37–40, 2010.
21. J.Y. Cha, J. Cha and S.H. Lee, Uncertainty analysis of two-step and three-step methods for deembedding on-wafer RF transistor measurements, *IEEE Transactions on Electron Devices* 55(8), 2195–2201, 2008.
22. X.S. Loo, K.S. Yeo, K.W.J. Chew and L.H.K. Chan, A new millimeter-wave fixture deembedding method based on generalized cascade network model, *IEEE Electron Device Letters* 34(3), 447–449, 2013.
23. K.H.K. Yau, I. Sarkas, A. Tomkins and P. Chevalier, On-wafer S-parameter de-embedding of silicon active and passive devices up to 170 GHz, *Microwave Symposium Digest (MTT)* 600–603, 2010, IEEE MTT-S International.
24. X. Wei, G. Niu, S.L. Sweeney and Q. Liang, A general 4-port solution for 110 GHz on-wafer transistor measurements with or without impedance standard substrate (ISS) calibration, *IEEE Transactions on Electron Devices* 54(10), 2706–2714, 2007.
25. M.-H. Cho, D. Chen, R. Lee and A.-S. Peng, Geometry-scalable parasitic deembedding methodology for on-wafer microwave characterization of MOSFETs, *IEEE Transactions on Electron Devices* 56(2), 299–305, 2009.
26. H. Cho and D.E. Burk, A three-step method for the de-embedding of high-frequency S-parameter measurements, *IEEE Transactions on Electron Devices* 38(6), 1371–1375, 1991.
27. S. Bousnina, C. Falt, P. Mandeville and A.B. Kouki, An accurate on-wafer deembedding technique with application to HBT devices characterization, *IEEE Transactions on Microwave Theory and Techniques* 50(20), 420–424, 2002.

28. B. Zhang, Y.-Z. Xiong, L. Wang and S. Hu, On the de-embedding issue of millimeter-wave and sub-millimeter-wave measurement and circuit design, *IEEE Transactions on Components, Packaging and Manufacturing Technology* 2(8), 1361–1369, 2012.
29. T.H. Lee, *The Design of CMOS Radio-Frequency Integrated Circuits*, Cambridge University Press, Cambridge, 1998.
30. B. Razavi, R.-H. Yan and K.F. Lee, Impact of distributed gate resistance on the performance of MOS devices, *IEEE Transactions on Circuits and Systems-I* 41, 750–754, 1994.

12 Silicon and Germanium Nanocrystals

Corrado Spinella and Salvo Mirabella

CONTENTS

12.1 Introduction 377
12.2 Synthesis and Imaging of Si Nanoclusters 378
12.3 Si Nanoclusters Formation: Effect of Deposition Methods 380
12.4 $Si-SiO_2$ Phase Separation: Phase Diagram and Energetics 386
12.5 Synthesis and Imaging of Ge Nanoclusters 388
12.6 Light Absorption in Si and Ge Nanoclusters 393
12.7 Conclusions 402
References 402

12.1 INTRODUCTION

Si nanoclusters (NCs) dispersed in SiO_2, synthesized by high-temperature anneal of substoichiometric silicon oxide (SiO_x) layer, are attracting increasing attention due to their application in fabricating optical and electronics devices, based on quantum confinement effects, by methods and processes fully compatible with the Si technology.[1,2] Furthermore, the reduced dimensionality of quantum dots (QDs) should help in fabricating innovative photovoltaic devices that can harvest the full solar energy, in a variety of modes, among which the multiple exciton generation (MEG),[3–5] the intermediate band formation (allowing the absorption of lower energy photon),[6,7] and the chance of modulating the absorption spectrum by changing the NC size, which should allow to tailor the light absorption onset for a better matching of the sun spectrum.[8,9] In this scenario, Ge QDs recently received considerable attention over Si QDs for various reasons, among which were the larger absorption coefficient of the bulk material, the easier bandgap tuning due to the larger exciton Bohr radius (24 nm),[10] and the lower synthesis temperature. Moreover, theoretical calculations estimate that an indirect-to-direct transition in the Ge spectra occurs with decreasing size, and oscillator strengths are expected to be larger in Ge NS than in Si NS.[11] This last effect could be attributed to stronger overlaps of the electron wave functions on Ge atoms,[12] which could enhance the light absorption in Ge NS.

To clearly understand these phenomena occurring in Si or Ge NCs, great care must be taken in the fabrication process in order to obtain well-isolated and size-controlled NCs. In the following sections, the Si and Ge QDs formation kinetics will be addressed while the last section will be devoted to light absorption phenomena in these confined systems for photovoltaic purposes.

12.2 SYNTHESIS AND IMAGING OF Si NANOCLUSTERS

Several methods to synthesize the Si NCs have been proposed and investigated in the last few years: ion implantation,[13–15] aerosol,[16,17] plasma-enhanced chemical vapor deposition (PE–CVD) from SiH_4 and N_2O gases,[18–21] and magneto co-sputtering (MS) from Si and SiO_2 targets.[22,23] Specifically, the latter two methods (PE–CVD or MS) have been demonstrated to be convenient techniques because of their immediate implementation in the very large-scale integrated (VLSI) processing and because of the good control on the deposition parameters, and, for this reason, in the following sections we will focus our discussion on these two methodologies.

As an example, the capability to finely control the SiO_x composition by PE–CVD technique is demonstrated in Figure 12.1. The ratio x between oxygen and silicon concentration, $x = C_O/C_{Si}$, increases by increasing the ratio between N_2O and SiH_4 fluxes (Figure 12.1a) or the chamber pressure (Figure 12.1b) during the PE–CVD process. In these measurements, Si and O concentrations were measured by coupling Rutherford backscattering spectrometry (RBS) analyses of the deposited layers (in order to determine the Si and O areal density) with transmission electron microscopy in cross-sectional configuration (X-TEM) to evaluate the layer thickness.

A key point for a full understanding of the optical and electrical properties of this system requires a clear picture of the thermodynamics regulating the phase separation between Si and SiO_2, induced by temperature anneal in the range 450–1250°C, which produces the formation of the Si NCs.

The control of the Si NCs mean radius and density plays an important role in the final device performance and is essential in understanding the kinetics and thermodynamics of the Si−SiO_2 phase separation. In this regard, several techniques have been employed to characterize Si NCs, namely, transmission electron

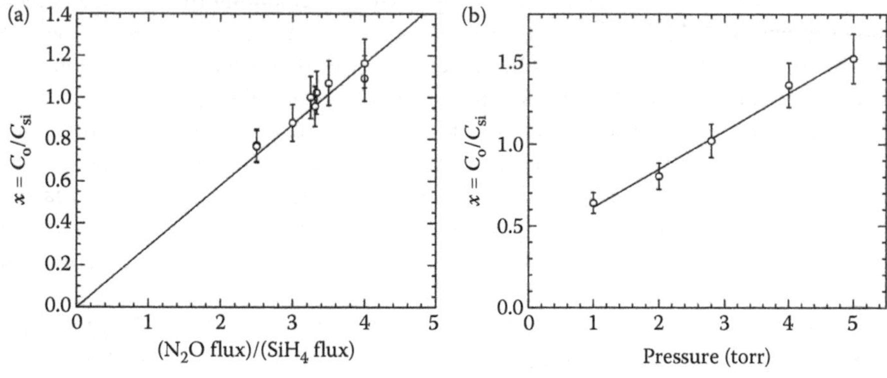

FIGURE 12.1 Composition, x, of SiO_x layers deposited by PE–CVD as a function of ratio between N_2O and SiH_4 fluxes (a), and as a function of chamber pressure (b).

microscopy (TEM),[24,25] x-ray diffraction (XRD),[14] and Raman spectroscopy.[13] All of these techniques provide some estimation of the Si ND mean radius. In particular, distribution of the Si NC size has been obtained by conventional TEM analysis (from dark-field or high-resolution measurements), but this technique is unable to provide a full quantitative picture of the system, as, for instance, the Si NCs density. Furthermore, all of these techniques are essentially blind to the presence of amorphous Si clusters. Very reliable determination of the cluster size distribution can be obtained by using energy-filtered transmission electron microscopy (EFTEM) bright-field images.[24,25] Indeed, the Si NCs can be imaged in the EFTEM configuration with electron energy loss tuned to the value of the Si bulk plasmon (17 eV), within an energy window of about 4 eV. Owing to the energy difference between the Si plasmon signal and the SiO_2 bulk plasmon signal at 23.5 eV, the contrast between the Si NCs and the surrounding silicon dioxide is clearly enhanced and the methodology successfully overcomes the problems related to the visibility of amorphous silicon dots. This is demonstrated in Figure 12.2 where a conventional X-TEM micrograph (Figure 12.2a) of a PE–CVD SiO_x layer ($x = 1.6$), 70 nm thick, annealed at 1100°C for half an hour, is compared to the corresponding EFTEM image on the Si plasmon peak (Figure 12.2b). The Si NCs are essentially invisible in Figure 12.2a while they are perfectly detected appearing as bright spots in Figure 12.2b.

Quantitative information on the average size of Si NCs, independent of their phase (crystal or amorphous), can be obtained by analyzing a large number of plan-view images obtained by energy-filtered technique, as shown in Figure 12.3a, which refers to the same sample described in Figure 12.2. Figure 12.3b shows the corresponding Si NCs radius distribution, n_R, that is, the number of Si NCs with a radius between R and $R + dR$ per unit volume in the layer. In order to obtain this information, it is necessary to have a good estimation of the thickness sampled by the EFTEM analysis. This is possible by collecting, under the same experimental conditions and in the same sample region, an electron energy loss spectrum (EELS), since the analyzed depth is proportional to the logarithm of the ratio between the integral intensity of the whole EELS and the integral intensity of the zero-loss peak.[26]

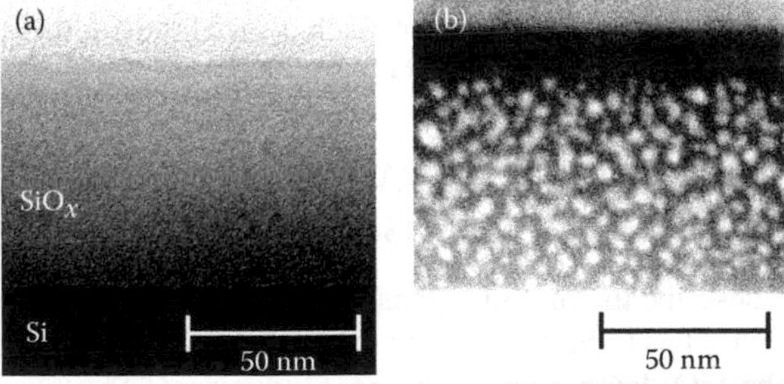

FIGURE 12.2 (a) Cross-sectional TEM micrograph of a PE–CVD SiO_x layer ($x = 1.6$) annealed at 1100°C for half an hour. (b) The same sample analyzed by EFTEM using the Si plasmon peak evidencing the presence of Si nanoclusters.

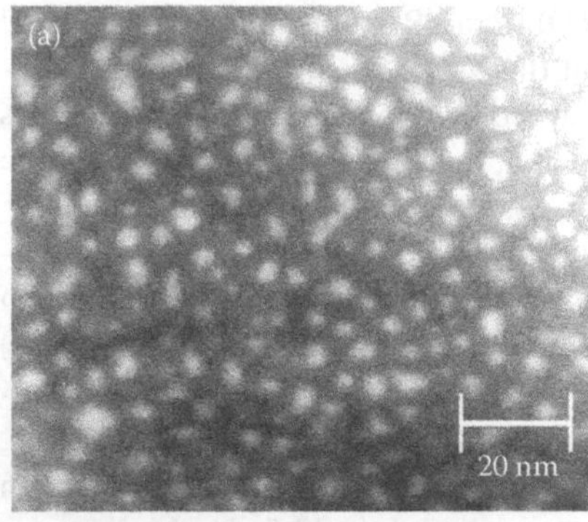

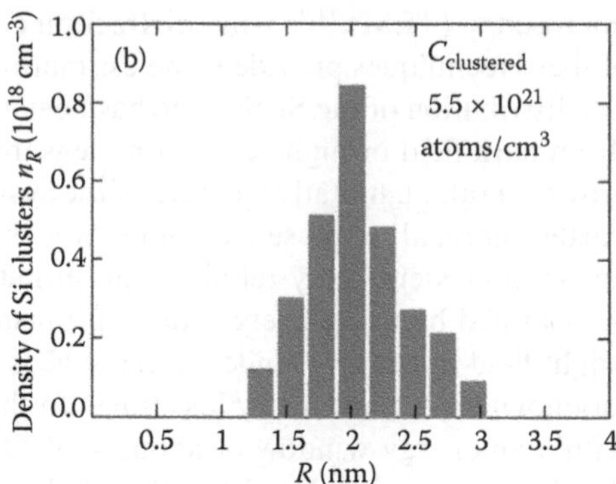

FIGURE 12.3 (a) Plan view EFTEM micrograph of a PE–CVD SiO_x layer ($x = 1.6$) annealed at 1100°C for half an hour. (b) Corresponding Si nanocluster size distribution. (Reprinted with permission from *Applied Physics Letters* 87, 044102. Copyright 2005, American Institute of Physics.)

From the Si NCs radius distribution of Figure 12.3b, we can obtain the average radius (2.2 nm for the example shown in Figure 12.3b) and the total Si concentration in the Si NCs, $C_{clustered}$, obtained as

$$C_{clustered} = \frac{4}{3}\pi\rho_{Si}\sum n_R R^3 \tag{12.1}$$

where $\rho_{Si} = 5 \times 10^{22}$ cm^{-3} is the Si atomic density.

A more reliable method to obtain $C_{clustered}$ is the one based on EELS analysis[25] in cross-sectional configuration (closed circles in Figure 12.4). This methodology allows to determine the total volume fraction, f_c, of clustered Si from fitting EELS by using the theoretical description proposed by Barrera and Fuchs.[27] The result of such a fitting procedure is the full line plotted in Figure 12.4 where several contributions to the energy loss probability coming from the host (dotted-dashed line), from the excitation of the interface modes (dotted curve), and from the Si NCs (dashed line), are also plotted. Once the volume fraction, f_c, is obtained, the clustered silicon concentration can be simply evaluated as $C_{clustered} = \rho_{Si} \cdot fc$.

12.3 Si NANOCLUSTERS FORMATION: EFFECT OF DEPOSITION METHODS

In order to fully describe the $Si-SiO_2$ phase separation process, the total Si concentration in the Si NCs has to be compared to the silicon excess concentration, C_{excess}, dissolved into the SiO_x layer, that is, the concentration of Si atoms exceeding the value corresponding to the SiO_2 stoichiometry:

$$C_{excess} = C_{Si} - C_O/2 \tag{12.2}$$

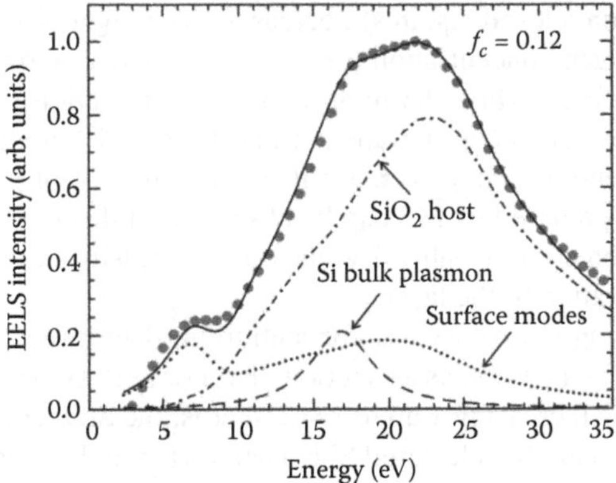

FIGURE 12.4 Electron energy loss spectrum from the SRO layer annealed at 1100°C for half an hour (closed circles). The full line is a fit to the experimental spectrum using the model discussed in the text. Dotted-dashed, dotted, and dashed curves are the contribution to the total energy loss intensity coming from the SiO_2 host, from the surface or interface modes, and from the Si nanoclusters, respectively. (Reprinted with permission from *Applied Physics Letters* 87, 044102. Copyright 2005, American Institute of Physics.)

where C_{Si} and C_O are the silicon and oxygen atomic concentrations in the SiO_x layer. According to the $Si-SiO_2$ phase diagram, we expect that all the excess Si (initially dissolved into the SiO_x layer) precipitates, forming Si NCs after a high-temperature anneal, since Si and SiO_2 are the stable phases.

As already mentioned, the Si and O concentrations can be accurately evaluated by dividing the surface atomic densities, evaluated from the analysis of RBS spectra, by the film thickness measured by conventional TEM. The results are shown in Figure 12.5a where C_{Si} and C_O are plotted as a function of the composition x of several SiO_x layers, prepared by PE–CVD or MS. In PE–CVD SiO_x, by increasing x the

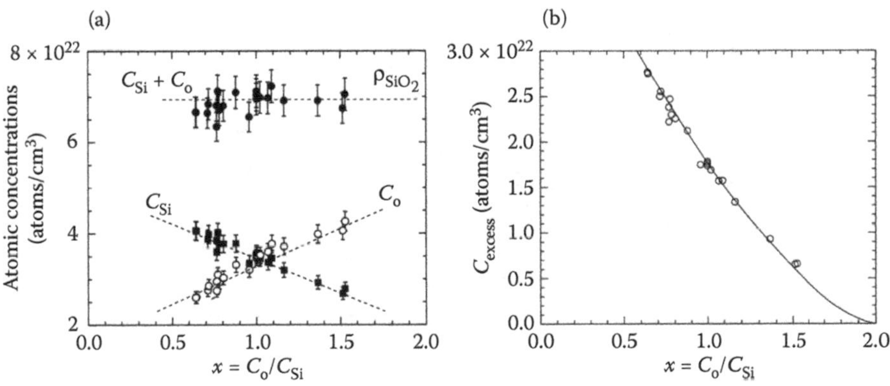

FIGURE 12.5 (a) Silicon (■), oxygen (○), and total (●) atomic concentrations as a function of the composition x of the SiO_x layer after annealing at 1100°C for half an hour. (b) Corresponding values of the excess Si concentration.

silicon concentration (closed squares) decreases from about 4.1×10^{22} to 2.7×10^{22} atoms/cm^3, the oxygen concentration (open circles) increases from about 2.6×10^{22} to 4.3×10^{22} atoms/cm^3, while the total atomic concentration $\rho = C_{Si} + C_O$ (closed circles) keeps almost constant and about equal to 7×10^{22} atoms/cm^3, really close to the atomic concentration $\rho_{SiO_2} = 6.9 \times 10^{22}$ characteristic of pure SiO_2. A similar behavior for layers prepared by MS can be observed. In that case, however, the total atomic concentration ρ appreciably decreases by decreasing x, that is, by increasing the silicon concentration in the layer.

The corresponding silicon excess concentration, determined by using Equation 12.2, is plotted in Figure 12.5b as a function of x. For both systems, it monotonically decreases to zero by increasing x up to $x = 2$, that is, the SiO_2 composition.

We can now compare the clustered Si concentration to the silicon excess concentration in SiO_x layers subjected to high-temperature anneal. It is straightforward to underline that these two quantities should be exactly equal if the SiO_x layer undergoes a full Si$-$$SiO_2$ phase separation. Extensive experiments have been carried out by using a constant thermal budget, corresponding to anneal at 1100°C for half an hour in N_2 atmosphere, and by using EELS analysis to evaluate the clustered Si volume fraction and, consequently, $C_{clustered}$.[23–28] The results are shown in Figure 12.6 where $C_{clustered}$ is plotted (open circles or open squares refer to PE–CVD or MS layers, respectively), jointly with C_{excess} (closed circles or closed squares refer to PE–CVD or MS layers, respectively), as a function of x. It is clear that in PE–CVD SiO_x layers, only about 30% of the excess Si agglomerates while the remaining part is dispersed in the matrix. In contrast, sputtered samples show an almost complete agglomeration of the excess Si within Si NCs, while the matrix appears

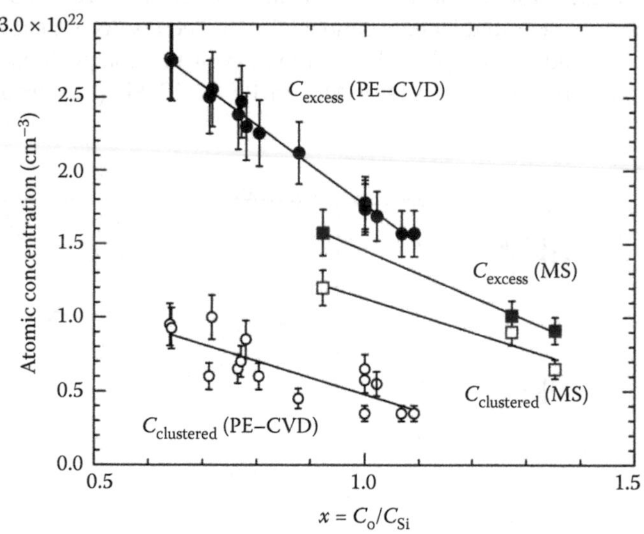

FIGURE 12.6 Comparison between excess and clustered Si concentrations after anneal at 1100°C for half an hour in PE–CVD (closed and open circles) and in MS samples (closed and open squares).

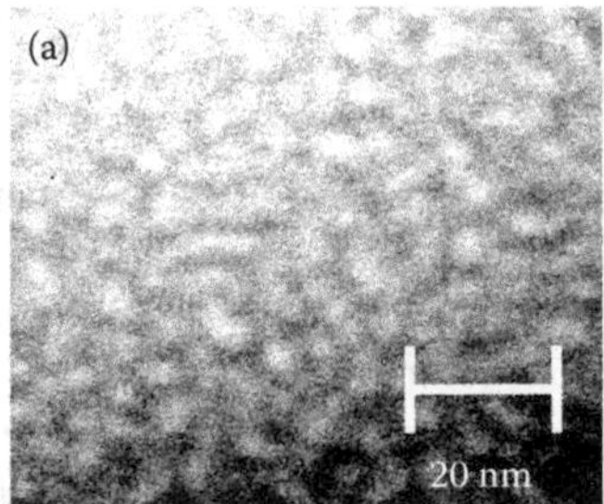

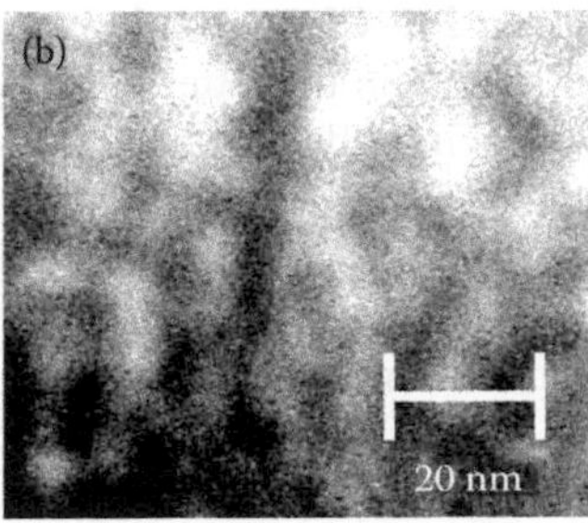

FIGURE 12.7 EFTEM images of SiO_x films having the same Si excess and prepared by (a) PE–CVD and (b) magnetron sputtering. The bright zones are associated with the presence of Si NCs. (Reprinted with permission from *Journal of Applied Physics* 104, 094306. Copyright 2008, American Institute of Physics.)

to be almost stoichiometric SiO_2. At a fixed value of the Si excess concentration (C_{excess} ~ 1.6×10^{22} atoms/cm^3), Si NCs are much larger in sputtered sample (Figure 12.7b) compared to PE–CVD NCs (Figure 12.7a). In fact, the cluster mean radius values are 4.0 and 2.0 nm, respectively.

The dielectric properties of the matrix surrounding the Si NCs were investigated by electron energy loss spectroscopy in a scanning transmission electron microscopy (STEM) configuration.[23] The electron probe, 0.5 nm in radius, was focused on a point about equidistant from two adjacent Si NCs, that is, in a region where only the matrix is visible in the micrographs of Figure 12.7. Figures 12.8a and 12.8b show the corresponding spectra of the PE–CVD (open squares) and of the sputtered sample (open circles). While the spectrum taken in the MS sample is identical to the one taken on a reference SiO_2 sample (continuous lines in Figure 12.8), the spectrum

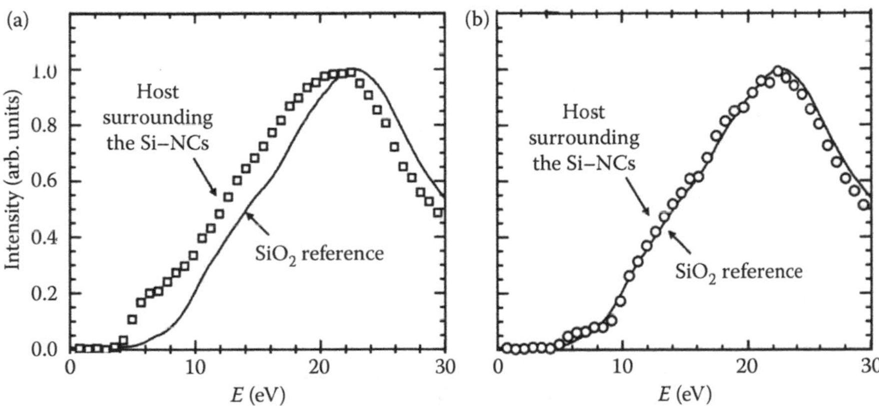

FIGURE 12.8 EELS spectra of the oxide matrix surrounding Si NCs taken by STEM in the center of the region between two adjacent NCs for (a) PE–CVD (open squares) and (b) sputtered (open circles) films. The continuous line in both panels is the EELS spectrum for pure SiO_2. (Reprinted with permission from *Journal of Applied Physics* 104, 094306. Copyright 2008, American Institute of Physics.)

coming from the PE–CVD sample is significantly different from the SiO_2 sample, thus confirming that in the latter case the matrix is only partially depleted from all the silicon excess initially dispersed in it.

A further confirmation that Si−SiO_2 phase separation is more effective in MS rather than in PE–CVD SiO_x system arises from photoluminescence (PL) measurements. Indeed, it has been widely proved by PL techniques that carrier quantum confinement effects produce room-temperature light emission in Si nanocrystals embedded in dielectric hosts.[20] The normalized room-temperature PL spectra relative to a PE–CVD SiO_x film (with Si concentration of 37%) are reported in Figure 12.9 for annealing temperatures of 1100°C, 1200°C, and 1300°C. Figure 12.9 shows that the wavelength of the PL peak (λ_{max}) increases with the annealing temperature; indeed, λ_{max} is found at 770 nm at 1100°C, and it shifts at 800 nm at 1200°C, and finally, at 880 nm at 1300°C. Light emission from silicon nanocrystals is due to the band-to-band radiative recombination of electron–hole pairs confined within the crystals. Indeed, the carrier quantum confinement theory predicts the progressive blueshift of the PL peaks by decreasing the crystal size due to the enlargement of the bandgap of the Si nanocrystals with respect to bulk crystalline silicon. This is clearly shown in Figure 12.10, where the energy of the experimental luminescence peaks is reported as a function of the crystal radius, as obtained by TEM analyses. It can be seen that the energy increases with decreasing mean radius of crystal; moreover, the bandgap upshift ΔE can be approximately described as $\Delta E \sim d^{-n}$ (with $n \sim 1.5$), where d is the Si NC diameter. The presence of a strong correlation between Si NC size and luminescence data is a clear evidence of the band-to-band radiative recombination of electron–hole pairs confined in Si NCs.

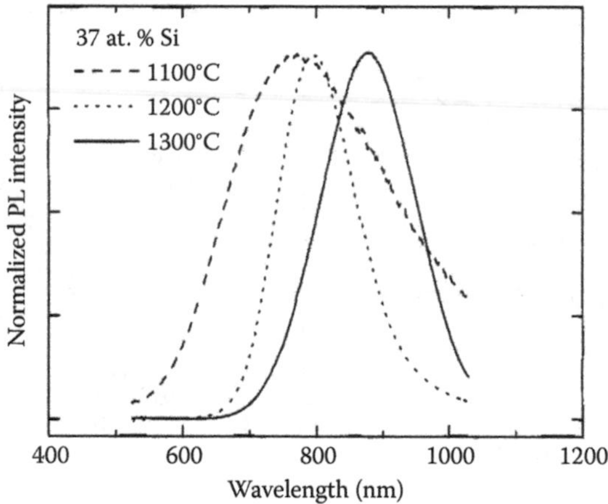

FIGURE 12.9 Normalized room-temperature PL spectra of a PE–CVD SiO_x film having a Si concentration of 37 at.% after thermal annealing processes performed at 1100°C, 1200°C, and 1300°C for 1 h. (Reprinted with permission from *Journal of Applied Physics* 87(3), 1295. Copyright 2000, American Institute of Physics.)

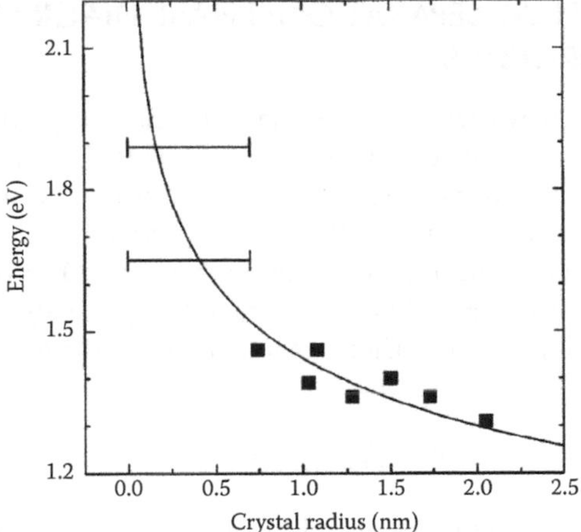

FIGURE 12.10 Energy of the luminescence peak versus Si nanoclusters mean radius, as obtained from TEM plan-view micrographs. The error bars refer to the PL peak of samples containing Si nanocrystals smaller than the TEM detection limit. The line is drawn to guide the eye. (Reprinted with permission from *Journal of Applied Physics* 87(3), 1295. Copyright 2000, American Institute of Physics.)

Owing to the different structural evolution, PL spectra of annealed sputtered samples are significantly different with respect to PE–CVD samples. Figure 12.11 reports the normalized PL spectra for sputtered and PE–CVD samples with a similar Si excess concentration of 1.6×10^{22} atoms/cm^3 and both annealed at 1250°C for 1 h. Sputtered samples show a redshift and a wider spectrum with respect to PE–CVD films, due to a larger Si NCs mean size and a more spread size distribution.

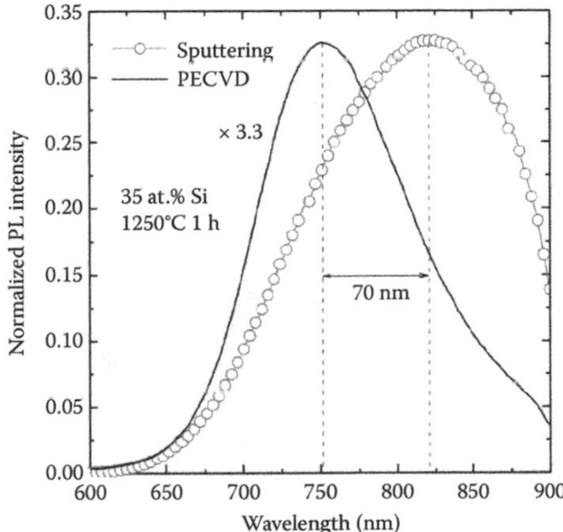

FIGURE 12.11 Normalized room temperature PL spectra for sputtered and PE–CVD SiO_x films with a similar Si content of 35 at.% and both annealed at 1250°C for 1 h. (Reprinted with permission from *Journal of Applied Physics* 104, 094306. Copyright 2008, American Institute of Physics.)

12.4 $Si-SiO_2$ PHASE SEPARATION: PHASE DIAGRAM AND ENERGETICS

A question arises on why there is this strong difference in agglomeration properties of the Si NCs in samples prepared by different methods. It is interesting to note that PE–CVD and MS SiO_x layers behave in a different way in terms of their Si and O concentration dependence as a function of composition x. It is interesting to compare this behavior to the one expected by assuming that the SiO_x system, after the high-temperature anneal, is composed of a mixture of two perfectly separated phases: Si clusters embedded in pure SiO_2. Under this hypothesis, the total atomic concentration will be given by

$$\rho = f\rho_{Si} + (1 - f)\rho_{SiO_2} \tag{12.3}$$

where f is the volume fraction occupied by silicon, $(1 - f)$ is the volume fraction occupied by SiO_2, and $\rho_{Si} = 5 \times 10^{22}$ atoms/cm^3 and $\rho_{SiO_2} = 6.9 \times 10^{22}$ atoms/cm^3 are the atomic concentrations of Si and SiO_2, respectively.

On the basis of the same assumption, the quantity $f\rho_{Si}$ in Equation 12.3 coincides with the silicon excess concentration. Equation 12.3 can then be inverted and used to find C_{Si} as a function of x:

$$C_{Si}(x) = \frac{2\rho_{Si}\,\rho_{SiO_2}}{3x\rho_{Si} + (2 - x)\rho_{SiO_2}} \tag{12.4}$$

The functions $C_{Si}(x)$, $C_O(x) = xC_{Si}(x)$, and $\rho(x)$, derived from Equation 12.4, are plotted in Figure 12.12 as dashed, dot-dashed, and continuous lines, respectively,

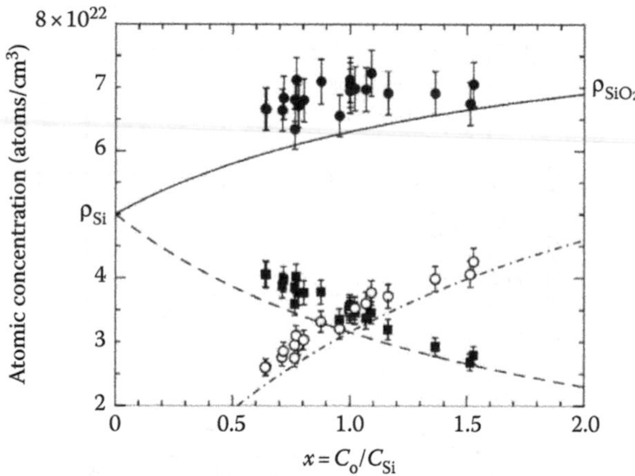

FIGURE 12.12 Silicon (■), oxygen (○), and total (●) atomic concentrations as a function of the composition x of the SiO_x layer after annealing at 1100°C for half an hour. Here dashed, dot-dashed, and continuous lines are the Si, O, and total atomic concentrations calculated under the hypothesis of a complete phase separation between Si and SiO_2. (Reprinted with permission from *Applied Physics Letters* 90, 183101. Copyright 2007, American Institute of Physics.)

jointly to the experimental concentration values already shown in Figure 12.5a. For any value of x, the experimental densities of the PE–CVD layers are significantly larger than the corresponding ones calculated from Equation 12.4, that is, by assuming a complete phase separation between Si and SiO_2 and by adopting for ρ_{Si} and ρ_{SiO_2} the known values for the bulk Si and SiO_2 atomic densities. This is a further clear evidence that Si−SiO_2 phase separation is partially inhibited in PE–CVD system compared to MS one. Phase separation and Si NCs formation require the nucleation of Si clusters and their growth through atomic mass transport. The starting properties of the as-deposited layers are profoundly different and give rise to a different evolution of the two systems. Indeed, the structure of as-prepared (by PE–CVD or MS) substoichiometric silicon oxide can be described as a simple pure Si/SiO_2 mixture [random mixture model (RMM)],[29] thus neglecting the existence of Si intermediate oxidation states, or, alternatively, as a mixture of all Si oxidation states able to respect the overall film stoichiometry [random bonding model (RBM)][30] with a negligible extent of Si−SiO_2 phase separation. SiO_x films can be "more" similar to an RMM or to an RBM and this should affect their thermal evolution.

The weak dependence on the composition x of the total atomic density ρ, which remains almost constant to a value different from that expected if a pure Si/SiO_2 mixture occurs (see Figure 12.5a), is a strong indication that the PE–CVD system tends to maintain the strain relaxation state of the as-deposited material in an RBM configuration. The experimental results suggest that pseudocoherency strain energy significantly influences the phase separation mechanism of the PE–CVD SiO_x layers.[31,32] The phenomenological description that makes plausible the hypothesis of a strain-driven incomplete phase separation of the PE–CVD SiO_x system is based on the observation that its free energy density can be approximated by a double minima function, $g^*(x,T)$, given by[31]

$$g^*(x, T) = \alpha(T)(x-1)^2 + \beta(T)(x-1)^4 \tag{12.5}$$

where T is the absolute temperature and $\alpha(T)$ and $\beta(T)$ are determined by imposing that $g^*(x,T)$ has minimum value at the points $x = 0$ (Si) and $x = 2$ (SiO_2), respectively, and a saddle point at $x = 1$ (SiO):

$$\alpha(T) = 2\beta(T) = 2\left\{g_{SiO}(T) - 0.5\left[g_{SiO_2}(T) + g_{Si}(T)\right]\right\} \tag{12.6}$$

where $g_{SiO}(T)$, $gSiO_2(T)$, and $g_{Si}(T)$ are the free energy of SiO, glass SiO_2, and diamond Si, respectively.

The function $g^*(x,T)$ is plotted for $T = 1100°C$ in Figure 12.13 as solid line. Obviously, the behavior of $g^*(x,T)$ predicts that the system should fully separate, in accordance with the Si−SiO_2 phase diagram.[33] In order to take into account the strain effect, $g^*(x,T)$ has to be locally corrected with the addition of a free strain energy contribution depending on the initial homogeneous composition x_0 of the as-deposited film:

$$\tilde{g}(x, x_0, T) = g^*(x,T) + \frac{1}{2}E\left[\varepsilon_0(x - x_{0^2})\right] \tag{12.7}$$

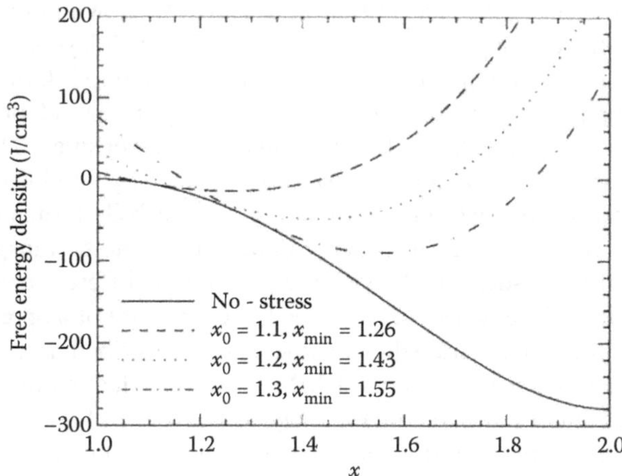

FIGURE 12.13 SiO_x free energy density as a function of the composition x calculated for the no-stress case (solid line), and by introducing the strain energy for different initial compositions: $x_0 = 1.1$ (dashed line), $x_0 = 1.2$ (dotted line), and $x_0 = 1.3$ (dot-dashed line). (Reprinted with permission from *Applied Physics Letters* 90, 183101. Copyright 2007, American Institute of Physics.)

where E is Young's modulus of the matrix and $\varepsilon_0 = (1/a)(\mathrm{d}a/\mathrm{d}x)$ is the composition expansion coefficient, a being the average atomic distance of the system. This correction is due to the cost paid by the evolving system to contrast the matrix stress field at the locations of the nuclei of the new (Si) phase. The function $\tilde{g}(x, x_0, T)$ is plotted in Figure 12.13 for different values of x_0: $x_0 = 1.26$ (dashed line), $x_0 = 1.43$ (dotted line), and $x_0 = 1.55$ (dot-dashed line). For the calculations, we used $E = 70$ GPa (the literature values of the SiO_2 Young's modulus span in the range 60–100 GPa), while ε_0 was approximated to be $\varepsilon_0 \approx (\rho_{SiO_2}^{-3} - \rho_{Si}^{-3})/\rho_{SiO_2}^{-3}$. The minima of the functions $\tilde{g}(x, x_0, T)$ are now significantly far from the point at $x = 2$ and located at $x_{min} = 1.26$, $x_{min} = 1.43$, and $x_{min} = 1.55$, respectively, thus confirming that stress prevents a complete phase separation of the PE–CVD SiO_x system. The Si NCs remain embedded in a substoichiometric silicon oxide whose final composition strongly depends on the initial composition of the as-deposited layer.

12.5 SYNTHESIS AND IMAGING OF Ge NANOCLUSTERS

In order to obtain Ge NCs, many techniques are now available, based both on chemical and physical approaches (e.g., sol–gel, chemical vapor deposition, physical vapor deposition, ion implantation). Most of the methods synthesize Ge NCs (amorphous or crystalline) as embedded in a suitable matrix, as oxides,[34] nitrides,[35] or carbides. As far as embedded Ge NCs are concerned, random or ordered configurations of NCs can be obtained by playing with different approaches.

Random configuration of Ge NCs in insulating matrix can be produced by cosputtering of SiO_2 and Ge targets upon fused silica substrates, obtaining SiGeO films with thickness ranging from 100 to 300 nm, and Si, Ge, and O atomic concentrations close to 24 at.%, 16 at.%, and 60 at.%, respectively. The formation of Ge NC is obtained through random nucleation and growth mechanisms induced by thermal annealing (600–800°C range, 1 h, N_2 ambient), which promotes the separation of SiGeO film into SiO_2, GeO_2, and Ge clusters (due to precipitation of excess Ge). Since the SiGeO deposition is performed at 400°C, some Ge NC can also be present in the as-deposited samples, acting as preferential sites for further Ge precipitation.

The formation of Ge NCs can be evidenced by cross-section transmission electron microscopy in high-resolution (HR–TEM) or scanning mode (STEM). A high density of Ge precipitates within the SiO_2 matrix is revealed by the STEM images (at the same magnification) in Figure 12.14, just after the deposition (Figure 12.14a), and after thermal annealing at 750°C (Figure 12.14b). The bright patches represent

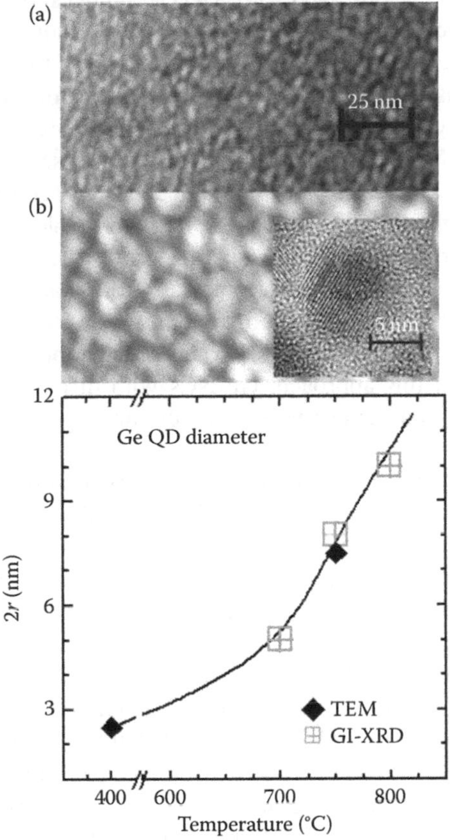

FIGURE 12.14 Cross-sectional dark-field STEM images of a SiGeO, as deposited (a) or after annealing at 750°C (b). The inset reports an HR-TEM of the annealed sample, showing the presence of a clear crystalline structure. Thermal evolution of the mean diameter (2*r*) of Ge nanostructures, measured by TEM (diamond) or GI-XRD (squares). Line is a guide for eyes (color online). (With kind permission from Springer Science+Business Media: *Nanoscale Res. Lett.*, The role of the surfaces in the photon absorption in Ge nanoclusters embedded in silica. **6**, 2011, 135, S. Cosentino et al.)

Ge NCs whose density and mean size noticeably change after annealing (the mean diameter increasing from 2.5 to 7.5 nm). Although Ge QDs are already present in the as-deposited films, as also found by Zhang et al.,[36] the deposition temperature was not high enough to induce the formation of crystalline QDs in our case. SiGeO film deposited by sputtering can be described as a mixture of Ge, GeO_2, and SiO_2 units, according to a random matrix model, similar to what occurs for silicon-rich oxide.[23] During annealing, Ge QDs undergo an Ostwald ripening mechanism, similar to the Si QD case,[37] leading to a size increase of precipitates with a concomitant amorphous-to-crystal (*a–c*) transition occurring in the 600–800°C range.[38] The inset in Figure 12.14b reports a high-resolution TEM image of the annealed sample, evidencing a clear crystalline phase for Ge QD with the fringes due to crystalline planes (indicated by red lines and separated by 0.33 nm, as the (111) planes of c–Ge bulk).

Ge clusters can also be observed with XRD techniques, if they are in the crystalline phase. To this aim, glancing incidence x-ray diffraction (GI–XRD) analysis can be employed. Basing on the Bragg diffraction peaks of the GI–XRD spectra, the average QD size is estimated by applying the Scherrer formula,[39] which correlates the nanocrystal size with the peak width. In Figure 12.14c, the mean QD diameter ($2r$) measured by TEM (diamond) and by GI–XRD (crossed squares, line is a guide for eyes) is reported as a function of the annealing temperature. Even if GI–XRD gives information only on crystalline QDs, the reasonable agreement between the two techniques observed at 750°C is supporting the idea that the size distribution of crystalline QDs does not significantly deviate from that of amorphous QDs. The overall variation of r can be extracted by joining the two techniques, showing a clear QD enlargement in the 400–800°C range compatible with an Ostwald ripening mechanism. With this approach, Ge NCs with different size distribution can be realized, by acting on the annealing temperature, which promotes the ripening mechanism.

Still, the distance among Ge NCs cannot be varied on purpose; thus a different approach must be used, based on *ordered* Ge NCs embedded in silica. Ordered Ge NC configuration can reveal new effects, on light emission or absorption processes, which are not visible in random configuration. In fact, it was evidenced that the atomic absorption cross section (σ_a, measuring the photon absorption probability per Ge atom) could depend on the QDs distribution within the film.[40] In order to clarify if and to what extent QD–QD interactions can enhance the light absorption, the dependence of light absorption on size and on QD–QD distance should be disentangled, based on properly ordered Ge NCs.[41] Multilayered samples (as shown in Figure 12.15) have been realized, repeating 15 times the film/barrier (SiO_2:Ge/SiO_2) structure, by magnetron sputtering deposition from SiO_2 and Ge targets, sequentially. After a SiO_2 buffer layer, thin (4 nm) films of SiO_2:Ge were alternated to pure SiO_2 barriers, whose thickness was varied to produce samples with tightly packed or fairly isolated Ge QDs films. The multilayer structure and thickness, as well as the presence of Ge QDs were evaluated by cross-sectional transmission electron microscopy (TEM) analysis.

Bright-field TEM images (Figure 12.15) show the multilayered structure of the films with SiO_2 barriers (brighter layers) embedding very thin SiO_2 films containing Ge QDs (darker layers). The thickness of the SiO_2 barrier ($d_{\perp}$) was 3 nm for the

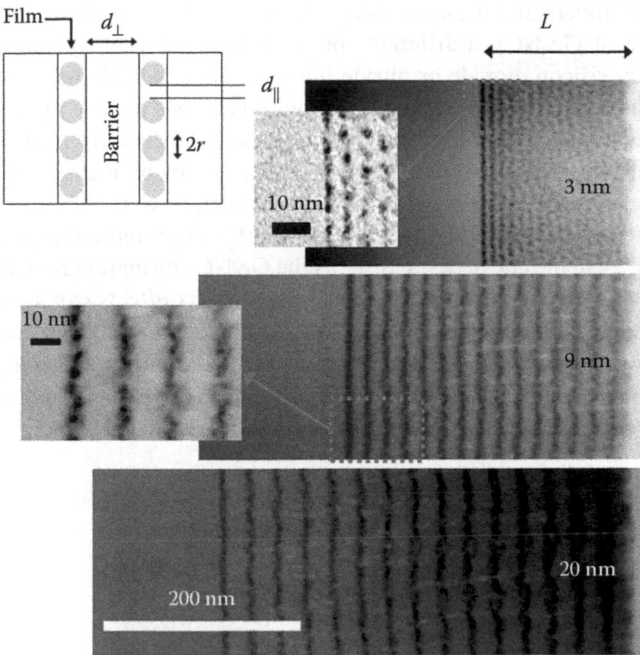

FIGURE 12.15 Schematic and cross-sectional bright-field transmission electron microscopy (BF-TEM) images of Ge QDs multilayered samples with different thicknesses of SiO_2 barrier. TEM images indicated by the arrows show the details of the deeper four QDs films in 3 and 9 nm samples. (Reprinted with permission from *Applied Physics Letters* 102, 193105. Copyright 2013, American Institute of Physics.)

tightest QDs configuration, 9 nm for the intermediate packaging, and 20 nm for the most spaced one. The multilayered configuration also allows a better control of the size and vertical order distribution of Ge QDs. To better show Ge nanostructures in each film, two TEM images at higher magnification (images indicated by the arrows in Figure 12.15) are reported, evidencing the presence of densely packed amorphous Ge QDs (3 nm in diameter).

In order to quantify the QD density and the QD–QD distance RBS was employed, giving the amount of Ge atoms embedded in the 15 layers. From RBS analysis, the Ge areal density (~4.3×10^{15} Ge/cm^2) within each QDs layer was estimated. Basing on TEM evidences, we can assume spherical QDs with an average diameter ($2r$) of 3–4 nm, thus a mean QDs areal density of ~7×10^{12} cm^2 in each layer, corresponding to a surface-to-surface distance ($d_\parallel$) of about 1 nm between adjacent Ge QDs in the same layer. As $d_\parallel$ is fixed and well lower than $d_\perp$, the multilayer approach used here allows to play only with the distance between Ge QDs films along the growth direction. In other words, in multilayer samples, the QD–QD distance can be varied only in the vertical direction, while it is fixed in the plane of the QDs film. This configuration will be used for evidencing the effect of QD–QD distance on the light absorption phenomena in Section 12.6.

In order to understand if and to what extent the embedding matrix has a role on the formation of Ge NCs, a different approach must be used to compare Ge NCs surrounded by silicon dioxide or nitride host medium. Stoichiometric silica or silicon nitride substrates have been implanted with Ge^+ ions at very high dose such as to have a Ge atomic concentration of few percent.[35] Differently from PE–CVD or sputter-based fabrication techniques, in this way, a stoichiometric and chemically stable matrix is used. After implantation, the matrices were subjected to furnace annealing processes (1 h, N_2 ambient, 600–900°C) to promote the precipitation of the excess Ge as in the above cases. Still, as the Ge NCs formation is affected both by the Ge diffusion and nucleation rates, important matrix effects can arise if comparing the silica or silicon nitride cases. In fact, large differences have been observed, as reported with the STEM images in Figure 12.16 (samples implanted with a dose of 7.3×10^{16} Ge/cm^2 and annealed at 850°C). Ge NCs appear as bright spots, due to the higher Z contrast of the NCs with respect to both the matrices.

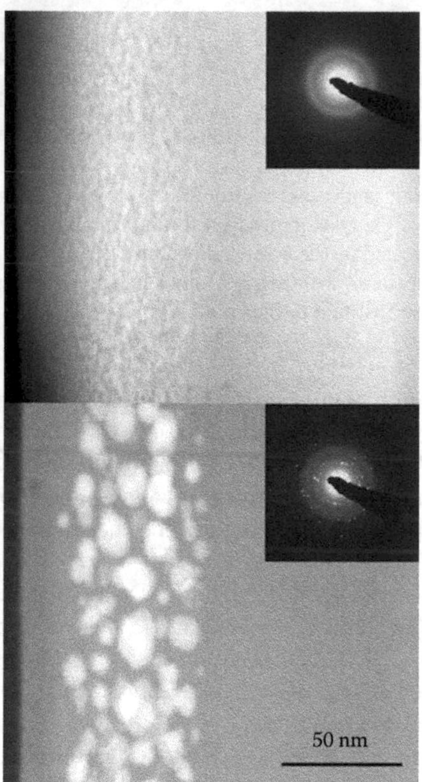

FIGURE 12.16 Cross-sectional high-angle annular dark-field STEM images of the Ge nanoclusters embedded in (top panel) Si_3N_4 or in (bottom panel) SiO_2 matrices, obtained after Ge ions implantation (100 keV, 7.3×10^{16} cm^{-2}) and annealing at 850°C, 1 h. Respective diffraction images are shown as insets. (Reprinted with permission from *Applied Physics Letters* 101, 011911. Copyright 2012, American Institute of Physics.)

The NC diameter ($2r$) is much larger in SiO_2 (bottom image, $2r \sim 3$–24 nm) than in Si_3N_4 (top image, $2r \leq 2$ nm). Moreover, as evidenced by electron diffraction analysis (shown in the insets), Ge NCs in SiO_2 are crystalline, contrary to the Si_3N_4 case, where no diffraction spots are observed. The same also holds after annealing at 900°C, which means that Ge formation in silicon nitride is quite challenging, also at very high temperature (Ge melting point is 938°C). This proves that the embedding matrix significantly affects the formation of Ge NCs. There are at least two reasons, related to the Ge migration ability in the two matrix and to the interface energy between Ge NC and the surrounding host. Actually, the Ge diffusivity in Si_3N_4 can be much smaller than in SiO_2, as it occurs for Si diffusivity ($\sim 3 \times 10^{-13}$ cm^2/s at 800°C in SiO_2, $\sim 10^{-24}$ cm^2/s at 840°C in Si_3N_4).[42,43] By using high-resolution RBS, it was verified that Ge diffusivity in Si_3N_4 at 850°C cannot be larger than 7×10^{-17} cm^2/s.[35] These results indicate that while at 850°C Ge easily migrates in SiO_2, leading to NC ripening in the inner part of the film and Ge out-diffusion in the surface near region, at the same temperature Ge diffusion in Si_3N_4 is very low, limiting the NC ripening in Si_3N_4. However, one should think that the nucleation step can be limited due to thermodynamic reason. In fact, according to the classical nucleation theory, a critical radius (r^*) exists above which the *a–c* transition lowers the free energy, since for large nuclei the extrainterfacial energy (γ) is compensated by the gain in the internal free energy (ΔG_{phase}) due to crystallization:

$$r^* = \frac{-2\gamma}{\Delta G_{phase}} \tag{12.8}$$

Recently, the Ge/Si_3N_4 interface was shown to have a larger γ in comparison to the Ge/SiO_2 interface.[44] This supports a larger critical radius for Ge NC in Si_3N_4 (since ΔG_{phase} is not affected by the matrix, at a first approximation) justifying the lack of crystalline Ge NCs in the Si_3N_4 samples. Thus, the larger interfacial energy and the reduced diffusivity of Ge in Si_3N_4 limit the NC ripening and crystallization. This means that the kinetics of NC formation and crystallization is much slower in Si_3N_4 than in SiO_2, evidencing as the hosting matrix is another critical factor in NC fabrication.

12.6 LIGHT ABSORPTION IN Si AND Ge NANOCLUSTERS

Semiconductor NCs are receiving increasing attention for boosting novel technologies with higher efficiency in the sunlight–electricity conversion.[45–48] Photon absorption in Si or Ge NCs is a promising way to enhance the efficiency of sunlight to electricity conversion. In fact, the light absorption spectra of bulk Si or Ge semiconductors (both in the amorphous or in the crystalline states) are widely compatible with the photon distribution coming from the sun, as shown in Figure 12.17. Indeed, as soon as Si or Ge are confined to nanostructures, one of the typical arising effects is the size-dependent tuning of the bandgap (E_G), which should tailor the light absorption onset for a better matching of the sun spectrum in multijunction

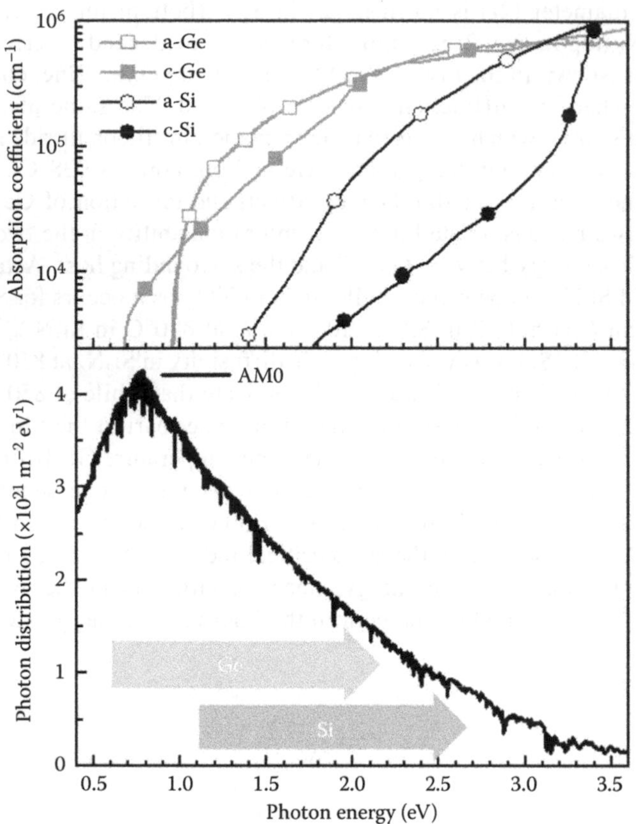

FIGURE 12.17 Top panel: Absorption coefficient for bulk Si (circles) or Ge (squares), in the amorphous (open symbols) or crystalline (closed symbols) phase. Bottom panel: Energy distribution of solar photons under AM0 conditions. Arrows refer to possible bandgap extensions for each material due to quantum confinement effects.

approaches. Assuming an infinite barrier for the matrix embedding NCs, the variation with size of E_G is given by[45,46]

$$E_G = E_G^{\text{bulk}} + A/(2r)^2 \tag{12.9}$$

where E_G^{bulk} is the not-confined bandgap, and A is the confinement parameter. In particular, Si or Ge NCs could have a profitable E_G tuning (as drawn with arrows in the lower panel of Figure 12.17), as it should cover the main part of the solar available spectrum. Still, an open question is whether the size of such nanostructures is the only parameter determining the photoresponse of PV devices. As a matter of fact, light absorption in semiconductor NC depends not only on the NC size, as relevant effects come out because of other factors as the embedding matrix,[47,48] the presence of surface or interface states,[34,49] the amorphous or crystalline structure of the NCs,[50,51] and the shape and layered structure of NCs.[40,52,53]

Thus, the mechanism of light absorption by NC appears to be quite complex and it cannot be fully described by the NC size alone. To focus this question, light absorption phenomenon in Si or Ge at the nanoscale must be investigated in detail, trying to separately treat significant effects related to the structural phase or density of the NC, to the embedding matrix and to the synthesis technique.

Si NCs fabrication can heavily affect the optical properties of nanostructures. In fact, despite several experimental works dealt with light absorption in Si or Ge NCs, contrasting results have been often published, which can be related to different recipes for the synthesis of QDs. The observed shift toward lower energies of the absorption edge of Si QDs, when increasing the Si content, has been attributed to a reduction of the QD energy bandgap due to a lower confinement effect.[19,54,55] However, the same redshift with Si content holds also in as-deposited samples,[19] where no Si QD has been observed up to now. Moreover, the effect of annealing itself is not univocal since a redshift has been reported for thermal processes up to 1100°C,[19,56] while a blueshift has been evidenced in the 1100–1250°C temperature range.[57] In addition, a size effect in the linear absorption coefficient of Si QDs has been argued,[58] by using a Si-rich SiO_2/SiO_2 multilayer approach, which ensures a greater control of the Si QD size.[41]

In Figure 12.18a, absorption spectra, extracted from spectrophotometry by combining transmittance and reflectance measurements,[51] have been reported for samples synthesized by PE–CVD (C– label) or magnetron sputtering (S– label) techniques. For all samples, α spectrum increases with the Si content. For PE–CVD samples, under annealing, α increases up to 900°C and then decreases, varying the intensity over about one decade. For MS samples, α spectra were insensitive to annealing up to 900°C, then decreasing at higher temperatures. Thus, the deposition technique determines different behaviors in the light absorption mechanism, inducing also different response to annealing processes.

To describe the light absorption in an amorphous semiconductor, the Tauc law is typically used,[59,60] according to which, for indirect transition (as it is the case for Si QD)[61] and for α higher than 1×10^4 cm^{-1}, we have

$$\alpha = \frac{B}{h\nu}\left(h\nu - E_g^{\mathrm{OPT}}\right)^2 \tag{12.10}$$

where $h\nu$, B, and E_g^{OPT} are the incoming photon energy, the Tauc constant, and the optical bandgap, respectively. The photon absorption leads to transitions between extended electronic states from the valence band toward the conduction band; thus, E_g^{OPT} is the energy difference, while B includes information on the convolution of the two energy bands and the matrix element of optical transition.[60] By plotting $(\alpha \cdot h\nu)^{1/2}$ versus $h\nu$ (Tauc plot) and fitting with a straight line (Figure 12.18b), E_g^{OPT} can be derived as its intercept with the abscissa axis. This approach is by far the most reliable to measure E_g^{OPT},[60] and it has been used for studying the absorption of embedded Si or Ge QDs. The fit always shows a very good agreement with experimental data.

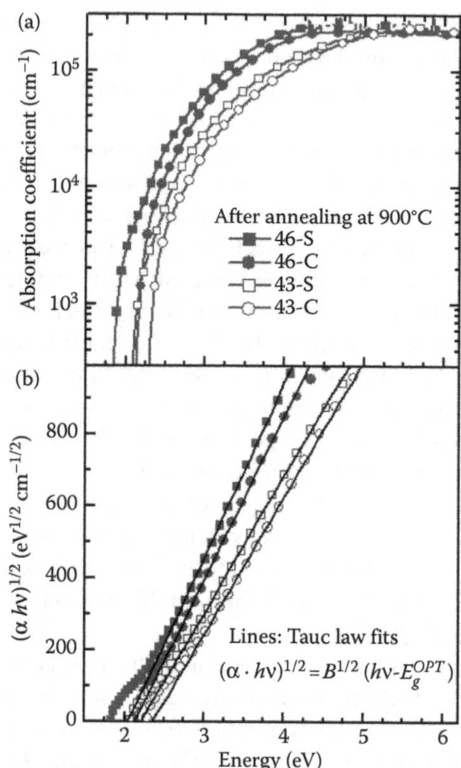

FIGURE 12.18 (a) Absorption spectra for SiO_x samples (deposited through PE–CVD or sputter deposition, with different x values, as reported in the legend) after annealing at 900°C, 1 h. (b) Tauc plot for the same samples and corresponding linear fit according to the reported Tauc law. (Reprinted with permission from *Journal of Applied Physics* 106, 103505. Copyright 2009, American Institute of Physics.)

Here, it should be noted that the slope of $(\alpha \cdot h\nu)^{1/2}$ is linear over a wide energy range (Figure 12.18b), justifying the Tauc approach for indirect transition.

Figure 12.19 reports the thermal behavior of E_g^{OPT}, for all samples (symbols plus dashed lines). The optical bandgap in Si NC samples is in the 3.2–2.1 eV range, clearly evidencing the quantum confinement effect in the photon absorption (as bulk Si shows at maximum an E_g^{OPT} of 1.7 eV, if amorphous material is considered). Having fixed the deposition technique, an increase of the Si content results in a weak reduction of the E_g^{OPT} at all temperatures. The PE–CVD films show a U-shaped trend for E_g^{OPT}, with a minimum at 900–1000°C and an overall variation of ~1 eV. This feature gives account for the discrepancy found in the literature concerning the thermal behavior of Si QDs.[19,56,57] The MS samples have a constant E_g^{OPT} up to 900°C, while above 900°C, their trend resembles the one for PE–CVD samples. It is clear that E_g^{OPT} is strongly affected by the deposition technique, at least at low temperatures.

Up to 900°C annealing, a memory of deposition technique is retained. In this thermal range, sputtered samples do not show large variation, while CVD samples

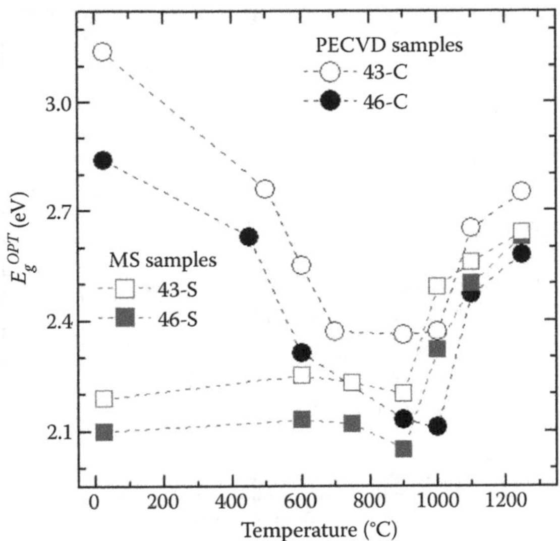

FIGURE 12.19 Optical bandgap versus annealing temperature for the SiO_x samples (deposited through PE–CVD or sputter deposition, with different x values, as reported in the legend).

exhibit a great reduction of optical bandgap. This is compatible with different Si–SiO_2 phase separation in the two sample families, since Si NC formation in SiO_2 is already present in sputtered samples while it needs relatively high thermal budget in CVD samples.[23] For T larger than 900°C, a common trend is evidenced with a surprising increase of the optical bandgap. In fact, in this thermal range, Si NC typically undergoes Ostwald ripening, which induces an increase of the mean size and an expected decrease of the quantum confinement effect with a consequent reduction of the optical bandgap.

Indeed, this evidence shows that QD size is not the only parameter that determines E_g^{OPT}. Instead, the increase in E_g^{OPT} can be associated to the a–c transition of QDs, which occurs at temperatures well higher than in bulk Si.[62] To confirm this, the effects on the optical bandgap of the a–c transition and of the Ostwald ripening have been disentangled (Figure 12.20) by ion implantation technique.[51]

By using Ge implantation, both amorphization of Si NC and damage in the embedding matrix occur, the last being resolved after 600°C annealing. Thermal annealing up to 1250°C induces the a–c transition of Si NC. In particular, Ge implantation has been performed on two samples: 900°C annealed (containing small and amorphous NC) and 1250°C annealed (containing large and crystalline NC). On 1250°C annealed samples, implantation plus 600°C annealing create large and amorphous Si NC, with an optical bandgap (2.4 eV) significantly lower than that in the starting sample (2.75 eV, with large and crystalline NC). It is clear that the phase transition from amorphous to crystalline affects the optical bandgap much more than the size variation. In particular, for the 1250°C annealed sample, the difference of 0.35 eV in the optical bandgap (before and after implantation plus 600°C annealing) appears to be related to the a–c transition of the Si QDs. These data show that the exact nature

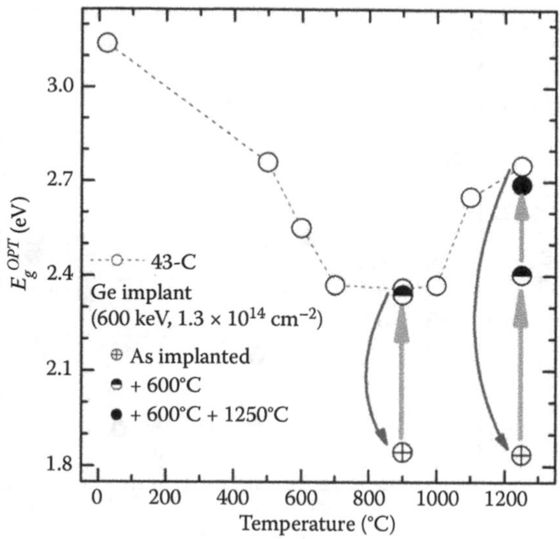

FIGURE 12.20 Variation of the optical bandgap of a SiO_x sample (grown by PE–CVD, with $x = 0.43$) after Ge implantation and annealing. (Reprinted with permission from *Journal of Applied Physics* 106, 103505. Copyright 2009, American Institute of Physics.)

of the phase of QD is crucial in the light absorption process, while the effect of the QD size is less relevant. This effect has also been found in theoretical calculations of the electronic bandgap of Si QDs.[63]

Thus, light absorption in Si NC is shown to be affected by many parameters, which can determine a large deviation from the simple QCE rule. It is clear that Si NC retain memory of the deposition techniques, and this affects the light absorption process, at least up to 900°C annealing. However, it has been shown that the structural phase of Si NC can influence the light absorption much more effectively than the NC size.

As the structural phase of NC plays a paramount role in photon absorption for Si QD, one can ask whether the same also holds for Ge NC. In Figure 12.21, the reported α spectra are associated with the photon absorption by Ge QDs in the crystalline or amorphous phase. The presence of crystalline QDs (as proved by TEM analyses) induces the two broad peaks at about 2.6 and 5 eV in the spectrum, recalling the E_1 and E_2 direct transitions (at 2.1 and 4.3 eV) of the crystalline Ge spectrum, but at a slightly larger energy. Such broad peaks in the 800°C annealed sample can be related to direct transitions within the crystalline Ge QDs having an energy band structure modified by the confinement. To investigate the role of the QD structural phase, the crystal-to-amorphous (*c–a*) transition of the Ge QDs in the sample annealed at 800°C was induced by ion implantation process followed by 550°C, 1 h annealing. A Tauc approach was employed to extract the optical bandgap of these two samples (both containing ~10 nm large Ge NC). Before and after the induced amorphization, the optical bandgap is found to be 1.6 eV, independent of the structural phase. Thus, *c–a* transition of Ge QDs does neither appreciably modify the onset of light

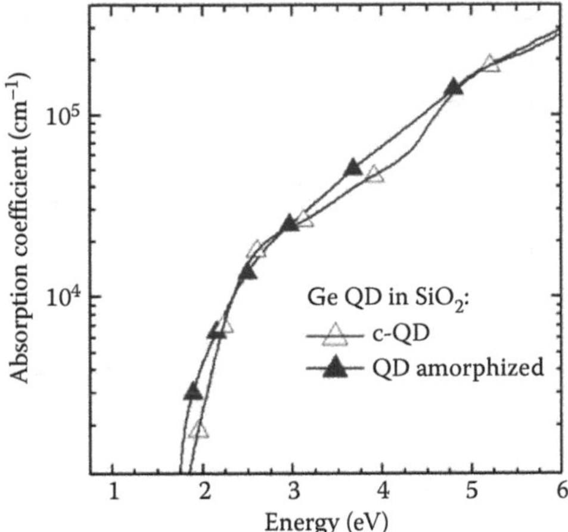

FIGURE 12.21 Absorption spectra of SiGeO sample annealed, containing crystalline (open symbols) or amorphous (closed symbols) Ge QDs embedded in SiO_2.

absorption nor the spectrum itself, except that for the disappearance of the direct resonance peaks as expected because of the lost crystalline order within the Ge QDs. Thus, this evidence reveals a net divergence in the behavior of Ge QDs with respect to Si QDs, which can be solved by considering that the *c–a* transition differently modifies the electronic bandgaps of Si and Ge. In fact, already in bulk, amorphous or crystalline Ge have similar light absorption spectra and optical bandgap, while this does not hold for Si, as evidenced in Figure 12.17.

Recently, it was shown that the atomic absorption cross section (σ_a, measuring the photon absorption probability per Ge atom) could depend on the QDs distribution within the film.[40] In particular, by comparing samples with Ge nanocrystals randomly distributed or multilayer of Ge nanocrystals (2.6–5.5 nm in diameter) separated by an 8 nm thick SiO_2 barrier layer, Uhrenfeldt et al. showed that σ_a for larger and close packed Ge nanocrystals exceeds that for similar concentration of randomly distributed nanocrystals.[40] Still, this interesting effect was not observed in smaller and less densely packed nanocrystals samples, and no optical bandgap values have been derived for Ge nanostructures. In order to clarify whether and to what extent QD–QD interactions can enhançe the light absorption, the Ge NC multilayer approach presented in Section 12.5 was employed, where QD–QD distance can be varied.

Figure 12.22 summarizes the optical parameters of the multilayer samples as a function of the barrier thickness. As expected, all the multilayer samples (Figure 12.22a) exhibit the same optical energy gap (1.9 ± 0.1 eV), mostly independent of the barrier thickness. This evidence is in agreement with the QCE of size tuning of E_g, as all the Ge QDs are similar in size ($2r \sim 3$–4 nm), and the E_g value is well larger than that of not-confined amorphous Ge. However, in Figure 12.22b, the Tauc coefficient B^* is reported, describing the efficiency in light absorption normalized to the volume of Ge.[53]

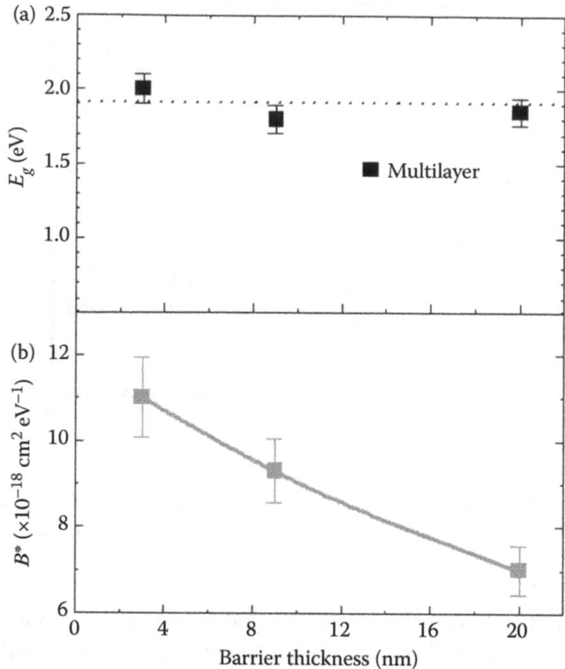

FIGURE 12.22 (a) Energy gap of Ge QDs embedded in SiO_2 and arranged in multilayers with different SiO_2 barrier thicknesses. (b) Absorption efficiency for the same samples. (Reprinted with permission from *Applied Physics Letters* 102, 193105. Copyright 2013, American Institute of Physics.)

A marked dependence of the absorption efficiency is revealed on the barrier thickness, clearly showing a strong decrease in the sample with the more spaced Ge QDs films. In particular, in the closest packed configuration, such absorption efficiency is almost doubled with respect to the case of the most spaced Ge QDs. Even if this experimental approach allows modification of the QD–QD distance only in the vertical direction, some general considerations can be drawn. By using well-spaced Ge QDs, an evident loss occurs in the efficiency of the absorption process. Such an effect has been observed in our samples up to 20 nm of QD–QD vertical spacing. To account for the effect of a three-dimensional (3D) QD–QD spacing on the absorption efficiency, a sample with a single layer (200 nm thick) of SiO_2:Ge was fabricated and characterized in the same way. Ge QDs of similar diameter have been found, with a surface-to-surface distance (now in 3D) of 3 nm. This single-layer sample can be compared with the multilayer sample with a barrier thickness of 3 nm, to account for the modulation of $d_{||}$, the in-plane QD–QD distance. The single-layer sample shows an absorption efficiency comparable with the largest barrier thickness. In other words, when Ge QDs are spaced by 3 nm in 3D, they absorb as much as in a multilayer configuration with 20 nm of 1D spacing and 1 nm of 2D spacing. All these data evidence that the QD–QD spacing plays a key role in the photon absorption process. Therefore, some long-range interaction between QDs has to be assumed to account for the observed effect. Actually, the presence of electronic coupling among semiconductor nanoparticles has been theoretically described,[64,65] for which energy transfer occurs between semiconductor nanocrystals up to 10–20 nm apart, mainly by means of dipole–dipole interactions. If some electron coupling occurs between closely packed Ge QDs, it should affect the electron transition probability and then the light absorption mechanism.

As the quantum confinement effect takes origin from the potential barriers offered by the large bandgap surrounding the NC, if a different matrix is used to embed NCs, significant variation of the optical bandgap is expected. Here, we will try to show this effect by comparing light absorption of Ge QDs in Si_3N_4 or SiO_2 matrices. Since it was previously shown that the fabrication techniques can affect the optical properties, the formation of Ge NCs was induced by implanting Ge ions in Si_3N_4 or SiO_2 matrices, and annealing.[35] As the NC formation is affected by the matrix, the comparison of light absorption is done after annealing at 700°C, which are expected to give comparable Ge NCs (amorphous and ~2 nm sized) in both matrices. In order to properly compare samples with different implant fluencies (D), the atomic absorption cross section, $\sigma = (\alpha \cdot d)/D$, giving the photon absorption probability per Ge atom, has been derived from α spectra. In Figure 12.23, σ spectra are reported for 700°C annealed Si_3N_4 or SiO_2 samples implanted with 100 keV, 7.3×10^{16} Ge/cm^2 (triangles or circles, respectively). For a fixed matrix, it was observed that the absorption coefficient (α) scales linearly with the amount of Ge atoms. By comparing Si_3N_4 and SiO_2 matrices, α is 3–4 times larger in the former, where the E_{04} threshold (energy for $\alpha = 10^4$ cm^{-1}) is reduced by 0.8 eV. This evidences a strong role of the matrix in the photon absorption, which can be profitably used for any light-harvesting application of Ge NCs. To extract the optical bandgap, the Tauc model can be used, taking care of the implant damage effects. In fact, optical transitions can occur involving electronic states in the band tails or in the midgap created by the implantation damage.

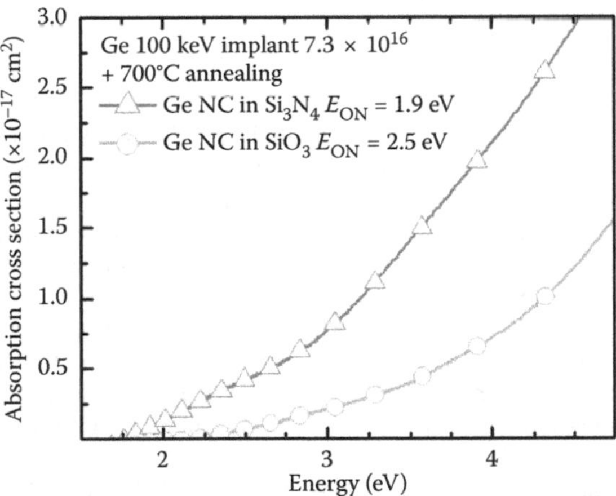

FIGURE 12.23 Atomic absorption cross-section spectra for Ge nanoclusters embedded in Si_3N_4 matrix (open triangles) or in SiO_2 matrix (open circles). (Reprinted with permission from *Applied Physics Letters* 101, 011911. Copyright 2012, American Institute of Physics.)

Ge NCs in Si_3N_4 (triangles) show a smaller onset energy for light absorption (E_{ON} =1.9 eV) than in SiO_2 (2.5 eV).

To explain the different E_{ON} of Ge NCs, the barrier heights seen by electrons and holes in the Ge NCs embedded in the high gap matrix should be considered. Assuming 2 nm sized Ge NCs in SiO_2, E_G is 2.7 eV, in good agreement with E_{ON} of Ge QDs in SiO_2 (whose barrier, ~8.2 eV, can be assimilated to an infinite one). Instead, Si_3N_4 offers a lower barrier to carriers (~4.5 eV) so that a finite barrier calculation is needed, where bandgap widening is reduced by the factor[8] $\left[1+\left(\hbar/\left(r\sqrt{2m*V_0}\right)\right)\right]^2$ (effective mass of exciton, m^*, is about 0.1 for Ge).[66] This factor lowers the expected E_G to 2.0 eV, compatible with the observed E_{ON} of Ge QDs in Si_3N_4. A crucial role of the embedding matrix is then demonstrated, pointing out that Si_3N_4 matrix allows Ge NCs to absorb light much more efficiently than in SiO_2.

12.7 CONCLUSIONS

In conclusion, Si and Ge nanoclusters can be fabricated with high precision and show really promising effects as the light absorption properties are concerned. Indeed, the synthesis of Si or Ge nanoclusters embedded in insulating matrices is clearly affected by the deposition techniques and by the matrix itself. Matrix surrounding the Si NCs shows different dielectric properties if a chemical (PE–CVD) or physical (sputter) deposition technique is employed, which, in turn, affects both the nanoclusters formation and the energy potential confining electrons and holes in the nanoclusters.

Light absorption phenomena in Si or Ge NCs have been presented and discussed, evidencing the paramount role of the quantum confinement of electrons and holes, which enlarges the optical bandgap with respect to the bulk materials. Nevertheless, clear evidences indicate that other effects take place, sometimes overwhelming in comparison to quantum confinement, as the structural phase of the NCs and the interaction among these confined systems.

REFERENCES

1. A. Irrera, F. Iacona, G. Franzò, S. Boninelli, D. Pacifici, M. Miritello, C. Spinella et al., Correlation between electroluminescence and structural properties of Si nanoclusters. *Opt. Mater. (Amsterdam, Neth.)* **27**, 1031, 2005.
2. I. Crupi, S. Lombardo, C. Spinella, C. Gerardi, B. Fazio, M. Vulpio, M. Melanotte, Y.G. Liao, and C. Bongiorno, Memory effects in MOS capacitors with silicon quantum dots. *Mater. Sci. Eng.* **C15**, 283, 2001.
3. R.D. Schaller and V. I. Klimov, High efficiency carrier multiplication in PbSe nanocrystals: Implications for solar energy conversion. *Phys. Rev. Lett.* **92**, 186601, 2004.
4. R.D. Schaller, M. Sykora, J.M. Pietryga, and V.I. Klimov, Seven excitons at a cost of one: Redefining the limits for conversion efficiency of photons into charge carriers. *Nano Lett.* **6**, 424, 2006.
5. R.J. Ellingson, M.C. Beard, J.C. Johnson, P. Yu, O.I. Micic, A.J. Nozik, A. Shabaev, and A.L. Efros, Highly efficient multiple exciton generation in colloidal PbSe and PbS quantum dots. *Nano Lett.* **5**, 865, 2005.
6. A. Luque and A. Martì, Increasing the efficiency of ideal solar cells by photon induced transitions at intermediate levels. *Phys. Rev. Lett.* **78**, 5014, 1997.

7. A. Martì, E. Antolìn, C.R. Stanley, C.D. Farmer, N. Lòpez, P. Dìaz, E. Cànovas, P.G. Linares, and A. Luque, Production of photocurrent due to intermediate–to–conduction–band transitions: A demonstration of a key operating principle of the intermediate-band solar cell. *Phys. Rev. Lett.* **97**, 247701, 2006.
8. G. Conibeer, M. Green, R. Corkish, Y. Cho, E.C. Cho, C.W. Jiang, T. Fangsuwannarak, et al. Silicon nanostructures for third generation photovoltaic solar cells. *Thin Solid Films* **511–512**, 654, 2006.
9. M.A. Green, G. Conibeer, I. Perez-Wurfl, S.J. Huang, D. König, D. Song, A. Gentle, et al. *Proceedings of the 23rd European Photovoltaic Solar Energy Conference,* Valencia, Spain, 1–5 September 2008 (unpublished).
10. Y.M. Niquet, G. Allan, C. Delerue, and M. Lannoo, Quantum confinement in germanium nanocrystals. *Appl. Phys. Lett.* **77**, 1182, 2000.
11. C.S. Garoufalis, Optical gap and excitation energies of small Ge nanocrystals. *J. Math. Chem.* **46**, 934 2009.
12. M. Yu, C.S. Jayanthi, D.A. Drabold, and S.Y. Wu, Enhanced radiative transition in Si_nGe_m nanoclusters. *Phys. Rev. B* **68**, 035404, 2003.
13. T. Shimizu-Iwayama, K. Fujita, S. Nakao, K. Saitoh, T. Fujita, and N. Itoh, Visible photoluminescence in Si+–implanted silica glass. *J. Appl. Phys.* **75**, 7779, 1994.
14. J.G. Zhu, C.W. White, J.D. Budai, S.P. Withrow, and Y. Chen, Growth of Ge, Si, and SiGe nanocrystals in SiO_2 matrices. *J. Appl. Phys.* **78**, 4386, 1995.
15. K.S. Min, K.V. Shcheglov, C.M. Yang, H.A. Atwater, M.L. Brongersma, and A. Polman, Defect-related versus excitonic visible light emission from ion beam synthesized Si nanocrystals in SiO_2. *Appl. Phys. Lett.* **69**, 2033, 1996.
16. E. Werwa, A.A. Seraphin, L.A. Chin, Chuxin Zhou, and K.D. Kolenbrander, Synthesis and processing of silicon nanocrystallites using a pulsed laser ablation supersonic expansion method. *Appl. Phys. Lett.* **64**, 1821, 1994.
17. L.N. Dinh, L.L. Chase, M. Balooch, L.J. Terminello, and F. Wooten, Photoluminescence of oxidized silicon nanoclusters deposited on the basal plane of graphite. *Appl. Phys. Lett.* **65**, 3111, 1994.
18. A.J. Kenyon, P.F. Trwoga, C.W. Pitt, and G. Rehm, The origin of photoluminescence from thin films of silicon-rich silica. *J. Appl. Phys.* **79**, 9291, 1996.
19. T. Inokuma, Y. Wakayama, T. Muramoto, R. Aoki, Y. Kurata, and S. Hasegawa, Optical properties of Si clusters and Si nanocrystallites in high-temperature annealed SiO_x films. *J. Appl. Phys.* **83**, 2228, 1998.
20. F. Iacona, G. Franzò, and C. Spinella, Correlation between luminescence and structural properties of Si nanocrystals. *J. Appl. Phys.* **87**, 1295, 2000.
21. F. Iacona, C. Bongiorno, C. Spinella, S. Boninelli, and F. Priolo, Formation and evolution of luminescent Si nanoclusters produced by thermal annealing of SiO_x films. *J. Appl. Phys.* **95**, 3723, 2004
22. F. Gourbilleau, X. Portier, C. Ternon, P. Voivenel, R. Madelon, and R. Rizk, Si-rich/SiO_2 nanostructured multilayers by reactive magnetron sputtering. *Appl. Phys. Lett.* **78**, 3058, 2001.
23. G. Franzò, M. Miritello, S. Boninelli, R. Lo Savio, M.G. Grimaldi, F. Priolo, F. Iacona, G. Nicotra, C. Spinella, and S. Coffa, Microstructural evolution of SiO_x films and its effect on the luminescence of Si nanoclusters. *J. Appl. Phys.* **104**, 094306, 2008.
24. G. Nicotra, S. Lombardo, C. Spinella, G. Ammendola, C. Gerardi, and C. Depuro, Observation of the nucleation kinetics of Si quantum dots on SiO_2 by energy filtered transmission electron microscopy. *Appl. Surf. Sci.* **205**, 304, 2003.
25. C. Spinella, C. Bongiorno, G. Nicotra, E. Rimini, A. Muscarà, and S. Coffa, Quantitative determination of the clustered silicon concentration in substoichiometric silicon oxide layer. *Appl. Phys. Lett.* **87**, 4102, 2005.

26. R.F. Egerton, *Electron Energy-Loss Spectroscopy in the Electron Microscope*, 2nd ed. (Plenum, New York, 1996).
27. R. G. Barrera and R. Fuchs, Theory of electron energy loss in a random system of spheres. *Phys. Rev. B* **52**, 3256, 1995.
28. A. La Magna, G. Nicotra, C. Bongiorno, C. Spinella, M.G. Grimaldi, E. Rimini, L. Caristia, and S. Coffa, Role of the internal strain on the incomplete Si/SiO_2 phase separation in substoichiometric silicon oxide films. *Appl. Phys. Lett.* **90**, 183101, 2007.
29. R.J. Temkin, An analysis of the radial distribution function of SiO_x. *J. Non-Cryst. Solids* **17**, 215, 1975.
30. H.R. Philipp, Optical and bonding model for non-crystalline SiO_x and SiO_xN_y materials. *J. Non-Cryst. Solids* **8–10**, 627, 1972.
31. D.J. Seol, S.Y. Hu, Y.L. Li, J. Shen, K.H. Oh, and L.Q. Chen, Computer simulation of spinodal decomposition in constrained films. *Acta Mater.* **51**, 5173, 2003.
32. J.W. Cahn, Coherent fluctuations and nucleation in isotropic solids. *Acta Metall.* **10**, 907 1962.
33. R.A. Seilheimer, in *Handbook of Semiconductor Silicon Technology*, edited by W.C. O'Mara, R. B. Herring, and L. P. Hunt (Noyes, Park Rodge, NJ, 1990), p. 776.
34. S. Cosentino, S. Mirabella, M. Miritello, G. Nicotra, R. Lo Savio, F. Simone, C. Spinella, and A. Terrasi, The role of the surfaces in the photon absorption in Ge nanoclusters embedded in silica. *Nanoscale Res. Lett.* **6**, 135, 2011.
35. S. Mirabella, S. Cosentino, A. Gentile, G. Nicotra, N. Piluso, L.V, Mercaldo, F. Simone, C. Spinella, and A. Terrasi, Matrix role in Ge nanoclusters embedded in Si_3N_4 or SiO_2. *Appl. Phys. Lett.* **101**, 011911, 2012.
36. B. Zhang, S. Shrestha, M.A. Green, and G. Conibeer, Size controlled synthesis of Ge nanocrystals in SiO_2 at temperatures below 400°C using magnetron sputtering. *Appl. Phys. Lett.* **96**, 261901, 2010.
37. L.A. Nesbit, Annealing characteristics of Si-rich SiO_2 films. *Appl. Phys. Lett.* **46**, 38, 1985.
38. J. Skov Jensen, T.P. Leervad Ledersen, R. Pereira, J. Chevallier, J. Lundsgaard Hansen, B. Bech Nielsen, and A. Nylandsted Larsen, Ge nanocrystals in magnetron sputtered SiO_2. *Appl. Phys. A* **83**, 41, 2006.
39. J.I. Langgord and A.J.C. Wilson, Scherrer after sixty years: A survey and some new results in the determination of crystallite size. *J. Appl. Crystallogr.* **11**, 102–13, 1978.
40. C. Uhrenfeldt, J. Chevallier, A.N. Larsen, and B.B. Nielsen, Near–infrared–ultraviolet absorption cross sections for Ge nanocrystals in SiO_2 thin films: Effects of shape and layer structure. *J. Appl. Phys.* **109**, 094314, 2011.
41. M. Zacharias, J. Heitmann, R. Scholz, U. Kahler, M. Schmidt, and J. Blasing, Size-controlled highly luminescent silicon nanocrystals: A SiO/SiO_2 superlattice approach. *Appl. Phys. Lett.* **80**, 661, 2002.
42. H. G. Chew, W. K. Choi, Y. L. Foo, F. Zheng, W. K. Chim, Z. J. Voon, K. C. Seow, E. A. Fitzgerald, and D. M. Y. Lai, Effect of germanium concentration and oxide diffusion barrier on the formation and distribution of germanium nanocrystals in silicon oxide matrix. *Nanotechnology* **17**, 1964, 2006.
43. H. Schmidt, U. Geckle, and M. Burns, Simultaneous diffusion of Si and N in silicon nitride. *Phys. Rev. B* **74**, 045203, 2006.
44. J.E. Chang, P.H. Liao, C.Y. Chien, J.C. Hsu, M.T. Hung, H.T. Chang, S.W. Lee, et al. Matrix and quantum confinement effects on optical and thermal properties of Ge quantum dots. *J. Phys. D: Appl. Phys.* **45**, 105303, 2012.
45. D.J. Lockwood, Z.H. Lu, and J.M. Baribeau, Quantum confined luminescence in Si/SiO_2 superlattices. *Phys. Rev. Lett.* **76**, 539, 1996.
46. E.G. Barbagiovanni, D.J. Lockwood, P.J. Simpson, and L.V. Goncharova, Quantum confinement in Si and Ge nanostructures. *J. Appl. Phys.* **111**, 034307, 2012.

47. W. Teng, J.F. Muth, R.M. Kolbas, K.M. Hassan, A.K. Sharma, A. Kvit, and J. Narayan, Quantum confinement of E_1 and E_2 transitions in Ge quantum dots embedded in an Al_2O_3 or an AlN matrix. *Appl. Phys. Lett.* **76**, 43, 2000.
48. S. Cosentino, M. Miritello, I. Crupi, G. Nicotra, F. Simone, C. Spinella, A. Terrasi, and S. Mirabella, Room-temperature efficient light detection by amorphous Ge quantum wells. *Nanoscale Res. Lett.* **8**, 128, 2013.
49. G.H. Shih, C.G. Allen, and B.G. Potter Jr., Interfacial effects on the optical behavior of Ge:ITO and Ge:ZnO nanocomposite films. *Nanotechnology* **23**, 075203, 2012.
50. R. Guerra and S. Ossicini, High luminescence in small Si/SiO_2 nanocrystals: A theoretical study. *Phys. Rev. B* **81**, 245307, 2010.
51. S. Mirabella, R. Agosta, G. Franzò, I. Crupi, M. Miritello, R. Lo Savio, M.A. Di Stefano, S. Di Marco, F. Simone, and A. Terrasi, Light absorption in silicon quantum dots embedded in silica. *J. Appl. Phys.* **106**, 103505, 2009.
52. C. Bulutay, Interband, intraband, and excited-state direct photon absorption of silicon and germanium nanocrystals embedded in a wide band-gap lattice. *Phys. Rev. B* **76**, 205321, 2007.
53. S. Mirabella, S. Cosentino, M. Failla, M. Miritello, G. Nicotra, F. Simone, C. Spinella, G. Franzò, and A. Terrasi, Light absorption enhancement in closely packed Ge quantum dots. *Appl. Phys. Lett.* **102**, 193105, 2013.
54. Z. Ma, X. Liao, G. Kong, and J. Chu, Absorption spectra of nanocrystalline silicon embedded in SiO_2 matrix. *Appl. Phys. Lett.* **75**, 1857, 1999.
55. L. Khriachtchev, M. Räsänen, S. Novikov, and L. Pavesi, Systematic correlation between Raman spectra, photoluminescence intensity, and absorption coefficient of silica layers containing Si nanocrystals. *Appl. Phys. Lett.* **85**, 1511, 2004.
56. A. Podhorodecki, G. Zatryb, J. Misiewicz, J. Wojcik, and P. Mascher, Influence of the annealing temperature and silicon concentration on the absorption and emission properties of Si nanocrystals. *J. Appl. Phys.* **102**, 043104, 2007.
57. G. Vijaya Prakash, N. Daldosso, E. Degoli, F. Iacona, M. Cazzannelli, Z. Gaburro, G. Puker, et al. Structural and optical properties of silicon nanocrystals grown by plasma-enhanced chemical vapor deposition. *J. Nanosci. Nanotech.* **1**, 159, 2001.
58. F. Gourbilleau, C. Ternon, D. Maestre, O. Palais, and C. Dufour, Silicon-rich SiO_2/SiO_2 multilayers: A promising material for the third generation of solar cell. *J. Appl. Phys.* **106,** 013501, 2009.
59. J. Tauc in *Amorphous and Liquid Semiconductors* (Ed. J. Tauc), Plenum Press, London and New York, p. 175, 1974.
60. S. Knief and W. von Niessen, Disorder, defects, and optical absorption in a–Si and a–Si:H. *Phys. Rev. B* **59**, 12940, 1999.
61. D. Timmerman, I. Izeddin, P. Stallinga, I.N. Yassievich, and T. Gregorkiewicz, Space-separated quantum cutting with silicon nanocrystals for photovoltaic applications. *Nat. Photonics* **2**, 105, 2008.
62. D. Pacifici, E.C. Moreira, G. Franzò, V. Martorino, F. Priolo, and F. Iacona, Defect production and annealing in ion-irradiated Si nanocrystals. *Phys. Rev. B* **65**, 144109, 2002.
63. R. Guerra, I. Marri, R. Magri, L. Martin–Samos, O. Pulci, E. Degoli, and S. Ossicini, Silicon nanocrystallites in a SiO_2 matrix: Role of disorder and size. *Phys. Rev. B* **79**, 155320, 2009.
64. G. Allan and C. Delerue, Energy transfer between semiconductor nanocrystals: Validity of Förster's theory. *Phys. Rev. B* **75**, 195311, 2007.
65. R. Baer and E. Rabani, Theory of resonance energy transfer involving nanocrystals: The role of high multipoles. *J. Chem. Phys.* **128**, 184710, 2008.
66. A.G. Cullis, L.T. Canham, and P.D.J. Calcott, The structural and luminescence properties of porous silicon. *J. Appl. Phys.* **82**, 909, 1997.

Index

A

A-center, *see* Vacancy–Oxygen center (VO center)
ABCD-matrices, 369, 370
ABCD-parameters (chain-or cascade-matrix), 366
Ab initio calculation, 120–121, 142, 339
  C_iO_i and hydrogen interaction, 279
  negative-U behavior, 280
  surface-induced charges, 143, 144
Absorbance (A), 325–326, 327
  for CZ silicon, 332, 333
  FZ silicon, 342
  of N-doped CZ silicon, 337
A defects, *see* Swirl-defects
AC, *see* Alternating current (AC)
Acceptor diffusion, 190–192
Acetic acid (HAc), 290
  as diluent, 298
  effect on *FS Cr-free SOI etch*, 296
  nitric acid dissociation, 293
AFM, *see* Atomic force microscopy (AFM)
Alternating current (AC), 353
  dopant profile, 362
  MOS varactors, 361, 362
  response, 361
Ampere/watt (A/W), 363
Annealed wafers (AWs), 120, 147; *see also* Epitaxial wafers (EWs)
  gettering technologies, 149
  OSF ring formation, 153
  oxygen precipitation, 151
  precipitate density depth profile, 148
  ramping-up rate, 152
  TEM, 150
Antireflection coating (AR coating), 242
APDs, *see* Avalanche multiplication photo diodes (APDs)
APT, *see* Atom-probe tomography (APT)
AR coating, *see* Antireflection coating (AR coating)
Arrhenius plot, 316, 357
Arsenic (As), 168
  diffusion, 175
  diffusion coefficient, 176, 177
  electronic intrinsic conditions, 177
Atom-probe tomography (APT), 160
Atomic force microscopy (AFM), 302
Avalanche multiplication photo diodes (APDs), 24
A/W, *see* Ampere/watt (A/W)
AWs, *see* Annealed wafers (AWs)

B

BC position, *see* Bond-centered position (BC position)
1,4-benzoquinone, 308
Benzoquinone (BZQ), 290, 308
BMD, *see* Bulk microdefect (BMD)
Boltzmann approximation, 354
Bond-centered position (BC position), 46, 219
  hydrogen, 219
  muonium, 229
  in Si, 44
  of Si–Si bonds, 45
  species, 46
Boron, 342
  diffusion, 172–173
  enhancement, 204
Boron-related complexes, 279–281; *see also* Carbon-related interstitial defects
Bourgoin–Corbett mechanism, 93, 96
BOX, *see* Buried oxide (BOX)
Bromine, 296
  etching mechanism and special role, 298–300
  oxidation mechanism of silicon, 300
B_s, *see* Substitutional boron (B_s)
Bulk crystals, 24
Bulk microdefect (BMD), 149, 293
Buried oxide (BOX), 296
BZQ, *see* Benzoquinone (BZQ)

C

Carbon (C), 2, 161, 193, 338
  concentrations in silicon melt, 10, 13
  impurities in silicon, 276, 339
  IR absorption spectra, 339
  isotopic effect, 335
  photoluminescence spectrum, 341
  X-and Y-bands, 340
Carbon-related interstitial defects, 276
  C_iO_i, B_iO_i, and C_iO_iI, 278
  DLTS spectra, 277
  ion implantation or particle irradiation, 276
  LVMs, 279
Carrier transport, 38
  Hall effect measurements, 39
  Hall mobilities, 40, 41

Cathodoluminescence, 327
Channel metaloxide semiconductor (CMOS), 24
Chemical etching techniques, 290
  activation energies and selectivities, 317
  AFM measurement, 319
  characteristic parameters, 315–319
  crystal defects, 313–315
  electron transfer from silicon to chromium, 315
  indentation with dislocation half loops, 318
  potential energy plot, 314
  of silicon, 291–293
Chemical vapor deposition (CVD), 62, 160
  Ge on Si epitaxy, 65
CMOS, *see* Channel metaloxide semiconductor (CMOS)
Conduction band, effective density of, 355
Copper (Cu), 161
Crystal growth, 25
  grown-in defects, 27–28
  grown crystals crystallinity, 26–27
Crystal-originated particles (COPs), 120, 289
Current gain, 371
Current–voltage characteristics (*I–V* characteristics), 354
  Boltzmann approximation, 354
  diffusion constant and length, 356
  effective density of conduction band, 355
  energy scheme for electron emission, 360
  ideality factor *vs.* voltage, 359
  p–n junction, 355, 361
  recombination current, 357
  SiGe heterostructure device, 357
  SRH theory, 360
CVD, *see* Chemical vapor deposition (CVD)
CZ process, *see* Czochralski process (CZ process)
Czochralski process (CZ process), 1, 24–25, 146–147, 256
  absorbance spectra, 327
  axial temperature gradients, 17
  C and O atoms distributions, 10, 13
  discrete system, 15
  gas flow fields, 5–8
  heat and mass transfer with magnetic fields, 11
  interface deflection, 17, 18
  light element transfer, 3–5
  melt flow, thermal field, and melt–crystal interface profiles, 16
  silicon materials, 289, 326
  SiO and CO distributions, 8–10, 12
  temperature fields, 5–8
  transverse magnetic field, 14

D

D-defects, 307
  in CZ bulk material, 308
DA, *see* Donor–acceptor (DA)
Dash etch, 293
Dash etchant, 301
DBS, *see* Doppler broadening spectroscopy (DBS)
DC, *see* Direct current (DC)
De-embedding, 366
Deep level transient spectroscopy (DLTS), 47, 88, 105, 186, 221, 222, 256
Defect engineering, 256
Density functional theory (DFT), 120, 186
Device realization, 353
Device structures, 363
Device under test (DUT), 366
  ABCD-matrix, 370
  with GSG contact pad and probe tips, 367
  S-parameters, 372
DFT, *see* Density functional theory (DFT)
Di-interstitial
  germanium, 114
  silicon, 111
Di-interstitial-related complexes, 281
Diffusion-oxygenated Float-zone material (DOFZ material), 260
Diffusion, 160; *see also* Dopant diffusion; Nonequilibrium diffusion; Self-diffusion
  in silicon–germanium alloys, 197–204
  temperature dependence, 162
Dilute Secco etching step, 290
Dilute Secco mixtures, 293
Direct current (DC), 353
Dislocations, 345
  DLTS and EPR measurements, 346
  EFG multicrystalline silicon sample, 347
  etch pit formation, 313
  PL spectra of dislocated FZ Si sample, 346
Divacancy center (V_2 center), 257, 268; *see also* Trivacancy center (V_3 center)
  absolute diffusivity values, 265
  as-grown and Fe-enriched samples, 267
  diffusion, 264–265
  DLTS spectra, 260
  electronic properties, 257–259
  Fe_iV_2 and VFeV atomic structures, 266
  H_2 molecular fraction, 264
  interaction with Fe, 265–268
  interaction with hydrogen, 261–264
  interaction with oxygen, 259–261
  simple atomic structure, 258, 259
Divacancy–oxygen complexes (V_2O complexes), 259
  annealing mechanism, 261
  electronic properties, 261
  thermal stability, 260
DLTS, *see* Deep level transient spectroscopy (DLTS)
DOFZ material, *see* Diffusion-oxygenated Float-zone material (DOFZ material)

Donor–acceptor (DA), 330
pair luminescence, 344
transitions, 330
Donor diffusion, 187, 188
coefficient, 190
doping dependence, 188, 190
in Ge, 189
Dopant–defect complex formation, 192–193
Dopant diffusion, 160, 187, 202
acceptor diffusion, 190–192
arsenic diffusion, 175–177
boron diffusion, 172–173
donor diffusion, 188–190
dopant–defect complexes formation, 192–193
in Ge, 187
I-mediated diffusion, 188
isotope structure, 170
n-type dopants, 187
phosphorus diffusion, 173–175
self-diffusion *vs*., 178
in Si, 169
in silicon–germanium alloys, 202–204
SIMS concentration profiles, 170, 171
Dopants, 342; *see also* Nondoping impurities and defects
DA pair luminescence, 344, 345
low-temperature silicon spectrum, 342
PL spectra, 344, 345
SoG, 343
Doppler broadening spectroscopy (DBS), 143
DS process, *see* Unidirectional solidification process (DS process)
DUT, *see* Device under test (DUT)

E

EDX, *see* Energy dispersive x-ray (EDX)
EELS, *see* Electron energy loss spectrum (EELS)
EFTEM, *see* Energy-filtered transmission electron microscopy (EFTEM)
EG, *see* Extrinsic gettering (EG)
EL, *see* Electroluminescence (EL)
Electrical levels, 89
excitonic effects, 90
in germanium, 95–96
self-interstitials, 89–90
in silicon, 89
Electroluminescence (EL), 61, 327, 363
Electromagnetic radiation (EM radiation), 365
Electron energy loss spectrum (EELS), 379
oxide matrix, 383
SRO layer, 381
total volume fraction, 380
Electron nuclear double resonance (ENDOR), 257
Electron paramagnetic resonance (EPR), 88, 219, 220, 257, 331
EM radiation, *see* Electromagnetic radiation (EM radiation)
ENDOR, *see* Electron nuclear double resonance (ENDOR)
Energy-filtered transmission electron microscopy (EFTEM), 379
micrograph of PE–CVD SiO_x layer, 380
of SiO*x* films, 383
Energy dispersive x-ray (EDX), 26
Entropy correction, 7
Epitaxial wafers (EWs), 147; *see also* Annealed wafers (AWs)
Epitaxy
GaAs-based solar cell, 63
of Ge on Si, 62, 65
thermal expansion coefficients, 65
thick-graded buffers, 63
EPR, *see* Electron paramagnetic resonance (EPR)
EWs, *see* Epitaxial wafers (EWs)
Excitation-enhanced diffusion, 93, 96
Ex situ experiments, 300
External stress, 130–133
Extrinsic gettering (EG), 149

F

Far Infrared spectroscopy (FIR), 326
FBI, *see* Forward bias injection (FBI)
FE, *see* Free-exciton (FE)
FeB pair, *see* Iron–boron pair (FeB pair)
FET, *see* Field effect transistors (FET)
FFC configuration, *see* Fourfold coordinated configuration (FFC configuration)
FIB, *see* Focused ion beam (FIB)
Fick's law, 180
Field effect transistors (FET), 182
FIR, *see* Far Infrared spectroscopy (FIR)
Float-zone (FZ), 326
absorbance spectra, 327
technique, 256
Flow pattern, 307
Focused ion beam (FIB), 303
Foreign atom diffusion, 161
Formation energy, 122
ab initio simulations, 122–123
in Si crystals, 124
Forward bias injection (FBI), 270
Fourfold coordinated configuration (FFC configuration), 256, 268–269
Fourier transform infrared (FTIR), 227
absorption, 227
annealing kinetics of VO(0), 272
H_2^* existence by, 229
spectroscopy, 256
FP, *see* Frenkel pairs (FP)

Frankfurt-SOITEC Cr-free SOI etches (FS Cr-free SOI etches), 290, 293
 with and without nitric acid, 299
 bromine, 296, 299
 optical micrographs, 297
 SEM, 297
 thin SOI etchants, 298
Free-exciton (FE), 328
 emission line, 343
 luminescence, 328
Free-to-bound transition, 329
Frenkel pairs (FP), 97, 103; *see also* Radiation damage; Self-interstitials
FS Cr-free SOI etches, *see* Frankfurt-SOITEC Cr-free SOI etches (FS Cr-free SOI etches)
FTIR, *see* Fourier transform infrared (FTIR)
Full-width at half-maximum (FWHM), 326
FWHM, *see* Full-width at half-maximum (FWHM)
FZ, *see* Float-zone (FZ)

G

Ge NC, *see* Germanium nanocluster (Ge NC)
Ge-on-Insulator (GOI), 144
Generalized gradient approximation (GGA), 120
Germanium (Ge), 61, 88, 113, 160, 223–224
 defect configurations, 94–95
 deposition on structured substrates, 70–72
 Di-interstitial, 114
 diffusion coefficient, 182, 183
 dopant diffusion, 187–193
 electrical levels, 95–96
 epitaxial crystal growth, 74
 epitaxy of, 62–65
 excitation-enhanced diffusion, 96
 high-purity Ge, 108–109
 lattice parameter, 72–73
 layers for light emission, 70
 MQW, 67–70
 n-type to *p*-type Ge, 109–110
 nonequilibrium diffusion, 193–197
 self-diffusion, 184–186
 self-interstitials in, 94
 as semiconductor, 62
 strained Ge QWs, 65–66
 thermally activated diffusion, 96
 tri-interstitial, 114
Germanium nanocluster (Ge NC), 388; *see also* Silicon nanocluster (Si NC)
 absorption coefficient, 394
 atomic absorption cross-section spectra, 401
 bright-field TEM images, 390, 391
 dark-field STEM images, 389
 Ge/Si_3N_4 interface, 393
 GI–XRD, 390
 high-angle annular dark-field STEM images, 392
 light absorption, 393–402
 RBS analysis, 391
 synthesis and imaging, 388–393
Gettering technologies, 149
GGA, *see* Generalized gradient approximation (GGA)
Glancing incidence x-ray diffraction (GI–XRD), 390
GOI, *see* Ge-on-Insulator (GOI)
Gold (Au), 161
Ground–signal–ground configuration (GSG configuration), 366
Grown-in defects, 27–28

H

H-parameters (hybrid-matrix), 365, 371
H_2O_2, *see* Hydrogen peroxide (H_2O_2)
Heterostructure bipolar transistor (HBT), 24
Hexagonal ring clusters (HRC), 126
High frequencies (HF), 363, 365
High-resolution transmission electron microscopy (HR–TEM), 389
HNO_3/HF etching system, 291
HRC, *see* Hexagonal ring clusters (HRC)
HR–TEM, *see* High-resolution transmission electron microscopy (HR–TEM)
Hydrofluoric acid (HF), 290, 298
Hydrogen (H), 46, 161, 217, 341–342; *see also* Interstitial hydrogen
 activation of impurities, 236–237
 band alignment diagram, 47
 C-rich Si materials, 237–240
 Ge-rich and Si-rich SiGe alloys, 48
 high-purity Si materials, 237–240
 hydrogenation of deep levels, 235–236
 with impurities and defects, 230–237
 interactions with native defects, 233
 molecules, 226–228
 Mu(+/–) level, 46
 O-rich Si materials, 237–240
 pairs, 226–230
 photovoltaic industry, 242–244
 platelets and smart-cut® process, 233–235
 Raman spectra, 228
 shallow impurities passivation, 230–233
 vibrational lifetimes, 240–242
Hydrogen-containing defects, 240
Hydrogen pairs, 226
 formation, 229
 hydrogen molecules, 226–228
 Si and Ge, 229–230
Hydrogen peroxide (H_2O_2), 290
Hydrogenation of deep levels, 235–236

I

I-mediated diffusion, 188
IG, *see* Internal gettering (IG)
Impact-ionized avalanche transit time diode (IMPATT-diode), 368
    real and imaginary parts, 368
Infrared spectroscopy (IR spectroscopy), 44, 324; *see also* Photoluminescence spectroscopy (PL spectroscopy)
    absorbance, 325
    impurities and defects, 324
    optical transmission measurements, 325
    vibrational mode, 325, 326
Insulators, 67
Interface deflection, 17, 18
Internal gettering (IG), 146–147
Internal quantum efficiency (IQE), 363
Internal stress, 130–133
Interstitial boron (B_i), 279
Interstitial defects, 276
    boron-related complexes, 279–281
    carbon-related interstitial defects, 276–279
    di-interstitial-related complexes, 281
Interstitial hydrogen, 219
    diffusion properties, 224–225
    diffusivity and solubility, 219–220
    germanium, 223–224
    silicon, 220–223
    vibrational properties, 225–226
Interstitial oxygen pairs (IOi pairs), 276
Intrinsic defects, 255
Iodine
    etching mechanism and special role, 298–300
    nitric acid to, 290
IOi pairs, *see* Interstitial oxygen pairs (IOi pairs)
IQE, *see* Internal quantum efficiency (IQE)
Iron (Fe), 161
Iron–boron pair (FeB pair), 265
IR spectroscopy, *see* Infrared spectroscopy (IR spectroscopy)
I–V pairs, 97; *see also* Frenkel pairs

J

Jahn–Teller distortion, 257
Junction space charge techniques (JSCT), 363

K

KI, *see* Potassium iodide (KI)

L

Laplace-transform deep-level transient spectroscopy (LDLTS), 105
Laser mass spectroscopy (LAMAS), 300
LDA, *see* Local density approximation (LDA)
LDLTS, *see* Laplace-transform deep-level transient spectroscopy (LDLTS)
LEPECVD, *see* Low-energy plasma-enhanced chemical vapor deposition (LEPECVD)
LF, *see* Low frequencies (LF)
Light-induced degradation (LID), 257
Local atomic structure, 36
    Ge and Si coordination numbers, 37
    of SiGe, 39
    x-ray fluorescence holography, 38
    XAFS results, 37
Local density approximation (LDA), 120
Local vibrational mode (LVM), 112, 239, 279
Low frequencies (LF), 362
Low-energy plasma-enhanced chemical vapor deposition (LEPECVD), 62, 65

M

Magneto co-sputtering (MS), 378
Maximum available gain (MAG), 373
Maximum oscillation frequency, 372, 374
Maximum stable gain (MSG), 373
Maxwell's equations, 365
MBE, *see* Molecular beam epitaxy (MBE)
mc-Si, *see* Multicrystalline silicon (mc-Si)
MCTS, *see* Minority carrier transient spectroscopy (MCTS)
MD, *see* Molecular dynamics (MD)
Medium Infrared spectroscopy (MIR), 326
MEG, *see* Multiple exciton generation (MEG)
Metal oxide semiconductor (MOS), 182, 354
    C–V characteristic, 362
    scheme, 362
    varactors, 361
Metal oxide semiconductor field effect transistor (MOSFET), 371
    fabrication, 197
    with gate resistance, 373
    Si, 62
    under VNA characterization, 371
Microwave measurements, 364
    ABCD-matrices, 369, 370
    de-embedding, 366
    DUT, 367
    equivalent lumped models, 367
    IMPATT diode, 368
    MAG, 373
    maximum oscillation frequency, 372, 374
    Moore's law, 364
    MOSFET, 371
    one-port network, 366
    real and imaginary parts, 368
    scattering matrix, 365
    transit frequency, 372
    vector network analyzer, 366

Microwave power source, 368
Minority carrier transient spectroscopy (MCTS), 256
MIR, *see* Medium Infrared spectroscopy (MIR)
ML, *see* Monolayers (ML)
Modulation-doped field-effect transistor (MODFET), 24
Molecular beam epitaxy (MBE), 62, 160
Molecular dynamics (MD), 119, 121–122, 168, 227
  ab initio and, 119–120, 281
  correlation factor calculation, 163
Monolayers (ML), 52
Moore's law, 364
Morphs, 96–97
MOS, *see* Metal oxide semiconductor (MOS)
MOSFET, *see* Metal oxide semiconductor field effect transistor (MOSFET)
MQW, *see* Multiple quantum-well (MQW)
MS, *see* Magneto co-sputtering (MS)
MSG, *see* Maximum stable gain (MSG)
Multi-interstitial clusters, 110
Multicrystalline silicon (mc-Si), 218
Multiple exciton generation (MEG), 377
Multiple quantum-well (MQW), 67; *see also* Quantum well (QW)
  optoelectronic devices, 68–70
  thermoelectric devices, 67–68
Muon spin rotation (μSR), 221

N

n-type germanium, 102; *see also* *p*-type germanium
  DLTS, 105
  equilibrium charge state, 103
  Fermi level dependence, 104
  FP, 103
n-type silicon, 98–99, 102; *see also* *p*-type silicon
Near-edge x-ray absorption fine structure spectroscopy (NEXAFS), 315, 316
Neutron reflectometry (NR), 160
Nickel (Ni), 161
Nitric acid (HNO_3), 290, 298
Nitrogen, 336–338
  IR absorbance spectrum, 337
  PL spectra from Si doped with, 338
Nondoping impurities and defects
  carbon, 338–341
  hydrogen, 341–342
  oxygen, 331–336
Nonequilibrium *ab initio* MD simulations, 240
Nonequilibrium diffusion, 178, 193
  B*I* pair formation entropy, 197
  Fick's law, 180
  in Ge, 194
  Ge surface property, 196
  P and As equilibrium diffusion, 195
  RESD comprehensive analysis, 179–180
  self-diffusion, 178–179
  temperature dependency, 181
  V concentrations, 182
Novel chromium-free chemistries; *see also* Silicon
  etching mechanism and special role, 298–300
  etching solutions with OOE, 308–313
  FS Cr-free SOI etching solutions, 293–298
  OPE, 300–308
NR, *see* Neutron reflectometry (NR)

O

ODMR, *see* Optically detected magnetic resonance (ODMR)
Old thermal donors (OTD), 331
One-port network, 366
OOE, *see* Organic oxidizing agent etches (OOE)
OPE, *see* Organic peracid etches (OPE)
Open-short (OS), 366
Optical
  properties, 363
  transmission measurements, 325
Optically detected magnetic resonance (ODMR), 257
Optoelectronic devices, 68
  MQW structures, 68, 69
  strain-induced splitting, 70
Organic oxidizing agent etches (OOE), 290
  etching solutions with, 308
  OSF densities, 313
  p-BZQ or *p-CA*, 311
  redox equilibrium, 310
  SOI surface after etching and HF dip, 312
Organic peracid etches (OPE), 290, 300
  CZ Si sample, 309
  D-defects, 307
  Dash etchant, 301
  defects in SOI, 304
  optical micrograph of SOI sample, 303
  OSF, 305, 306
  PAA and decreasing H_2O_2 concentration, 302
  properties, 301
  removal rate, 303
  swirl-defects, 308, 309
OS, *see* Open-short (OS)
OSF, *see* Oxidation-induced stacking fault (OSF)
Ostwald ripening, 397
OTD, *see* Old thermal donors (OTD)
Oxidation-induced stacking fault (OSF), 120, 289
  characteristic of, 305
  densities, 313
  ring formation, 153
  SEM image, 311, 312
  silicon substrates, 289
  in SOI material, 306

Oxygen, 44, 331
atoms in SiGe alloys, 45
concentrations in silicon melt, 10, 13
IR absorbance spectrum, 332, 333
luminescence lines, 334–336
O-related defects, 331
PL spectrum of CZ silicon samples, 335
precipitates, 3
precipitation, 146–147
Si–Si bond, 46
Si_xGe_{1-x} alloys IR spectra, 44
TD-bound-exciton photoluminescence lines, 334

P

p-MODQW, *see* *p*-type modulation-doped QW (*p*-MODQW)
p-type germanium, 105
P1-EPR center, 106–107
positive charge state, 108
RT, 106
stable spin-0 state, 107
two-state defect, 106, 107
p-type modulation-doped QW (*p*-MODQW), 65
p-type silicon, 97
AA12 center, 101, 102
double-positive charge state, 101
e-irradiation, 98–99, 100
self-interstitial, 97
PAA, *see* Peracetic acid (PAA)
PAC, *see* Perturbed angular γ–γ correlation (PAC)
Para-benzoquinone (*p*-BZQ), *see* 1,4-benzoquinone
Para-chloranil (*p*-CA), *see* 2,3,5,6-tetrachloro-1,4-benzoquinone
PAS, *see* Positron annihilation spectroscopy (PAS)
Passivation, 46
PECVD, *see* Plasma-enhanced, chemical-vapor deposition (PECVD)
Peracetic acid (PAA), 301
Permeation experiments, 220
Perpropanoic acid (PPA), 301
Perturbed angular γ–γ correlation (PAC), 88, 186
neutrino-recoil, 114
spectroscopy, 109–110
trapping experiments, 104
Phonon-aided energy, 93
Phosphorus (P), 168
diffusion, 173, 175, 176
interstitial, 174
in isotope multilayers, 173
Photocurrents, 363
Photoexcitation process, 328
Photoluminescence spectroscopy (PL spectroscopy), 52, 257, 327
advantages, 330
DA transitions, 330
defects, 331
free-to-bound transition, 329
luminescence decay processes, 328
PE–CVD SiO*x* system, 384
photoexcitation process, 328
silicon-related research, 324
W PL line, 112
Photonic properties, 363–364
Photovoltaic cells (PV cells), 1, 255
deposition of nanocrystalline Si, 65
hydrogenation of mc-Si, 218
hydrogen in, 242–244
microelectronic and, 323, 331
Si co-doped with Ga, 345
PL spectroscopy, *see* Photoluminescence spectroscopy (PL spectroscopy)
Plasma-enhanced, chemical-vapor deposition (PECVD), 243, 378
EFTEM micrograph, 380
films, 396
room-temperature PL spectra, 384, 385
SiO*x* system, 378, 384
TEM micrograph, 379
Platelets, 233–235
Platinum (Pt), 161
Point defect complexes in silicon, 256
concentration, 256
experimental techniques, 257
interstitial defects, 276–281
V_2 center, 257–268
V_3 center, 268–271
VO, 271–276
Poisson's law, 354
Polished wafer (PW), 147
Portevin–LeChatelier phenomenon, 52
Positron annihilation spectroscopy (PAS), 143, 257
Potassium iodide (KI), 298
PPA, *see* Perpropanoic acid (PPA)
Precipitation hardening, 151
Preferential etching solution, selectivity of, 317
Proton implantation, 234
PV cells, *see* Photovoltaic cells (PV cells)
PW, *see* Polished wafer (PW)

Q

Quantum dot (QD), 377
Quantum-confined Stark effect (QCSE), 61
Quantum well (QW), 24, 61
p-MODQWs, 65
strained Ge QWs, 65–66

R

Radiation damage, 97; *see also* Self-interstitials
high-purity Ge, 108–109
n-type germanium, 102–105
n-type silicon, 102
n-type to *p*-type Ge, 109–110
p-type germanium, 105–108
p-type silicon, 97–102
in Si and Ge, 97
Radiation-enhanced self-diffusion (RESD), 179
Raman spectroscopy, 324
Random bonding model (RBM), 387
Random mixture model (RMM), 387
Rapid thermal annealing (RTA), 120
Rapid thermal processing (RTP), 151
RBA, *see* Reverse bias anneals (RBA)
RBM, *see* Random bonding model (RBM)
RBS, *see* Rutherford backscattering spectrometry (RBS)
Reactive ion etching (RIE), 120
Recombination current, 357
RESD, *see* Radiation-enhanced self-diffusion (RESD)
Reverse bias anneals (RBA), 231
RIE, *see* Reactive ion etching (RIE)
RMM, *see* Random mixture model (RMM)
Room temperature (RT), 37
FS Cr-free SOI etch, 296
transformations, 106
vacancy-related defects in silicon, 257
RT, *see* Room temperature (RT)
RTA, *see* Rapid thermal annealing (RTA)
RTP, *see* Rapid thermal processing (RTP)
Rutherford backscattering spectrometry (RBS), 378

S

S-parameters, 144, 365
Scanning electron micrograph (SEM), 296
Scanning transmission electron microscopy (STEM), 383, 389
Scattering matrix, 365
Secco etching, 306, 307
Secondary ion mass spectroscopy (SIMS), 45, 160, 220
concentration profiles, 170, 171, 198
depth profile, 221
detection limit, 244
diffusion studies, 184
Seebeck coefficient, 43
Self-diffusion, 160, 162, 184, 197
activation enthalpy and preexponential factor, 200
Arrhenius expression, 199
correlation factor, 163
diffusion studies, 186, 202
dopant diffusion *vs*., 178
in Ge, 184, 185
I-mediated mechanism, 164
MD, 168
p-type doping on, 185
phosphorus in silicon, 169
in Si, 162
in silicon–germanium alloys, 197
temperature dependences, 168, 198–199
thermal equilibrium conditions, 167
uncorrelated contribution, 164–165
V-mediated self-diffusion, 166
V contribution, 165
Vegard's law, 200
Self-interstitials, 88
aggregation of, 110
charge-dependent structures, 89
di-interstitial, 111
experimental formation energies, 110
in germanium, 94–96
morphs, 96–97
silicon, 88–93, 110–111
tetra-interstitial, 113
tri-interstitial, 111–113
SEM, *see* Scanning electron micrograph (SEM)
Semiconductor Equipment and Materials International (SEMI), 343
Semiconductor junctions, 354
SF, *see* Stacking fault (SF)
Shallow impurities passivation, 230, 233
[B, H] complex, 232
depth profiles dopant concentration, 232
P and Sb donors, 231
spreading resistance profiles, 231
Shockley–Read–Hall theory (SRH theory), 360
Short-open (SO), 367
Si–SiO_2 phase separation, 386–388
SiC, *see* Silicon carbide (SiC)
SiGe alloys, *see* Silicon–germanium alloys (SiGe alloys)
SiGeO, *see* Silicon Germanium oxides (SiGeO)
Silicon (Si), 88, 110–111, 160, 220–223, 323
chemistries for etching, 291–293
classical chemistries for preferential etching, 293, 294–295
configuration-coordinate diagram, 90
defect configurations, 88–89
electrical levels, 89–91
nitride, 243
oxidation mechanism, 300
self-interstitials in, 88
substrates, 289
thermally activated diffusion, 91–93
Silicon carbide (SiC), 2

Silicon Germanium oxides (SiGeO), 388
absorption spectra, 399
dark-field STEM images, 389
Silicon–germanium alloys (SiGe alloys), 23, 197
applications, 24, 52–53
carrier transport, 38–41
critical growth velocity, 30
crystal growth, 25–28
dislocations, 48–49
dopant diffusion, 202–204
$Ge_{1-x}Si_x$ system phase diagram, 25
heteroseeding, 28–29
hydrogen defect, 46–48
impurities and defects, 44
impurities distribution coefficient, 33–36
local atomic structure, 36–38
mechanical strength, 49–52
oxygen, 44–46
Seebeck coefficient, 43
self-diffusion, 197–202
Si_xGe_{1-x} alloys, 25
spatial variation of composition, 30–33
thermal conductivity, 41–43
Silicon nanocluster (Si NC), 377
absorption coefficient, 394
deposition method effect, 380–385
EELS, 381, 383
EFTEM micrograph, 380, 383
excess and clustered Si concentrations, 382
light absorption, 393–402
luminescence peak energy *vs.* Si nanoclusters mean radius, 385
normalized room-temperature PL spectra, 384, 385
SiO*x* composition, 378, 381
synthesis and imaging, 378–380
TEM micrograph, 379
Silicon on insulator (SOI), 218, 289, 304, 305
defect delineation process, 296, 297
defects in, 304
OSF, 305, 306
Silicon oxides (SiO*x*), 395
absorption spectra, 396
optical bandgap variation, 398
optical bandgap *vs.* annealing temperature, 397
SIMS, *see* Secondary ion mass spectroscopy (SIMS)
Si NC, *see* Silicon nanocluster (Si NC)
Single crystal silicon wafers, 324
SiO*x*, *see* Silicon oxides (SiO*x*)
Sirtl etch, 293
Smart-cut® process, 218, 233–235
Smart-Cut™ technology, 289
SO, *see* Short-open (SO)
SoG, *see* Solar-grade silicon (SoG)
SOI, *see* Silicon on insulator (SOI)
Solar-grade silicon (SoG), 343
SPC, *see* Spherically shaped clusters (SPC)
Spectroscopic studies, 323
Spherically shaped clusters (SPC), 126
Split configuration, 264
SRH theory, *see* Shockley–Read–Hall theory (SRH theory)
sSOI, *see* strained Silicon-on-insulator (sSOI)
Stacking fault (SF), 126
STEM, *see* Scanning transmission electron microscopy (STEM)
Stillinger–Weber (SW), 121
strained Silicon-on-insulator (sSOI), 289
Substitutional boron (B_s), 276
Substitutional carbon (C_s), 276
Sulfur (S), 161
Surface-normal configuration, 68
Surface(s), 137
ab initio calculation, 142
intrinsic point defect formation energy, 140
melt–solid interface, 143
microscopic mechanism, 143
S-parameter, 144, 145
self-interstitial Si, 146
Si and Ge, 139, 144
SW, *see* Stillinger–Weber (SW)
Swirl-defects, 308, 309

T

T-parameters (chain-transfer-or chain-scattering-matrix), 366
Tauc law, 395, 396
TD, *see* Thermal donors (TD)
TED, *see* Transient enhanced diffusion (TED)
TEM, *see* Transmission electron microscopy (TEM)
2,3,5,6-tetrachloro-1,4-benzoquinone, 310
Tetra-interstitial
germanium, 114
silicon, 113
Thermal conductivity, 41
compositional dependences, 41
difference in, 15
SiGe alloys, 42, 43
Thermal donors (TD), 334
H-enhanced, 238
shallow, 279
Thermalization process, 328
Thermally activated diffusion, 91
activation energies, 92
excitation-enhanced diffusion, 93
in germanium, 96
long-range migration paths, 91
migration energies and diffusion paths, 93

Thermoelectric devices, 67–68
heavily impurity-doped SiGe crystals, 25
SiGe, 43
Thick-graded buffers, 63
Thin SOI etchants, 298
3*d* transition metal (3*d* TM), 229
Total variation diminishing (TVD), 6
Davis–Yee symmetric, 6
Transient enhanced diffusion (TED), 88, 276
Transmission electron microscopy (TEM), 150
bright-field TEM images, 390, 391
cross-sectional, 390
PE–CVD SiO*x* layer, 379
SEM and, 303
Si NCs, 378–379
Transverse magnetic field, 14
application of, 14
CZ-Si growth in, 16, 17
Tri-interstitial, 111
configurations, 111
lowest-energy forms, 113
models for, 112
Trivacancy center (V_3 center), 268; *see also* Vacancy–Oxygen center (VO center)
DLTS spectra, 269, 270
electronic properties, 268–271
FBI, 270
planar and FFC atomic structures, 269
thermal stability, 271
TVD, *see* Total variation diminishing (TVD)
Two-port network, 368, 369
ABCD-matrices, 369
HF characterization, 371
S-parameters, 373
voltage amplitudes of reflected wave, 369

U

Unidirectional solidification process (DS process), 1
carbon and oxygen incorporation, 2
impurity transfer in DS furnace, 11

V

V_2 center, *see* Divacancy center (V_2 center)
V_2O complexes, *see* Divacancy–oxygen complexes (V_2O complexes)
V_3 center, *see* Trivacancy center (V_3 center)
Vacancy cluster formation, 124
cluster growth modes, 126
continuum models, 127–128
germanium divacancy, 129
MD simulations, 125, 129–130
power law results, 128
vacancy formation energy in Ge, 125
Vacancy engineering, 146
annealed wafers, 147–153
oxygen precipitation, 146–147
Vacancy properties, 119
ab initio calculation, 120–121
annealed wafers, 147–153
capping layers, 137–146
external stress, 130–133
formation energy, 122–124
impurities, 133–137
interfaces, 137–146
internal stress, 130–133
MD, 121–122
oxygen precipitation, 146–147
surfaces, 137–146
vacancy cluster formation, 124–130
vacancy engineering, 146
Vacancy–Oxygen center (VO center), 271
atomic structure, 272
electrical, optical, and annealing properties, 271–274
growth, 274
interaction with hydrogen, 274–276
interaction with oxygen, 274
logarithm of absorption coefficient, 273
metastable complex formation, 274
Valence band maximum (VBM), 185
van der Waals interaction (VdW interaction), 330
Vector network analyzer (VNA), 366
MOSFET under, 371
S-parameter measurement, 368
Vegard's law, 24, 36, 200, 204
Very large-scale integrated processing (VLSI processing), 2, 378
Vibrational
disorder, 42
excitations, 218
frequency for H_2, 228
lifetimes, 240–242
mode, 113, 325, 326
properties, 225–226
spectrum, 225
Virtual substrate (VS), 63
SiGe, 52
VLSI processing, *see* Very large-scale integrated processing (VLSI processing)
VNA, *see* Vector network analyzer (VNA)
VO center, *see* Vacancy–Oxygen center (VO center)
VS, *see* Virtual substrate (VS)

W

Wilke's mixing rule, 6

X

X-ray absorption fine-structure (XAFS), 37
    Ge and Si coordination numbers, 37
    investigations, 45
X-ray diffraction (XRD), 98, 99, 379
X-ray fluorescence holography, 38
X-ray photoelectron spectroscopy (XPS), 292, 300

Y

Y-parameters (admittance-matrix), 365

Z

Z-parameters (impedance-matrix), 365
Zinc (Zn), 161

Milton Keynes UK
Ingram Content Group UK Ltd.
UKHW021835071024
449327UK00021B/1501